Excipient Development for Pharmaceutical, Biotechnology, and Drug Delivery Systems

Excipient Development for Pharmaceutical, Biotechnology, and Drug Delivery Systems

Edited by

Ashok Katdare
NeuroMolecular Pharmaceuticals, Inc.
Emeryville, California, U.S.A.

Mahesh V. Chaubal
Baxter Healthcare
Round Lake, Illinois, U.S.A.

CRC Press
Taylor & Francis Group
Boca Raton London New York

CRC Press is an imprint of the
Taylor & Francis Group, an **informa** business

CRC Press
Taylor & Francis Group
6000 Broken Sound Parkway NW, Suite 300
Boca Raton, FL 33487-2742

Version Date: 20130723

International Standard Book Number-13: 978-0-8493-2706-3 (Hardback)

Visit the Taylor & Francis Web site at
http://www.taylorandfrancis.com

and the CRC Press Web site at
http://www.crcpress.com

Preface

To facilitate the development of novel drug delivery systems and biotechnology-derived drugs, the need for new excipients continues to increase. This book *Excipient Development for Pharmaceutical, Biotechnology, and Drug Delivery Systems* serves as a comprehensive source to improve understanding of excipients and forge new avenues to promote independent regulatory review and development of novel excipients. In addition, this book presents in-depth information on various aspects of excipient development, safety/toxicology testing, regulatory processes, quality, manufacturability, and the utility of excipients for various drug delivery systems. We have relied on numerous experts and thought leaders from all over the world who have shared their expertise and time to prepare the chapters included in this book. Each chapter also provides a wealth of useful references that should prove to be invaluable for the reader.

This book is intended for formulation scientists, analytical scientists and engineers, regulatory and compendia personnel, procurement personnel, preclinical scientists, excipient manufacturers, quality control and assurance personnel, and distributors.

What makes this book so timely? In recent years, an awareness and understanding of excipients has increased based upon several important factors.

First, as pharmaceutically active ingredients continue to become more "potent," the effective doses have become smaller. As a result, excipients now often constitute the major portion of many pharmaceutical dosage forms and as such can have profound impact on the reproducibility of manufacture and overall quality of the dosage forms.

Second, regulatory authorities, especially the U.S. Food and Drug Administration, have clearly set an expectation that quality should be built in drug products from the beginning of development and manufacture rather than simply testing quality of the finished product (quality for 21st century initiative). This stance has forced the industry and academia to develop a thorough understanding of the functionalities and modalities of excipients, as well as to develop and adopt testing methodologies from other industries to refine the characterization of excipients. Also, increased use of process analytical technologies has helped excipient manufacturers and users to develop improved in-process controls and better-controlled manufacturing processes. These efforts should enhance building quality in the manufacture of drug products.

Third, the technical complexities associated with drug development have increased due to challenges such as poor drug solubility, complex drug actives, and, in cases of biotech products, stabilization of the active ingredient. Often times, the current array of excipients in approved products are not sufficient to formulate challenging molecules, forcing pharmaceutical scientists to explore new excipients. The development and testing of new excipients require a multidisciplinary understanding of technical, safety, quality, and regulatory aspects, which, prior to this effort, has not been available in a single resource.

Finally, the drug development business has become truly global, especially in the area of procurement of components, outsourcing of manufacture, and global commercialization. Numerous guidances issued by the International Council on Harmonization have led the groundwork and have had a far-reaching effect in accomplishing globalization. As the regulatory standards on efficacy and, especially, safety of drug products become higher and higher, the pace of drug discovery and launch of new products has slowed considerably. As a consequence, cost conservation has forced excipient users to look for less expensive alternative sources of excipients without sacrificing quality. This broadening of sourcing base has further necessitated improved understanding and control of excipients sourced from multiple global sources.

Although the increased attention to excipients has followed with more academic and industrial activity in the area of excipients, published literature on excipients has greatly lagged behind. Although the industry has benefited handsomely from the seminal book *Handbook of Pharmaceutical Excipients*, there is little published literature on preclinical testing, regulatory processes for novel excipients, and a 'best practice' guide for the use of excipients in various dosage forms. This is the area where this book clearly distinguishes itself.

The chapters in this book can be broadly categorized into four major themes:

Global regulatory processes (Chapters 2, 4, 5, and 7): This section provides a regulatory perspective and reviews existing global regulatory processes . It also proposes new and innovative ways for regulatory review of excipients, which, if adopted, should promote innovation. This section also provides a status update on the global compendial harmonization, which should eliminate non–value-added testing that manufacturers and users of excipients currently have to perform.

Preclinical testing and development and development of new and coprocessed excipients (Chapters 3, 6, 9, and 20): This section describes the type of preclinical testing that is required in support of the development and registration of new excipients and presents a case study for successful development of a novel excipient. Lastly, Chapter 20 looks to the future and identifies excipients needed for innovative biotechnologically derived dosage forms.

Excipient interactions and best practice guide for use of excipients and types of interactions possible in different dosage forms (Chapters 8, 10–19): These chapters should be extremely useful for formulators and regulatory reviewers. They suggest types of excipients that are suitable for various dosage forms and "what to do and more importantly what not to do" when selecting a suitable excipient for a specific dosage form.

Quality, manufacture and distribution of excipients (Chapters 21, 22, and 23): These chapters provide a perspective on quality assurance considerations for the testing of excipients and describe unique characteristics for use, manufacture, and distribution of excipients.

We certainly hope that this book will encourage regulatory authorities to develop new regulatory processes for independent review and use of excipients. The availability of independent review will encourage innovation and development of commercially viable new excipients. Ultimately, all this should help quickly develop lifesaving drug delivery systems benefiting humans.

Ashok Katdare
Mahesh V. Chaubal

Contents

Contributors

Meagan Anderson Aptuit, Kansas City, Missouri, U.S.A.

Gopal Anyarambhatla Research and Development, Akorn, Inc., Decatur, Illinois, U.S.A.

Shireesh P. Apte Chemologic LLC, Mansfield, Texas, U.S.A.

Jane P. F. Bai ZyxBio, Cleveland, Ohio, U.S.A.

Paul Baldrick Scientific and Regulatory Consulting, Covance Laboratories Ltd., Harrogate, North Yorkshire, U.K.

Arvind K. Bansal Department of Pharmaceutical Technology (Formulations), National Institute of Pharmaceutical Education and Research (NIPER), Punjab, India

Lokesh Bhattacharyya Department of Standards Development, United States Pharmacopeia, Rockville, Maryland, U.S.A.

Glenn T. Carlson Pfizer, Inc., Groton, Connecticut, U.S.A.

Mahesh V. Chaubal Baxter Healthcare, Round Lake, Illinois, U.S.A.

Harold Davis Division of Drug Information (HFD-240), Center for Drug Evaluation and Research (CDER), United States Food and Drug Administration, Rockville, Maryland, U.S.A.

D. Delmarre Morton Grove Pharmaceutical, Vernon Hills, Illinois, U.S.A.

Christopher C. DeMerlis Colorcon Inc., West Point, Pennsylvania, U.S.A.

Raafat Fahmy Office of New Animal Drug Evaluation, Center for Veterinary Medicine, Food and Drug Administration, Rockville, Maryland, U.S.A.

Yatin R. Gokarn Department of Pharmaceutics, Amgen Inc., Thousand Oaks, California, U.S.A.

Jian-Hwa Guo Wyeth Consumer Healthcare, Richmond, Virginia, U.S.A.

Piyush Gupta Department of Pharmaceutical Technology (Formulations), National Institute of Pharmaceutical Education and Research (NIPER), Punjab, India

Bruno C. Hancock Pfizer, Inc., Groton, Connecticut, U.S.A.

Stephen W. Hoag School of Pharmacy, University of Maryland, Baltimore, Maryland, U.S.A.

Julia C. Howell J.C. Howell Consulting, LLC, Powder Springs, Georgia, U.S.A.

Amit Jain Department of Chemical Engineering, University of California, Santa Barbara, California, U.S.A.

Pankaj Karande Department of Chemical Engineering, University of California, Santa Barbara, California, U.S.A.

James Kipp Baxter Healthcare, Round Lake, Illinois, U.S.A.

Andrew Kosky Department of Pharmaceutics, Amgen Inc., Thousand Oaks, California, U.S.A.

Eva Kras Department of Pharmaceutics, Amgen Inc., Thousand Oaks, California, U.S.A.

J. Lane United States Pharmacopeia, Rockville, Maryland, U.S.A.

Arnold McAuley Department of Pharmaceutics, Amgen Inc., Thousand Oaks, California, U.S.A.

Samir Mitragotri Department of Chemical Engineering, University of California, Santa Barbara, California, U.S.A.

R. Christian Moreton Idenix Pharmaceuticals, Inc., Cambridge, Massachusetts, U.S.A.

Satish K. Nachaegari Department of Pharmaceutical Technology (Formulations), National Institute of Pharmaceutical Education and Research (NIPER), Punjab, India

Derek O'Hagan Vaccine Delivery Group, Chiron Vaccines, Chiron Corporation, Emeryville, California, U.S.A.

F. Opawale Formulation Development, NovaDel Pharma Inc., Flemington, New Jersey, U.S.A.

Robert G. Pinco Buchanan Ingersoll, P.C., Washington, D.C., U.S.A.

Barrett Rabinow Baxter Healthcare, Round Lake, Illinois, U.S.A.

M. Rao Morton Grove Pharmaceutical, Vernon Hills, Illinois, U.S.A.

Richard L. Remmele, Jr. Department of Pharmaceutics, Amgen Inc., Thousand Oaks, California, U.S.A.

Stefan Schuber Department of Standards Development, United States Pharmacopeia, Rockville, Maryland, U.S.A.

Victoria M. Shaheen Mutchler Pharmaceutical Ingredients, Inc., Harrington Park, New Jersey, U.S.A.

Catherine Sheehan Department of Standards Development, United States Pharmacopeia, Rockville, Maryland, U.S.A.

Irwin Silverstein IBS Consulting in Quality LLC, Piscataway, New Jersey, U.S.A.

Manmohan Singh Vaccine Delivery Group, Chiron Vaccines, Chiron Corporation, Emeryville, California, U.S.A.

Hugh Smyth Pharmaceutical Sciences, College of Pharmacy, University of New Mexico, Albuquerque, New Mexico, U.S.A.

Prashant Srivastava Focused Health Solutions, Deerfield, Illinois, U.S.A.

Theodore M. Sullivan Buchanan Ingersoll, P.C., Washington, D.C., U.S.A.

Diane O. Thompson CyDex Inc., Lenexa, Kansas, U.S.A.

Sydney O. Ugwu NeoPharm Inc., Waukegan, Illinois, U.S.A.

Roger William Department of Standards Development, United States Pharmacopeia, Rockville, Maryland, U.S.A.

1
Excipients: Background/Introduction

Lokesh Bhattacharyya, Stefan Schuber, Catherine Sheehan, and Roger William
Department of Standards Development, United States Pharmacopeia, Rockville, Maryland, U.S.A.

Almost all therapeutic products, including therapeutic products for human and veterinary use, include excipients—indeed, the total amount of excipients frequently used is greater than the amount of the active drug substance(s) in a dosage form. As with drug substances, excipients are derived from natural sources or are synthesized either chemically or by other means. They range from simple, usually highly characterized, organic, or inorganic molecules to highly complex materials that are difficult to fully characterize.

In earlier days, excipients were considered inactive ingredients. Over time, pharmaceutical scientists learned that excipients are not inactive and frequently have substantial impact on the manufacture and quality, safety, and efficacy of the drug substance(s) in a dosage form. Further, variability in the performance of an excipient—both batch to batch within a single manufacturer as well as between batches from different manufacturers—came to be understood as a key determinant of dosage form performance. Excipients are now known to have defined functional roles in pharmaceutical dosage forms. These include (i) modulating solubility and bioavailability of the active ingredient(s); (ii) enhancing stability of the active ingredient(s) in finished dosage forms; (iii) helping active ingredients maintain a preferred polymorphic form or conformation; (iv) maintaining pH and osmolarity of liquid formulations; (v) acting as antioxidants, emulsifying agents, aerosol propellants, tablet binders, and tablet disintegrants; (vi) preventing aggregation or dissociation; and (vii) modulating the immunogenic response of active ingredients (e.g., adjuvants) and many others. *United States Pharmacopeia 28–National Formulary 23* lists 40 functional categories of excipients for pharmaceuticals, and many more are expected as new—and usually increasingly complex—drug-delivery systems emerge and evolve. Approximately 800 excipients are currently used in the marketed pharmaceutical products in the United States. This number is also expected to grow with new therapeutic categories, such as gene therapy and cell therapy, and new drug-delivery technologies.

In these various contexts, excipients and issues associated with them can be considered in the following different areas. "Functionality": An excipient interacts with the active in the formulated dosage form and/or provides a matrix that can affect critical quality attributes of the drug substance, including stability and bioavailability. Given an excipient's potential influence on the finished dosage form, manufacturers will execute careful characterization studies, with due attention to final specifications and change control, in order to ensure consistent performance of the dosage form. Many examples have demonstrated that limited understanding of excipient functionality can compromise process control and product quality. As a general rule, the more complex the dosage form and/or its ingredients, the greater is the impact of excipient functionality. "Safety and efficacy": Excipients can themselves affect safety and efficacy outcomes. Excipients, or their impurities, can be associated with adverse events, either by direct action or by formation of undesirable adducts. By modifying absorption and, for parenteral products, distribution, excipients can change exposure patterns and thus influence both safety and efficacy outcomes. Excipients are well known to affect the safety and efficacy profiles of locally acting products. As adjuvants, excipients required for protein and conjugate vaccines play a crucial role in the immunogenic properties of vaccines. "New excipients": These may require careful and, not uncommonly, extensive safety studies, with corresponding careful attention to characterization and specification setting. At present, new excipients in the United States do not undergo separate approval but attain market access frequently via a regulatory process in association with the new drug application process for a dosage form. "Processability": Manufacturers increasingly rely on a good understanding of the characteristics and functional contributions of excipients to aid in the day-to-day manufacture of a dosage form. "Evolving regulatory and compendial approaches and harmonization": Regulatory agencies and compendia now fully realize the value of careful attention to the safety and quality attributes of excipients and their impact on dosage form performance and safety/efficacy outcomes. This has led to an increasing number of regulatory and compendial documents, many of which are in active harmonization. "Excipients and food additives": The relationship between excipients and food additives, in their manufacture, and regulatory control, is complex and evolving. They are frequently identical in character, yet are controlled according to different regulatory requirements and compendial standards. In the United States, food additives are the "excipients" used in a dietary supplement. Many excipients arise in the manufacture of food-grade material, a point that poses special challenges in terms of achieving pharmaceutical-grade material and regulatory control.

In the rapidly evolving world of excipient manufacture, with attendant challenges of regulatory control and compendial standards-setting, the need for a timely, comprehensive, and thoughtful publication is clear. This need is filled by the following text, prepared with talented editorial oversight from Dr. Ashok Katdare and Dr. Mahesh Chaubal. The author list developed by these editors is composed of distinguished experts with a broad range of skills, experience, and geographical representation. The topics covered are broad and challenging. The text fulfills a critical need for up-to-date and comprehensive information about a rapidly evolving topic for which regulatory guidance is only now emerging. We encourage readers to learn from this text and to consider themselves challenged in helping pharmaceutical scientists, excipient and dosage form manufacturers, and regulatory and compendial experts understand how to advance the field. Careful consideration of the many issues discussed in this book will help talented experts advance to the next stage of understanding of the importance of excipients and food additives in the manufacture of therapeutic products. The need is clear—and the benefit to patients and practitioners is unquestionable.

2

Food and Drug Administration Perspective on Regulation of Pharmaceutical Excipients

Harold Davis
Division of Drug Information (HFD-240), Center for Drug Evaluation and Research (CDER), United States Food and Drug Administration, Rockville, Maryland, U.S.A.

The Food and Drug Administration (FDA) is generally recognized as one of the, if not the, premier therapeutic agent gatekeepers among nations. Consequently, the pharmaceutical and medical library stacks are laden with journals and manuals devoted to drug development and instructions on how to run the FDA gauntlet to reach the jackpot of drug approval. However, little attention is paid to the regulation of excipients. A number of standard texts on the subject are exhaustive in their reviews, although they offer little on how this agency regulates excipients, an integral and essential part of drug development in the review process for drugs. We trust the following provides a window on our actions and thinking in this area.

The regulation of drug inactive ingredients was an outgrowth of the regulation of food colors (1). That began with the Pure Food and Drugs Act of 1906. The adulteration of foods and drugs was prohibited. Seven synthetic organic colors, chosen to give the required range of color, and because no mention of their causing unfavorable effects on humans and animals could be found in the scientific literature, were permitted for food use. A procedure was set up for voluntary certification of the identity and purity of these seven colors, and the use of artificial coloring other than these colors could be grounds for prosecution. This list was revised in subsequent years.

The Elixir of Sulfanilamide disaster, in which 107 people died as a result of the use of a toxic inactive ingredient, dramatized the need to establish drug safety before marketing and provided the impetus to pass the pending Federal Food, Drug, and Cosmetic Act of 1938. Certification of colors became mandatory, with all coal-tar colors used in foods, drugs, and cosmetics required to be from a certified batch. The law also created, out of less than 20 colors, three categories of certified colors: food, drugs, and cosmetic (FD&C) colors acceptable for food, drug, and cosmetic use, drugs and cosmetics (D&C) colors allowed in drugs and cosmetics only, and external D&C colors intended for external use only (2). The 1938 Act required that the presence of an uncertified coal tar be shown to prove that a food, drug, or cosmetic was adulterated, whereas under the 1906 Act, a color was considered to be in compliance until it was shown that its addition to a food rendered that food "injurious to health" (3). Most importantly from our standpoint, the 1938 Act required the submission of a New Drug Application (NDA) for drugs wherein the drug product was considered in its entirety, active and inactive ingredients together. This remains in effect for all drugs subject to an NDA or an abbreviated NDA (ANDA). Inactive ingredients in nonprescription drugs subject to a monograph as described in Title 21 Code of Federal Regulations Part 330.1 and 330.10 (21CFR 330.1 and 330.10) are considered separately from active ingredients and need to be suitable and "safe in the amounts administered and do not interfere with the effectiveness of the preparation or with suitable tests or assays to determine if the product meets its professed standards of identity, strength, quality, and purity. Color additives may be used only in accordance with Section 721 of the Act and subchapter A of this chapter" (4).

Chronic toxicity studies showed that most color additives were toxic when fed at high levels. The position of the FDA was that it lacked authority under the 1938 Act to permit the certification of a coal-tar color that was not harmless when fed to animals in any amount or to impose tolerances or limitations on the use of such colors; this position was confirmed by the U.S. Supreme Court. It appeared that in the future, few, if any, coal-tar colors would be permitted to be certified. The passage of the Color Additive Amendments of 1960 solved the problem of permitting the safe use of colors in foods, drugs, and cosmetics. All color additives had to be listed, regardless of their nature, by regulation (only after a complete showing of safety was made) and also required defining in the regulation the necessary conditions of safe use of the color additive. These amendments placed the burden of proof upon the party interested in obtaining the listing of the color additive (5). Colors derived primarily from plant, animal, and mineral (other than coal and petroleum) sources are exempt from FDA certification.

An inactive ingredient is defined by the FDA as "any component of a drug product other than an active ingredient" [Title 21 Code of Federal Regulations [21CFR Part 218.3C(b)(8)]. While the agency regulations are consistent in using this perhaps obsolescent term, an FDA guidance document (6) defines "new excipients" as "any ingredients that are intentionally added to therapeutic and diagnostic products, but which, we believe, (i) are not intended to exert any therapeutic effects at the intended dosage (although they may act to improve product delivery, e.g., enhancing absorption or controlling release of the drug substance) and (ii) are not fully qualified by existing safety data with respect to the currently proposed level of exposure, duration of exposure, or route of administration. Examples of current ingredients include fillers, extenders, diluents, wetting agents, solvents, emulsifiers, preservatives, flavoring agents, absorption enhancers, sustained-release matrices, and coloring agents." This definition is very much in line with those offered by numerous researchers in the field.

Compendia that describe excipients used for various formulations such as parenterals, vaginal formulations, and antibiotics are offered in a number of publications (7–9). The FDA publishes on its internet site, www.fda.gov, the downloadable "Inactive Ingredient Database." The components of proprietary inactive ingredients are not always included. All inactive ingredients that are present in currently approved final dosage form in drug products are listed. Whenever included, one may need to search for such data under individual component entries.

Synonyms of many ingredients do not appear in the database. Inactive ingredients are listed as specifically intended by the manufacturer. Some of these ingredients could also be considered as active ingredients under different circumstances.

Radiopharmaceutical kit reactants, and inactive ingredients, which chemically or physically combine with active ingredients to facilitate drug transport, are considered as inactive ingredients for the purposes of the database.

The inactive ingredients are updated quarterly, by the fifth working day of April, July, October, and January. To search for the excipient, one can enter any portion of the name of an excipient, of at least three characters. Search results are displayed alphabetically, sorted first by ingredient, then by the route of administration and dosage form. Routes of administration and dosage forms are derived from current approved labeling. Refer to the IIG query search results' column headers for data field definitions.

Industry can use this information to assist in developing drug products. Once an inactive ingredient has appeared in an approved drug product for a particular route of administration, the inactive ingredient is no longer considered new and may require a less extensive review the next time it is included in a new drug product. If, for example, a particular inactive ingredient has been approved in a certain dosage form at a given potency, a sponsor could consider it safe for use in a similar manner for a similar type of product.

Another source of very useful excipient data is the United States Pharmacopeia-National Formulary (USP-NF). Despite certain limitations, it appears that this compendium may become more useful in the years to come.

There are over 400 excipient monographs listed in the current USP 28-NF23. It is of interest to note that 32 new monographs were admitted this year (2005), 10 new monographs approved to USP 28-NF23 (Supplement 1 to USP 28), and four new monographs proposed to USP 28-NF23 (Supplement 2). These contrast sharply with, in chronological descending order, the 12, 4, and 3 new monographs admitted in earlier years.

Informational guidelines, Chapter 1024 in the USP, provides a scientifically based protocol for the safety assessment of new excipients intended for use in any dosage form. The USP has moved beyond addressing identity and purity concerns (9). The issues of physical characteristics are being examined by excipient committees. Methods have been and are being developed to incorporate (quality standards) basic physical characteristics such as particle size, density, and surface area into monographs. Such characterization can aid in identifying differences in materials manufactured in different locations by different suppliers. The point is that by focusing on physical characterization, further assurance is given that functionality will be maintained for a specific intended application. For example, this label claim approach now assures that different physical properties deliver different functionalities, such as liquid retention or ease of compressibility, which may be because of a change in particle shape. These could be appropriately defined. Methodology can be standardized so that the manufacturer and supplier are following the same rules. However, Moreton (10) cautions that variability is an inherent part of any production process. One concern is the extent to which improvement of an excipient's quality can be made without pricing it out of the market. Pharmacopeial monographs should include tests that establish excipient safety. Tests that are needed to differentiate between available pharmaceutical grades should be included, and placed in a labeling section allowing the flexibility to include all the various grades in the monograph.

"... Requests for Revision of the USP-NF, Chapter 3" at the USP Web site www.usp.org offers guidance on various tests useful for new monograph excipients. Details as to what should be included in the submission package are given. Assuming all the required data are present, the package is sent to the expert committees on excipients for review. If, after a thorough evaluation, the submission package is accepted, it will be incorporated in the Pharmacopeial Forum (PF), published every two months. This allows for public review and comment.

After the comments are received and considered, the complete package is sent back to the committee. If no comments are received, the committee may allow the monograph proposal to become an official monograph 60 to 90 days after its PF publication. If comments are made, the committee may reject them or revise the monograph. A revised monograph must be published in the PF. It then can be voted upon to become official 60 days after publication. If a monograph requires only one publication in the PF, it can become official in about six to eight months. The process can take 15 months or longer, should a second publication cycle be needed.

Excipient manufacturers have a number of reasons for wanting their novel excipients to be included in the USP/NF. The NF publishes the highest quality standard publicly available for the product. Drug manufacturers then have confidence in product quality, with corresponding higher excipient sales. The USP has a document disclosure policy, subject to negotiation, which serves to protect confidential, proprietary information and intellectual property rights. The company that submits a new monograph has a dominant role in developing the various tests, procedures, and acceptance criteria that should be performed when evaluating substance quality. Drug manufacturers who purchase compendial grade materials for inclusion in their products are assured that the appropriate tests and procedures have been used with appropriate quality standards.

Compendial grade materials also give FDA inspectors a high degree of confidence, and they do not generally question the tests and acceptance criteria used. Indeed, FDA chemistry reviewers ordinarily do not review the manufacturing of compendial excipients. A new or inadequately qualified inactive ingredient proposed for use in any product pursuant to an NDA, Biological License Application, or ANDA should be supported by adequate data, which may be placed in the application directly or in a Drug Master File (DMF) (11). For compendial excipients that have an unusual use (e.g., lactose for inhalation products), FDA expects to see complete Chemistry, Manufacturing, and Controls (CMC) information (12), which is usually submitted in a DMF.

There are a few concerns about inclusion of an excipient monograph, however. The excipient can only be considered if it has been used in at least one FDA-approved product, or is on the generally recognized as safe (GRAS) list. Under 21CFR211, excipients, as with active drug substances, are required to be manufactured under current good manufacturing practices. Often, the excipient may be used primarily in other applications such as food or non–FDA-regulated products not requiring the same level of manufacturing standards. Significant additional costs may be incurred to meet Good Manufacturing Practice (GMP) requirements. The FDA does not review excipients separately from formulations. They are only approved as part of an NDA or Investigational New Drug Applications (IND). For novel excipients, the manufacturer must essentially develop the same amount of safety data required for a new active ingredient. A strong need for a certain characteristic may make such an investment worthwhile.

FDA guidances[a] serve as a flexible approach to assist compliance with FDA's requirements. Safety testing of novel and potential excipients is addressed in the

---

[a] The Center for Drug Evaluation and Research List of Guidnces, which includes ICH Guidances for Industry, can be accessed at http://www.fda.gov/cder/guidance/index.htm. All the documents can be downloaded.

FDA's 2002 draft Guidance for Industry "Nonclinical Studies for Development of Pharmaceutical Excipients." This guidance lists safety-related issues that should be addressed under an IND or NDA in support of proposals to use excipients in new drug products. The safety-related topics that must be considered under different exposure conditions are given. All pivotal toxicological studies should be performed in accordance with state-of-the-art protocols and good laboratory practice regulations. These excipients should be appropriately evaluated for pharmacological activity using a battery of standard tests. Osterberg and See (13) have reviewed this guidance and discussed in some detail specific development strategies to support marketing of new excipients in drug products.

Some safety issues for excipients with a history of use may be addressed by citations of the clinical and nonclinical database, marketing history, or regulatory status of the compound, e.g., "GRAS" status as a direct food additive may support oral administration of that product up to the levels allowed in foods.

For antibacterial liquid dosage forms, preservative stability and effectiveness require thought. The sterilization method and its effects on the active pharmaceutical ingredient (API) and excipients of ophthalmic liquid dosage forms take on significance. Assurance of sterility for parenterals is paramount, and the effect of the method of sterilization on excipients, API, and preservative (when applicable) stability need investigation. Antimicrobial properties of the preservative require investigation to assure preservative effectiveness. Compendial tests (antimicrobial preservative effectiveness test, microbial limits test, and sterility test, and biological assay tests for antibiotics) appropriate to a specific dosage form should be tested to evaluate the microbiological component during preformulation studies (14). Control of composition and impurities in excipients are briefly discussed (15).

Genotoxicity or carcinogenicity potential may need to be addressed. The FDA's Center for Drug Evaluation and Research (CDER) uses a "cause for concern" approach when determining the scope of the database needed to support a given use of an excipient. The International Conference on Harmonisation (ICH-S1A) (1996) document should be consulted for an analogous approach.

Mitigating circumstances may affect the decision. Duration of exposure, levels of local and systemic exposure, patient population (pediatric, geriatric, debilitated, and healthy), route of administration, knowledge of excipient congeners, and earlier studies that point to areas needing further study are examples. All are part of the risk–benefit assessment. If one can show that an excipient provides benefits to the product, such as promoting absorption of the active ingredient or affecting its release rate, or if it can be shown that the excipient provides some unique and critical property, that therapeutic enhancement (benefit) will be weighed against any risk to the patient. Each proposed use of an excipient must be considered on a case-by-case basis consistent with a positive risk–benefit ratio. Similar to new drug substances, the potential pharmacological activity of the new excipient must be delineated. The ICH guidance S-7A (2001) should be followed with the focus on testing for effects on the central nervous, cardiovascular, and respiratory systems. The ICH M-3 (1997) document identifies these as vital functions. Any activity found could involve performance of detailed investigations to more precisely determine excipient effects on the affected system(s) and the no-observed-effect levels and to calculate acceptable daily intakes.

Silverberg and See also point out that often proper planning will allow assessment of an excipient's toxicity in a relatively efficient manner. A less expensive "study within a study" can be conducted by developing new excipients concurrently with the development of new drugs. Satellite groups of animals receiving an excipient may be added to studies that would have been conducted anyway to develop a drug substance.

Other examples are given. Suitable safety data may be present in DMFs and NDAs. It may be necessary, however, to document a right to reference such data by submitting written permission, from the owners of the data, to the agency, thereby allowing the agency to review the information.

The September 2000 draft guidance considers excipient databases associated with drug products with three different therapeutic durations. For a drug product intended for a 14-day therapy or less, and for infrequent use, the excipient should be tested in acute toxicity studies and in one-month, repeat-dose toxicity studies in two mammalian species (one being a nonrodent), using the intended route of therapeutic administration.

Pharmacokinetic profiling (ICH-S3B 1995) may prove useful. Review of the battery of genetic toxicity tests ICH-S2B (1997) and the ICH reproduction toxicity guidances (S5A) (1994) and S5B (1996, 2000) are valuable.

All of the above studies should be performed if the intended therapeutic duration is less than or equal to 90 days. In addition, two 90-day, repeat-dose studies, with the procedure as previously mentioned, need to be conducted. An intended use of more than 90 days requires all of the above studies plus chronic toxicological studies in both a rodent species (usually six-month duration) and an appropriate nonrodent species (usually nine-month duration). The agency will request, under certain circumstances, chronic toxicology studies of different duration [ICH S4A (1999)]. Excipients intended for use in chronically administered drug products should have a carcinogenicity evaluation. The sponsor has the option of conducting a two-year bioassay in rats and an alternative assay as per the ICH documents S1A (1996) and S1B (1997) or two 2-year bioassays in rodents. The need for such data can be waived (see ICH-S1A), if the sponsor can adequately document that carcinogenicity data are unnecessary. As usual, these decisions will be reviewed on a case-by-case basis. The appropriate division-level staff will make the evaluation together with the center's Pharmacology and Toxicology Coordinating Committee's (PTCC) Executive Carcinogenicity Assessment Committee. The sponsor's decisions will be reviewed from the following aspects:

- Any previous demonstration of carcinogenic potential in the relevant excipient class
- Structure–activity relationships suggesting a carcinogenic risk
- Evidence of preneoplastic lesions in repeated-dose toxicity studies
- Long-term tissue retention of the excipient or a metabolite of the excipient, resulting in local tissue reaction or other pathophysiological responses that are suggestive
- Genetic toxicity data

Sponsors may need data generated from all of the above tests for excipients used in drugs administered by topical or inhalation routes. Data on sensitization potential by either route would be needed. Data obtained from a parenteral or oral (if supported by toxicokinetic data) study may be needed to evaluate the excipient's potential for producing systemic toxicity if systemic exposure is identified in the pharmacokinetic studies. Safety evaluation of the excipient should also include its ability to absorb ultraviolet and visible light. If such a capacity is obtained, the phototoxicity potential could be evaluated using the FDA Guidance for Photosafety Testing (16). Other guidelines provide information on, for example, Liposome Drug

Products, as do Kumi and Booth (17). De George et al. offer guidance on excipients used in inhalation drug products (18).

Toxicological test results may cause the agency to request further studies to examine the toxicity in question to understand the level of risk that the compound may pose. Thus, special studies may be requested to clarify some adverse effect or finding. On the other hand, during the course of product development, some studies could conceivably be eliminated. A decision from the appropriate FDA division can be rendered upon consultation. The division responsible for a given drug product can answer information requests regarding use in the product. Questions are typically posed in pre-IND meetings or in an IND or NDA submission, depending on the product's regulatory status. Guidance on general excipient issues that do not pertain to a specific drug product or questions that pertain to potential excipients not yet associated with a drug product should be directed to the Inactive Ingredient Subcommittee of the PTCC of CDER.

To sum up, the issues and recommendations discussed in the guidance for industry relating to the nonclinical development of excipients, as with other agency guidances, are flexible and open to discussion and modification, as long as any change can be validated. The issues and recommendations should be viewed as a series of topics that should be addressed in an acceptable manner. Again, information or guidance specific to a particular excipient or drug product concerning the development of a safety database is usually available from CDER.

Pharmaceutical manufacturers may wish to change an excipient in a marketed drug. The reasons are several. For example, there may be a change in compendial standards. The USP does revise excipient monographs. Those changes can force a firm to reevaluate and change the excipient used in a formulation to meet the compendial requirements, especially when it comes to grades of excipients. An excipient on occasion may become unavailable due to a loss of source—for example, natural disasters (fire, war, etc.). Some excipients are available only in limited geographic areas, much like many other natural resources. Firms may make formulation modifications tailored to a specific patient population—pediatrics for example. Some changes are driven by the specialty excipient manufacturer—often excipients are also foodstuffs and food additives. Certainly, economics plays a role. Specialized excipients tailored to pharmaceutical market are a small portion of the total excipient market. The demand for excipients in vitamins and food supplements can cause pharmaceutical manufacturers to reduce or reevaluate their use of those excipients (19).

It is requested, but not required, that drugs listed according to 21CFR207.20 qualitatively list the inactive ingredients in the format given in Form 2656 (Drug Product Listing). An external color change of a drug product requires the submission of a new National Drug Code [21CFR35 (4)(i)]. Neither the Act nor the regulations mention that the wholesaler or retailer be notified if an excipient change is made. This is often done in practice, however.

If the product is the subject of an NDA or an ANDA, a supplemental NDA must be filed [21CFR314.70(b)(2)]. It must be shown that the change does not affect the bioavailability of the active ingredient(s). CMC information for drug substances used in over-the-counter (OTC) products covered by an OTC monograph (e.g., calcium carbonate) are not reviewed. Therefore, a DMF need not be filed. The fact that there are existing DMFs for calcium carbonate does not mean that they are reviewed. CMC information for OTC products not covered by an OTC monograph (e.g., famotidine) does need to be reviewed. A DMF is an appropriate mechanism to submit such information.

Section 502(e) of the Act requires that the drug label bear the "established name of each inactive ingredient (and also be) listed on the outside container of the retail package." This includes any quantity of alcohol. Based on this section, 21CFR201.10(c)(4) does not allow "the featuring ... of ... inactive ingredients in a manner that creates an impression of value greater than their true functional role in the formulation." If for other than oral use, the names of all inactive ingredients must be listed [21CFR201.100 (a) (5)]. The members of the Pharmaceutical Manufacturers Association (now the Pharmaceutical Research and Manufacturers of America) voluntarily agreed to list inactive ingredients in Rx drugs for oral use (20). Generic manufacturers followed suit. As a result, a regulation to this effect was never issued.

Whenever data demonstrating a relationship between inactive ingredients in drugs and possible adverse reactions come to the FDA's attention, appropriate steps are taken by the agency. These changes include requiring labeling to contain information about the relationship or prohibiting the use of the ingredient. Thus, the labeling for Rx drugs containing aspartame and sulfites, except epinephrine, for injection, when intended for use in allergic or other emergency situations, requires specific warning statements (21CFR201.21 and 22, respectively).

Section 706(b) (3) of the Act provides that regulations for the listing of a color additive shall "prescribe the conditions under which such additive may be safely employed for such use or uses (including but not limited to... and directions or other labeling or packaging requirements for such additive)." The FDA's position then is that the name of a color additive will not routinely be required on the labels of all foods and drugs unless its declaration is necessary for safety reasons. The presence of FD&C Yellow #5 and/or FD&C Yellow #6, potential sensitizing agents for many individuals, must be declared on the label of foods and certain drugs (21CFR201.20).

In 1984, the FDA welcomed a voluntary program, adopted by the Proprietary Association, now the Consumer Health Products Association, to identify on the product label the inactive ingredients used in OTC drug products (21). The listing of these ingredients was on an alphabetical basis instead of in the descending order of predominance.

The voluntary program was mooted by the 1997 FDA Modernization Act [see FDC Act Section 502(e) (1) (A) (iii)].

Nonprescription drug labels are required by law to identify all active ingredients and to identify and list quantities of certain ingredients, such as alcohol, whether active or not. Sodium content per dosage unit of oral OTCs is required (21CFR201.64). Terms that may be used, such as low sodium, very low sodium, and sodium-free, are defined. Inactive ingredient–labeling requirements are discussed in 21CFR201.66, both for drugs and for drugs that may also be considered as cosmetics. A number of Guidances for Industry that describe OTC labeling are available (22–24).

Interest in facets of excipient development is growing and in some cases is forced upon us. The agency has published an Interim Final Rule and proposals regarding the use of materials derived from cattle in human food and cosmetics (25). This addresses the potential risk of bovine spongiform encephalopathy in human food, including dietary supplements and cosmetics. Registration of all manufacturing sites and prior notification of all food ingredient imports will be required. It is a certainty that comparable systems for drug excipients will follow. Of course, many pharmaceutical excipients are used in food products. Thus, excipients may be required to be registered if used in food products. The excipient supplier then is under the gun and may face charges. Such a regulation would affect animal-derived excipients, including tallow, gelatin, stearyl alcohol, lactose, and glycerin. It appears then that the status of the generally cheaper offshore sourcing of pharmaceutical excipients may change or they may adapt themselves to the regulations (26).

There are a number of review refinements in the works that should streamline the review of some excipients. These were discussed at the October 2004 Generic Pharmaceutical Association meeting. Among these is a "fast-track" system for handling changes-being-effected (CBE) supplements. If either a CBE-0 or a CBE-30 supplement arrives at the office and is reviewed, and a determination is made that the proposed change is acceptable and no additional review is needed, the project manager will draft and send a letter notifying the company immediately. This action obviates the need for the supplement to be placed in a queue for review by the chemists, as had been the case earlier. The agency will undoubtedly seek other methods to speed review time. New DMFs are almost always found deficient on review. More information contained in the file can mean a quicker acceptance, but it can also mean more fodder for questions from the FDA.

Dr. John Kogan (27), speaking at a January 2005 International Pharmaceutical Excipients Council (IPEC) conference, said he believed that, because of downward price pressure, a lack of innovation, and rising costs of new product development, the excipient industry will diversify into two groups: one, focusing on high-tech excipients with greater functionality and high prices—developed in partnership with drug companies and in a manner akin to an API—and the other, a commodity sector. Helping in driving this split is the development of pharmaceuticals without the need for excipients, with the exception of diluents to provide bulk. Work on identifying the best physical or crystalline form of an API is already doing away with the need for wet/dry binders and making APIs more compressible. Next in line could be lubricants, dissolution agents, and disintegrants. A second problem facing the industry is that, on the whole, the 1200 plus marketed excipients fulfill the needs of most of the finished drug products, at least for immediate-release dosage forms. The big exception is for modified-release dosage forms.

A different view is taken by Apte and Ugwu (28), who focus on predicting trends and classifying delivery systems for parenterals, especially biotechnology products. The need to deliver drugs to specified therapeutic targets is a major driver for investigating the use of new excipients. They contend that in the near future, kilogram quantities of fusion proteins, polylysine, fibronectin, or alpha hemolysin could become available as "off-the-shelf" excipients or as designer excipient kits.

Apte and Katdare (29) aver that new mechanisms in the form of guidelines and procedures are needed to regulate the functionality of new and emerging excipients. In the examples below, the pharmacological effectiveness of a drug can be influenced by the excipient. These new excipients may be antigens, viral vectors, microbial products, or other complex proteins. Their pharmacological activities are not completely independent of their excipient functionality and straddle the line between excipients and APIs. For example, paclitaxel bound to albumin (30) (Abraxane) improves breast cancer therapy.

Solvents are no longer needed and the albumin passes into the body. More pertinent examples include the pegylated interferons (31). Polyethylene glycol (PEG) is attached in a random fashion and at variable numbers of sites on each molecule. A single dose of the combination in each cycle of chemotherapy is as effective as the original version, which required daily injections for up to two weeks. The PEG moiety is essential for the increased effectiveness, yet at the same time is inactive by itself. The U.K. company Biocompatibles reports (32) that it has developed a system of bioinert etched microspheres that not only block blood vessels supplying tumors, but also deliver a payload of chemotherapeutic drugs. A device then serves as an active weapon against a disease at the same time that it serves as a drug.

Apte and Katdare question whether a molecule classified as both an excipient and an API can be regulated as both. Excipients are only reviewed as part of an NDA. Including a new excipient is a gamble on a new drug approval that includes a heavy financial investment. Vital issues that must be addressed include expanding the definition of excipients—but they must still wend their way as part of an NDA. Perhaps there should be an independent excipient review—possibly by outside experts. How can excipient innovation and creativity be promoted by government policies? Of course, one problem is that regulatory guidance always trails innovation.

Osterberg (33) comments that our draft excipient guideline be consulted together with the procedures outlined by Steinberg and Silverstein (34). The FDA stands ready to consult with innovators. He also suggests that an expert panel could be developed to pass on the safety of excipients.

As discussed in Chapter 20 by Apte and Ugwu, the future for new, unusual excipients that have exotic properties is hot and sunny. A quick scan of pharmaceutical science and pharmacology journals demonstrates very active research that could bear fruit unimaginable at this time.

REFERENCES

1. Anderson OE Jr. Pioneer Statute: The Pure Food and Drugs Act of 1906. J Pub Law 1964; 13:189–196.
2. Young JH. The Government and the Consumer Evolution of Food and Drug Laws. The 1938 Food, Drug, and Cosmetic Act. J Pub Law 1964; 13:197–204.
3. Toxicological Principles for the Safety Assessment of Direct Food Additives and Color Additives Used in Food, 1993, 21CFR 330.1(e) (draft-updated 2003).
4. Rumoe MM, Strauss S, Kothari AB. Regulatory aspects of color additives. Pharm Tech 1992; 68–82.
5. FDA Guidance for Industry, Nonclinical Studies for Development of Pharmaceutical Excipients, September, 2002. Guidance documents, published by the FDA, represent the agency's current thinking on a particular subject. They can be accessed at www.fda.gov/cder/guidance/index.htm.
6. Von Behren DA. Physical characterization of excipients in practice. Pharm Tech 1996; 85–88.
7. Nema S, Washkuhn RJ, Brendel RJ. Excipients and their use in injectable products. PDA J Pharm Sci Tech 1997; 51:166–170.
8. Garg, S. Tambwekar K, Vermani K, et al. Compendium of pharmaceutical excipients for vaginal formulations. Pharm Tech 2001; 25:14–24.
9. Kumar A, Weatherly MR, Beaman DC, et al. Sweeteners, flavorings, and dyes in antibiotic preparations. Pediatrics 1991; 87:352–360.
10. Moreton RC. Excipient functionality. Pharm Tech 2004; 28:98–119.
11. FDA Guidance for Drug Master Files September 1, 1989. Guidance documents, published by the FDA, represent the agency's current thinking on a particular subject. They can be accessed at www.fda.gov/cder/guidance/index.htm.
12. FDA Guidance for Industry Drug Product Chemistry, Manufacturing and Controls Information Draft Guidance, 2004. Guidance documents, published by the FDA, represent the agency's current thinking on a particular subject. They can be accessed at www. fda.gov/cder/guidance/index.htm.

13. Osterberg RE, See N. Toxicity of excipients—a Food and Drug Administration perspective. Int J Toxicol 2003; 22:377–379.
14. Montgomery ER, Manu-Tawiah W. Microbiological considerations when selecting excipients during product development. Am Pharmaceut Rev 2004; 34–39.
15. Erickson M. Identification and control of impurities in excipients. Am Pharm Rev 2005:88–94.
16. FDA Guidance for Industry Photosafety Testing, 2003. Guidance documents, published by the FDA, represent the agency's current thinking on a particular subject. They can be accessed at www.fda.gov/cder/guidance/index.htm.
17. Kumi KI, Booth BP. A regulatory view of liposomal drug product characterization. New Drug Development. Marcel Dekker, Inc., 2004.
18. De George JJ, Ahn CH, Andrews PA, et al. Considerations for toxicology studies of respiratory drug products. Regul Pharm Toxicol 1997; 25:189–197.
19. Sam T. Regulatory implications of excipient changes in medicinal products. Drug Info J 2000; 34:875–894.
20. PMA News Release. Feb 8, 1984.
21. OTC Inactive Ingredients. FDA Talk Paper, T84–31, May 14, 1984. Talk papers are issued by the FDA press office and are available upon written request from HFI-35, FDA, Rockville MD. The federal register, is an official newspaper of the federal government.
22. FDA Guidance for Industry-Labeling OTC Human Drug Products Using a Column Format, December 19, 2000. Guidance documents, published by the FDA, represent the agency's current thinking on a particular subject. They can be accessed at www. fda. gov/cder/guidance/index.htm.
23. FDA Guidance for Industry-Labeling OTC Human Drug Products-Updated Labeling in ANDA's (1), February 2, 2001. Guidance documents, published by the FDA, represent the agency's current thinking on a particular subject. They can be accessed at www.fda. gov/cder/guidance/index.htm.
24. FDA Guidance for Industry-Labeling OTC Human Drug Products; Submitting Requests for Exemptions and Deferrals. December 19, 2000. Guidance documents, published by the FDA, represent the agency's current thinking on a particular subject. They can be accessed at www.fda.gov/cder/guidance/index.htm.
25. Federal Register, 69, 422565, 42275, 42288, July 14, 2004.
26. Rowlands SS. Transmissible spongiform encephalopathy and medicinal products in the United States and Europe. Drug Info J 2001; 35:993–1001.
27. Excipient industry's future under debate. Pharma Technologist.com, February 2, 2005.
28. Apte SP, Ugwu SO. A review and classification of emerging excipients in parenteral medications. Pharm Tech 2003; 27:46.
29. Apte SP, Katdare A. Emerging excipients demand new regulations. Pharm Tech 2005; 122–124.
30. Biocompatibles plans next generation microsphere. Pharma Technologist.com, February16, 2005.
31. Protein modification technology patented by Bolder. Pharma Technologist.com, February 4, 2005.
32. Albumin delivery improves breast cancer therapy. Pharma Technologist.com, January 10, 2005.
33. Osterberg RE. Speech quoted. Pharm Tech 2004; 124.
34. Steinberg M, Silverstein F. The use of unallowed excipients. Int J Toxicol 2003; 22:373.

3

Pharmaceutical Excipient Development—A Preclinical Challenge

Paul Baldrick
Scientific and Regulatory Consulting, Covance Laboratories Ltd., Harrogate, North Yorkshire, U.K.

INTRODUCTION

The development of excipient materials for use in drug formulations represents a growing area of interest (and of invested time and cost) for pharmaceutical companies. Such development has been fuelled by the increasing need for more sophisticated excipients and/or new uses for established ones. However, a key consideration is how safe the material is. Answering such a question is vital, especially because pharmaceutical excipients can no longer be regarded as totally inert/inactive substances within the formulation of pharmacologically active drugs. New drug development itself involves a range of preclinical studies to show efficacy (pharmacology investigations) and safety (kinetic and toxicology studies) to support clinical trial work and eventual product licensing. Safety studies can include adsorption, distribution, metabolism, and excretion (ADME)/pharmacokinetic (PK), general toxicity, reproduction toxicity, genotoxicity, and carcinogenicity investigations. Additionally other specific studies, for example, local tolerance investigations for drugs administered by the topical or inhalation route, or immunological evaluation for biological drugs may be needed. Safety pharmacology studies (which examine for unexpected high-dose pharmacological effects) can also be considered as part of the safety package. Obviously, pharmacological evaluation per se is not the norm for excipient materials. However, evaluation for potential toxicity is vital, and this chapter examines the safety evaluation process for excipients (new, "essentially" new, and established) from a preclinical perspective and shows that the role of the toxicologist is indeed a challenging one.

PRECLINICAL TESTING RECOMMENDED BY REGULATORY SITUATION

Until recently, there has been a paucity of regulatory agency guidance relating to the safety evaluation (and indeed development in general) of excipients, both established and new. Furthermore, even knowing which excipients are readily "acceptable" to the regulators is not necessarily clear. As a rule of thumb, a regulatory situation of acceptance can be assumed for an excipient when regulatory approval is obtained for a new product license, of which the excipient is a component of the formulation (1,2). Such a system, however, does not address stand-alone excipient development. Examination of drug approvals (especially perusal of the associated product label or summary basis of product characteristics information) by the U.S. Food and Drug Administration (FDA), Center for Drug Evaluation and Research, and the European Agency for the Evaluation of Medicinal Products (EMEA) can reveal information on the constituents in the formulation (3,4). Although now becoming outdated, the FDA has also published a listing of inactive ingredients in drug approvals; a FDA online information service on ingredients (updated quarterly) is also available (5). Information on "approved" excipients in Japan has been published (6). Various recent textbooks also contain information on the regulatory status of some excipients (7).

A general lack of knowledge of excipients has proved an effective barrier to the development of novel materials, and companies have tended to opt for the less complicated and less expensive solution of using well-known (but not necessarily the most effective) excipients. Thus, excipients in use about 100 years ago are still in common use today (8). The lack of specific regulatory guidance to assist any development of new excipients led the International Pharmaceutical Excipients Council (IPEC), an industry association, which champions excipients, to publish safety evaluation guidance (9,10). This guidance covers a whole range of preclinical testing considerations. In 1999, a paper relating to considerations for safety evaluation of new excipients in Japan was published and includes studies on acute, subacute, and chronic toxicity, mutagenicity, and effects on reproduction and carcinogenicity (11). In Europe, although it is a requirement that new excipients need to undergo a full safety evaluation, no detail is given on what is needed (12). Some clarity on expectations has recently occurred in that "the toxicology and pharmacokinetics as of an excipient used for the first time in the pharmaceutical field shall be investigated" and the same pivotal studies as for a new active drug substance are expected (13).

Possibly as a response to all this uncertainty, the FDA has released a guidance document entitled "Nonclinical Studies for Development of Pharmaceutical Excipients," which was finalized in May 2005 (a draft version of this document first appeared in September 2002) (14). Among other things, the guidance is intended to foster and expedite the development of new excipients and to communicate agency expectations to industry. A key message is that excipients are potential toxicants and need to be evaluated accordingly, and so the document proposes a range of preclinical studies, in a manner similar to those of IPEC.

PRECLINICAL TESTING FOR A NEW EXCIPIENT

Essentially, a new (novel) excipient is a material that has not been previously used in a pharmaceutical formulation. New proposed excipients cover a range of functions from conventional use to active roles of enhanced drug uptake and specific drug delivery. Indeed, the "activating" of older drug formulations by inclusion of new excipients for a range of pharmaceutical classes is an ongoing process (15). Most of the emerging excipients have been categorized as natural products (e.g., polymers and derivatives), synthetic polymers, small molecules, natural products modified with synthetic polymers (or vice versa), natural products modified by small molecules (or vice versa), and synthetic polymers modified with small molecules (or vice versa) (16).

The preclinical safety evaluation of a new excipient generally commences after initial in vitro pharmacy work to demonstrate the material's proposed role. Additionally, some in vivo investigations (often a short exposure study in the rodent) may occur, for example, comparing the new proposed material in a drug formulation versus a marketed drug formulation. Enhanced drug exposure and/or a reduced toxicity profile (through the use of lower-dose levels or excipient protection) may be a study end point.

Further development of a new excipient may then take the form of a "stand alone" material for potential inclusion in a range of drug formulations or solely as part of a specific drug formulation. In the latter case, the testing package may be reduced, but any inherent toxicity that the excipient may possess needs to be established. It would be foolhardy to develop a new medicinal product without first checking that any toxicity findings (which could slow down or even terminate its development) are not, in fact, related to the active drug substance. A possible approach in toxicity studies is to add groups of animals that receive the excipient alone as well as the drug-treated groups, as mentioned in the FDA guidance (14,17). However, this approach can make the size of the study enormous, especially if more that one excipient-only group is included. Another concern would be excipient-related toxicity (especially using materials with "activity"), which compromises findings seen in all drug-treated groups. Thus, a case-by-case approach is needed for the safety evaluation of new excipients.

As mentioned earlier, the testing strategies proposed by IPEC and the FDA offer a useful starting point for preclinical excipient testing. The essentials of these strategies are summarized in Table 1. IPEC has proposed guidance from both a European and a U.S. perspective, reflecting single or limited human exposure (<two two weeks), limited or repeated human exposure (two to six weeks for IPEC-US and <four weeks for IPEC-Europe), and long-term human exposure (>six weeks for IPEC-US and >four weeks for IPEC-Europe) for a new excipient (9,10). Proposed study types are given for a range of dose routes, including oral, topical, parenteral, and inhalational. The FDA has divided testing requirements into those needed to support maximum clinical duration of up to 14 consecutive days (short-term use), more than two weeks but three months or less (intermediate use), and more than three months of use (long-term use) (14,17).

Although some differences occur among the proposed testing strategies, a great deal of commonality is apparent. Thus, recommended toxicity studies for initial human use of the new material include single-dose toxicity, repeat-dose toxicity, and genotoxicity studies; the toxicity studies need to reflect the proposed clinical dose route, with repeated dosing for one month in a rodent (usually the rat) and non-rodent (usually the dog) species. The latter studies are routinely performed for new drug substances and have end points of clinical observations, body weights, food consumption, clinical pathology, and organ weights plus macroscopic and histological examination. Dose levels are usually related to multiples of the proposed human drug use. As such a situation is not directly relevant to a new excipient per se, study dose level selection is vital. It is likely that for totally nontoxic excipients, a high-dose level of 2000 mg/kg/day is appropriate. Such a level will give large safety margins over the levels used by the industry for the majority of excipients. The final FDA guidance now also suggests a high limit dose of 2000 mg/kg/day (or 2% in the diet), which is more sensible than the draft FDA document, which suggested consideration of a heroic high-dose level of 5000 mg/kg/day (or 5% in the diet) (14). The latter level of testing is unnecessary because very high doses of materials by oral gavage

Table 1 Summary of Available Literature Guidance Relating to Preclinical Testing Strategies

		Recommended preclinical study		
Guidance	Initial	Short-term clinical use	Midterm clinical use	Longer-term clinical use
IPEC-US (intended clinical route)[a]	–	Acute oral and dermal toxicity, skin and eye irritation, and skin sensitization. Bacterial gene mutation and chromosome damage. ADME (intended route). 28-day toxicity (2 species by intended clinical route)	Short-term use studies. 90-day toxicity (most appropriate species). Teratology (rat and/or rabbit). Genotoxicity assays. Additional assays (conditional)[d]	Short-/midterm studies. One-generation reproduction. Chronic toxicity (rodent and nonrodent) and carcinogenicity (conditional)
IPEC-Europe (intended clinical route)[b]	ADME	Acute toxicity (intended route) and skin sensitization. Ames, chromosome damage and micronucleus. Four weeks toxicity (2 species by intended route)	Short-term use studies. Three-month toxicity (most appropriate species). Teratology (rat and rabbit). Genotoxicity assays	Short-/midterm studies. Segment I reproduction. Six to nine months toxicity (rodent and nonrodent), segment III reproduction, and carcinogenicity (conditional)
FDA (intended clinical route)[c]	Standard safety pharmacology battery	Acute toxicity (rodent and nonrodent by intended route, although option of not	Short-term use studies (although option of not performing 1-month studies).	Short-/midterm studies (although option of not performing 1 and 3 mos

	performing these studies if sufficient dose levels are used in repeat-dose studies). ADME (intended route). Standard genotoxicity battery. One-month toxicity (rodent and nonrodent by intended route). Single-study rodent assay to evaluate all phases of reproductive toxicity and a teratology study in a nonrodent	Three-month toxicity (rodent and nonrodent species by appropriate route). Parenteral use studies (conditional)	studies). Six-month toxicity in rodent and chronic toxicity in nonrodent (by appropriate route). Carcinogenicity in 2 rodent species or 1 rodent species plus, for example, a transgenic model (conditional)

[a]Additional considerations for inhalation/intranasal route: acute inhalation, application site, and pulmonary sensitization studies; for parenteral route: acute parenteral toxicity and application site studies; mucosal use: application site evaluation; transdermal and topical drugs: application site and phototoxicity/photoallergy evaluation. Photocarcinogenicity is a conditional option for transdermal and topical excipients.

[b]Additional considerations for mucosal, transdermal, dermal/topical, parenteral, inhalation/intranasal, and ocular use: skin and eye irritation and application site studies; for parenteral route: acute parenteral toxicity study. Pulmonary sensitization is a conditional option for inhalation/intranasal excipients and phototoxicity/photoallergy plus photocarcinogenicity are conditional options for transdermal and dermal/topical materials.

[c]Additional considerations for topically (dermal, intransal, introral, ophthalmic, rectal or vaginal) or pulmonary adminstered excipients are ocular irritation, sensitisation, oral or parenteral route toxicity studies; additional considerations for injectable excipients are an in vitro hemolysis study, measurement of creatinine kinase and protein binding evaluation; where appropriate new excipients should also be examined for photosafety.

[d]Studies specific to the nature of the excipient, e.g., screening for endocrine modulators.

Abbreviations: FDA, Food and Drug Administration; IPEC, International Pharmaceutical Excipients Council; ADME, adsorption, distribution, metabolism, and excretion.

Source: From Refs. 9, 10 and 12.

or in the diet in repeat-dose toxicity studies can lead to "expected" findings of altered body weight and food consumption and "local" effects such as cecal enlargement, purely due to the presence of large amounts of unabsorbed material in the gastrointestinal tract or the fact that the material is nonnutritive. Genotoxicity evaluation would normally involve an in vitro bacterial gene mutation (Ames) test and either a mouse lymphoma or chromosome aberration assay as well as an in vivo rodent micronucleus test. All three sets of guidance mention ADME assessment with IPEC-Europe indicating that it may be useful to perform such studies before other testing begins (9). Thus, for example, information on whether absorption has occurred following oral use is important in designing the testing package (9). In vitro metabolism studies (e.g., with hepatocytes) can be used to examine potential differences across species. Excipient exposure can also be assessed from measurements in blood samples from toxicology studies (toxicokinetic evaluation). However, ADME/PK measurements may not be possible for a new excipient due to technical reasons, for example, difficulty in finding a suitable molecular site for labeling or insufficient sensitivity of the method to detect very low levels. Following administration, a number of excipient materials (e.g., fatty acids and glycerol) are quickly metabolized into the normal components of the body's cellular system. Some workers have successfully overcome such problems; for example, absorption of polyethylene glycols (PEGs) can be followed using urinary excretion measurements (18). A final area involving kinetics may be a need to fully characterize potential excipient–drug interactions. Interactions occur more frequently between excipient and drug than between excipient and excipient and take the form of either a physical interaction (which can modify the speed of dissolution or uniformity of the dose form) or a chemical interaction (which can lead to drug degradation and/or the formation of degradation impurities) (19,20).

Some of the studies suggested by IPEC across all dose routes, such as acute dermal toxicity plus assessment for the potential for skin and eye irritation (IPEC-US) and sensitization studies (IPEC-US and IPEC-Europe) are probably not necessary. However, skin and eye irritation testing across all routes of administration is defended in the literature as data necessary to protect researchers during the research and production life of the excipient (21). Neither of the IPEC proposals make any provision for unexpected high-dose pharmacological activity from the new excipient, for example, effects on the central nervous system or cardiovascular/respiratory system outside those examined by toxicity studies. However, such safety pharmacology studies are suggested in the FDA guidance (14).

Midterm clinical use for a new excipient involves the need for three-month repeat-dose toxicity studies. The IPEC approach indicates examination of the findings from the one-month data and selection of only one (the most appropriate) species for such a study (9,10). Unless marked toxicological findings occur, the rat is likely to be selected. The FDA guidance indicates a different approach in that if the excipient is to be used for a period ranging from more than two weeks to three months or less, it may be possible to perform three-month toxicity studies in two species without the need for one-month studies (14). However, it is highly unusual (risky and costly) to follow this strategy, because most one-month studies are vital markers for potential target organ toxicity at high-dose levels. An alternative strategy might be to perform "preliminary" two-week repeat-dose studies before embarking on three-month studies. All three sets of guidance mention reproduction toxicity. In new drug development, such studies are needed to allow the inclusion of women of child-bearing potential in clinical trial work. The earlier reproduction toxicity studies are performed, the earlier such a population can be enrolled in these studies.

Probably to reflect this situation, the FDA has included such studies at an earlier stage than the IPEC guidance. It should be pointed out that if such an assessment showed a potentially new excipient to be teratogenic, it is highly likely that further development (never mind additional reproduction toxicity studies) would not occur.

Chronic toxicity studies need to be considered for longer-term clinical use of a new excipient. IPEC has suggested that such studies should be conditional and only performed if evaluation of available data indicates a need (9,10). The FDA guidance suggests that such studies are needed, namely rodent (for six months) and nonrodent (for 6–12 months) toxicity studies (14). The option is given of performing these without the need for one- or three-month studies, although this would appear to be a highly risky strategy. All three sets of guidance indicate that assessment for carcinogenicity is conditional, based on other data. Thus, it is unlikely that such studies would be needed for an excipient with little or no toxicity at high-dose levels, limited systemic exposure, and negative genotoxicity findings, and in a class of noncarcinogenic materials. Classical carcinogenicity testing has involved dosing in the mouse and rat, daily for up to two years, with assessment of survival and tumor incidence. A recent consideration has been to replace the mouse bioassay with an alternative assay such as the use of transgenic animals. Because any such assessment will have a large cost, it has been suggested by the FDA that a possible option is to include an excipient-alone group (using the maximum tolerated or maximum feasible dose), when performing bioassays with the new drug substance (14).

As mentioned earlier, new excipients are being developed to improve and make formulations more economic and alter bioavailability (to produce more favorable drug exposure) and as specific drug delivery materials (e.g., for large molecules and gene therapies). A massive array of published literature is available in this field and only a few examples will be discussed here. Thus, drug delivery in cochleates (phospholipid-cation precipitates usually composed of phosphatidylserine and calcium) for conventional drugs and in gene therapy is being evaluated (22,23). These materials appear to be nontoxic and do not result in the development of an immune response, which is a disadvantage of viral vector–based delivery systems. In recent years, liposomes (phospholipid-based vesicles) have been examined as drug delivery systems, and the recent literature has many examples of these materials with proposed/actual use, largely in cancer therapy (24–26). Liposomes have the ability to greatly increase circulation time of the drug, protect the drug from enzymatic or chemical degradation, and reduce side effects from high-potency drugs. A major forerunner was the stealth liposomal form of the anticancer drug doxorubicin, which has been successfully marketed as Caelyx (in Europe) and Doxil (in the United States) (27). Various toxicology studies were performed to show the safety of this lipid excipient (Tables 2 and 3). Modification of liposomes by the addition of the well-known excipient PEG has occurred to increase hydrophilicity and therefore reduce interactions with reticuloendothelial cells responsible for their systemic elimination; furthermore, liposomes have been conjugated to antibodies or ligands to enhance target-specific drug therapy (24,61). In addition, a range of other PEGylated candidate drugs are under investigation or are marketed (e.g., PEGylated interferons) (62). Polymeric micelles, including those made from PEG–phospholipid conjugates, are also being evaluated (61,63). Another area of major excipient interest is in the use of polymers, including those derived from glycolic and lactic acids (PLGAs), polyglycolic acid, or poly(lactic acid) (PLAs) for use in drug delivery micro- or nanospheres. Marketed products using these materials include the luteinizing hormone–releasing analogue Lupron Depot and Zoladex (64). PLA–PEG

text continues on page 29

Table 2 Preclinical Studies for Recent Excipients Under Development or Used in Marketed Drug Products

Excipient	Proposed/actual use	Toxicology studies	Remarks	References
Ac-Di-Sol (croscarmellose sodium)	Dissolution aid and disintegrant	Repeat-dose toxicity with routine end points (90 days—diet rat) and embryo-fetal study (rat)	No adverse toxicity or embryo toxicity	28
Aquacoat ECD (ethylcellulose polymer, acetyl alcohol, and sodium lauryl sulfate in water)	Coating for tablets and capsules	Repeat-dose toxicity with routine end points (90 days—oral rat) and reproduction toxicity (embryo-fetal study in rat)	No adverse findings for general toxicity or reprotoxicity	29, 30
Aquatic aqueous enteric coating (cellulose acetate phthalate, distilled acetylated monoglycerides, and poloxamer 188)	Film coating for tablets and capsules	Repeat-dose toxicity with routine end points (90 days—diet rat), reproduction toxicity (embryo-fetal study in rat) and genotoxicity (2 in vitro and 1 in vivo studies)	No adverse toxicity, reprotoxicity, or genotoxicity	31, 32
Chitosan	Controlled release tablets, dissolution aid and disintegrant	Repeat-dose toxicity (10 days in rabbit)	No toxicity reported	33
Liposomes (DOTAP: DOPE 1:1/DDAB: DOPE 1:1)	Drug delivery systems for hydrophobic drugs	Repeat-dose toxicity with routine parameters (every 3 days for 3 wks—intravenous rat)	Low-level toxicity (clinical observations)	34
Ethylene glycols	Formulation aid for nasal delivery	Single and repeat-dose nasal toxicity (up to 14 days—intranasal rabbit)	Mild local toxicity; likely to be acceptable in clinical nasal formulations for short-term use	35

HP-β-CD	Formulation vehicle for poorly soluble drugs	Single and repeat dose toxicity studies (with later mainly in the rat and dog by oral or intravenous route and up to 1 year duration), reproduction toxicity (embryo-foetal studies in the rat and rabbit), battery of genotoxicity assays, carcinogenicity studies (by diet route in mouse and rat) plus ADME studies (single and multiple dosing)	Well tolerated (Some high dose effects seen – see Table 3)	36
HPMC	Constituent of oral and topical pharmaceuticals	Repeat-dose toxicity with routine end points (3 mos—oral rat)	No adverse effects	37
Labrasol/Labrafil/Transcutol (mixture of mono-, di-, and triglycerides with mono- and diesters of polyethylene glycerol and fatty acids and diethylene glycol monoethyl ether)	Bioavailability enhancer and solubilizer	Repeat-dose toxicity with routine end points (4 wks—oral rat)	High-dose effects of renal and adrenal changes (related to ethylene glycol) and hepatic enzyme induction	38
Liposome (sphingomyelin and cholesterol—55:45)	Used in liposomal-encapsulated vincristine sulfate (antitumor drug)	Repeat-dose toxicity with routine end points (once a wk for 6 wks—intravenous rat)	No toxicity seen	26

(*Continued*)

Table 2 Preclinical Studies for Recent Excipients Under Development or Used in Marketed Drug Products (*Continued*)

Excipient	Proposed/actual use	Toxicology studies	Remarks	References
Me PEG/PCL nanospheres	Injectable drug carrier	Single and short-term (7 days—intraperitoneal) toxicity in mice	No toxicity reported	39
MPL	Vaccine adjuvant	Cardiovascular/respiratory function safety pharmacology study, repeat-dose toxicity in rat (up to 4 wks—subcutaneous), rabbit and dog, reproduction (embryo-fetal studies in rat and rabbit) and 2 in vitro genotoxicity studies	No adverse effects	40
PVA copolymer	Bioavailability enhancer	Single-dose toxicity (in rat and dog) and maximum tolerated dose/short-term (2 wks—oral) (in rat and dog) toxicity as well as 2 in vitro and 1 in vivo genotoxicity studies. ADME studies with ^{14}C-labelled material are underway and a 3–6 mo toxicity study in the rat is planned	No adverse effects seen to date	41

PVAP	Tablet coatings/ink component for capsules	Single dose in various species, repeat dose (oral gavage or diet and up to 2 yrs in duration in rat and dog) and reproduction (fertility study in rat, embryo-fetal studies in rat and rabbit, and peri-postnatal study in rat) toxicity	Well tolerated [limited extreme high-dose effects seen (Table 3)]	42
Stealth liposomes (HSPC: MPEG-DSPE: cholesterol—5.5:56.4:38.3)	Drug delivery system for stealth liposomal doxorubicin	Examined in cardiovascular safety pharmacology study, single- and multiple-dose toxicity studies (intravenous—rat and dog), embryo-fetal study (rat), genotoxicity package (3 in vitro and 1 in vivo studies), and a local tolerance study (subcutaneous—rabbit)	No adverse findings seen (transient effects in dogs—see Table 3)	27
Surelease (aqueous ethylcellulose dispersion)	Modified release and taste-masking applications	Repeat-dose toxicity with routine end points (3 mos—diet rat) and genotoxicity package (2 in vitro and 1 in vivo studies)	No adverse toxicity or genotoxicity seen	43

Abbreviations: DOTAP: DOPE, dioleoyltrimethylammonium propane: dioleoylphosphatidylethanolamine; DDAB: DOPE, dimethyldioctadecylammonium bromide: dioleoyl-phosphatitidylethanolamin E; HP-β-CD, 2-hydroxypropyl-β-cyclodextrin; HPMC, hydroxypropyl methylcellulose; MePEG/PCL, methoxy poly(ethylene glycol)/poly (epsilon-caprolactone); MPL, monophosphoryl lipid A; PVA copolymer, polyvinyl alcohol acrylic acid methyl methacrylate polymer; PVAP, polyvinylacetate phthalate; HSPC, hydrogenated soy phosphatidylcholine; MPEG-DSPE, N-(carbomoyl-methoxypolyethylene glycol 2000)-1,2-distearoyl-sn-glycero-3-phosphoethanolamine sodium salt. ADME, adsorption, distribution, metabolism, and excretion.

Table 3 Examples of Recently Reported Excipient Toxicity

Excipient	Toxicology findings	Explanation	References
BZC and PS	Nasal lesions of inflammatory nature in rat	A low concentration (0.01% for BZC and 0.1% for PS) of these materials can lead to nasal lesions in the rat; however, this level is known as safe for human nasal mucosa exposure	44
Corn oil	Maternal toxicity (reduced body weight gain, food consumption, and renal pathology) and reduced pup viability in rat reproduction toxicity study	Related to the composition of the diet along with stress of pregnancy, parturition, and lactation (not seen with same diet in males and nonpregnant females); daily gavage administration of 10 mL/kg of corn oil is not recommended to pregnant rats	45
α- and β-CDs	Renal toxicity in rats from parenteral administration	Cause of toxicity is not clearly understood but, in part, may be related to an adaptive response due to excretion of osmotic agents at extremely high concentration; parenteral use of α- and β-CDs is not recommended	46, 47
HP-β-CD	Minor clinical pathology changes with urinary tract, liver and pancreas histopathology at 2000 mg/kg/day and above following chronic oral administration in rats, clinical pathology changes plus renal, urinary tract, lung, spleen and liver histopathology at 200 mg/kg/day and above with intravenous dosing in rats. Urinary tract changes and increased incidence of tumours in the pancreas and intestine seen in dietary carcinogenicity study in rats	Due to repeat dose toxicity study findings being restricted to high dose levels with reversibility demonstrated, it is concluded that HP-β-CD is a well tolerated excipient. For carcinogenicity study, urinary tract changes were reported as due to osmotic "necrosis," intestinal tumours were related to increased osmotic activity and pancreatic tumours were shown to be due to rat-specific hormonal stimulation	36
	Renal toxicity in rats from intraperitoneal dosing		48
	Renal, cardio and lung toxicity in monkeys from intravenous administration		49

Dibasic sodium phosphate	Nephrotoxicity in the form of proteinuria and glomerular calcification following intravenous (bolus) administration at 284 and 408 mg/kg/day for 14 days. These findings were not seen at 1 and 28 mg/kg/day	It is concluded that high-dose repeated use results in an overload of the glomerular epithelium during filtration through glomerular capillaries and produces insoluble calcium salt and glomerular lesions, resulting in proteinuria	50
HPMCAS	Low-level incidence of fetal clubfoot in older (1980s) rat teratology study following oral administration	Not seen in modern embryo-fetal rat study; earlier finding concluded to be a misdiagnosis or artifact	51
Menthol/peppermint oil	Evidence of genotoxicity in some in vitro assays	May be related to different components present in the oil; it is concluded that a genetic risk assessment is "very complicated or even impossible"	52, 53
Miglyol 812	Rats dosed orally with 10 mL/kg/day of 100% Miglyol for 4 wks showed soft and/or mucoid stools, reduced body weight gain, altered clinical pathology (decreased blood urea nitrogen, total protein, and globulins plus increased cholesterol and triglycerides), increased urine specific gravity, decreased thymus weight, and increased alveolar histiocytosis with focal interstitial inflammation; these changes reversed during a 4 wks non–dose recovery period	It was concluded that 100% Miglyol may not be innocuous when used orally in long-term toxicology studies in rats	54
PEG-linked proteins	Marked renal cortical tubular vacuolation in mice and rats following parenteral administration	Related to the combination of PEG and protein and the configuration of the PEG side chain and was seen at molecular weights of <70 kDa and at high doses; the clinical significance remains unknown	55, 56
P-407	Hyperlipidemia (raised serum triglycerides and cholesterol) seen in rabbits injected with 137.5 mg/kg/day of 22% P-407 for up to 14 days. No effect was seen at lower doses of 5.5 and 27.5 mg/kg/day	Finding is considered by study authors to be the result of "supraphysiologic doses"; lower doses may be used in controlled release drug delivery without the untoward hyperlipidemic effect	57
Propylene glycol	Maternal and reproductive toxicity in embryo-fetal studies in rabbit	Not recommended as a vehicle in embryo-fetal toxicity studies by oral administration	58

(*Continued*)

Table 3 Examples of Recently Reported Excipient Toxicity (*Continued*)

Excipient	Toxicology findings	Explanation	References
PVAP	Gastrointestinal tract irritation in rat and dog and embryo toxicity in rat and rabbit	Very high-dose effects only following oral administration	42
Stealth liposomes	Liposomal infusion reaction (hypoactivity, flushing, diarrhea, emesis, and decreased blood pressure seen following intravenous infusion in dogs)	Transient effect (resolved within 1–2 hr). Biological significance is not apparent but the finding has been related to histamine release due to the infusion of a large amount of lipid	27
L-Tartaric acid	Nephrotoxicity seen with intravenous infusion in monkeys	High-dose effect; caution is recommended in the use of this excipient	59
Various excipients with parenteral use	Hemolytic effects with rat blood (e.g., >1%-hydroxypropyl-β-CD, >2.5%-PEG 400, >5%-propylene glycol, >0.125%-Tween 80)	May be related to high-concentration effect	60

Abbreviations: BZC, benzalkonium chloride; PS, potassium sorbate; CDs, cyclodextrins; HP-β-CD, 2-hydroxypropyl-β-cyclodextrin; HPMCAS, hydroxypropyl methylcellulose acetate succinate; PEG, polyethylene glycol; P-407, poloxamer 407; PVAP, polyvinylacetate phthalate.

nanoparticles as drug carriers across mucosal surfaces are also in development (65). From a toxicological perspective, it should be noted that although these drug delivery systems appear to have little or no toxicity, they might be associated with a low level of immunological activity in the clinic. It has recently been questioned whether the immunotoxicological activity of polymers used in drug delivery (e.g., PEG and PLGAs) has been fully assessed (66). Indeed, it has been shown that the hypersensitivity reactions (including anaphylaxis) that have occurred in patients are associated with the polymer content of Zoladex. Furthermore, the PEG-grafted liposomal carrier for Caelyx/Doxil, polyethoxylated ether cremophor EL (a solubilizing agent for paclitaxel and cyclosporin), and poloxamer 188 (a stabilizer for various drugs) are all reported to produce low levels of immunological reactions. It is likely that such rare findings would not be predicted from routine toxicology studies, and more specialized immunotoxicological evaluation may be needed for these types of excipients. Finally, the clearance of these drug-delivery systems (e.g., through the kidneys) and any potential to accumulate in the body and/or biodegradability will need specific consideration during preclinical evaluation.

Overall, an important need with newer excipients, which are included to enhance activity of the formulation, is to clearly indicate their proposed mechanism of action and/or their relationship to the active drug in regulatory documentation (2). To take the point on activity further, it is interesting to note that a formal definition as an "inert" formulation constituent for some of the newer excipient materials is becoming impossible. Thus, it has been questioned whether materials such as attenuated adenoviruses and retroviruses (used as vectors for cell nuclei delivery), bacterial protein components, monoclonal antibodies, bacteriophages, fusion proteins, and molecular chimera are excipients, parts of a prodrug, or something in between (16,19). Other materials in development, which may be considered as difficult to classify, are the topical penetration enhancers [e.g., chitosan and soft enhancement of percutaneous absorption (SEPA or 2-*n*-nonyl-1,3-dioxolone)] (33,67–69). Also, "inactive" excipients such as cellulose acetate phthalate and hydroxypropyl methylcellulose phthalate, which are used in the production of tablets, have been shown to have antiviral activity in their own right and are being tested for efficacy in various animal models as "microbicides" for the prevention of infection (70).

A final consideration when testing a new excipient is its impurity profile. The specifications of the excipient material used in any preclinical testing package should be as close as possible, if not identical, to that proposed in drug formulations. Impurities may be toxic in their own right or can interact with active ingredients, leading to degradation and loss of efficacy (19,64,71). Thus, it is crucial to clarify the excipient impurity profile as even established pharmacopeia-listed materials such as the commonly used magnesium stearate has had questions raised on the safety and toxicity of its impurities (72).

Various materials have recently been evaluated in extensive preclinical programs to allow for regulatory "approval" as stand alone excipients. These include the cyclodextrins (CDs) and the hydrofluoroalkanes (HFAs) (46,47,73). Published literature on preclinical studies performed for other excipient materials under development is not extensive (presumably as developers do not want to aid competitors by allowing them to reference the available data), but information on some materials is available. Various examples of toxicology assessment work for an excipient that is under development or has been used in approved drugs for a range of applications are given in Table 2. Generally, excipient developers summarize their preclinical data in a Drug Master File, which remains confidential but can be made available to the regulators.

PRECLINICAL TESTING FOR AN ESSENTIALLY NEW EXCIPIENT

Materials that have had prior human use/exposure in food and cosmetics, or from the chemical industry, can be categorized as essentially new excipients. Such previous exposure is likely to be of help for oral and topical use excipient development. Excipients that have had established medicinal product use but are being used by a different dose route and/or chemically modified to enhance their properties are also likely to belong to this category.

Materials used in the food industry may be generally recognized as safe (GRAS) for human use and/or have established acceptable daily intakes (ADIs) based on toxicological data as established by, e.g., the Joint Food and Agriculture (FAD) and World Health Organization (WHO) Expert Committee (JECFA). Use of this information in a robust expert literature review may reduce the need for preclinical testing if the material is being considered as an excipient, although such data is of little use for nonoral products. Also, the reviewed toxicity data may be old and unreliable and, indeed, raise specific toxicological concerns, or the new proposed level of use may be higher than the oral ADI, all necessitating new investigations. Thus, although the CDs had a well-established history of use in food products, a full package of preclinical investigations, including metabolism work, short- to long-term rodent and nonrodent toxicity, reproduction toxicity, genotoxicity, and carcinogenicity studies were performed to support the excipient use of these materials as drug delivery systems (46,47). Interestingly, this work highlighted that α- and β-CDs are unsuitable for parenteral administration as they cause kidney tubule damage upon intravenous and subcutaneous administration in rats (46,47). In Japan, food additives or cosmetic substances used in drug formulations need to be treated as new excipients (11).

A case-by-case approach is needed for materials with previous medicinal product use but with a proposed route change. However, some testing is likely as indicated from an unofficial FDA perspective (74). Thus for inhalation excipients, evaluation of the toxic potential of the excipient after repeated inhalation dosing is recommended for materials with previous use in humans but with limited inhalation information. Furthermore, in Japan, a change of an excipient already used in orally or intravenously administered products to an externally applied product necessitates additional testing, including acute and subacute toxicity and local irritation investigations (11). A final consideration is for materials that have a fairly conventional excipient use but are being examined for a new, more active role. A good example is chitosan, which is under investigation for drug delivery potential as well as absorption-enhancing effects (68). As the material has a well-established history of low toxicity, further preclinical safety studies to support such use are likely to be minimal.

PRECLINICAL TESTING FOR AN ESTABLISHED EXCIPIENT

Established excipients are those with a history of use in drug formulations and thus are known to the regulators. Indeed, many appear in pharmacopeias and can be referenced as such in new drug applications in which they occur in the formulation. Thus, in theory, preclinical testing should not be needed. However, there is a common misconception that once an excipient is used in an approved product, it is automatically assumed to be safe for use in any product thereafter that involves the same route of administration and level of exposure (17). In reality, even in such cases, the level of toxicity information may need to be brought up to current standards for inclusion in new products. Toxicity testing may be needed if the known use is from limited exposure to the excipient (e.g., short-term therapy) and the new drug product is for long-term use (e.g., chronic therapy) or the excipient is used at higher levels than those currently known.

Even for established excipients, regulators will look carefully at their presence in new drug formulations because they are not necessarily inert materials and some have well-established activity and/or "toxicity." Clinically relevant adverse reactions are known for well-known excipients and the subject is covered elsewhere in the published literature (2,20,75–81). Findings tend to be uncommon compared to the overall prevalence of adverse drug reactions and often involve hypersensitivity reactions that are not likely to be predicted by conventional toxicity studies.

Cross-reference to published scientific reviews of the safety of materials used as excipients in a drug formulation is acceptable to the regulators. More and more reviews are becoming available for materials used as excipients. Examples of recently reviewed materials are the CDs (36,46,47), the HFAs (73), lactose (82), methyl and propyl paraben (83,84), peppermint oil, and menthol (52); PEG (85); polysorbates (Tweens) (86); and polyvinylacetate phthalate (42), polyvinylpyrrolidone (PVP) (87), propylene glycol (88), sodium metabisulfite (89), and trehalose (90). Even a new vaccine adjuvant monophosphoryl lipid A, which in terms of a constituent of a drug formulation can be considered an excipient, has been recently reviewed (40). The cross-reference process needs to involve a robust scientific appraisal of the published data, with comment on the relevance of any animal findings to humans in the new proposed formulation, together with the establishment of safety margins. Animal toxicity study findings have been reported for a number of established excipients such as lactose, maltodextran, mannitol, menthol, and PVP; however, these findings are generally minor and not relevant to human use (1,2). Some further more recent examples are given in Table 3. Again, the reported findings do not necessarily affect clinical use. Thus, overall, little or no extra preclinical studies are normally required for well-known excipients.

THE CHALLENGE

Some guidance on testing strategies for new excipients is available (9,10,14). However, as pointed out elsewhere (2), although useful as a starting point for development, such proposed packages of studies are extensive and generally no different from that of a new drug substance itself. Thus, they should not be viewed as a concrete list of preclinical studies that must be submitted to regulatory bodies but a series of topics that should be examined (17). The challenge to the toxicologist is what is the minimal, yet most scientifically robust, set of studies needed to support safe inclusion of an excipient in a drug formulation to be used in humans. Although still fragmentary, the literature would appear to suggest that developers have taken a fairly conservative approach with a range of preclinical studies for new excipients. An interesting challenge will be the design of such studies to support the expanding use of drug delivery systems.

CONCLUSION

Overall, a wide range of testing considerations are needed for new excipient materials, although the actual package of study types still remains a case-by-case approach.

On some occasions, a full program of studies may be needed to confirm a risk–benefit situation, whereas in others (e.g., lifesaving therapy), it may be acceptable to have reduced toxicity data (17). Similarly, the extent of studies needed to support the safe use of essentially new excipients, and indeed well-known materials, needs careful consideration based on available knowledge.

REFERENCES

1. Baldrick P. Pharmaceutical excipient development: the need for preclinical guidance. Reg Tox Pharm 2000; 32:210.
2. Baldrick P. Pharmaceutical excipient testing—a regulatory and preclinical perspective. In: Swarbrick J, Boylan JC, eds. Encyclopedia of Pharmaceutical Technology. 2nd ed. New York: Marcel Dekker, Inc., 2002:2141.
3. European Agency for the Evaluation of Medicinal Products (EMEA). www.emea.eu.int/index/indexh1.htm.
4. United States Food and Drug Administration, Center for Drug Evaluation and Research (FDA, CDER). www.fda.gov/cder/drug/default.htm.
5. U.S. Food and Drug Administration. FDA Inactive Ingredients Guide. Rockville, Maryland: Division of Drug Information Resources, 1996.
6. Japanese Pharmaceutical Excipient Directory. Japanese Pharmaceutical Excipient Council (JPEC), ed. Tokyo: Yakuji Nippon Ltd, 1996.
7. Rowe RC, Sheskey PJ, Weller PJ. Handbook of Pharmaceutical Excipients. 4th ed. American Pharmaceutical Association and The Pharmaceutical Press, 2003.
8. Shangraw R. Developments in tablet excipients since 1960. Manufact Chem 1986; 22.
9. International Pharmaceutical Excipients Council (IPEC). The proposed guidelines for the safety evaluation of new excipients, The IPEC Europe Safety Committee. Eur Pharm Rev 1997.
10. Steinberg M, Borzelleca JF, Enters EK, et al. A new approach to the safety assessment of pharmaceutical excipients. Reg Toxicol Pharm 1996; 24:149.
11. Uchiyama M. Regulatory status of excipients in Japan. Drug Inform J 1999; 33:27.
12. European Commission. Rules relating to marketing authorisation of medicinal products for human use. Council Directive 75/318/EEC. Luxembourg: Office for Official Publications of the European Communities, 1998.
13. Committee for Proprietary Medicinal Products (CPMP). Note for guidance on repeat dose toxicity. Final guidance issued by EMEA CPMP, October 2000, www.emea.eu.
14. Center for Drug Evaluation and Research (CDER). Guidance for Industry: Nonclinical Studies for Development of Pharmaceutical Excipients. Final guidance issued by FDA CDER, May 2005, www.fda.gov/cder/guidance.
15. Kalinkova GN. Studies of beneficial interactions between active medicaments and excipients in pharmaceutical formulations. Int J Pharm 1999; 187:1.
16. Apte SP, Ugwu SO. A review and classification of emerging excipients in parenteral medication. Pharm Tech 2003; 27(3):46.
17. Osterberg RE, See NA. Toxicity of excipients—A Food and Drug Administration perspective. Int J Tox 2003; 22:377.
18. Donovan MD, Flynn GL, Amidon GL. Absorption of polyethylene glycols 600 through 2000: the molecular weight dependence of gastrointestinal and nasal absorption. Pharm Res 1990; 7(8):863.
19. Crowley PJ, Martini LG. Excipients for pharmaceutical dosage forms. In: Swarbrick J, Boylan JC, eds. Encyclopedia of Pharmaceutical Technology. 2nd ed. New York: Marcel Dekker, Inc., 2002:1151.
20. Pifferi G, Restani P. The safety of pharmaceutical excipients. Il Farmaco 2003; 58:541.

21. Steinberg M, Kinoshita FK. Excipients—safety testing. In: Swarbrick J, Boylan JC, eds. Encyclopedia of Pharmaceutical Technology. 2nd ed. New York: Marcel Dekker, Inc., 2002:1143.
22. Perlin DS. Amphotericin B cochleates: a vehicle for oral delivery. Curr Opin Investig Drugs 2004; 5(2):198.
23. Zarif L, Mannino RJ. Cochleates: lipid-based vehicles for gene delivery-concept, achievements and future development. In: Kluwer H, ed. Cancer Gene Therapy: Past Achievements and Future Challenges. New York: Academic/Plenum Publishers, 2000:83.
24. Drummond DC, Kirpotin D, Benz CC, et al. Liposomal drug delivery for cancer therapy. In: Brown DM, ed. Drug Delivery Systems in Cancer Therapy. Totowa, New Jersey: Humana Press Inc., 2004:191.
25. Cattel L, Cerutti M, Dosio F. From conventional to stealth liposomes a new frontier in cancer chemotherapy. Tumori 2003; 89:237.
26. Tam PM, Oshanek R, Yasuda N, et al. Comparative repeated-dose toxicity of liposome-encapsulated vincristine sulfate and free vincristine sulfate in rats. Tox Sci 2004; 78(1-S).
27. Working PK, Dayan AD. Pharmacological-toxicological expert report Caelyx™ (Stealth® liposomal doxorubicin HCL). Hum Exp Tox 1996; 15:752.
28. Freeman C, Weiner ML, Kotkoskie LA, et al. Subchronic and developmental toxicity studies in rats with Ac-DiSol croscarmellose sodium. Int J Tox 2003; 22:149.
29. Kotkoskie LA, Freeman C. Subchronic oral toxicity study of Aquacoat® ECD ethylcellulose aqueous dispersion in the rat. Food Chem Tox 1998; 36:705.
30. Palmieri MA, Freeman C, Kotkoskie LA. Developmental toxicity study of Aquacoat® ECD ethylcellulose aqueous dispersion administered orally to rats. Food Chem Tox 2000; 38:71.
31. Kotkoskie LA, Freeman C, Palmieri MA. Subchronic toxicity and developmental toxicity studies in rats with Aquatic® aqueous enteric coating. Int J Tox 1999; 18:109.
32. Batt KJ, Kotkoskie LA. An evaluation of genotoxicity tests with Aquatic® aqueous enteric coating. Int J Tox 1999; 18:117.
33. Illum L. Chitosan and its use as a pharmaceutical excipient. Pharm Res 1998; 15(9):1326.
34. Tosca P, Grossi I, Brooker M, et al. Toxicity comparison of liposomes comprised of dioleoyltrimethylammonium propane: dioleoylphosphatidylethanolamine (DOTAP: DOPE) or dimethyldioctadecylammonium bromide: dioleoylphosphatitidylethanolamin E (DDAB: DOPE) following multiple IV injections in Fischer 344 rats. Tox Sci 2003; 72(S–1):303.
35. Hjortkjaer RK, Bechgaard E, Gizurason S, et al. Single- and repeated-dose local toxicity in the nasal cavity of rabbits after intranasal administration of different glycols for formulations containing benzodiazepines. J Pharm Pharmacol 1999; 51(4):377.
36. Gould S, Scott RC. 2-Hydroxypropyl-12-cyclodyctoin (HP-β-CD)—a toxicological review. Ford Chem Tox 2005; 43:451.
37. Obara S, Muto H, Shigeno H, et al. A three-month repeated oral administration study of a low viscosity grade of hydroxypropyl methylcellulose in rats. J Toxicol Sci 1999; 24(1): 33–43.
38. Spire C, Lepage JF, Vermeil de Conchard G, et al. Assessment of Labrasol®/Labrafil®/ Transcutol® (4/4/2, v/v/v) as a vehicle for hydrophobic compounds after a 4-week oral toxicity study in Wistar rats. Tox Lett 2003; (suppl 1):144.
39. Kim SY, Lee YM, Baik DJ, et al. Toxic characteristics of methoxy poly[ethylene glycol/ poly(epsilon-caprolactone)] nanospheres: in vitro and in vivo studies in normal mice. Biomat 2003; 24(1):55.
40. Baldrick P, Richardson D, Elliott G, et al. Safety evaluation of monophosphoryl lipid A (MPL): an immunostimulatory adjuvant. Reg Tox Pharm 2002; 35:398.
41. Hoshi N, Ogura T, Shimamoto T, et al. Development of PVA copolymer capsules. Pharm Tech Eur 2004; 37.
42. Schonecker DR, DeMerlis CC, Borzelleca JF. Evaluation of the toxicity of polyvinylacetate phthalate in experimental animals. Food Chem Tox 2003; 41:405.

43. DeMerlis CC, Schoneker DR, Borzelleca JF. A subchronic toxicity study in rats and genotoxicity tests with an aqueous ethylcellulose dispersion. Food Chem Tox 2005; 43: 1355.
44. Cho J-H, Kwun Y-S, Jang H-S, et al. Long-term use of preservatives on rat nasal respiratory mucosa: effects of benzalkonium chloride and potassium sorbate. Laryngoscope 2000; 110(2 Part 1):312.
45. Sato M, Wada K, Marumo H, et al. Influence of corn oil and diet on reproduction and the kidney in female Sprague–Dawley rats. Tox Sci 2000; 56:156.
46. Mosher G, Thompson D. Complexation and cyclodextrins. In: Swarbrick J, Boylan JC, eds. Encyclopedia of Pharmaceutical Technology. 19. Basel, New York: Dekker Inc., 2000:49.
47. Thompson DO. Cyclodextrins-enabling excipients: their present and future use in pharmaceuticals in drug formulations. Crit Rev Ther Drug Carrier Sys 1997; 14(1):1.
48. Wild SL, Cosenza M, Sullivan L, et al. Comparative pathology of vehicles used for rapid in vivo screening of small molecules. Tox Sci 2001; 60(1):375.
49. Oliver J, Crofts F, Mangipudy R, et al. Organ toxicity in cynomolgus monkeys following intravenous administration of the vehicle 2-hydroxypropylbetacyclodextrin. Tox Path 2003; 31(suppl):144.
50. Tsuchiya N, Matsushima S, Takasu N, et al. Glomerular calcification induced by bolus injection with dibasic sodium phosphate solution in Sprague–Dawley rats. Tox Path 2004; 32:408–412.
51. Cappon GD, Fleeman TL, Rocca MS, et al. Embryo/fetal development studies with hydroxypropyl methylcellulose acetate succinate (HPMCAS) in rats and rabbits. Birth Defects Res (Part B) 2003; 68:421.
52. Anonymous. Final report on the safety assessment of *Mentha piperita* (peppermint) oil, *Mentha piperita* (peppermint) leaf extract, *Mentha piperita* (peppermint) leaf and *Mentha piperita* (peppermint) leaf water. Int J Toxicol 2001; 20(3):61.
53. Luzutka JR, Mierauskiene J, Slapsyte G, et al. Genotoxicity of dill (*Anethum graveolus* L.), peppermint (mentha x piperita L.) and pine (*Pinus sylvestris* L.) essential oils in human lymphocytes and *Drosophila melanogaster*. Food Chem Tox 2001; 39:485.
54. Sellers R, Antman M, Khan N, et al. Effects of Miglyol 812 on rats after 4-weeks of gavage as compared to methylcellulose/Tween 80. Tox Path 2004; 32(suppl 1):154.
55. Bendele A, Seely J, Richey C, et al. Short communication: renal tubular vacuolation in animals treated with polyethylene-glycol-conjugated protein. Tox Sci 1998; 42:152.
56. Bolon B, Faust J, Storm N, et al. Cytoplasmic vacuoles in renal tubular epithelium of mice given polyethylene glycol (PEG)-conjugated proteins are reduced by alternating PEG size and conformation. Tox Sci 2001; 60(1):376.
57. Blonder JM, Baird L, Fulfs JC, et al. Dose-dependent hyperlipidemia in rabbits following administration of poloxamer 407 gel. Tox Sci 2000; 54(1):409.
58. Frieling WJAM, Heijink E, Tesh SA, et al. Embryo-foetal development toxicity of propylene glycol in NZW rabbits. Repro Tox 2000; 14:562.
59. Ortega S, Mounho B, Kelley S, et al. L-Tartaric acid-induced nephrotoxicity. Tox Sci 2001; 60(1):324.
60. Sawant SG, Greene S, Cosenza ME. Hemolytic potential of vehicles used for rapid in vivo screening of small molecules, Tox Sci 2002; 66(1):248.
61. Sahoo SK, Labhasetwar V. Nanotech approaches to drug delivery and imaging. Drug Dis Today 2003; 8(24):1112.
62. Harris JM, Chess RB. Effects of PEGylation on pharmaceuticals. Nat Rev Drug Dis 2003; 2:214.
63. Torchilin VP, Lukyanov AN. Peptide and protein drug delivery to and into tumors: challenges and solutions. Drug Dis Today 200; 8(6):259.
64. Nema S, Brendel RJ, Washkuln RJ. Excipients—their role in parenteral dosage forms. In: Swarbrick J, Boylan JC, eds. Encyclopedia of Pharmaceutical Technology. 2nd ed. New York: Marcel Dekker, Inc., 2002:1164.

65. Vila A, Sanchez A, Perez C, et al. PLA-PEG nanospheres: new carriers for transmucosal delivery of proteins and plasmid DNA. Poly Adv Tech 2003; 13(10–12):851.
66. Hunter AC, Moghimi SM. Therapeutic synthetic polymers: a game of Russian roulette? Drug Dis Today 2002; 7(19):998.
67. Alonso MJ, Sanchez A. The potential of chitosan in ocular drug delivery. J Pharm Pharmaco 2003; 55:1451.
68. Singla AK, Chawla M. Chitosan: some pharmaceutical and biological aspects—an update. J Pharm Pharmacol 2001; 53(8):1047.
69. Zhong WZ. Application of solid-phase extraction in the method development for determination of SEPA, an acronym for soft enhancement of percutaneous absorption in human, rat and rabbit serum using GC-FID method. J Pharm Biomed Anal 1998; 16(7):1179.
70. Neurath AR, Strick N, Li Y-Y, et al. Design of a "microbicide" for prevention of sexually transmitted diseases using "inactive" pharmaceutical excipients. Biologicals 1999; 27:11.
71. Jakel D, Keck M. Purity of excipients. In: Weiner ML, Kotkoskie LA, eds. Excipient Toxicity and Safety, Drugs and the Pharmaceutical Sciences. 103. Basel, New York: Marcel Dekker, Inc., 2000:21.
72. Chowhan ZT. A rational approach to setting limit tests and standards on impurities in excipients. Pharm Tech Eur 1995; 7(10):66.
73. Alexander DJ, Libretto SE. An overview of the toxicology of HFA-134a (1,1,1,2-tetrafluoroethane). Hum Exp Toxicol 1995; 14(9):715.
74. DeGeorge JJ, Ahn CH, Andrews PA, et al. Considerations for toxicology studies of respiratory drug products. Regul Toxicol Pharmacol 1997; 25(2):189–193.
75. American Academy of Pediatrics (AAP), Committee on Drugs. "Inactive" ingredients in pharmaceutical products: update. Pediatrics 1997; 99(2):268.
76. Barband A. Place of excipients in drug-related allergy. Clin Rev Aller Immun 1995; 13:253.
77. Golightly LK, Smolinske SS, Bennett ML, et al. Pharmaceutical excipients. Adverse effects associated with inactive ingredients in drug products (Part I). Med Tox 1988; 3:128.
78. Golightly LK, Smolinske SS, Bennett ML, et al. Pharmaceutical excipients. Adverse effects associated with inactive ingredients in drug products (Part II). Med Tox 1988; 3:209.
79. Martindale. The Complete Drug Reference. Sweetman SC, ed. 33rd ed. London: Pharmaceutical Press, 2002.
80. Napke E, Stevens DGH. Excipients and additives: hidden hazards in drug products and in product substitution. Can Med Assoc J 1984; 131:1449.
81. Winek CL. History of excipient safety and toxicity. In: Weiner ML, Kotkoskie LA, eds. Excipient Toxicity and Safety, Drugs and the Pharmaceutical Sciences. Vol. 103. Basel, New York: Marcel Dekker, 2000:59.
82. Baldrick P, Bamford DG. A toxicological review of lactose to support clinical administration by inhalation. Food Chem Tox 1997; 35:719.
83. Soni MG, Burdock GA, Taylor SL, et al. Safety assessment of propyl paraben: a review of the published literature. Food Chem Tox 2001; 39:513.
84. Soni MG, Taylor SL, Greenberg NA, et al. Safety assessment of methyl paraben: a review of the published literature. Food Chem Tox 2002; 40:1335.
85. Anonymous. Final report on the safety assessment of PEG-7, -30, -40, -78 and -80 glyceryl cocoate. Int J Toxicol 1999a; 18(suppl 1):33.
86. Anonymous. Final report on the safety assessment of PEG-20 sorbitan cocoate; PEG-40 sorbitan diisostearate; PEG-2, -5, and -20 sorbitan isostearate; PEG-40 and -75 sorbitan lanolate; PEG-10, -40, -44, -75 and -80 sorbitan laurate; PEG-3, and -6 sorbitan oleate; PEG-80 sorbitan palmitate; PEG-40 sorbitan perisostearate; PEG-40 sorbitan peroleate; PEG-3, -6, -40, and -60 sorbitan stearate; PEG-20, -30, -40, and -60 sorbitan tetraoleate; PEG-60 sorbitan tetrastearate; PEG-20 and -160 sorbitantriisostearate; PEG-18 sorbitan trioleate; PEG-40 and -50 sorbitol hexaoleate; PEG-30 sorbitol tetraoleate laurate; and PEG-60 sorbitol tetrastearate. Addendum to the final report on the safety assessment of

polysorbates. [Elder RL. Final report on the safety assessment of polysorbates 20, 21, 40, 60, 61, 65, 80, 81 and 85. J Am Coll Toxicol 1984; 3:1; Int J Toxicol 2000; 19(suppl 2):43.]
87. Anonymous. Final report on the safety assessment of polyvinylpyrrolidone (PVP). Int J Toxicol 1998; 17(4):95.
88. Anonymous. Final report on the safety assessment of propylene glycol (PG) dicaprylate, PG dicaprylate/dicaprate, PG dicocoate, PG dipelargonate, PG isostearate, PG laurate, PG myristate, PG oleate, PG oleate SE, PG dioleate, PG dicaprate, PG diisostearate and PG dilaurate. Int J Toxicol 1999b; 18(suppl 2):35.
89. Anonymous. Final report on the safety assessment of sodium sulfite, potassium sulfite, ammonium sulfite, sodium bisulfite, ammonium bisulfite, sodium metabisulfite and potassium metabisulfite. Int J Toxicol 2003; 22(suppl 2):63.
90. Richards AB, Krakowka S, Dexter LB, et al. Trehalose: a review of properties, history of use and human tolerance, and results of multiple safety studies. Food Chem Tox 2002; 40:871.

4

Regulation of Pharmaceutical Excipients

Robert G. Pinco[a] **and Theodore M. Sullivan**[b]
Buchanan Ingersoll, P.C., Washington, D.C., U.S.A.

INTRODUCTION

The development of innovative pharmaceuticals is critical to improving the health care and standard of living for billions of people around the globe. Bringing novel pharmaceuticals to market requires the expenditure of very significant resources by both drug companies and government regulators. Most of the development efforts and government regulatory expenditures are directed toward the discovery, testing, and oversight of novel pharmaceutical "active" ingredients, as these ingredients are seen as the key to making new drugs available to the world's population. As a result, the process by which new active ingredients are investigated and regulated is well developed and understood by all parties involved in the process. In contrast, the regulation of pharmaceutical excipients, the "inactive" ingredients used in drug products, presents significant challenges for both government regulators and industry. Historically, excipients were inert substances that were used mainly as fillers, coatings, manufacturing aids, and diluents. Commonly used excipients such as cornstarch, lactose, talc, and sucrose did not present significant questions of safety, and were largely ignored by the regulatory community. Advancements in pharmaceutical technology have rendered this view of excipients as simple inert pharmaceutical fillers obsolete. Pharmaceutical companies and government regulators are slowly developing mechanisms to effectively develop and regulate these pharmaceutical ingredients.

Traditional excipient ingredients of the type mentioned above remain quantitatively the most important and widely used, and for these ingredients, the dictionary definition of excipient is adequate. Webster's defines an excipient as "an inert substance used as diluent or vehicle for a drug" (1). However, sophisticated, high technology excipients, which are critical to the quality and bioavailability of some modern drug products and novel dosage forms, do not fit within the traditional definition. Government agencies have begun to recognize this change even though regulatory mechanisms have not yet evolved to adequately address the regulation

---

[a] Robert G. Pinco is a Senior Counsel at the law firm Buchanan Ingersoll, P.C.

[b] Theodore M. Sullivan is an associate at Buchanan Ingersoll, P.C.

of such excipients. The U.S. Food and Drug Administration (FDA) does not have a formal regulatory definition of "excipient"; however, recent guidance on nonclinical safety studies for excipients provides some indication of FDA acceptance of excipients as more than fillers. The background section of that guidance states:

> In this guidance, the phrase *new excipients* means any inactive ingredients that are intentionally added to therapeutic and diagnostic products, but that: (i) we believe are not intended to exert therapeutic effects at the intended dosage, although they may act to "improve product delivery" (e.g., enhance absorption or control release of the drug substance); and (ii) are not fully qualified by existing safety data with respect to the currently proposed level of exposure, duration of exposure, or route of administration. Examples of excipients include fillers, extenders, diluents, wetting agents, solvents, emulsifiers, preservatives, flavors, absorption enhancers, sustained-release matrices, and coloring agents (emphasis added) (2).

This definition begins to capture the wide potential range of excipient usage in modern pharmaceutical products. It is, however, a long distance from recognizing the many potential uses of excipients to the development of a rational regulatory scheme that can facilitate market and regulatory acceptance of nontraditional novel ingredients.

Despite widespread agreement that the current system is woefully inadequate in its ability to review new excipients, the regulation of excipients remains mired in the traditional system that relies upon approval of excipients only as components of a drug product, with no true independent review. In this chapter, we discuss the existing regulatory process for excipient review, as well as some potential alternatives to the existing process, some of which have been attempted, without much success, in the past. This article focuses on the regulatory environment in the United States. Where the discussion is applicable to other countries, it will be specifically mentioned in the text.

NO INDEPENDENT STATUS FOR EXCIPIENTS

There is no process or mechanism currently in place within the FDA to independently evaluate the safety of pharmaceutical excipients. Instead, for drugs subject to FDA premarket approval, excipients are only "approved" as components of new drugs.

A number of proposals (some of which are discussed elsewhere in this article) have been made over the years to attempt to provide some mechanism for independent review or approval of inactive ingredients, but such proposals have been largely unsuccessful for a variety of reasons. In large measure, it is because excipients are not included in recent User Fee legislation, which provides funds to the FDA in exchange for rapid review and clearance of drugs, and medical devices. It is a political truism that "regulation follows the money." Excipients, because they generally do not pose safety concerns, remain largely ignored. As will be discussed below, in the absence of direct legislation, Congressional oversight, and funding, there has been no major effort to increase FDA regulation and oversight over excipients. Further, the worldwide pharmaceutical industry is somewhat ambivalent about increased government regulation and/or premarket clearance of excipients.

Formulation of finished pharmaceuticals has always been somewhat of an art, rather than totally a science. Attempts to completely objectify regulation of with excipients have been largely ignored by the regulators. The pharmaceutical industry, particularly the large multinational companies, have been searching for mechanisms to standardize worldwide regulation of excipients by means of a harmonization process which combines the efforts of the United States Pharmacopeia (USP), the Council of Europe (European Pharmacopeia), and the Japanese Pharmacopeia. These efforts have generally been along the lines of creating harmonized standards for excipient identity, safety, and manufacturing.

Despite the harmonization efforts, the pharmaceutical industry is somewhat leery of potential excipient regulatory systems (such as regulatory premarket review and approval requirements for excipients) that may create increased barriers to the development of new pharmaceutical biologics, or medical devices. The industry is concerned that any efforts to require that excipients face formal independent premarket review prior to use in a drug could result in significant delays in bringing novel excipients to market. Various major regulatory agencies charged with the regulation of drugs, biologics, and medical devices are burdened by legislative measures, as well as market changes, which result in significantly increased responsibilities, while agency budgets have not increased proportionally. Thus, where there are many competing legislative mandates for limited funds, there is, and has been, little incentive to create a major government regulatory effort specifically targeted at excipients. Even where such efforts are begun, as with the FDA's over-the-counter (OTC) inactive ingredient proposal in the mid-1970s (discussed in detail infra), there has been virtually no incentive for FDA, nor its European counterparts, to develop a specific regulatory plan to more adequately define the legal status for excipients.

EXCIPIENTS FOR OVER-THE-COUNTER DRUGS

In contrast to the requirement that excipients in new drugs be approved as a component of that new drug, for OTC drugs that are regulated under the FDA's OTC monograph system, there is no required approval for excipients. FDA's monograph system is a series of regulations that define what active ingredients and claims are permitted for a variety of OTC drugs. These regulations do not define specific formulations, but rather set forth dose ranges for acceptable active ingredients. For inactives, FDA regulations provide that the drugs must contain only suitable inactive ingredients which are safe in the amounts administered and do not interfere with the effectiveness of the preparation or with suitable tests or assays to determine if the product meets its professed standards of identity, strength, quality, and purity (3).

Therefore, excipients that comprise the bulk of OTC drug products must only be deemed safe and suitable by the drug manufacturer prior to use.[c]

However, in the 1970s, there was an attempt to create a more ingredient-specific review for excipient ingredients for OTC drugs. The proposed (but never finalized) OTC inactives regulations were issued in the Federal Register on April 12, 1977 (4) and grew out of overreaching by several of the 17 expert Advisory Panels formed during the first years of the OTC review process. This process began in 1972, when the FDA tasked expert panels with reviewing and rendering advice on the general recognition of safety and effectiveness of what was expected to be approximately 40 therapeutic classes of drugs that included more than 300,000 marketed products

[c] Except in the case of the few OTC drugs that are the subjects of approved New Drug Applications.

(eventually the number of therapeutic classes reached approximately 80). During this review, the FDA attempted to keep the task at a manageable level by not reviewing specific marketed products, but rather by creating various monographs (standards) for OTC "active ingredients" in numerous therapeutic classes. Expert Panels were advised only to consider "excipients" (or complete product formulations) when such excipients or formulations materially impacted the drug's efficacy or diminished its safety. In therapeutic classes such as antacids, laxatives, antidiarrheals, expectorates, antitussives (cough suppressants), sleep aids, and numerous oral products, excipients such as lactose, starch, methylcellulose, magnesium stearate, etc., had long been used in the manufacturing of OTC finished products without raising safety concerns, and were therefore not examined. However, there were a number of exceptions where inactive ingredients had a noticeable (and usually detrimental) impact on safety or effectiveness. For example, slight variation in the formula (including changes to excipient ingredients) of antiperspirants rendered the final formulated product ineffective. Certain antimicrobials were rendered ineffective as a result of changes to pH (e.g., quarternary ammonium compounds) or by the addition of certain color ingredients. Chlorhexidine, for example, would bind irreversibly with certain red color ingredients and be rendered largely ineffective.[d] Cough and cold remedies when compounded in time-release formulations were prone to dose-dumping.[e] The problems with time-release dose-dumping was deemed serious enough that it caused the FDA to remove all time-release drugs from the monograph review, and require that such drugs be individually approved through new drug applications (NDAs).

In some therapeutic classes, particularly topical drugs, Expert Panels focused extensively on the excipients and began developing monographs for these ingredients. When this occurred, as happened with the Hemorrhoidal Panel, the FDA management that provided oversight to the Advisory Panel, including the FDA Commissioner and the Commissioners' OTC Steering Committee (which essentially ran the entire OTC review project), tried to convince the Panel that ingredients that did not contribute to a product's stated therapeutic use should be dropped from the review. Despite this advice, the Panel continued its review of excipients that were used in hemorrhoidal drug products. The FDA allowed the Panel to continue its review of these ingredients, but did not incorporate any Panel comments on the inactives into its draft findings for hemorrhoidal drugs. This ultimately led to a somewhat contentious meeting between the Panel members and FDA Commissioner Schmidt, Chief Counsel Peter Hutt, the Director of the Bureau of Drugs, Dr. Richard Crout, and Mr. Robert Pinco, Head of the OTC review.[f] During this meeting, the Panel was persuaded to omit any inactive ingredient which the Panel did not believe had some therapeutic activity.

The controversy with the Hemorrhoidal Panel was mainly the result of the fine line between active and inactive ingredients in some therapeutic classes. Ingredients that were generally used as inactive ingredients in most drugs could potentially be considered as the active ingredient in products indicated for the relief of such minor

---

[d] ICI's Hibiclens™ stated it was a 4% solution of chlorhexidiene but, in fact, only 1% was available. This became evident when the first ANDA (generic) versions went on the market in 1985, and were, much to the chagrin of ICI, more effective than the pioneer product.

[e] Time-release technology from the time of the OTC review (1970s) was prone to dose-dumping. Significant improvements have been made in time-release formulation in the last several decades.

[f] The Bureau of Drugs was the predecessor of the Center for Drug Evaluation and Research.

maladies as hemorrhoids, irritation of the skin, windburn, and diaper rash. In these cases, traditionally inactive ingredients such as petrolatum (Vaseline®) or baby powder (talc) could be considered as active ingredients.

Because of the problems with certain expert panels reviewing topical products, the FDA proposed a regulation identifying those classes or groups of excipients that could be included in OTC products without specific FDA premarket review. The Agency expanded upon the general requirement that excipients be "safe and suitable" in a proposed regulation published in the Federal Register on April 12, 1977. In that document, the Agency began by referring to 21C.F.R.210.3(d)(5), in defining an active ingredient as "any component that is intended to furnish pharmacological activity or other direct effect in the diagnosis, cure, mitigation, treatment, or prevention of disease, or to affect the structure or function of the body of man ...". The Agency then explained that they considered all other components of a finished drug product to be inactive (excipients).

The Agency's main concern in promulgating the proposed regulation was to make it clear to the industry that the ingredients heretofore considered active, and that failed to be included in a final OTC monograph (i.e., considered either category II—unsafe or ineffective or category III—needs more study) would have to be removed from the market at the conclusion of the administrative process, if these ingredients did not serve a useful purpose at the product's final formulation. The Agency was particularly concerned that consumers of products long on the U.S. market would be defrauded if the nonmonographed ingredients that did not serve a useful purpose in the product were merely recharacterized as inactive excipients.

The Agency set out six criteria for an excipient in an OTC product as follows:

1. The ingredient is listed in an official compendium [e.g., USP/National Formulary (NF)] as a pharmaceutical aid or performed certain physical or technical functions in the final formulations (as will be set forth below).
2. The inactive ingredient is used at a level no higher than reasonably required to achieve its physical or technical function. For example, an antimicrobial excipient ingredient could only be used at a level consistent with preservation of the finished product (not at therapeutic levels), and a sunscreen ingredient could only be used at levels that protected the product from breaking down if the top of the jar was left open, not for protecting the user.
3. It is safe when at levels used as inactive ingredient.
4. If it is a color, it must meet appropriate color additive regulations.
5. It does not interfere with the effectiveness of the product. For example, fluoride toothpastes have to be formulated carefully, as the various fluorides react with certain excipients, rendering the product ineffective.
6. The inactive ingredient does not interfere with suitable tests or assays used to assure the identity, quality, strength, or purity of the finished product.

The FDA then set forth a list of 23 physical or technical functions these excipients perform as follows:

1. Air displacement agents: substances that displace air.
2. Color additives: as defined in Section 201(t) of the Act.
3. Denaturing agents: substances added to alcohol to render it unfit for use as an intoxicating beverage.
4. Dispersing agents: substances that promote even distribution throughout a liquid, gaseous, or solid medium with the formulation of a two-phase system.

5. Emollients: bland, fatty, or oleaginous substances that may be applied locally, particularly to the skin, and also to mucous membranes or abraded tissue.
6. Emulsifiers and emulsifying salts: substances that modify surface tension in the component phase of an emulsion to establish a uniform dispersion or emulsion.
7. Flavors and flavoring adjuncts: substances added to impart or help impart a taste or aroma to a product.
8. Fragrances: substances, extracts, or preparations for fragrance diffusing or imparting an agreeable or attractive odor.
9. Humectants: hygroscopic substances incorporated in a product to promote retention of moisture, including moisture retention agents and antidusting agents.
10. Identifiers: substances incorporated in a product to aid manufacturers to distinguish their products from similar or counterfeit products.
11. Levigating agents: substances that aid in reducing another substance to an extremely fine state of subdivision after that other substance has been made into a paste with some suitable liquid in which it is insoluble; also, nonsolid vehicles used to disperse a solid substance to a paste.
12. Ointment bases: vehicles to permit topical application of active medicinal substances.
13. pH control agents: substances added to change or maintain active acidity or basicity, including buffers, acids, alkalies, and neutralizing agents.
14. Preservatives: substances that are added to preparations to prevent or retard deterioration or degradation in a product; such substances include antifungal agents, antioxidants, antimicrobial agents, mold and rope inhibitors, and agents having the effects listed by the National Academy of Sciences—National Research Council under "preservatives." For the purpose of this section, "antioxidant" is defined as a substance that inhibits oxidation and is used to prevent rancidity of oils or fats or the deterioration of other materials through oxidative processes, including color changes.
15. Propellants, aerating agents, and gases: gases used to supply force to expel a product or used to reduce the amount of oxygen with the product in packaging.
16. Solvents and vehicles: substances used to dissolve or extract another substance or used as carriers of other substances.
17. Stiffening agents: substances that increase the viscosity of certain pharmaceutical preparations, especially ointments.
18. Suppository bases: pharmaceutical bases that are solid at room temperature but melt at body temperature.
19. Surface-active agents: substances used to modify the surface properties of liquids for a variety of effects. The definition includes solubilizing agents, dispersants, detergents, wetting agents, dehydration enhancers, whipping agents, foaming agents, and defoaming agents.
20. Suspending agents: substances required to overcome agglomeration of the dispersed particles and increase the viscosity of the medium so that the particles settle slowly.
21. Tablet and capsule diluents: inert substances incorporated to increase the bulk, to make the tablet or capsule of practical size.

22. Tablet binders.
23. Tablet-coating agents.

Unfortunately, this proposed regulation was never finalized. However, it remains as an active proposal and forms the Agency's enforcement policy with regard to OTC drugs that were relabeled to declare formally active ingredients that were excluded from a monograph as inactive excipients. It also gives the industry a good idea of acceptable excipient categories, at least for older, established products and product types.

EXCIPIENTS IN NEW DRUGS

Most OTC monograph drugs are ones with long histories of use. They are generally unsophisticated, and contain relatively simple and well-characterized excipients. Aside from some issues discussed above that arose during the OTC monograph process, there are relatively few regulatory issues regarding these ingredients. Problems that do occur generally have to do with the quality and purity of the excipient rather than use of novel excipients.

Most novel excipient issues are related to use of the excipients in new drugs. New drugs in the United States may only be marketed after approval of the drug in either a NDA or an Abbreviated New Drug Application (ANDA). NDAs generally concern drugs that contain either a new active pharmaceutical ingredient (API) or a new dosage form of an existing API. ANDAs are applications for drugs that are essentially copies of other approved new drugs, but may generally have different excipients.[g]

The FDA requires that NDA and ANDA applicants submit information about all components of the drug, including all excipients. Information on the safety of excipients used in a drug product has been required by federal law since the enactment of the first modern national food and drug legislation, the Food, Drug, and Cosmetic Act, in 1938. This law was, in fact, precipitated by an incident involving an unsafe excipient. In 1937, more than 100 people in the United States died as a result of poisoning from ingesting "Elixir Sulfanilamide," a liquid dosage form of a common drug, Sulfanilamide, used to treat streptococcal infections. A chemist at S.E. Massengill Co. found that sulfanilamide would dissolve in diethylene glycol, and the company formulated and produced the liquid form of the drug. No toxicology or safety testing whatsoever was performed on the drug and none was required by the law of the time (5). As a result of this tragedy, the new drug law required that all new drugs, including their components, must be evaluated for safety.

NDA applicants are required to submit a list of all excipients (as well as other drug components), used in the manufacture of a proposed new drug. Additionally, the applicant must provide sufficient information to establish that the use of each excipient is safe for its intended use, at its intended quantity. This information includes safety data, a statement of the composition, specifications, and any analytical methods used for the excipient. When a USP/NF monograph exists for an excipient, the applicant may state that the excipient in the drug will comply with the standards in the monograph instead of providing composition, specification, and analytical method information. The required safety information includes

[g] For some dosage forms, e.g., parenteral drug products, the ANDA applicant must use the same excipients as are found in the original NDA.

manufacturing information, and full toxicology studies to demonstrate that the excipient is safe for the intended use (6). In some cases, there may be sufficient assurances of safety from other sources, such as documented human exposure, or professional/scientific review of the excipient through some other mechanism (as will be discussed later in this chapter), to adequately demonstrate the safety of the excipient.

In the absence of existing human exposure data or other review of the excipient, FDA recommends in its guidance on Pharmaceutical Excipients (2) that the excipient be evaluated using a battery of standard nonclinical tests (7). Which tests are appropriate depends upon the likely patient exposure given the intended use of the drug if approved. Table 1 provides a summary of the necessary tests. This test paradigm will likely be considered as setting the standard for data requirements for any new excipients, whether or not approved by the FDA in an NDA or ANDA, or reviewed by some future alternative review/approval mechanism.

The required information and data about an excipient is submitted to the FDA by the drug product sponsor in the NDA. Many excipient manufacturers choose to protect the proprietary parts of their excipient manufacturing and safety data through use of a drug master file (DMF). A DMF is a mechanism that permits drug component (including excipient) manufacturers to submit information to the FDA in a document that is held as confidential by the Agency. The information in the DMF can be reviewed by the FDA, but is not disclosed to other parties. When an NDA is submitted to the FDA for a drug that is manufactured with an excipient that

Table 1 Recommended Pharmacology and Toxicology Testing for New Excipients

For all new excipients	Pharmacology studies battery as per ICH S7A guidance
Max. exposure of 14 consecutive days or less	Acute toxicology testing in a rodent and mammalian nonrodent species; 1 mo repeat dose toxicology studies in rodent and nonrodent mammalian species; reproductive toxicology testing as per ICH S5A and S5B guidance; and absorption, distribution, metabolism, and excretion studies; standard genetic toxicology testing as per ICH S2B guidance
Max. exposure of 14 to 90 consecutive days	All the tests above, except 3 mo repeat dose toxicology studies should be performed in place of, or in addition to the 1 mo studies. Other studies may be required depending on the data gathered from the completed studies
Exposure of more than 3 mo	All the tests above, except a 6 mo repeat dose toxicology study in a rodent species should be performed in place of, or in addition to the 3 mo and 1 mo studies. Additionally, a chronic toxicology study of 6 mo or 1 yr in a nonrodent mammalian species should be performed. Carcinogenicity studies may be required. Other studies may be required depending on the data gathered from the completed studies, or other factors
Pulmonary or topical products	All the tests above, as appropriate. Additional tests will depend on the route of administration, and may include a sensitization study, and an ocular irritation study. Other studies may be required depending on the data gathered from the completed studies

Abbreviation: ICH, International Conference on Harmonisation.
Source: From Ref. 8.

is the subject of a DMF, the NDA applicant will include a right of reference to that DMF instead of the actual data and information relating to that excipient. This right of reference is given to the NDA applicant by the excipient manufacturer that owns the DMF. This right of reference permits the FDA to review the excipient safety information and data, without disclosure of that information to the NDA applicant.

As discussed above, the FDA will consider the review of excipient ingredients through mechanisms other than the NDA review as indicative of the safety of the ingredient. Traditionally, the FDA has generally accepted as safe for oral dosage forms excipient ingredients that have been reviewed and approved or acknowledged as safe for use in foods. Food ingredients that are generally recognized as safe (GRAS), subjects of approved food additive petitions, or reviewed through the United Nations (U.N.) [Food and Agriculture Organization (FAO)/World Health Organization (WHO)] Joint Expert Committee on Food Additives (JECFA) have generally been accepted as safe for use as excipients in oral dosage form of drugs. The FDA has also generally accepted as safe excipients that comply with USP/NF monographs.

Once an excipient has been reviewed by the FDA during the NDA approval process, that ingredient is then listed in the FDA's database for drug excipients, the Inactive Ingredient Guide (9). This guide lists the ingredient, the use(s) of that ingredient in approved drugs, the number of approved drugs containing that ingredient, the date of first approval of a drug containing the ingredient, and a quantity range for the ingredient. The guide is periodically updated by the FDA, and available through the FDA's Internet website (10).[h] This guide provides a fairly reliable indicator that any excipient listed in the guide will be acceptable to the FDA in an NDA where the use, route of administration, and concentration are consistent with the specifications provided in the guide. Note that this does not represent FDA "approval" of an excipient for those uses, routes of administration, or concentrations, because (as noted previously) excipients do not have any formal status with the FDA, and are approved only as a component of an NDA-approved drug product. Nevertheless, FDA reviewers are not likely to give close scrutiny to use of an excipient that complies with use and quantity specified in the guide.

INFORMAL MECHANISMS TO PROMOTE EXCIPIENT ACCEPTANCE

As mentioned above, while the FDA does not recognize any formal status for the independent review of excipients, the FDA does give consideration to, and will generally permit the use of, ingredients that have been determined to be safe for food use, and those ingredients that meet the conditions of a USP/NF monograph. Below is a summary of the various mechanisms that the FDA accepts as providing some indicia of acceptability for an excipient ingredient.

Food Additive Status

The FDA will generally accept as safe for use in an oral dosage form those ingredients that are considered to be safe for use as a food ingredient. According to the Federal Food, Drug, and Cosmetic Act, food ingredients are safe for food use if they are either approved by the FDA as a food additive or are considered GRAS.

---

[h] As of the date of this chapter, the online version of the Inactive Ingredient Guide is located at <http://www.fda.gov/cder/drug/iig/>.

Food additive approval is obtained through a formal rulemaking process (11),[i] where a company submits a petition to the FDA to have the agency issue a regulation allowing use of a particular ingredient in foods at a specified level for a specified purpose. The process is fairly long and involved, and can require a significant effort on the part of the petitioner. The petitioner will prepare a petition that identifies the chemical composition of the proposed ingredient, the proposed use of the ingredient, the amount of the ingredient to be used, proposed analytical methods for the ingredient, and full safety reports for the ingredient. A report of the environmental impact of the use of the ingredient will occasionally be required. Once the petition is prepared and submitted to the FDA, it is subject to, at a minimum, a chemistry review, safety review, and environmental review. After the reviews are completed, the FDA publishes the food additive regulation specifying the permitted use of the ingredient. Between preparation of the petition and FDA review, obtaining approval for a new food additive generally takes a number of years, and a substantial commitment on the part of the petitioner.

Generally Recognized as Safe Food Status

Exempt from the food additive regulations and process are food ingredients considered to be GRAS. Under FDA regulations (12), a food ingredient is GRAS when (i) there is a GRAS of the ingredient among qualified experts and (ii) that recognition is based on scientific procedures, or experience with the ingredient in foods prior to 1958. If the GRAS status is to be determined on the basis of scientific procedures, this requires studies of the same sort as are required for food additive petitions. GRAS substances that are formally recognized by the FDA are listed in the regulations at 21C.F.R.§§ 182, 184, and 186. GRAS status does not depend on formal FDA recognition, and as FDA notes in its regulation, "it is impractical to list all such substances that are GRAS" (13). Nevertheless, for pharmaceutical excipient purposes, it is unlikely that an ingredient that does not have some formal status with the FDA will be accepted by the agency as safe for use in a new drug without the full studies that are necessary for novel excipients.

Obtaining formal FDA recognition of GRAS status is called GRAS affirmation (14), a process that is not dissimilar from the food additive petition process. A petitioner, or the FDA on its own initiative, starts the process. The GRAS process is an open rulemaking procedure where the data supporting the GRAS status is placed in a public docket for comment. Once all comments are evaluated, the FDA will make a determination on whether the ingredient should be considered GRAS. The GRAS affirmation process requires a considerable effort on the part of a company to acquire toxicological data (often long-term data) that is acceptable to the FDA. The FDA then often takes years to review the data and issue the regulation. As a result, this process is rarely used.

GENERALLY RECOGNIZED AS SAFE NOTIFICATION

Instead of the formal GRAS affirmation petition, the FDA initiated a new procedure for GRAS ingredients, the GRAS notification. Under the notification process, a

---

[i] The Regulation for the petition process are found at 21 C.F.R. § 171.

manufacturer makes a determination that an ingredient is GRAS, and instead of petitioning the FDA to affirm this determination by formal rulemaking, it submits a notification to the FDA of its GRAS determination. Within 90 days, the FDA responds to the manufacturer that either (i) the agency does not question the basis for the manufacturer's GRAS determination, or that (ii) FDA concludes that the notice does not provide a sufficient basis for a GRAS determination. In any event, the FDA does not formally recognize the GRAS status of the ingredient as it did under the affirmation process. The notification process has been valuable to food and food ingredient manufacturers, as it is much faster and less burdensome than the GRAS affirmation process; however, its utility for pharmaceutical excipient manufacturers is less clear. Under the affirmation process, the ingredient had a formal recognition as GRAS in FDA regulations. Under the notification process, the FDA does not make any finding that the ingredient is GRAS, and as a result, ingredients subject to the notification process are potentially less acceptable to the FDA as pharmaceutical excipients.

For many years, chemical manufacturers (makers) and pharmaceutical firms (users) as well as FDA reviewers informally used these FDA food clearance mechanisms to give new drug reviewers (e.g., toxicologists and pharmacologists) a level of comfort about the safety of an excipient contained in a finished pharmaceutical product.

WORLDWIDE FOOD ADDITIVE STATUS

For many years, international chemical companies and related food and pharmaceutical companies have had no formal legal mechanism to provide some indicia of safety for ingredients, especially pharmaceutical excipients. As a result, these companies turned to the U.N. in a manner similar to the way that U.S. companies informally used the FDA food additive and GRAS processes to give an indicia of government acceptance regarding the safety of ingredients in the United States. The JECFA was established in 1956 under the auspices of the U.N. and is made up of committees from two U.N. constituent organizations, the FAO and the WHO, as an international committee of experts (primarily toxicologists) to evaluate food additives. The WHO part of JECFA would review all of the available toxicological data on an ingredient, and, where appropriate, establish acceptable daily intake levels. The FAO part of JECFA would establish chemical specifications for the ingredient. The recommendations of JECFA were then used as the basis for decisions on the safety of that ingredient as a pharmaceutical excipient. This subject is discussed in greater detail in Chapter 6, by DeMerlis and Howell.

JECFA originally met annually, but in recent years began meeting biennially. This program was intended to obtain international agreement on the acceptability and usage levels of food additive chemicals, in order to protect consumer safety and facilitate international trade.

EXCIPIENT DEVELOPMENT STAGNATION

The existing structure for review of excipient ingredients has several structural defects that result in significant disincentives in the development of novel excipients. Currently, only excipients that have been approved in an NDA are subject to a USP/NF monograph, or excipients that have an FDA- or a JECFA-sanctioned food status are likely to be acceptable to pharmaceutical manufacturers. Novel ingredients without any prior review or status represent a risk to pharmaceutical manufacturers involved in product development. Novel excipients, particularly, those that are not suitable for food use (and thus not appropriately reviewed under food clearance mechanisms) present a situation where the drug manufacturer, who has usually spent millions of dollars on the development and testing of the active pharmaceutical substance, has no reasonable assurance that the novel excipient will not cause further delays in the regulatory review and approval process. As a result of the concern over potential regulatory clearance delays caused by new excipients, pharmaceutical manufacturers are likely to use existing and accepted excipients despite potential technical advantages of novel excipients.

The lack of acceptance of new excipients by regulatory agencies, and therefore pharmaceutical manufacturers, creates disincentives for companies to develop innovative new excipients. The risks of not obtaining acceptance by potential buyers for new excipients make development of these ingredients economically unacceptable.

INDUSTRY INITIATIVES

In an attempt to increase the likelihood of regulatory and pharmaceutical industry acceptance of new excipients, excipient manufacturers have made several attempts to foster a regulatory environment that provides mechanisms for acceptance of new and novel excipients. Attempts to create a preapproval review or formal status for excipients have not been successful, and are likely not to be in the interests of the industry nor economically feasible for regulatory agencies.

Efforts to create mechanisms for obtaining some indicia of regulatory acceptance of excipients have been more successful. One notable effort by industry along these lines was the creation of the International Pharmaceutical Excipients Council (IPEC) in 1991. IPEC was formed to represent the interests of both excipient manufacturers and users (pharmaceutical companies), and is notable for a number of initiatives on behalf of the industry; these include the following:

1. Efforts to harmonize the international standards for excipients.
2. Drafting of Safety Evaluation Guidelines, which have been reviewed by various regulatory agencies, and that currently represent industry standard for excipient use evaluation. During the creation of these guidelines, IPEC met with FDA and USP extensively. The FDA has informally accepted these guidelines as the basis for their review of excipient safety, and USP has published a modified version of these guidelines as General Chapter ⟨1074⟩.
3. Drafting of the IPEC good manufacturing practices (GMP) standards for the manufacture of bulk pharmaceutical excipients. The GMP standards were included in USP as General Chapter ⟨1078⟩, and represent the international industry standard for GMPs for bulk pharmaceutical excipients (15).

All of these efforts were directed at standardizing excipient manufacture and use, and have been important in creating a regularized regulatory environment for excipients. Nevertheless, these steps are primarily directed toward the safety and use evaluation of existing excipients, and do little to provide any indicia of acceptability for novel ingredients.

Recently, IPEC has taken the initiative in evaluating different models for evaluation of novel excipient ingredients in an attempt to obtain some acceptance of new and novel ingredients by regulatory agencies. The latest effort is a proposal to create a system where independent experts will review toxicology and safety information for novel excipients, and issue a report that will be made available to regulatory agencies. The system is similar in concept to the FDA's GRAS (food) notification process, and it is hoped that such a system will provide new excipients with some imprimatur of regulatory acceptability. It remains to be seen whether this system will be implemented and have the desired effect of encouraging development of new excipients to meet the needs of a rapidly evolving pharmaceutical industry.

FOOD AND DRUG ADMINISTRATION EXCIPIENT GUIDANCE

While the FDA still has no formal mechanisms for providing independent approval status to pharmaceutical excipients, the agency has recently provided guidelines for what types of data it will require for new excipients. In May 2005, the FDA issued its "Guidance for Industry: Nonclinical Studies for the Safety Evaluation of Pharmaceutical Excipients" (2). This guidance provides detailed recommendations on the types of testing required for excipients depending on the length of exposure for the ingredient and on the route of administration. A summary of these recommendations is provided in Table 1. This guidance provides some clarity on what the FDA will require of such ingredients, and may increase the willingness of excipient manufacturers to develop and pharmaceutical manufacturers to use novel excipients.

CONCLUSION

Pharmaceutical excipients have no official regulatory status independent of the finished dosage form in which they are used. As a result, the mechanism for regulation of these ingredients is uncertain and variable. For excipients found in OTC drug products regulated under the FDA's OTC monograph system, the agency operates under a set of proposed rules (still pending) that provide the general considerations for acceptable excipients and their functions. For prescription drugs (as well as any OTC drugs approved pursuant to a new drug application), excipients are reviewed as a part of the drug application, and not given any independent review. While in theory the FDA examines every ingredient in a new drug application, in practice, excipients long used in drugs or as food ingredients are given only cursory review. The FDA looks to several sources to identify these previously reviewed excipients, including food additive or food GRAS status, favorable review by JECFA, inclusion in the USP/NF, or prior review in other new drug applications. These previously used/acceptable excipients are identified in the FDA's Inactive Ingredient Guide. Inclusion of an ingredient in this guide provides the FDA a reasonable assurance that an ingredient, used within the scope of the usage provided in the guide, will be acceptable. Therefore, regulation of well-known excipients in both OTC and prescription drugs are subject to a reasonably certain set of procedures. However, the lack of an independent review for excipients creates significant issues for companies that wish to use new or novel excipients in their drug products. Lack of any independent review means that there is no formal mechanism for providing indicia of acceptability by regulatory agencies, and this creates uncertainties that reduce incentives to develop and use novel ingredients. Despite this lack of formal review, there has been some movement toward mechanisms to address this problem, but official independent status and review of excipients is not likely, or necessarily in the best interests of industry or the regulatory agencies.

REFERENCES

1. Webster's II New College Dictionary. Boston, Massachusetts: Houghton Mifflin Company, 1995:391.
2. U.S. Department of Health and Human Services. Guidance for Industry: Nonclinical Studies for the Safety Evaluation of Pharmaceutical Excipients. Food and Drug Administration, May 2005.
3. 21 C.F.R. § 330.1(e).
4. 42 Fed. Reg. 19156 (April 12, 1977).
5. Carol Ballentine. Taste of Raspberries, Taste of Death: the 1937 Elixir Sulfanilamide Incident. FDA Consumer magazine, June 1981.
6. 21 C.F.R. § 314.50(d) (1) (ii).
7. ICH Guideline S7A: Safety Pharmacology Studies for Human Pharmaceuticals. Add citation for ICH S7 guidance.
8. Guidance for Industry: Nonclinical Studies for Development of Pharmaceutical Excipients. Food and Drug Administration 2005.
9. U.S. Department of Health and Human Services. Inactive Ingredient Guide. Food and Drug Administration.
10. http://www.fda.gov/cder/drug/iig/.
11. 21 C.F.R.§ 171.
12. 21 C.F.R. § 170.30.
13. 21 C.F.R. § 170.30(d).
14. 21 C.F.R. § 170.35.
15. PhEur. Vol. 14. No 2, April 2002.

5

Cyclodextrins—Enabling Excipients: A Case Study of the Development of a New Excipient—Sulfobutylether β-Cyclodextrin (CAPTISOL®)

Diane O. Thompson
CyDex Inc., Lenexa, Kansas, U.S.A.

CYCLODEXTRINS PROVIDE CASE STUDIES OF NEW EXCIPIENT DEVELOPMENT

Cyclodextrins (CDs) (Fig. 1) have been used in drug development only since the mid-1970s. The CD experience highlights the two different routes for the development of and the associated hurdles facing the introduction of new inactive ingredients in the pharmaceutical industry. The two approaches involve either obtaining acceptance as a generally recognized as safe (GRAS) food additive or developing the appropriate preclinical and clinical safety and current Good Manufacturing Practices (cGMP) manufacturing and quality control package. This chapter will summarize the properties and status of the parent CDs (α-, β-, and γ-CD) and the proprietary-modified CDs [hydroxypropyl (HP)- and sulfobutylether (SBE)-β-CD (SBE7-β-CD) (CAPTISOL®)]. In general, the parent CDs were introduced as GRAS food additives, and the modified CDs were introduced as proprietary pharmaceutical ingredients. The development story of SBE-β-CD will highlight the latter approach of generating an industry-defined safety package and manufacturing quality [chemistry, manufacturing, and quality control (CMC)]. As the safety and quality standards are set high for any ingredient (active or inactive) incorporated into a drug product, the cost and time to meet these regulatory requirements follow suit and necessarily affect the ingredient's cost of goods.

New Excipient Cost of Goods: Novelty to Commodity Status

Four decades of experience with different CDs has provided a general pricing history (Fig. 2) that should be expected for the introduction of a totally new pharmaceutical excipient. The price is high initially due to a combination of factors involving low volume productions, the cost of cGMP manufacturing standards expected by the

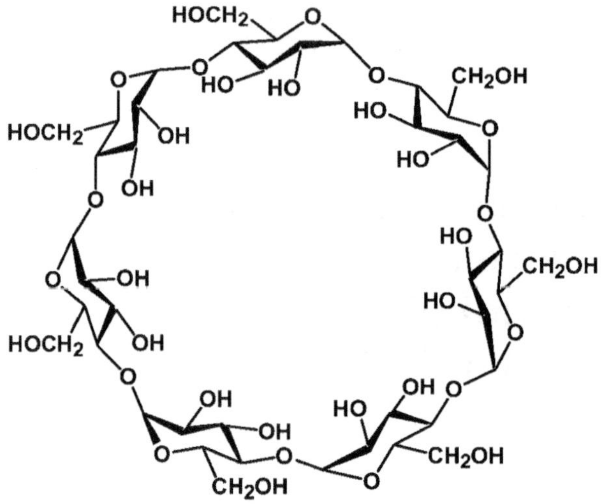

Figure 1 Chemical structure of β-cyclodextrin.

pharmaceutical customer, the expense of safety packages to underwrite the new material, and the potential proprietary nature of a new chemical agent. The price decreases only as the manufacturer recoups expenses and production volumes rise to access economies of scale in manufacturing. A kilogram of β-CD that cost US \$1500 in 1975, depending on the grade of the bulk, can now be purchased for US \$5.25, 31 years later. In addition, whereas, the supply of α- and γ-CD was limited in the early 1980s, these materials are now being produced in multiton quantities, and the cost of goods is continually decreasing.

Commodity prices, therefore, will be realized only several decades after the introduction of a novel excipient and only when the bulk is generic and/or when there are other excipients that are interchangeable for the function the excipient

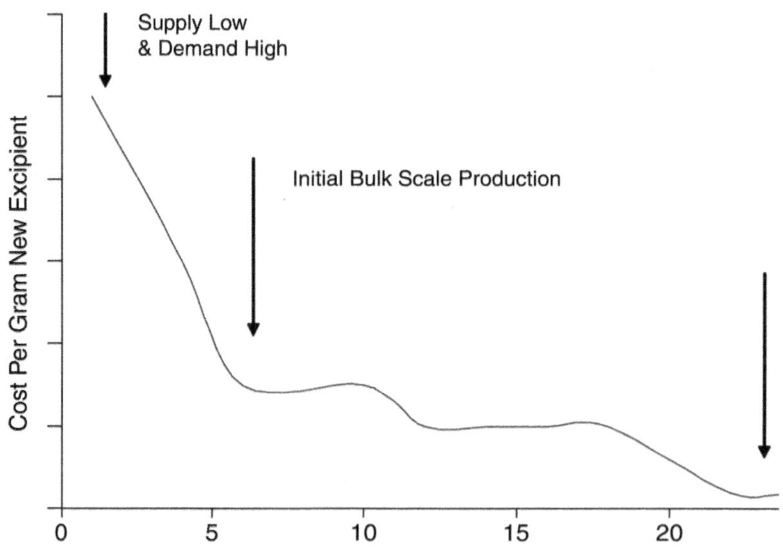

Figure 2 Anticipated pricing history for a totally "new" excipient.

provides the dosage form. The cost of a novel excipient decreases only with the increased call for the bulk material, which in turn only occurs as more drug products using the new ingredient successfully pass the regulatory approval process.

Timing to First Drug Approval Using a New Excipient

The introduction of the first pharmaceutical products using a novel excipient will typically take the same period of time as the development of the pharmaceutical ingredient requiring the novel excipient (Fig. 3). As new excipients, CDs have provided viable dosage forms, which were not accomplishable with any other formulation, and the development course was therefore linked to the development course for the new drug substance. For all three CDs, which have been incorporated in multiple marketed pharmaceutical products, the first products introduced with these agents took between 12 and 16 years, corresponding to the development time of the drug product.

CDs: Enabling Excipients

CDs are enabling excipients used to address solubility, stability, and bioavailability issues in a manner not possible with other inactive ingredients. Enabling excipients are involved in the functionality of the dosage form—ingredients recognized as essential for the appropriate delivery of the drug from the dosage form. Unlike "interchangeable" commodity excipients, enabling excipients are unique in that no other additive can accomplish the desired effect. Such agents enable the creation of a viable dosage form, for without their inclusion the drug would not have a formulation suitable for the market place.

Complexation of a drug with a CD enables the creation of formulations for water-insoluble drugs typically difficult to formulate with more traditional additives. The development of a CD formulation (Fig. 4) may be as simple as mixing the

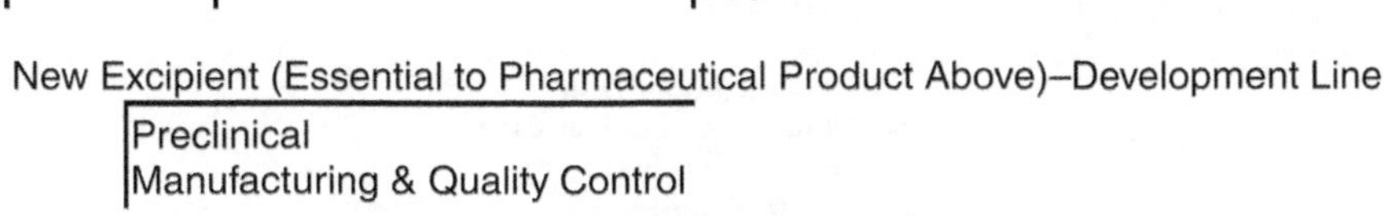

	Start	1st Marketed Product		Time To Market
β-CD	~1970	~1986	Prostandin—Japan	~16 yrs
HP-β-CD	~1982	~1997	Sporanox—USA	~15 yrs
SBE-β-CD	~1990	~2002	Vfend—USA	~12 yrs

Figure 3 Three CDs provide case studies of the development time of a new excipient. *Abbreviations*: CD, cyclodextrin; HP, hydroxypropyl; SBE, sulfobutylether.

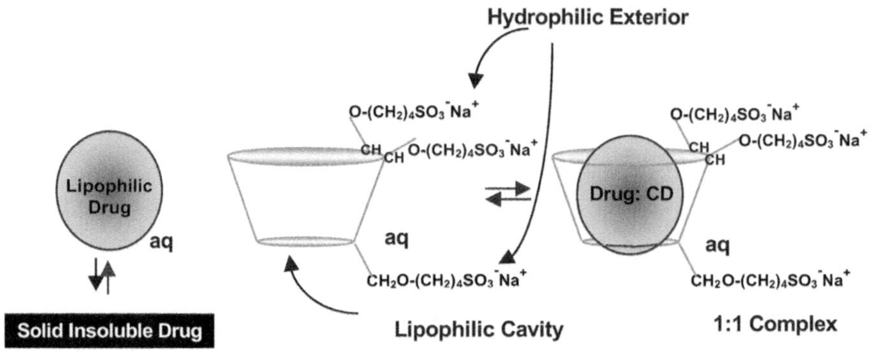

Figure 4 Equilibrium solubilization of a water-insoluble drug by a CD. *Abbreviations*: CD, cyclodextrin; aq, aqueous.

insoluble drug solid with an aqueous solution of the CD and allowing for equilibration of the system to the maximum amount of drug solubilized in the form of a drug–CD complex. The saturated mixture is filtered, and the resulting solution contains a mixture of the drug–CD complex, the level of free drug molecules soluble without the CD, and a given amount of free CD molecules necessary for the equilibrium solubility.

When formulated under these conditions, the drug will dissociate from the complex upon dilution without any precipitation issues (Fig. 5). This is quite different than the solubilization of drugs by co-solvents, where solubility often shows a positive deviation from linearity, and hence precipitation may occur with dilution. Although the strength of the binding will vary with drug and CD, due to the characteristics of linear equilibrium solubility, for properly formulated drugs, the binding strength should not affect the dissociation and delivery of the drug from the CD complex (1).

In addition to improved water solubility, the CDs can provide a myriad of other formulation benefits (2–4). These formulation benefits provided the justification for the incorporation of these CDs into pharmaceutical products and their establishment as new excipients. Two different development approaches for new

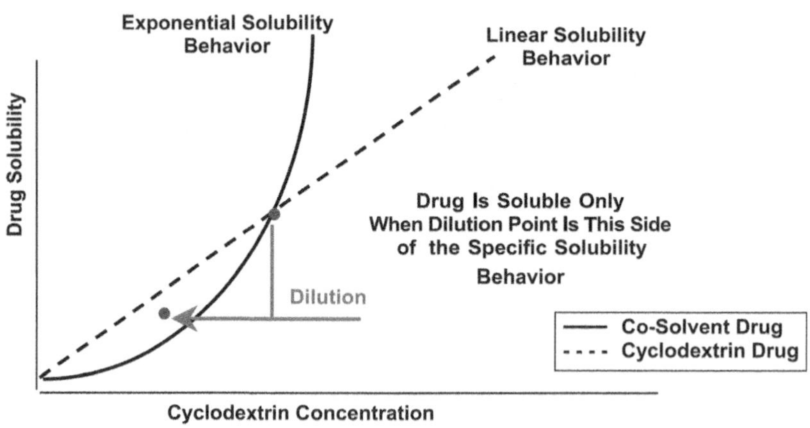

Figure 5 Solubilization and effect of dilution: drug–cyclodextrin complex or drug cosolvent formulation.

excipients are highlighted by the CD stories: the development as a GRAS food additive and the development of a proprietary pharmaceutical safety and CMC package for the excipient during the course of the development of a drug using the excipient.

PARENT CDs

Functionality and Limitations

The parent CDs are cyclic carbohydrates consisting of a variable number of glucopyranose units linked by 1,4-glycosidic bonds. The chemical structure of β-CD (Fig. 1) shows the cyclic nature and the three hydroxyl groups on each glucopyranose unit. Two of the hydroxyls are secondary alcohols and are located at the C2 and C3 positions of the glucopyranose unit. The third hydroxyl is a primary alcohol at the C6 position.

The conformation of the glucopyranose units results in a three-dimensional (3-D) structure best represented by a segment of a hollow cone (Fig. 4). The 3-D structure of the CD provides a cavity that is hydrophobic relative to an aqueous environment and that varies in size with α-CD being the smallest and γ-CD the largest. The hydroxyls or substituents of the modified CDs provide the hydrophilic exterior responsible for the aqueous solubility of the CDs. The properties of the parent CDs that affect their use in drug complexation are (i) their maximum aqueous solubilities (Table 1) and (ii) the differences in complexation due to differences in cavity dimensions.

The extent of solubilization of a drug will be determined by the maximum amount of CD that can be dissolved in water, the binding constant for the complex, and the intrinsic solubility of the drug. On a molar basis, α- and γ-CD are approximately 7 to 14 times more soluble than β-CD at their maximum aqueous solubilities and hence have a potentially better solubilizing capacity than β-CD. α-CD has been utilized in the stabilization of prostaglandins, but it has limited application in the solubilization of therapeutic agents due to the small size of the cavity. γ-CD has the best water solubility and the largest hydrophobic cavity suitable for complexation with hydrophobic small molecule therapeutics; however, many drugs exhibit B phase solubility behaviors (Fig. 6) with γ-CD, the other two parent CDs limiting the full exploitation of their solubilizing capacity. As of 2005, γ-CD has not yet been incorporated into an approved pharmaceutical drug product.

Table 1 Comparison of CD Water Solubility and Theoretical Solubilizing Capacity

	α-CD	β-CD	γ-CD	HP-β-CD[a]	SBE-β-CD[b]
Maximum achievable CD solution concentration					
% wt/vol. (g/100 mL)	14.5	1.85	23.2	60[c]	80[c]
Molecular weight (g/mol)	973	1135	1297	1402	2160
CD molarity (mol/1000 mL)	0.149	0.016	0.179	0.428	0.370
Maximum theoretical achievable drug concentration[d]					
Drug concentration (mg/mL)	76	8	90	214	186

[a]HP-β-CD: Encapsin®—DS, 4.6.
[b]SBE-β-CD: CAPTISOL®—DS, 7.
[c]Viscosity limiting.
[d]Assumptions: molecular weight of drug = 500 g/mol; 1:1 molar ratio of CD to drug and all CD cavities are occupied.
Abbreviations: CD, cyclodextrin; HP, hydroxypropyl; SBE, sulfobutylether; DS, degree of substitution.

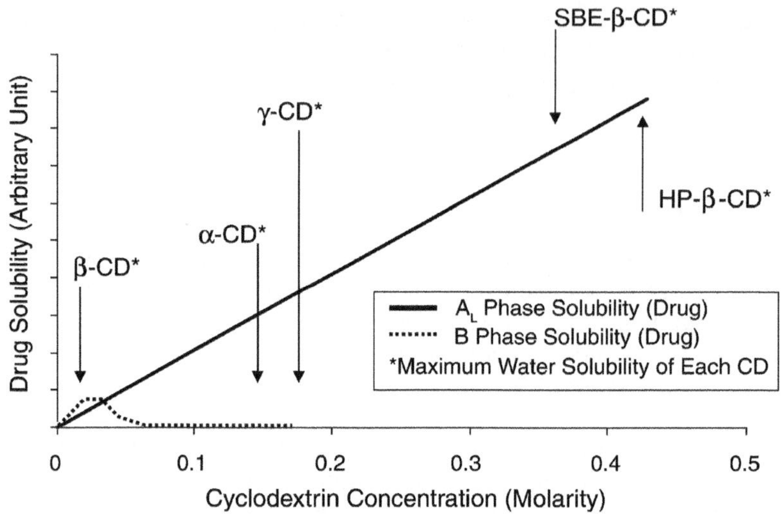

Figure 6 Theoretical maximum drug solubilization by different CDs at their maximum water solubilities (assuming 100% 1:1 complexation). *Abbreviations*: CDs, cyclodextrin; HP, hydroxypropyl; SBE, sulfobutylether.

The cavity of β-CD is appropriate for many of the insoluble small molecule therapeutics, but the low intrinsic water solubility (~1.85% wt/vol., 0.016 M) results in the lowest solubilizing capacity. At the maximum theoretical solubility and assuming all of the CD molecules are occupied, only 8 mg/mL of drug could be solubilized by β-CD. Even with this limitation, the majority of the marketed pharmaceutical CD formulations (Table 2) have used β-CD both due to its cost and the availability in bulk quantities. The use of β-CD is limited to oral and topical products due to its renal toxicity (5) when administered parenterally. The renal toxicity of β-CD is not understood, but it is thought to be due to accumulation of the CD in the renal tubule cells. Although not proven, the hypothesis has been presented that once in the renal cells, either β-CD or a β-CD–cholesterol complex precipitates as acicular crystals due to their low water solubility. How, or even if, these crystals disrupt cellular function is unknown; but exposure of the renal tubule cells to β-CD causes progressive degeneration of the cellular organelles with ultimate mortality.

Developed as GRAS Food Additives

The first CD-based pharmaceuticals were introduced into the market in Japan. The Japanese regulatory agency had granted the natural parent CDs the status of a natural starch, which allowed the utilization of these materials as food additives, a source of agents often embraced by the oral formulator. In Europe and the United States, the natural CDs, however, were not considered as natural starches, and safety data were generated (Table 3) to support the oral consumption of these agents. The natural CDs have now received approval by various European regulatory authorities as food additives, and in the United States, the Food and Drug Administration (FDA) has accepted the Notification of GRAS status filed by Wacker Chemie (6) for each of the natural CDs. The parent CDs demonstrate the use of the GRAS food additive approach to establishing a new pharmaceutical excipient. The details of the GRAS Notification Process are summarized by the FDA Center for Food Safety and Applied Nutrition (7).

Table 2 A Selection of Marketed Pharmaceuticals with CD-Based Formulations

Cyclodextrin	Drug	Route	Market	Trade name
Oral and topical products				
α-CD	OP-1206	Oral	Japan	Opalmon®
	Cefotiam hexetil HCl	Oral	Japan	Pansporin T®
β-CD	Piroxicam	Oral, rectal	Europe	Brexin®
	PGE2	Buccal	Japan	Prostarmon E®
	Benexate	Oral	Japan	Ulgut®, Lonmiel®
	Iodine	Topical	Japan	Mena-Gargle®
	Dexamethasone Glyteer	Dermal	Japan	Glymesason®
	Nitroglycerin	Buccal	Japan	Nitropen®
	Nimesulide	Oral	Europe	Nimedex®
	Tiaprofenic acid	Oral	Europe	Surgamyl®
	Omeprazole	Oral	Europe	Ombeta®
	ME 1207 Cephalosporin	Oral	Japan	Meiact®
HP5-β-CD (Encapsin®)	Itraconazole	Oral	United States, Europe	Sporanox®
	Cisapride	Rectal	Europe	Prepusid®
Parenteral products				
α-CD	PGE1 prostaglandin	IV	United States, Europe, Japan	Prostandin®
HP5-β-CD (Encapsin®)	Itraconazole	IV	United States, Europe	Sporanox®
SBE7-β-CD (CAPTISOL®)	Voriconazole	IV	United States, Europe, Japan	Vfend®
	Ziprasidone mesylate	IM	United States, Europe	Geodon®/ Zeldox®

Abbreviations: CD, cyclodextrin; HP, hydroxypropyl; SBE, sulfobutylether; PGE, prostaglandin; IV, intravenous; IM, intramuscular.

The parent CDs as GRAS food additives are suitable for oral pharmaceutical use when used at the levels approved for foods. The GRAS estimated daily mean oral exposures for β-CD, α-CD, and γ-CD were reported as 300, 1700, and 4000 mg/day, respectively. These levels provide reasonable quantities for consideration in oral pharmaceutical products. As stated earlier, the parent CDs, however, are not suited for intravenous (IV) use due to the early reports of their renal toxicity, and this limitation led researchers to introduce chemical modifications to provide new, systemically safe CDs for use in parenteral pharmaceuticals.

MODIFIED CDs

The ideal, modified CD for use as an excipient should be safe for all routes of delivery, have high water solubility, chemical and metabolic stability, no pharmacological

Table 3 Partial Listing of Reported Preclinical Safety Studies for Parent and Modified CDs

Preclinical safety	α-CD (6)	β-CD (6)	γ-CD (6)	HP-β-CD (9)	SBE-β-CD (25)
Parenteral					
IV bolus					
1 day	Y		Y	Y	Y
7 or 14 day					Y
1 mo			Y		Y
3 or 6 mo			Y	Y	Y
Fertility					Y
Teratology					Y
Peri-postnatal development					Y
IV continuous infusion					
1 or 4 day					Y
14 day					Y
Subcutaneous					
1 mo					Y
6 mo					Y
Oral					
1 day	Y		Y	Y	Y
7 or 14 day	Y		Y		Y
1 mo	Y				Y
3 or 6 mo	Y			Y	Y
9 or 12 mo			Y	Y	Y
Carcino		Y[a]		Y[b]	
Fertility					
Teratology	Y		Y		
Peri-postnatal development					
3-Generation + teratology		Y			
Inhalation					
1 day					Y
7 day					Y
28 day					Y

[a]No tumors were observed in either the rat or the mouse study with β-CD.
[b]Pancreatic neoplasms were observed in the rat but not in the mouse oral carcinogenicity study for HP-β-CD.
Abbreviations: CD, cyclodextrin; HP, hydroxypropyl; SBE, sulfobutylether; IV, intravenous.

activity, complexation behavior not less than β-CD's, and an economical manufacturing process capable of generating a quality bulk suitable for use in pharmaceutical products.

Chemical modifications of β-CD were explored to improve the systemic safety of this valuable solubilizing agent. HP and SBE substituents were introduced to provide new CDs for use in parenteral drug delivery.

Hydroxypropyl CDs

The first modified CD to be explored significantly in parenteral delivery was the HP-β-CD. The development of this proprietary CD has been previously presented (8) and will be briefly summarized.

Improved Functionality and Safety

The introduction of the HP substituent did increase the water solubility of the CD 40-fold over the parent β-CD while retaining much of the solubilizing power of the β-CD cavity. The new derivative did demonstrate improved renal safety allowing for the utilization of this CD in both oral and parenteral product development.

Developed with a Proprietary Pharmaceutical Data Package

HP-β-CD was developed by Janssen as a proprietary inactive ingredient (Encapsin®) and used in both the oral and IV formulation of their anti-fungal itraconazole (Sporanox®). The preclinical safety studies (9) conducted on HP-β-CD in support of Sporanox® are highlighted in Table 3.

The value of the anti-fungal product and the patent protection for HP-β-CD (10) justified Janssen's development of Encapsin®. Janssen further justified the development of this new excipient by offering licenses to Encapsin® as long as the licensee's drug product was not a competitor. In the 1980s and 1990s, this licensing restriction and the existence of a competitive U.S. National Institutes of Health (NIH) patent (11) for HP-β-CD impeded the expansive use of this new excipient. The situation changed due to several events. The Janssen patent protection has expired in European countries, and as of April 1998, Janssen Biotech closed Encapsin® operations allowing the introduction of generic (12) HP-β-CD. The Janssen litigation filed against the NIH HP-β-CD patent position resulted in disallowance of the NIH patent position (13) and issuance in 2002 of the U.S. Janssen patent for HP-β-CD (14). Therefore, although HP-β-CD is generic in Europe, until 2022, a license is still necessary to use HP-β-CD in commercial drug products in the United States.

This unexpected necessity to obtain a license to use HP-β-CD in the United States may continue to slow down the use of HP-β-CD, already further constrained due to the unexpected observation of pancreatic tumors in an oral rat carcinogenicity study (9). These tumors were unexpected because there had been no observations of tumors in the carcinogenicity studies with β-CD, and genotoxicity studies with all CDs have shown negative results. In addition, HP-β-CD produced pancreatic tumors only in the rat but not in the mouse oral carcinogenicity study. The unusual observation in the rat study with HP-β-CD was hypothesized (15) to be due to sensitivity of the rat to increased levels of circulating cholesystokinin (CCK), a pancreatic mitogen in the rat. The increased levels of CCK were a secondary effect to the dosing with HP-β-CD due to the ability of this very water-soluble CD to bind bile acids in the intestine, preventing their reabsorption. The fecal elimination of the bile acids was postulated to stimulate the production of CCK to replenish the bile acids. As CCK is a pancreatic mitogen in the rat but not in other species (16,17), this was proposed to account for the unusual observation. Although CCK-enhancing agents such as cholestyramine have been used chronically in humans without the observation of neoplastic effects on the pancreas, until further evidence is gathered to support the validity of this hypothesis, HP-β-CD may be limited to acute use and life-saving therapies.

Sulfoalkylether CDs

Even with the introduction of HP-β-CD, there was still a need for other oral and systemically safe CDs. SBE7-β-CD (CAPTISOL) was developed in the mid-1990s to provide another improved CD excipient.

Improved Functionality and Safety

The introduction of the anionic SBE substituent (Fig. 4) increased the water solubility of the CD 40-fold over the parent β-CD while retaining much of the solubilizing power of the β-CD cavity. The new derivative demonstrated renal safety allowing for the utilization of this CD in both oral and parenteral product development.

Developed with a Proprietary Pharmaceutical Data Package

SBE7-β-CD (CAPTISOL®) was developed by CyDex Inc., as a proprietary inactive ingredient. An extensive global patent estate exists for SBE7-β-CD (18), and CAPTISOL® must be licensed for use in commercial pharmaceutical products. Patent protection of this new excipient, however, provided the proprietary protection for the bulk material that is sufficient to warrant the expense associated with the development of the preclinical and CMC data packages to establish this new ingredient. The development of this new excipient was spearheaded by the need for a new solubilizing agent to enable the IV formulation of a very insoluble anti-fungal in development by Pfizer.

Pfizer licensed the use of CAPTISOL® and developed for CyDex the initial bolus parenteral safety data and manufacturing scale-up. CyDex and other clients expanded the safety package to include continuous infusion, subcutaneous, oral, ophthalmic, and nasal safety data. The preclinical safety studies (9) conducted on CAPTISOL® are highlighted in Table 3. These data have supported the introduction of two Pfizer products, the IV formulation of the anti-fungal voriconazole (Vfend®) and the intramuscular (IM) formulation of the anti-pyschotic agent, ziprasidone mesylate (Geodon®). Additional *Captisol-Enabled®* drugs are in development not only for parenteral but also for oral, ophthalmic, nasal, and inhalation delivery. The detailed story of the development of CAPTISOL® will demonstrate the second approach to establishing a new excipient.

A CASE STUDY OF THE DEVELOPMENT OF A NEW ENABLING EXCIPIENT—SBE-β-CD (CAPTISOL®)

From 1975 to 1990, scientists at the University of Kansas utilized a rational synthetic design for the definition of a new excipient, the SBE derivative of β-CD (SBE7-β-CD; CAPTISOL®). Designing renal safety into the CD was approached by introducing anionic substituents onto the CD structure. This approach capitalized on the increased water solubility that would be realized with the introduction of an ionic substituent. Higher intrinsic water solubility was expected to help minimize the potential precipitation of the CD, if concentrated in the kidney cell, and the charged substituent was expected to capitalize on the ability of the kidney to efficiently excrete ionic compounds into the urine, thus reducing residence time and exposure of the kidney cells to the CD.

Selecting the Anionic Substituent

Anionic substituents that are considered are the salts of carboxylic acids, phosphoric acids, and sulfur acids. The salts of sulfur acids are chosen based on their low pK_a values that allow these derivatives to remain un-protonated (anionic) throughout the pH range used in and experienced by pharmaceutical formulations. Figure 7 shows the chemical structures of the three different families of anionic CD derivatives

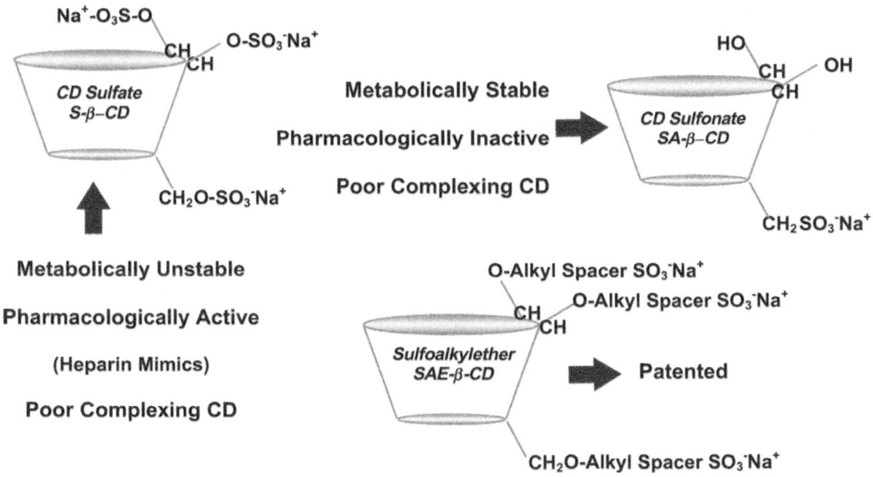

Figure 7 Various substituents evaluated in the rationale design study of CAPTISOL® (sulfobutylether7-β-CD). *Abbreviations*: CD, cyclodextrin; S, sulfate; SA, sulfonate; SAE, sulfoalkylether.

(sulfate, sulfonate, and alkyl sulfonates) (19) evaluated in the search of a safe but functional CD derivative.

Sulfate Derivatives

The first family is a directly sulfated CD. These molecules are easy to produce chemically and have the anionic sulfate group randomly distributed at the C2, C3, and C6 positions. The substituent is attached to the carbohydrate via a sulfate–ester linkage that may be metabolically unstable in vivo. One feature regarding this family is the negative charge of the substituent in close proximity to the carbohydrate backbone.

Sulfonate Derivatives

The directly sulfonated CDs were the second family of compounds studied. The substituent was introduced, after multiple synthetic steps, at the C6 position with a metabolically stable C–S bond. Like the sulfate derivatives, the sulfonates have the negative charge of the substituent in proximity to the carbohydrate backbone.

Alkyl Sulfonate (Sulfoalkylether) Derivatives

In the last family, the sulfonate anion is attached to a neutral alkyl spacer unit [$-(CH_2)_n-$, where $n = 2$ to 6] that links to the CD structure by a metabolically stable ether linkage. This group of compounds differs from the first two families in that the anionic charge is spaced away from the carbohydrate backbone by the alkyl group.

All of the anionic CD derivatives were shown to exhibit water solubilities 20 to 40 times greater than β-CD, and this property was independent of the degree of substitution (DS). The sulfated CDs were shown to be chemically and enzymatically unstable due to the sulfate–ester linkage, and these derivatives were ultimately shown to exhibit pharmacological activity as anticoagulants. The sulfonated CDs were chemically and metabolically stable, but they were not anti-coagulants. However, both the sulfated and sulfonated CDs were poor solubilizing agents. The introduction of the anionic charge close to the CD cavity appears to have disrupted the thermodynamics driving the complexation of drugs. Therefore, the sulfoalkylether derivatives are the best candidates for the new CD excipient.

Identifying the Specific Substitution Type

The introduction of the alkyl spacer chain between the CD and the sulfonate ion reestablished the favorable complexation of hydrophobic drugs (Fig. 8). Spacing the negative charge away from the CD backbone through the use of the 3-carbon propyl spacer unit verifies the impact of the proximity of the charge on complexation. When the derivative bears only one substituent [sulfoproylether 1 (SPE1)-β-CD], the binding constants for the drugs rival that observed with the parent β-CD. However, as the DS increases to 4 (SBE4-β-CD) and 7 (SPE7-β-CD), the binding constants decrease. This may result from the propyl spacer not effectively lessening the increase in charge density that occurs when going from a mono- to tetra- to hepta-anion, or from the physical bulk of the SPE group that sterically hinders the entry of the drug into the cavity.

The bulkier 4-carbon SBE derivatives, however, do not display a significant change in the complexation capability with a change in the DS. The steric effect does not seem to be operative, and the electronic effect seems to be minimized by further distancing the charge from the CD cavity. The SBE groups may possibly orient themselves up and away from the cavity (Fig. 9). The hydrophobic butyl changes may align to minimize interactions with the aqueous solution similar to the process seen in micelle formation. If the butyl groups align, the sulfonate anions would be brought together; but due to electrostatic repulsions, the anionic sulfonates should spread out, providing an elongated hydrophobic cavity with an unobstructed opening.

The SBE derivative was chosen for development as a new excipient, because the material demonstrated high water solubility and excellent complexation capacity relatively unaffected by the substitution level, and the raw materials were reasonably available for commercial scale manufacturing. The only remaining decision was the level of substitution to introduce.

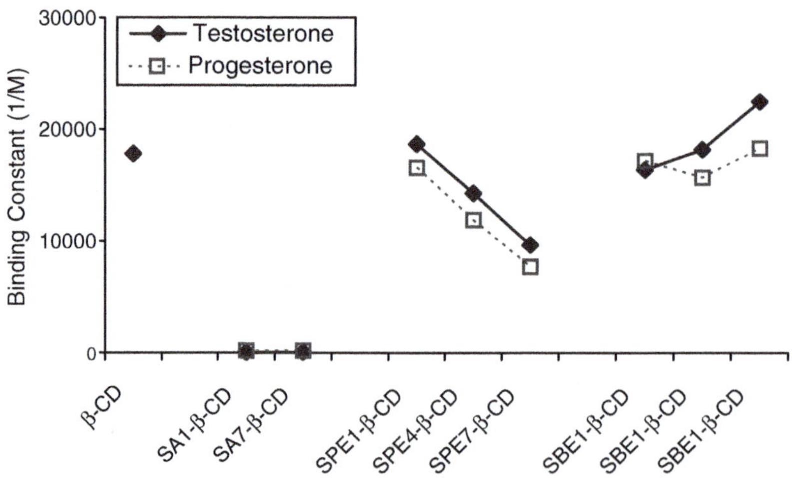

Figure 8 Effect of CD anionic substituents on the complexation of water-insoluble steroids. *Abbreviations*: SA, sulfonate; SPE, sulfopropylether; SBE, sulfobutylether; CD, cyclodextrin. *Source*: From Ref. 19.

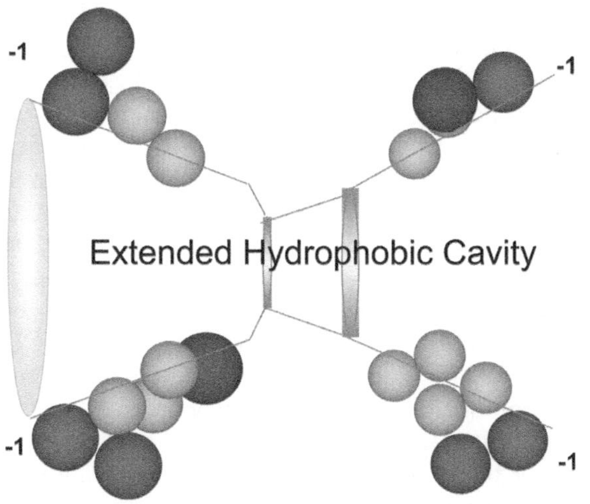

Figure 9 Proposed 3-D structure of sulfobutylether-cyclodextrin.

Selecting the DS

One primary factor that led to the selection of the DS for the SBE-β-CD was the need to economically produce a bulk that is safe for parenteral administration. The safety of the bulk required that it be devoid of any residual β-CD and that the SBE-β-CD exhibit no systemic toxicity. The most economical approach to the elimination of the unreacted β-CD was to derivatize all of the parent CD, a feat accomplished by reaching a DS of approximately 6.5 (SBE7-β-CD). A preliminary evaluation of the potential safety of the SBE7-β-CDs was demonstrated by its minimal involvement in membrane destabilization (20).

The SBE7-β-CD bulk is prepared through the reaction of butane sultone and β-CD and is a mixture of SBE derivatives of different levels of substitution (Fig. 10), with the average composite having a DS of 7. In establishing the manufacture of SBE7-β-CD, the requirements for a rugged and consistent process were high as it was expected that the quality control department of the pharmaceutical company using the new excipient would audit for manufacturing consistency, expecting product quality similar to that of an active drug substance.

cGMP MANUFACTURING—ANALYSIS, STABILITY, AND QUALITY

The quality of the new excipient, SBE7-β-CD, was built in by the design of the manufacturing process. The product's quality was created by understanding the synthesis and work-up procedures, the requirements for the raw materials, and the parameters that affected the reaction and isolation procedures. Validation of the process and cleaning procedures ensured a quality product, and this was confirmed by the analytical characterization of the product.

The reproducibility of the composite nature of modified SBE7-β-CD preparations was a quality issue that was addressed by the management of the manufacturing process. The consistency of the manufacturing process and the composition of the modified CDs were confirmed using analytical methods that were developed and validated to characterize the modified CDs and to evaluate the purity profile and micro-burden. The SBE7-β-CD composite is characterized by (i) the average DS, (ii) the fingerprint pattern of the substitution bands, and (iii) the distribution of the substituents at different regional and positional sites (21).

A number of methods were used to determine the average DS for a modified CD. Nuclear magnetic resonance (NMR) spectroscopy (22) is the most common method used and involves comparing the NMR signal for the anomeric C1 or its respective hydrogen to signal(s) for atom(s) distinct to the substituent. To calculate the DS for the SBE derivatives, the signal(s) for the methylene units in the butyl spacer is compared to the signal for the anomeric hydrogens. The DS can also be determined by additional methods, for example, elemental analysis of the SBE CD preparations can be used to determine the DS. Each substituent contains a sulfur and a sodium atom, and the percent composition of sulfur to carbon or sodium to carbon can define the extent of substitution.

The average DS as determined by these methods provides only the simplest characterization of these derivatives, and further analysis was necessary to characterize the mixture of SBE bands of different levels of substitution. Due to the presence of the anionic sulfonate substituent, it is possible to use capillary electrophoresis (23), to separate the SBE-CD substitution bands and to characterize the fingerprint of the composition (Fig. 10). Anion exchange chromatography (24) was utilized to isolate separate substitution bands (mono- to deca-derivatives) that were subsequently identified by NMR and fast atom bombardment mass spectroscopy (FAB-MS).

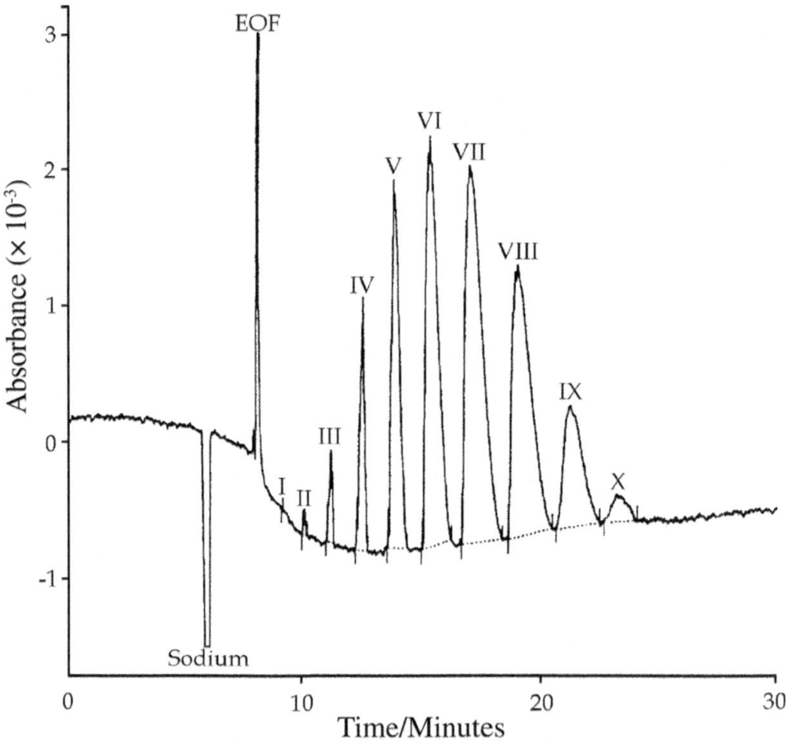

Figure 10 Capillary electrophoresis characterization of the composite nature of sulfobutylether7 (SBE7)-β-cyclodextrin CD. Roman numerals indicate the degree of substitution of each SBE band (I = SBE1-β-CD, ..., IX = SBE9-β-CD).

In addition to the assay and characterization of the SBE-CD mixture, analytical specifications were established for all potential residual raw materials and impurities. As SBE7-β-CD was to be used in parenteral products, multiple microbial and endotoxin specifications were established.

PRECLINICAL SAFETY PACKAGE

Table 3 indicates the preclinical safety studies for CAPTISOL® (25) conducted as of 2005. The strategic safety plan for CAPTISOL® was designed based on the guidelines discussed in the 1990s by the International Pharmaceutical Excipients Council which resulted in the May 2005 issuance of the FDA Guidance (26): Nonclinical Studies for the Safety Evaluation of Pharmaceutical Excipients. These studies and others in the CAPTISOL® Drug Master File have delineated the safety of CAPTISOL® (SBE7-β-CD) for parenteral, ophthalmic, oral, nasal, and inhalation administration.

THE COST TO DEVELOP A NEW EXCIPIENT

The extensive basic research, the analytical method development and validation, the establishment of cGMP manufacturing, the extensive International Conference on Harmonization (ICH) stability studies, and the nonclinical safety data package necessary for the establishment of the new excipient, SBE7-β-CD, are estimated to have cost well over US $30 million dollars. This expense is only justified for a new excipient which is proprietary and which enables the formulation to overcome difficulty in drug delivery. The reality of new excipient development is that only when the material performs a function not achieved by any other agent will the pharmaceutical industry be willing to pay the cost of a proprietary excipient. However, as new drug therapies are discovered, there will continue to be a need for new specialty inactive ingredients, and the CD development stories demonstrate two different pathways to establish these future new excipients.

REFERENCES

1. Stella VJ, Rao VM, Zannou EA, et al. Mechanisms of drug release from cyclodextrin complexes. Adv Drug Del Rev 1999; 3(1):36.
2. Uekama K, Otagiri M. Cyclodextrins in drug carrier systems. CRC Crit Rev Ther Drug Carrier Syst 1987; 3(1):1.
3. Thompson DO. Cyclodextrins-enabling excipients: their present and future use in pharmaceuticals. CRC Crit Rev Ther Drug Carrier Syst 1997; 14(1):1.
4. Loftsson T, Brewster ME. Pharmaceutical applications of cyclodextrins. I. Drug solubilization and stabilization. J Pharm Sci 1996; 85(10):1017.
5. Frank DW, Gray JE, Weaver RN. Cyclodextrin nephrosis in the rat. Am J Pathol 1976; 83:367.
6. γ-CD: 2000 GRAS Notice No. GRN 000046, β-CD: 2001 GRAS Notice No. GRN 000074, α-CD: 2004 GRAS Notice No. GRN 000155; further information on each GRAS Notification can be obtained from the Food and Drug Administration, Office of Management Programs, Division of Freedom of Information (HFI-35), 5600 Fishers Lane, Rockville, MD 20857. http://www.fda.gov/.Food and Drug Administration.

7. http://www.cfsan.fda.gov.
8. Pitha J, et al. Hydroxypropyl-β-cyclodextrin: preparation and characterization, effects on solubility of drugs. Int J Pharm 1986; 29:73.
9. Approval Documents for NDA 20-657 SPORANOX® (itraconazole) Oral Solution Oropharyngeal and Esophageal Candidiasis, Division of Antiviral Drug Products submitted by Janssen Pharmaceutical, February 21, 1997. Available from Food and Drug Administration, Office of Management Programs, Division of Freedom of Information (HFI-35), 5600 Fishers Lane, Rockville, MD 20857. http://www.fda.gov/.Food and Drug Administration.
10. Brauns U, Muller BWW, inventors. Pharmaceutical compositions containing drugs which are instable or sparingly soluble in water and methods for their preparation. Janssen Pharmaceutica N.V. assignee, DE 3346123, December 21, 1983.
11. Pitha J, inventor. Pharmaceutical preparations containing cyclodextrin derivatives. Department of Health and Human Services, United States of America, assignee; US 4,727,064, February 23, 1988.
12. Research Diagnostics website, http://www.researchd.com/janssen/410200.htm.
13. http://www.uspto.gov/web/offices/dcom/bpai/decisions/jd102413.pdf or by request from USPTO FOIA Officer, United States Patent and Trademark Office, PO Box 1450, Alexandria, VA 22313–1450, efoia@uspto.gov.
14. Muller BW, Brauns U, inventors. Pharmaceutical compositions containing drugs which are instable or sparingly soluble in water and methods for their preparation. Janssen Pharmaceutica N.V. assignee; US Patent 6,407,079, June 18, 2002.
15. Van Cauleren H, et al. Monitoring pharmacodynamic effects in toxicology studies. In: Sundwall A, Johansson B, Lindbom L, Lindgren E, and Sjoberg P, eds. Workshop on the Use of Pharmacology Studies in Drug Safety Assessment—Present Situation and Future Perspectives, Sweden, 1994.
16. Sundaram S, Dayan AD. Effects of a cholecystokinin receptor antagonist on rat exocrine pancreatic response to raw soya flour. Human Exp Toxicol 1991; 10:179.
17. Patten JR, Richards EA, Pope H. The effect of raw soybean on the pancreas of adult dogs. Proceedings of the Society for Experimental Biology and Medicine, Cambridge, 1971:37.
18. Stella VJ. Polar drug or prodrug compositions with extended shelf-life storage and a method of making thereof. United States Patent 6,133,248, 2000; Stella VJ, Rajewski RA, McGinity J. Sulfoalkylether cyclodextrins based solid pharmaceutical formulations and their use. US Patent 5,874,418, 1999; Stella VJ, Rajewski RA, Rao V, McGinity J, Mosher GL. Sulfoalkyl Ether Cyclodextrin Based Solid Pharmaceutical Formulations and their use. US Patent 6,046,177, 1999; Stella VJ, Rajewski R. Derivatives of cyclodextrins exhibiting enhanced aqueous solubility and the use thereof. US Patent 5,376,645, 1994; Stella VJ, Rajewski R. Derivatives of cyclodextrins exhibiting enhanced aqueous solubility and the use thereof. US Patent 5,134,127, 1992.
19. Rajewski RA. Development and Evaluation of the Usefulness and Parenteral Safety of Modified Cyclodextrins. Lawrence, KS, U.S.A., Univ. Kansas, 251 pp. Available Univ. Microfilms Int. Order No. DA9110909 from: Diss Abstr Int B 1991; 51(11):5332.
20. Rajewski R, et al. Preliminary safety evaluation of parenterally administered sulfoalkylether β-cyclodextrin derivatives. J Pharm Sci 1995; 84(8):927.
21. Rogmann N, Jones P, Mischnick P. Determination of substituent distribution in *O*-sulfonylbutyl-(1-4)-glucans. Carbohydr Res 2000; 327:275.
22. Luna EA, et al. Isolation and characterization by NMR spectroscopy of three monosubstituted 4-sulfobutylether derivatives of cyclomaltoheptaose (β-cyclodextrin). Carbohydr Res 1997; 299:111.
23. Tait RJ, et al. Characterization of sulphoalkylether derivatives of β-cyclodextrin by capillary electrophoresis with indirect UV detection. J Pharm Biomed Anal 1992; 10:615.

24. Luna EA, et al. Fractionation and characterization of 4-sulfobutylether derivatives of cyclomaltoheptaose (b-cyclodextrin). Carbohydr Res 1997; 299:103.
25. CyDex Inc., Drug Master File No. 14364, 10513 West 84th Terrace, Lenexa, Kansas 66214, U.S.A.
26. FDA Guidance Documents at www.fda.gov.

6

The Use of Food Additive Safety Evaluation Procedures as a Basis for Evaluating the Safety of New Pharmaceutical Excipients

Christopher C. DeMerlis
Colorcon Inc., West Point, Pennsylvania, U.S.A.

Julia C. Howell
J.C. Howell Consulting, LLC, Powder Springs, Georgia, U.S.A.

INTRODUCTION

This chapter examines the procedures used to review the safety and specifications of a new food additive by the Joint Food and Agriculture Organization (FAO)/World Health Organization (WHO) Expert Committee on Food Additives (JECFA), the data requirements, and procedures used for the approval of a new food additive in the United States and the European Union (EU), to determine their use as a basis for evaluating the safety of new pharmaceutical excipients. In addition to the food additive petition process, the procedure whereby a substance can be recognized so Generally Recognized As Safe (GRAS) in the United States is discussed. This chapter deals only with food additives directly added to food (direct food additives) and not food additives indirectly added to food (food contact substances), secondary direct food additives, or prior sanction substances.

The U.S. Food and Drug Administration (FDA) defines novel (new) pharmaceutical excipients as those substances used in the United States for the first time in a human drug product or by a new route of administration (1). The International Conference on Harmonization of Technical Requirements for Registration of Pharmaceuticals for Human Use (ICH) includes sections in its Common Technical Document (CTD) that details the information required for the approval of novel (new) excipients. Information on the control of excipients is included in Section P.4 of the CTD, and any additional information that may be required should be included in Appendix A.3 of the CTD.

A new excipient can only be approved for use within the review process for a new drug application (NDA) under the current system of drug product approval in the United States. There is no separate approval procedure for a new excipient that is to be used in drug products. This approval process is also applicable in the EU and other countries of the world. In this chapter, it is argued that the approval procedures used to evaluate the safety of new food additives could be considered as methods to evaluate the safety of a new excipient, provided the new excipient fulfills a technological function in food.

The evaluation of the safety of a new excipient as a food additive could be accomplished by the submission of a food additive petition to the United Nations expert panel operating under the auspices of the FAO and the World Health Organization known as the JECFA. A JECFA review could serve as a separate independent safety review to support the new excipient for potential use in drug products. Alternatively, the safety of a new excipient could be evaluated through the food additive petition processes, as currently in practice both in the United States and the EU, assuming that the new excipient can be demonstrated to have a technological function as a food additive.

The U.S. FDA states in the Guidance for Nonclinical Studies for Development of Pharmaceutical Excipients (2) that they will continue to consider factors such as use in previously approved products, GRAS-status, or a food additive to evaluate the safety of a new excipient. The FDA states "... an excipient with documented prior human exposure under circumstances relevant to the proposed use may not require evaluation in a full battery of toxicology studies ... " FDA also states "under some circumstances (e.g., similar route of administration, level of exposure, patient population, and duration of exposure) other factors can adequately qualify an excipient (2)." The sponsor of a new excipient should meet with the FDA to provide information regarding the toxicology, chemistry, manufacturing, and controls necessary to evaluate a potential new excipient.

Although the approval of a new food additive or the determination of an ingredient as GRAS are potential mechanisms available for the evaluation of the safety of a new excipient, food additive approval or GRAS approval is no guarantee that an ingredient will be accepted and approved as a new excipient in a drug product, because such an approval can take place only within the context of an NDA. However, a food additive approval or a GRAS determination may indicate an adequate measure of safety to a pharmaceutical manufacturer and may eliminate some of the uncertainty associated with the use of a new excipient for the oral route of administration.

The use of food additive petitions and GRAS procedures to evaluate the safety of a proposed new excipient would apply to the oral route of administration for the excipient and would not generally apply to other routes of administration. Some routes of administration (e.g., inhalation) result in unique toxicological requirements, and data would have to be developed for the specific route of administration. While toxicological data from systemic studies are important for excipients used for nonoral applications, separate data would be needed for the specific route of administration. Nevertheless, the amount of safety data, specifications, and intake information required for a food additive review is extensive, and therefore could provide a firm basis of safety for a new excipient.

SAFETY EVALUATION PROCEDURES FOR THE REVIEW OF FOOD ADDITIVES

Joint Food and Agriculture Organization/World Health Organization Expert Committee on Food Additives

The JECFA is a scientific committee administered by the FAO of the United Nations and the WHO. JECFA is a very important supranational organization responsible for the evaluation and assessment of the safety, specifications, and intake analysis of food additives and contaminants. Procedures for the preparation of a toxicological monograph, an intake assessment, and the specifications for the FAO and WHO are discussed in the FAO and WHO procedural guidelines (3,4) and on their Web sites (5,6).

JECFA provides scientific advice to FAO, WHO, member governments, and the Codex Alimentarius Commission operating under the Joint FAO/WHO Food Standards Programme, which was created to develop food standards, guidelines, and related texts such as codes of practice for the protection of health of consumers, ensure fair trade practices in food trade, and promote coordination of all food standards' work undertaken by international governmental and nongovernmental organizations. JECFA provides advice to the Codex Alimentarius Commission and its specialist committees. The Commission works with many aspects of food involving the protection of consumers and fair trade practices. JECFA provides expert advice on all scientific matters to the Codex Committee on Food Additives and Contaminants (CCFAC), which is responsible for endorsing food additive levels in all Codex food standards. Decisions within CCFAC are based on the safety assessments and recommendations of JECFA. In addition, CCFAC recommends priority food additives and contaminants for evaluation by JECFA for safety and specification review. More information regarding the responsibilities of Codex is explained in a document published in 1999 (7).

Many countries do not have the expertise and funds to conduct risk assessments of food additives and contaminants. JECFA performs a critical function in providing these risk assessments, and many countries use evaluations from JECFA to establish national regulatory programs for food additives and contaminants.

WHO selects JECFA committee members (experts) to conduct the toxicological evaluation of food additives and contaminants and establishing an acceptable daily intake (ADI). FAO selects JECFA experts to establish specifications for the identity and purity of food additives and to assess their intake. Scientists are chosen for their expertise to serve on JECFA committees according to the types of compounds on the agenda, and membership is on an ad hoc basis using individual scientists from all regions of the world.

WHO provides experts with specialized skills in the following areas: toxicology, pharmacology, metabolism, microbiology, pathology, epidemiology, molecular biology, and chemistry. FAO provides experts with specialized skills in the following areas: manufacturing, quality control, analytical chemistry, food technology, and good manufacturing practice. WHO and FAO attempt to balance the experts between academic and regulatory experience and geographical distribution, and the experts are invited as independent members and do not represent their employers or governments. Conflicts of interest must be disclosed in writing, and if one exists, JECFA will decide whether the expert can participate in the evaluation of a particular substance.

A list of potential experts can be found on the JECFA Web site. The members of JECFA are responsible for making decisions based on their experience and the scientific information submitted, and they are assisted by the JECFA Secretariat consisting of the FAO and WHO Joint Secretaries and the WHO temporary advisors. It should be noted that JECFA is not a standing committee and can only make decisions during the time of the meeting. Final decisions of JECFA are published by the Joint Secretariat and are made available to the FAO, WHO, CCFAC, and other interested parties. Once a meeting of JECFA has been finalized, the Joint Secretariat cannot modify or amend the interpretation or decisions of JECFA in the written report, but it may make editorial changes.

The FAO and WHO Joint Secretaries have the responsibility for organizing the JECFA meetings, inviting the experts, preparing the documents, and publishing the meeting report. The substances evaluated by JECFA are selected based on priorities set by CCFAC, requests from FAO and WHO, and requests from governments (FAO and WHO member governments). The Secretariat also prepares the agenda and distributes a call for data prior to each meeting. The Joint Secretaries assign substances on the agenda to a temporary advisor to draft the relevant information necessary for the JECFA evaluation, and an expert is assigned as a reviewer.

The International Programme on Chemical Safety (IPCS) is a joint venture of the United Nations Environment Programme, the International Labour Organization, and the World Health Organization. The IPCS evaluates the effects of chemicals on human health and the environment. A joint publication by the IPCS and JECFA, "Principles for the Safety Assessment of Food Additives and Contaminants in Food" (8), discusses the testing of chemicals used in foods, the evaluation of the test results, and the general basic set of data requirements necessary to evaluate food additives and contaminants.

An important outcome of the JECFA evaluation is the establishment of an ADI for a food additive. The ADI is based on the available toxicological data and the no adverse effect level in the relevant species. JECFA defines the ADI as "an estimate of the amount of a food additive, expressed on a body weight basis, that can be ingested daily over a lifetime without appreciable health risk" (8). JECFA utilizes animal data to determine the ADI based on the highest no-observed-adverse-effect level (NOAEL), and a safety factor is applied to the NOAEL to provide a margin of safety when extrapolating animal data to humans. JECFA typically uses safety factors of 50, 100, or 200 in the determination of an ADI. The NOAEL is divided by the safety factor to calculate the ADI. The food additive is considered safe for its intended use if the human exposure does not exceed the ADI on a chronic basis. This type of information may potentially be used to help assess the safety of a pharmaceutical excipient that is also used as a food additive, based on a comparison of the ADI to the estimated daily intake of the excipient.

Specifications for food additives are established by JECFA to ensure that the commercially used product is of consistent quality and is equivalent to the product evaluated in the toxicological studies.

JECFA publishes reports and monographs based on their review and evaluation of food additives and contaminants. A summary of each meeting is published on the FAO and the WHO Web sites, providing information on the outcome of the meeting. WHO then publishes the detailed conclusions of the meeting in the WHO Technical Report Series. Toxicological and intake monographs are published in the WHO Food Additive Series and are available on the WHO Web site. Specifications are published in the Compendium of Food Additive Specifications (9) and are available on the FAO Web site.

An outline of the review process for a food additive by JECFA is shown in Table 1. Procedures for placing food additives and contaminants on the JECFA agenda are discussed in Annex I of the FAO guideline (3). Table 2 lists the criteria for the inclusion of additives on the JECFA priority list. FAO and WHO member governments may also request the review of a food additive or contaminant by JECFA. Industry must provide a request for evaluation through a member governments, and the request must include the information listed in Table 3. The Joint Secretariat includes the substance in the call for data 10 to 12 months before the meeting and any interested party may then submit data to JECFA. The member governments must

Table 1 Steps in the Review of a Food Additive by Joint Food and Agriculture Organization/World Health Organization Expert Committee on Food Additives

Member governments request to WHO or FAO Secretariats or by request from member states at CCFAC meeting
Member governments provides commitment to supply data package to JECFA
Call for data issued by Joint Secretariat 10–12 mo before the meeting
Prepare and submit dossier 6–7 mo before the meeting
Joint Secretariat assigns responsibility and provides dossier to temporary advisors to prepare the working papers for the toxicology, specifications, and intake analysis
Evaluation of food additive by JECFA experts. Possible outcome of meeting: Decision on the ADI, specifications prepared, and intake analysis
Summary of JECFA meeting published on WHO and FAO Web sites after meeting
Specifications published by FAO in Food and Nutrition Paper
Specifications reviewed and either revised or accepted as Codex specifications at CCFAC meeting and INS number assigned
Technical Report Series containing the evaluation is published by WHO
Food Additive Series containing the toxicological monograph is published by WHO

Abbreviations: WHO, World Health Organization; FAO, Food and Agriculture Organization; CCFAC, Codex Committee on Food Additives and Contaminants; JECFA, Joint Food and Agriculture Organization/World Health Organization Expert Committee on Food Additives; ADI, acceptable daily intake; INS, International Numbering System.

provide a commitment to JECFA that a dossier with the supportive data and information will be submitted six to seven months before the JECFA meeting.

Information provided to JECFA should be submitted in the format detailed in the WHO and FAO procedural guidelines and the WHO guidelines for the preparation of toxicological and intake working papers for JECFA (10,11).

The JECFA expert assigned the responsibility for preparing the specifications for identity and purity for the food additive prepares a Chemical and Technical Assessment (CTA) document. The CTA contains the chemistry, manufacturing process, technological justification, and intended use of the food additive (FAO CTA Guideline, 2003) (12). The CTD guideline should be used for submission of chemical and technical information to JECFA.

Table 2 Criteria for the Inclusion of Food Additives on the Joint Food and Agriculture Organization/World Health Organization Expert Committee on Food Additives Priority List[a]

Use of the compound in accordance with the general principles for the use of food additives. Technological justification and need shall be indicated
Commodities in which the compound will be used are in international trade and represent a significant portion of the diet
Use of the compound will have potential to cause public health and/or trade problems
The compound is commercially available
Commitment that a dossier will be available for evaluation by JECFA

[a]Criteria adopted by Codex Committee on Food Additives and Contaminants.
Abbreviation: JECFA, Joint Food and Agriculture Organization/World Health Organization Expert Committee on Food Additives.

Table 3 Information Required for a Food Additive to Be Evaluated by Joint Food and Agriculture Organization/World Health Organization Expert Committee on Food Additives[a]

Name of sponsor submitting the proposal for inclusion
Name of compound
Trade name
Chemical name
Names and addresses of basic producers
Justification for use
Food products in which the compound is used
Has the compound been registered in two or more countries?
Has the manufacturer made a commitment to provide data?
List of data (toxicology, metabolism, specifications) available
Date on which data could be submitted to JECFA

[a]Information requirement adopted by Codex Committee on Food Additives and Contaminants.
Abbreviation: JECFA, Joint Food and Agriculture Organization/World Health Organization Expert Committee on Food Additives.

United States

The Regulation of Food Additives

In the 1950s, the increased use of food additives in foods became a concern to the U.S. government. Prior to 1958, the burden of proof was on the FDA to show that a food was adulterated by the misuse of food additives and, therefore, unsafe for consumption. As a result of this concern, Congress passed the 1958 Food Additives Amendment to the Federal Food, Drug, and Cosmetic Act (Act). A premarket clearance system was set up requiring that a food additive be shown to be safe for its intended use and that the food additive be approved by the FDA before it could be used in food.

In the United States, food additives are classified into two categories:

1. Additives that can be added directly to food (see 21 CFR 172)
2. Additives that can be added indirectly to food through contact of the food with packaging materials, processing equipment, or other food-contact materials (see 21 CFR 174–178)

The Federal Food, Drug, and Cosmetic Act and the Code of Federal Regulations contain the basic framework for the regulation of food additives in the United States. Section 409 (c)(3)(A) of the Act requires that food additives must be safe for their intended uses before they can be intentionally added to food. 21 CFR 170.3 (i) defines "safe" or "safety" as:

> A reasonable certainty in the minds of competent scientists that the substance is not harmful under the intended conditions of use. It is impossible in the present state of scientific knowledge to establish with complete certainty the absolute harmlessness of the use of any substance; therefore, safety may be determined by scientific procedures or by general recognition of safety. In determining safety, the following factors shall be considered:

1. The probable consumption of the substance and of any substance formed in or on food because of its use.
2. The cumulative effect of the substance in the diet, taking into account any chemically or pharmacologically related substance or substances in such diet.

3. Safety factors, which, in the opinion of experts qualified by scientific training and experience to evaluate the safety of food and food ingredients, are generally recognized as appropriate."

Section 201(f) of the Act defines the term "food" as (i) articles used for food or drink for man or other animals, (ii) chewing gum, and (iii) articles used for components of any such article. The definition of a food in the Act is a broad term and includes food components, including both food additives and GRAS substances. Section 402 of the Act states that a food is adulterated if it contains an unsafe food additive as defined in the Act. Section 201(s) of the Act defines the term "food additive" as "any substance the intended use of which results or may reasonably be expected to result, directly or indirectly, in its becoming a component or otherwise affecting the characteristics of any food (including any substance intended for use in producing, manufacturing, packing, processing, preparing, treating, packaging, transporting, or holding food; and including any source of radiation intended for any such use), if such substance is not generally recognized, among experts qualified by scientific training and experience to evaluate its safety, as having been adequately shown through scientific procedures (or, in the case as a substance used in food prior to January 1, 1958, through either scientific procedures or experience based on common use in food) to be safe under the conditions of its intended use; except that such term does not include

1. a pesticide chemical in or on a raw agricultural commodity; or
2. a pesticide chemical to the extent that it is intended for use or is used in the production, storage, or transportation of any raw agricultural commodity; or
3. a color additive; or
4. any substance used in accordance with a sanction or approval granted prior to the enactment of this paragraph pursuant to this Act, the Poultry Products Inspection Act (21 U.S.C. 451 and the following) or the Meat Inspection Act of March 4, 1907, (34 Stat 1260) as amended and extended (21 U.S.C. 71 and the following);
5. a new animal drug; or
6. an ingredient described in paragraph (ff) in, or intended for use in, a dietary supplement."

Therefore, a food additive is a substance that is intentionally added to foods and does not include substances that are GRAS. A food additive petition must be filed in order to obtain an authorization from FDA specifying the conditions under which a food additive may be safely used. The FDA conducts a very comprehensive review of the safety of a food additive when a petition is submitted.

The Act describes the information that a food additive petition must contain for the chemistry, safety, and functionality of the additive. Food additive petitions are submitted to the FDA under the provisions of section 409(b) of the Act and must include the following information:

1. Chemical identity and composition
2. Proposed use
3. Intended technical effect
4. Methods to determine the amount in the finished food
5. Full safety reports

FDA may request a description of the methods, facilities, and controls used for the production of the food additive, and it may also request samples of the food additive and the foods in which the additive will be used. The requirements for a food additive petition are discussed in 21 CFR Part 171.

The FDA Center for Food Safety and Applied Nutrition (CFSAN), Office of Food Additive Safety (OFAS) is responsible for reviewing food and color additive petitions and GRAS Notifications (13). OFAS has four divisions:

- Division of Petition Review
- Division of Food Contact Substance Notification Review
- Division of Biotechnology and GRAS Notice Review
- Division of Chemistry Research and Environmental Review

OFAS provides guidance for potential petitions and has provided for electronic submissions.

The FDA must act upon a food additive petition within 180 days. It can take many years to obtain an approval for a new food additive due to the lengthy review process and the interaction with the petitioner. When approved, FDA issues a specific regulation for the food additive, which specifies the amount of the food additive that may safely be used and the conditions of use for the additive. The conditions of use may be related to levels of use, types of foods, and application of use.

CFSAN has published guidelines for the data requirements for food additive petitions such as "Recommendations for Submission of Chemical and Technological Data for Direct Food Additive and GRAS Food Ingredient Petitions" (14), which describes the type of chemical and technological data that the FDA considers necessary for the evaluation of a petition. The guideline includes information on the identity, manufacturing process, and specifications for food grade material, stability of the added substance, intended technical effect and use, methodology for analysis of the added substance in food, and consumer exposure.

In addition, FDA has published a guideline "Estimating Exposure to Direct Food Additive and Chemical Contaminants in the Diet"(15), which details the databases and methodologies used by the FDA to estimate exposure to food additives found in the diet and that are very important considerations in assessing the safety of a food ingredient.

Information and data from toxicological tests are essential requirements of the food additive petition. The Redbook 2000 Toxicology Principles for the Safety Assessment of Food Ingredients (16) provides guidance to industry concerning the appropriate tests for the determination of safety. The Redbook discusses concern levels as a method to determine recommended toxicology tests for food and color additives. A level of concern can be assigned based on the potential health risk of the food additive.

During the review of a food additive, the FDA makes a determination of the NOAEL from the toxicological studies, selects an appropriate safety factor, and calculates the ADI for the food additive.

Sections of the Redbook have been updated in phases in October 2001, November 2003, and April 2004, and the updating process is continuing. Alternative approaches can be used but should be discussed with the OFAS. All toxicological studies must be in compliance with Good Laboratory Practice regulations published in 21 CFR Part 58.

The FDA maintains an inventory of more than 3000 total substances used in food, referred to as "Everything Added to Food in the United States" (EAFUS) (17). The database is useful in the determination of the regulatory status of an ingredient for use in food, and is maintained under an ongoing program known as the

Priority-based Assessment of Food Additives (PAFA). It contains administrative, chemical, and toxicological information for over 2000 substances regulated by the FDA as direct, "secondary" direct, color additives, GRAS substances, and prior-sanctioned substances. In addition, the database contains only administrative and chemical information for less than 1000 such substances.

Although the EAFUS database includes some GRAS substances added to food, it is only a partial list because under U.S. federal law, some ingredients may be added to food by a GRAS determination made independently of the FDA. Therefore, the EAFUS database contains many, but not all, of the substances subjected to independent GRAS determinations; however, the EAFUS database is still a useful resource to consult to determine the food additive or GRAS status of a proposed ingredient that may potentially need to be evaluated as a new excipient.

Voluntary Notification Program for "Generally Recognized as Safe" Additives

The Federal Food, Drug, and Cosmetic Act provides that substances to be added to food are subject to a premarket approval requirement, but the definition of the term "food additive" provides an exception to premarket approval for substances that are GRAS, because substances that are GRAS are exempted from the food additive definition. The general requirements for a GRAS assessment of flavor ingredients have been examined in detail by Hallagan and Hall (18), who have provided a basic reference for information regarding the GRAS process in the United States. In addition, information regarding the GRAS evaluation process can be found on the FDA Web site (19).

Requirements that must be met in order to classify a substance as GRAS are described in 21 CFR 170.30. According to 21 CFR 170.30 (b), the same quantity and quality of scientific evidence is required to obtain general recognition of safety based upon scientific procedures as is required to obtain approval of a food additive. General recognition of safety through scientific procedures can be based on published or unpublished studies and additional supportive data and information; however, the data and information relied on to establish a substance as GRAS must be generally available and, by definition, cannot be held confidential. The usual procedure to establish that scientific data and information is generally available is by publishing in a peer-reviewed journal.

Hallagan and Hall (18) identified four requirements that must be met for a substance to be determined GRAS.

1. Qualified experts must conclude that there is a general recognition of safety for the substance.
2. The experts must be qualified by scientific training and experience.
3. The experts must base their opinion on scientific procedures or the experience that the substance was used in foods prior to 1958.
4. The conclusion that a substance is GRAS is based on specific intended use or uses.

Some GRAS substances have been published in 21 CFR 182, 184, 186; however, FDA makes it clear that additional substances independently determined to be GRAS are not listed in 21 CFR regulations (20).

In the past, petitioners filed petitions with FDA to review the GRAS status of a substance and affirm the substance as GRAS. In 1997, FDA proposed to replace this GRAS affirmation petition process with a proposed voluntary notification procedure where any interested party may notify FDA of the determination that a substance is

GRAS. FDA must acknowledge the date of receipt of a GRAS notice in writing within 30 days, and respond on the status of the notice within 90 days. Although there is no formal regulation issued, the notifier receives a letter that is either neutral meaning "FDA has no objection to the GRAS notification" or cites a concern. While FDA has not yet published a final regulation on its voluntary GRAS notification procedure, the agency operates as if the 1997 proposal were a final regulation.

Appropriate documentation must be maintained for ingredients subject to a GRAS notification. The proposed Dietary Supplement Good Manufacturing Practices (GMPs) regulations recently published by FDA outlines clearly the type of supporting information required to use a GRAS substance as a component in a dietary supplement product (Ref.(21), proposed section 111.35). A key section of this proposal related to components used in dietary supplement products states:

> "For those substances that are GRAS, proposed 111.35(d)(4) would require the manufacturer [of the dietary supplement] to have documentation for the basis for why such a substance, that is not a 'dietary ingredient' within the meaning of section 201(ff) of the act, is approved for use or is GRAS for use in a dietary ingredient or dietary supplement."

The proposed regulation describes the specific types of documentation that are needed to support the use of nondietary ingredients and what types of information would not be appropriate (i.e., simple reference to an FDA GRAS Notification in the Fed. Register, etc., is not considered to be acceptable without additional data). Documentation supporting the GRAS status of an ingredient will, therefore, also be required to be held by the user of the GRAS substance in the dietary supplement product.

European Union

The Regulation of Food Additives

The objective of the EU community legislation is to protect human health and to prevent different national legislation that may hinder free trade within the EU.

This section discusses the regulation of food additives, colors, and sweeteners in the EU that are subject to three separate Directives. Three important aspects of the approval process for food additives in the EU are that (i) there is a technological need for their use; (ii) they are not misleading to the consumer; and (iii) they present no health hazard to the consumer (22). The EU Web site contains detailed information for the use of food additives and should be consulted for food additive information (23).

The Framework Directive 89/107/EEC deals with food additives used as ingredients during the manufacture of food and that become a part of the finished food product. It describes the criteria by which food additives are evaluated and specifically states that directives should be established for the list of permitted additives, which are authorized to the exclusion of all others. Alternatively, the directive does not deal with

- processing aids,
- substances used in the protection of plants and plant products in conformity with Community rules relating to plant health,
- flavorings for use in foodstuffs, falling within the scope of Council Directive 88/388/EEC, and
- substances added to foodstuffs as nutrients (for example, minerals, trace elements, or vitamins).

A "food additive" is defined in Article 1 of Directive 89/107/EEC as:

> "any substance not normally consumed as a food in itself and not normally used as a characteristic ingredient of food whether or not it has nutritive value, the intentional addition of which to food for a technological purpose in the manufacture, processing, preparation, treatment, packaging, transport, or storage of such food results, or may be reasonably expected to result, in it or its by-products becoming directly or indirectly a component of such foods."

The general criteria for the use of food additives in the EU are described in Annex II of the Framework Directive 89/107/EEC. The authorization of a new food additive in the EU involves a two-step procedure: a safety evaluation is completed by the European Food Safety Authority (EFSA) and the food additive is included in the appropriate Directive by the Commission and the Commission adopts a specification of the purity criteria for the food additive.

EFSA is the newly established scientific body of the EU that provides objective scientific information and evaluations on all food safety issues, including food additives. EFSA is an independent European agency located in Parma, Italy, which provides risk assessments to the European Commission, European Parliament, and Council and operates various scientific panels including the panel on food additives, flavorings, processing aids, and materials in contact with food.

Petitioners for the use of food additives can find information on the authorization of new additives, revisions of existing provisions, or approval of a new additive source or manufacturing method in Guidance on Submissions for Food Additive Evaluations published by the Scientific Committee on Food (24). This document describes the required administrative and technical data, toxicological tests, and submission format, and should be consulted for the detailed information required for the preparation of a dossier on a new food additive.

Directive 89/107/EEC also includes conditions whereby a member governments can grant provisional authorization of two years for the marketing of an unlisted additive. In addition, the Directive specifies the requirements for the labeling and packaging of food additives for sale to the consumer and the manufacturer.

Three separate Directives must be consulted for the positive list of approved food additives, colors, and sweeteners, including the detailed listing of the food categories and the maximum level of use permitted within each food category.

1. European Parliament and Council Directive 95/2/EC of 20 February 1995 on food additives other than colors and sweeteners for use in foodstuffs, amended by Directives 96/85/EC, 98/72/EC, 2001/5/EC, 2003/52/EC, and 2003/114/EC
2. European Parliament and Council Directive 94/36/EC of 30 June 1994 on colors for use in foodstuffs
3. European Parliament and Council Directive 94/35/EC of 30 June 1994 on sweeteners for use in foodstuffs, amended by Directives 96/83/EC and 2003/115/EC

In addition, three separate directives detail the purity criteria (specifications) for approved food additives, colors, and sweeteners.

1. Commission Directive 96/77/EC for food additives other than colors and sweeteners, amended by Directives 96/86/EC, 2000/63/EC, 2001/30/EC, 2002/82/EC

2. Commission Directive 95/45/EC for colors, amended by Directives 99/75/EC and 2001/50/EC
3. Commission Directive 95/31/EC for sweeteners, amended by Directives 98/66/EC, 2000/51/EC, and 2001/52/EC

Member governments must monitor sweetener consumption. The Commission may change the conditions of use for sweetener, based on the information submitted.

CONCLUSION

The evaluation as a food additive or GRAS substance may provide relevant safety information to support the use of a proposed new excipient in a drug product. The evaluation of a new excipient that is to be used orally as a food additive and that is evaluated independent of the drug product approval process may serve as a review of the safety of the excipient by a recognized regulatory authority.

In a recent article, the FDA stated "For excipients with a history of use, it may be possible to adequately address some or all of the safety issues through citation of the existing nonclinical and clinical database, marketing history, or regulatory status of the compound (e.g., "GRAS" status as a direct food additive may adequately support oral administration of that compound up to the levels permitted in foods)" (25). Therefore, it appears reasonable that the FDA would consider the use safety data based on the food additive regulatory status of an excipient to evaluate the safety of the excipient. If a new excipient has undergone a food additive safety review, this may reduce the perceived risks associated with the development and use of a new excipient.

If a petitioner for a new excipient plans to use a food additive evaluation procedure, the appropriate toxicological testing program must comply with those specified in the guidelines for excipients. The FDA Guidance for the Safety Evaluation of Pharmaceutical Excipients should be consulted to determine the necessary safety testing. The FDA Guidance discusses relevant ICH requirements that should be considered for an excipient. Safety studies conducted on a food additive using toxicological study protocols developed using the FDA "Redbook" and the JECFA Safety Assessment Principles will provide very useful data for potential new excipients. In addition, the safety guide of the International Pharmaceutical Excipient Council (IPEC) should also be consulted for guidance (26).

In the United States and the EU, obtaining food additive approval of a new excipient must be carefully evaluated. It can take many years to obtain approval for a new food additive. A GRAS determination may be a more practical method to ascertain the safety of a new excipient especially in the United States, because the determination of an ingredient as GRAS can be performed independent of a review by the FDA.

The intake of a food additive or GRAS substance used in foods may generally be significantly higher than that of an excipient used in pharmaceutical products. Generally, food additives and GRAS substances will be ingested over a lifetime, whereas excipients are ingested with drug products for a defined period of time in controlled amounts. Intake of a new excipient should be compared to that ingested as a food additive to demonstrate that it falls within the maximum daily intake of the food additive or GRAS substance. Therefore, the maximum permitted amount of a food additive or GRAS substance can be used as a guide when establishing a safe use level for an excipient that is also approved as a food additive or GRAS substance.

Many pharmaceutical excipients are food additives or GRAS substances that have been used in foods for many years. The Handbook of Excipients provides information in the regulatory status section for the accepted uses of excipients in foods (27). In addition, Appendix II of the Handbook lists the "E" number for excipients that are approved as food additives in the EU.

A timely and systematic approach is needed for the independent review of excipients to encourage the development of new excipients. A number of independent review models are used in other industries, such as food, cosmetics, and medical devices, and could be adapted to the review of excipients. IPEC is currently surveying its members to determine which system might be most useful (28), and IPEC has developed an Excipient Master File Guide to standardize and harmonize the information needed to review a new excipient (29). The format of the master file is modeled after the electronic ICH CTD for presenting chemistry, manufacturing, and controls and safety information.

In conclusion, the JECFA evaluation process for food additives is a thorough, comprehensive review of safety, intake, and specifications resulting in an assignment of an ADI. An evaluation by JECFA could serve as a credible review of the safety of a new excipient. Likewise, the safety evaluation by FDA or the EFSA of an ingredient for use as a food additive could also be very useful to support the potential use of a new excipient for the oral route of administration.

REFERENCES

1. FDA Draft Guidance for Industry: Drug Product Chemistry, Manufacturing, and Controls Information. January 2003.
2. FDA Guidance for Industry: Nonclinical Studies for the Safety Evaluation of Pharmaceutical Excipients. May 2005.
3. FAO Procedural Guidelines for the Joint FAO/WHO Expert Committee on Food Additives. Rome, February 2003.
4. WHO Procedural Guidelines for the Joint FAO/WHO Expert Committee on Food Additives. Geneva, January 2001.
5. FAO website: www.fao.org/es/esn/jecfa.
6. WHO website: www.who.int/pcs/jecfa.htm.
7. Understanding the Codex Alimentarius. Food and Agriculture Organization of the United Nations World Health Organization. Rome: Editorial Group, FOA Information Division, 1999.
8. Principles for the Safety Assessment of Food Additives and Contaminants in Food. Environmental Health Criteria 70. World Health Organization. Geneva, 1987.
9. Compendium of Food Additive Specifications. FAO Food and Nutrition Series No.52 and Addendum.
10. WHO Guidelines for the preparation of toxicological working papers for the Joint FAO/WHO Expert Committee on Food Additives. Geneva, December 2000.
11. WHO Guidelines for the preparation of working papers on intake of food additives for the Joint FAO/WHO Expert Committee on Food Additives. Geneva, January 2001.
12. FAO guidelines on the structure and content of the document called "Chemical and Technical Assessment (CTA)". Rome, February 2003.
13. Mitchell Cheeseman, James Wallwork. FDA's Office of Food Additive Safety. Food Safety Magazine; 2002.
14. Recommendations for Submission of Chemical and Technological Data for Direct Food Additive and GRAS Food Ingredient Petitions. Office of Premarket Approval. U.S. Food and Drug Administration Center for Food Safety and Applied Nutrition, May 1993.

15. Estimating Exposure to Direct Food Additives and Chemical Contaminants in the Diet. Office of Premarket Approval. U.S. Food and Drug Administration Center for Food Safety and Applied Nutrition, September 1995.
16. Redbook 2000. Toxicology Principles for the Safety of Food Ingredients. U.S. Food and Drug Administration: Center for Food Safety and Applied Nutrition. October 2001, April 2004.
17. FDA Database. "Everything" Added to Food in the United States (EAFUS). Department of Health and Human Services. Food and Drug Administration, March 2004 (available from: http://www.cfsan.fda.gov/~dms/eafus.html).
18. John Hallagan, Richard Hall. FEMA GRAS A GRAS Assessment Program for Flavor Ingredients. Regul Toxicol Pharmacol; 21:422–430.
19. FDA Inventory of GRAS Notifications. Department of Health and Human Services. Food and Drug Administration, August 2004 (available from: http://www.cfsan.fda.gov/~rdb/opa-gras.html).
20. Federal Register Volume 62, Number 74. Department of Health and Human Services. Food and Drug Administration. 17 April, 1997.
21. Federal Register Volume 68, Number 49. Department of Health and Human Services. Food and Drug Administration. 13 March, 2003, p. 12194–12196.
22. European Union Food Additive Legislation (available from: http://europa.eu.int/comm/food/food/chemicalsafety/additives/comm_legisl_en.htm).
23. General Scheme for Authorization of Food Additive in the European Union (available at : http://europa.eu.int/scadplus/leg/en/lvb/l21067.htm).
24. Guidance on Submissions for Food Additive Evaluations. Scientific Committee on Food. Brussels, July 2001.
25. Osterberg, Robert E, et. al. Toxicity of excipients—A Food and Drug Administration perspective. Int J Toxicol 2003; 22:377–380.
26. Steinberg M et al. A new approach to the safety assessment of pharmaceutical excipients. Regul Toxicol Pharmacol 1996; 24(2).
27. Rowe RC et al. Handbook of Pharmaceutical Excipients. London: Pharmaceutical Press, 2003.
28. DeMerlis CC, Goldring JM. Marketing new excipients, clearing regulatory hurdles. Pharmaceut Technol 2003:102–108.
29. DeMerlis, Christopher C. The IPEC-Americas Excipient Master File Guide. Pharmaceut Technol 2002:36–44.

7
Pharmacopeial Harmonization

J. Lane
United States Pharmacopeia, Rockville, Maryland, U.S.A.

Catherine Sheehan
Department of Standards Development, United States Pharmacopeia, Rockville, Maryland, U.S.A.

INTRODUCTION

The United States Pharmacopeia (USP) was established in 1820 by medical practitioners to ensure quality and correct nomenclature of therapeutic preparations. Today, over 180 years after the first USP was published, USP is the oldest continuously published pharmacopeia in the world and is the only nongovernment compendia. In 1974, USP merged with the National Formulary (NF) and the current *USP–NF* contains about 385 monographs for excipients. The USP mission statement states that, "the United States Pharmacopeia promotes the public health by establishing and disseminating officially recognized standards of quality and authoritative information for the use of medicines and health care technologies by health professionals, patients, and consumers." Although USP is a private organization, federal and state laws in the United States have allowed adoption of USP standards for many purposes. According to the provisions of the 1938 Federal Food, Drug, and Cosmetic Act, USP standards are enforceable by the Food and Drug Administration for drugs manufactured and sold in, or imported into, the United States. Standards established by USP are also recognized by law in Canada and by practice in many other countries.

As the pharmaceutical industry becomes more globalized, harmonization holds the key to effective international pharmaceutical commerce. USP is actively involved in international harmonization. When manufacturers have to comply with only a single harmonized standard, it reduces or eliminates the need to duplicate the testing carried out during the global production of medicines. The following table illustrates how the monographs of the three pharmacopeias compare to the harmonization draft and the reduced testing results (Table 1).

The table shows that without a harmonized monograph, material that is marketed in Europe, Japan, and the United States would require 37 individual tests to be compliant. However with the harmonization draft, industry may comply with compendial requirements in the three regions, with only 12 tests. In some cases, the

Table 1 Carboxymethylcellulose Calcium

United States pharmacopoeia monograph	Japanese pharmacopoeia monograph	European pharmacopoeia monograph	Harmonization draft
Identification A	Identification A	Identification A	Identification A
Identification B	Identification B	Identification B	Identification B
Identification C	Identification C	Identification C	Identification C
Identification D	Identification D	Identification D	Identification D
Alkalinity	Alkali	Alkalinity	Alkalinity
Chloride	Chloride	Chlorides	Chloride
Sulfate	Sulfate	Sulfates	Sulfate
Silicate	Silicate	Silica	
Heavy metals	Heavy metals	Heavy metals	Heavy metals
	Arsenic		
Starch	Starch		
Loss on drying	Loss on drying	Loss on drying	Loss on drying
Residue on ignition	Residue on ignition	Sulfated ash	Residue on ignition
Organic volatile impurities			

differences between each compendial monograph is only a slight difference, but because of regulatory requirements, industry is obligated to perform all testing, unless there is appropriate data to show equivalence. Harmonization also helps to avoid unnecessary delays in the regulatory process and consequently in the availability of medicines, while ensuring their quality, safety, and efficacy. USP continues to work with the European and Japanese pharmacopoeias, through the Pharmacopoeial Discussion Group (PDG), toward the harmonization of content in the world's major pharmacopoeias.

PDG was established in 1989, in response to requests from industry. The PDG was formed with representatives from the European Directorate for the Quality of Medicines in the Council of Europe, the United States Pharmacopeial Convention, Inc., and the Japanese Pharmacopoeia (JP) in the Ministry of Health and Welfare—now the Ministry of Health, Labor, and Welfare (MHLW). Since that time, the PDG generally meets twice a year to work on pharmacopeial harmonization topics. In May 2001, the PDG welcomed the World Health Organization as an observer. While not part of the International Conference on Harmonization (ICH) of Technical Requirements for Registration of Pharmaceuticals for Human Use, the PDG usually meets in conjunction with the ICH and provides the ICH Steering Committee with reports of its progress. To facilitate harmonization of some ICH Quality guidelines and the Quality section of the Common Technical Document, the PDG representatives sometimes attend ICH expert working group discussions as observers. Pharmacopeial harmonization amplifies the work of the ICH, particularly for Quality topics. While the PDG is not part of the ICH, the PDG periodically provides updates to the ICH Steering Committee, and in the past participated in a joint task force. This task force focused on harmonization of general chapters considered important to the ICH harmonized document *Specifications: Test Procedures and Acceptance Criteria for New Drug Substances and New Drug Products: Chemical Substances (Q6A)*. USP also participates in the International Cooperation on Harmonization of Technical Requirements for Registration of Veterinary Products (VICH). As with the ICH, some of the quality guidelines developed in VICH depend upon harmonization of pharmacopeial general chapters. A major difference between the PDG and ICH/VICHs is that the ICH/VICH guidelines are generally applicable only to ingredients and drug products not previously registered in an ICH/VICH region or nation, whereas the PDG harmonization applies to all marketed products in the applicable region or nation. The Group's primary focus is on the harmonization of pharmaceutical excipient monographs and some general chapters. This will reduce manufacturers' burden of performing analytical procedures in different ways, using different acceptance criteria.

Harmonization may be carried out retrospectively for existing monographs or chapters or prospectively for new monographs or chapters. The three pharmacopeias have a commitment to respect the agreed working procedures and the associated time deadlines as an essential part of the harmonization procedure. The PDG has defined harmonization of a pharmacopeial monograph or general chapter as follows:

> A pharmacopeial general chapter or other pharmacopeial document is harmonized when a pharmaceutical substance or product tested by the document's harmonized procedure yields the same results, and the same accept/reject decision is reached."

When using a fully harmonized pharmacopeial monograph or general chapter, an analyst will perform the same procedures and reach the same accept/reject decisions irrespective of which PDG pharmacopeia is referenced. This approach is called interchangeability, and each pharmacopeia will identify, in an appropriate manner, such a monograph or general chapter.

When full harmonization of a pharmacopeial monograph or general chapter is not possible, the PDG works to harmonize it using an approach termed harmonization by attribute. In this approach, some elements of a monograph or general chapter may be harmonized, but others may not. When a monograph is harmonized by attribute, a combination of approaches is needed. For nonharmonized elements, reliance on the individual PDG pharmacopeia is necessary. The PDG works transparently in many ways, but principally through the public notice and comment procedures of each pharmacopeia. Where necessary, meetings of experts are held to identify potential solutions to difficult problems.

In all, 61 excipient monographs and some general chapters (Table 2) are in various stages of the seven-stage harmonization process that are described below.

STAGE 1: IDENTIFICATION

On the basis of an inquiry among its users, the PDG identifies subjects to be harmonized among PDG pharmacopeias and nominates a coordinating pharmacopeia for each subject.

The PDG distributes the work by consensus among the three pharmacopeias and strives for a balance in the distribution of assignments to coordinating pharmacopeias.

STAGE 2: INVESTIGATION

For a subject to be harmonized retrospectively, the coordinating pharmacopeia collects the information on the existing specifications in the three pharmacopeias, on the grades of products marketed, and on the potential analytical procedures.

The coordinating pharmacopeia prepares a draft monograph or chapter, accompanied by a report giving the rationale for the proposal with validation data.

Table 2 Status of Harmonization

	Harmonization item	Stage
General methods relevant to Q6A		
Q-01	Dissolution	6
Q-02	Disintegration	6
Q-03/04	Uniformity of content/mass	6
Q-05	Microbial contamination	
Q-05a	Tests for specified microorganism	5A
Q-05b	Microbial enumeration	5A
Q-05c	Microbial contamination limits for nonsterile products	5A
Q-06	Bacterial endotoxins	6
Q-07	Color (and clarity of solutions)	2
Q-08	Extractable volume of parenterals (Rev. 1)	6
Q-09	Test for particulate contamination: subvisible particles (Rev. 1)	6
Q-10	Residue on ignition (Rev. 2)	6
Q-11	Sterility test	6
General chapters		
G-01	Analytical sieving	6
G-02	Bulk density and tapped density	4
G-03	Conductivity	2
G-04	Density of solids	4
G-05	Flow ability (powder flow)	6
G-06	Tablet friability	6
G-07	Heavy metals	3
G-08	Inhalation	4
G-09	Optical microscopy	6
G-10	Powder fineness	4 Rev.
G-11	Specific surface area	6
G-12	Porosimetry by mercury intrusion	4
G-13	Laser diffraction measurement of particle size	3
G-14	X-ray powder diffraction	3
G-15	Gravimetric water sorption of powders	2
G-16	Thermal behavior of powders	2
B-01	Amino acid determination	6
B-02	Capillary electrophoresis	6
B-03	Isoelectric focusing	6
B-04	Protein determination	6
B-05	Peptide mapping	6
B-06	Polyacrylamide gel electrophoresis	6
Excipients		
E-01	Alcohol (Rev. 1)	6
E-02	Dehydrated alcohol (Rev. 1)	6
E-03	Benzyl alcohol	6
E-04	Calcium disodium edetate	5A3
E-05/06	Calcium phosphate dibasic (and anhydrous)	5A
E-07	Carboxymethylcellulose calcium (Rev. 1)	6
E-08	Carboxymethylcellulose sodium	4
E-09	Croscarmellulose sodium	6
E-10	Microcrystalline cellulose	6
E-11	Cellulose, powdered	6

(*Continued*)

Table 2 Status of Harmonization (*Continued*)

	Harmonization item	Stage
E-12	Cellulose acetate (Rev. 1)	6
E-13	Cellulose acetate phthalate	6
E-14	Citric acid, anhydrous (Rev. 1)	6
E-15	Citric acid, monohydrate (Rev. 1)	6
E-16	Crospovidone	4
E-17	Ethylcellulose	6
E-18	Hydroxyethylcellulose	4–2
E-19	Hydroxypropylcellulose	4
E-20	Hydroxypropylcellulose, low substituted	4
E-21	Hydroxypropylmethylcellulose	6
E-22	Hydroxypropylmethylcellulose phthalate	5A
E-23	Lactose, anhydrous (Rev. 2)	5A
E-24	Lactose, monohydrate	6
E-25	Magnesium stearate	4 Rev.
E-26	Methylcellulose	6
E-27	Methyl paraben	6
E-28	Petrolatum	4
E-29	Petrolatum, white	4
E-30	Polyethylene glycol	4
E-31	Polysorbate 80	3
E-32	Povidone	5A
E-33	Saccharin	6
E-34	Saccharin, sodium (Rev.1)	6
E-35	Saccharin, calcium	6
E-36	Silicon dioxide	4 Rev.
E-37	Silicon dioxide, collodial	4 Rev.
E-38	Sodium chloride (Rev. 2)	6
E-39	Sodium starch glycolate (Rev.1)	6
E-40	Starch, corn (Rev. 1)	6
E-41	Starch, potato	6
E-42	Starch, rice	5A
E-43	Starch, wheat	6
E-44	Stearic acid	4
E-45	Sucrose	4
E-46	Talc	6
E-47	Titanium dioxide	5A2
E-48	Ethyl paraben	6
E-49	Propyl paraben	6
E-50	Butyl paraben	6
E-51	Glycerin	3
E-52	Carmellose	3
E-53	Calcium carbonate	2
E-54	Copovidone	3
E-55	Gelatin	2
E-56	Glucose monohydrate	2
E-57	Glyceryl monostearate	2
E-58	Mannitol	2
E-59	Propylene glycol	3
E-60	Sodium laurylsulfate	3
E-61	Starch, pregelatinized	2

Stage 2 ends with the proposal draft, which is mentioned in this procedure as a Stage 3 draft. The Stage 3 draft, accompanied by supporting comments or data that explain the reasons for each test procedure or limit proposed, is sent by the coordinating pharmacopeia to the secretariats of the other two PDG pharmacopeias.

STAGE 3: PROPOSAL FOR EXPERT COMMITTEE REVIEW

The three pharmacopeias forward the Stage 3 draft to their expert committee (through meetings or consultation by correspondence).

Comments by the experts resulting from this preliminary survey are sent to their respective pharmacopeial secretariat, preferably within two months. However, the comment period should not exceed four months. Within two months of receipt of the comments, the pharmacopeial secretariat should consolidate the comments and forward them to the coordinating pharmacopeia.

The coordinating pharmacopeia reviews the comments received and prepares a harmonized document (Stage 4 draft) accompanied by a commentary discussing comments received about the previous text and providing reasons for action taken in response to those comments.

The Stage 4 draft, as far as possible written in global style—a style easily understood by a variety of readers—together with the commentary, are sent to the secretariats of the other pharmacopeias (end of Stage 3).

STAGE 4: OFFICIAL INQUIRY

The Stage 4 draft and the commentary are published in the revision document of each pharmacopeia in a section entitled International Harmonization. The draft is published in its entirety.

The corresponding secretariats may have to add information essential to the understanding of the implementation of the texts (e.g., the description of an analytical procedure or of reagents that do not exist in the pharmacopeia) and a translation is added by the European and Japanese Pharmacopoeias. The style may be adapted to that of the pharmacopeia concerned or global style may be used. A pharmacopeia can add text, either to amplify some of the requirements with additional information or because national requirements and compendial policy dictate that the addition is necessary. However, there must be a clear indication that this additional information is not part of the harmonized document. This will avoid additional text being included after the harmonization process is completed, but will allow interested parties to review a complete text. The three pharmacopeias endeavor to publish the drafts simultaneously or as close together as possible.

Comments regarding this draft are sent by readers of the revision document to their respective pharmacopeial secretariat, preferably within four months and at most within six months of its publication.

Each pharmacopeia analyzes the comments received and submits its consolidated comments to the coordinating pharmacopeia within two months of the end of the review or comment period.

The coordinating pharmacopeia reviews the comments received and prepares a draft harmonized document (Stage 5A draft), accompanied by a commentary discussing comments received regarding the previous text and providing reasons for action taken in response to those comments.

The Stage 5A draft, together with the commentary, is sent to the secretariats of the other two PDG pharmacopeias.

STAGE 5: CONSENSUS

Provisional

The Stage 5A draft is reviewed and commented on by the other two PDG pharmacopeias within four months of receipt. The three pharmacopeias shall do their utmost to reach full agreement at this stage to obtain a final consensus document.

If a consensus has not been reached, the coordinating pharmacopeia prepares a revised version (Stage 5A/2), taking into consideration relevant, substantiated comments on the Stage 5A document from the two other pharmacopeias. The revised document (Stage 5A/2), together with the commentary is sent to the secretariats of the other two PDG pharmacopeias. The revised document is reviewed and commented on by the other two PDG pharmacopeias, preferably within two months of receipt. This review or comment and revision process of the 5A document is repeated (Stage 5A/*n*) until the three PDG pharmacopeias reach a consensus or until the coordinating pharmacopeia considers that harmonization by attribute should be applied.

If the coordinating pharmacopeia considers certain attributes in the monograph or provisions in a general chapter (especially for retroactive harmonization) are such that it will not be possible to harmonize within a reasonable time period, harmonization by attribute will be applied. If harmonization by attribute is applied, a special cover page indicating harmonization is included with the draft. The text contains harmonized attributes and provisions, and nonharmonized and local attributes are not included. The nonharmonized attributes are clearly indicated in the text as such. The table is prepared as follows: if the three pharmacopeias agree on the attribute, there will be a (+) in all columns; if two pharmacopeias agree that the attribute should be included and have agreed on the method and limit, there will be a (+) in the column for those two pharmacopeias, and a (–) in the column for the pharmacopeia that will not stipulate the test.

For nonharmonized or local requirements, if the three pharmacopeias agree that the attribute should be included, but have not come to agreement on the method or limit: state attribute under "nonharmonized attributes." If only one pharmacopeia will include an attribute: state under "local requirement."

If the Stage 5A draft is substantially different from the Stage 4 draft, the PDG may decide that it should be published again in the revision documents; the draft then reverts technically to Stage 4, revised.

Draft Sign-Off

When agreement is reached, the 5B draft is sent by the coordinating pharmacopeia to the other pharmacopeias no later than four weeks before a PDG meeting for final confirmation. The document is then presented for sign-off at the PDG meeting. This document includes nonharmonized attributes clearly marked as such.

STAGE 6: REGIONAL ADOPTION AND IMPLEMENTATION

The last two stages of the implementation of the "harmonized" chapters and monographs take place independently according to the procedures established by each pharmacopeial organization.

Regional Adoption

The document is submitted for adoption to the organization responsible for each pharmacopeia. Each pharmacopeia incorporates the harmonized draft according to its own procedures. Stylistic and editorial differences may occur.

Adopted texts are published by the three pharmacopeias in their supplements, or where applicable, in a new edition.

If necessary, the Stage 5B draft may be adopted with some amendments (local requirements) corresponding to a general policy in the national or regional (European) area. If a pharmacopeia includes a local attribute after the sign-off of a text, it will inform the PDG. It is, however, preferred to include the nonharmonized text in Stage 5B as an alert to the other pharmacopeias that there will be some differences in text in the final document.

Users of the pharmacopeias are appropriately informed of the harmonization status of monographs and general chapters. In the *European Pharmacopoeia* (*EP*) and *USP–NF*, for general chapters, this is done via a preliminary paragraph. For the *JP*, a notification is made by the MHLW and information is given in a general chapter.

Implementation

The pharmacopeias will inform each other of the date of implementation in their particular region.

The date of implementation of a harmonized document varies in the three PDG regions depending on their legal requirements, need of translation, and publication schedules. Each pharmacopeia generally allows some period of time after publication for implementation to allow manufacturers and other users to achieve conformity. Harmonization is not achieved until the text becomes official in all the three pharmacopeias.

STAGE 7: INTERREGIONAL IMPLEMENTATION

When a harmonized text has become official in all the three pharmacopeias, EP and USP publish a statement indicating the harmonization status of the text; JP publishes a statement to the same effect at Stage 6B. These statements are intended to promote regulatory acceptance of interchangeability of harmonized monographs and general chapters.

Because input from industry is valuable, the PDG has been working closely with TriPEC, the coalition of the International Pharmaceutical Excipient Councils (IPEC) of Europe, Japan, and the United States, to further expedite the harmonization of excipient monographs. These groups are trade organizations that consist of manufacturers and users of pharmaceutical excipients; thus, they play an important role in providing industry input related to various aspects of the harmonization process, including drafting early stage documents and providing analytical testing support. More than 180 multinational excipient manufacturers and users are members of one or more of the three IPEC associations.

Currently, 29 excipient monographs have reached Stage 6 in the process and have been signed-off and accepted by the three pharmacopeias (Stage 6). To proceed to the next stage in the process, it was necessary for USP to create a new general chapter. The new chapter ⟨1196⟩, *Pharmacopeial Harmonization*, explains the role of the PDG, defines the PDG process, elaborates on the definitions of harmonized, nonharmonized, and harmonized by attribute, and details the procedure for users to identify harmonized monographs and general chapters. As more monographs reach Stage 6 in the process (Table 2), it was necessary for the PDG to add 10 new excipient topics to the process, to bring the total number of excipients undergoing the harmonization process to 61. Because it has taken nearly 15 years to reach this point, it has become necessary to reevaluate the working procedures of the PDG. Several steps have been initiated to ensure a more efficient and timely harmonization process, which is beneficial to all interested parties. (i) The PDG proposes to work more closely with IPEC to gain an insight in to the industry viewpoint. IPEC is being asked to contribute to the compilation of early stage drafts that will be gathered from multiple parties and will therefore, by incorporating a wide array of opinions, be closer to harmonized from conception. (ii) Harmonization by attributes has been introduced to settle issues within a monograph that cannot be resolved between the three pharmacopeias. If a monograph or general chapter is not completely harmonized with the corresponding texts of the JP and the EP, it is considered to be harmonized by attributes. Only certain attributes of the text can be considered harmonized with the indicated attributes of the JP or the EP. (iii) A newly revised PDG harmonization procedure has been initiated that will streamline the process. The new process entails reaching a general pharmacopeial consensus early in the process before the draft is open for public comment. It also provides for a shortened process for those monographs where a consensus is reached early. With the acceptance and implementation of the new process, collaboration with IPEC, and utilization of harmonization by attributes, the PDG harmonization of pharmaceutical excipients will take on a new shape and results will be more quickly realized.

While the harmonization process has undergone revision to make it more streamlined, there are difficulties that will continue to impede the process. Although the monographs of the three pharmacopeias contain standards that are universal for the quality of excipients, there are several differences that can be attributed to regional requirements. One example is the monographs for Magnesium Stearate. It is a requirement in the United States, through the Code of Federal Regulations (CFR), that this material be manufactured from "edible sources." Because USP feels that it would be inappropriate to have two standards, one related to the CFR and one compendial, it will retain this requirement in the USP monograph for Magnesium Stearate. The European and Japanese Pharmacopoeias do not have this specific regional requirement and will not include the use of "edible sources" in their monographs. Although there are cultural, regulatory, historical, and other issues that make harmonization of compendial standards difficult, the PDG is committed to resolving differences, with the help of industry.

8
Excipient Interactions

R. Christian Moreton
Idenix Pharmaceuticals, Inc., Cambridge, Massachusetts, U.S.A.

INTRODUCTION

Today we enjoy the benefits of an expanding and increasingly sophisticated range of treatments for illnesses and diseases. A significant part of the advance in medical science has been the successful development of a wide range of medicines, a revolution in therapy. This revolution began slowly with, for example, Paul Erhlich's work that led to the discovery of the arsenical, neosalvarsan, for the treatment of syphilis. The revolution continued through the 1930s with the discovery of the sulfonamides and penicillin, and on to today's ever-widening range of medicinal products to treat a wide range of clinical conditions. The development of new treatments for an expanding group of diseases and conditions continues. But patients are not interested in the drug substance; they want a product they can use to make them better, or alleviate their symptoms, and thereby allow them an enhanced quality of life. Excipients help transform a drug substance [active pharmaceutical ingredient (API)] into a medicine; a form of the drug that can be administered to or taken by the patient, and that is acceptable to them.

On their own, most bulk APIs are not particularly convenient for the patient. Ignoring taste concerns, etc., we might be able to give the patient a bag of acetaminophen powder with the instruction to take one level teaspoonful four times a day. But how would the patient cope with digoxin presented in a similar manner? Does the average patient understand the concept of a microgram? We formulate drugs to make them suitable and convenient for use by the patient.

In order to develop and manufacture a medicine, we need to consider three main components:

- The API—its properties and *limitations*
- The excipients—their properties and *limitations*
- The manufacturing process—its advantages and *limitations*

For some types of product we may also need to consider the primary packaging. Very often, it can be as important to understand the limitations of these three components as it is to understand their properties or advantages. Beyond these three components in isolation, the formulation scientist also needs to understand how they interact and combine to produce the finished medicine.

Excipients are thus one of the three components that in combination produce the medicine that the patient will take. In therapeutic terms, the API is of primary importance because without it there is no treatment and no product. However, in terms of the development and manufacture of the product, all three components are equally important, and we neglect any one of them at our peril. The annals of formulation development in most companies, both large and small, are probably littered with examples where some aspect of one of these three components has been neglected in some way, with unfortunate consequences for the project. The interactions between excipients and the other two components (the API and the manufacturing process), and/or between two or more excipients, are fundamental to the transformation of an API into a medicinal product.

In a sense, formulation science and pharmaceutics may be described as the investigation and application of interactions between excipients, the API, and the manufacturing process. The formulation scientist brings expertise in the use of excipients and pharmaceutical processing, and then adds an understanding of the API. In this discussion we will be considering only the interactions of excipients, but we must remember that the other two components, the API and the manufacturing process, can also interact with each other.

Excipients are those "other" materials used in formulation science. In short, excipients comprise everything in the formulation other than the API (1). IPEC-Americas and IPEC-Europe have developed a more detailed definition:

> Pharmaceutical excipients are any substance other than the active drug or pro-drug that has been appropriately evaluated for safety and is included in a drug delivery system to either aid processing of the system during manufacture, or protect, support or enhance stability, bioavailability or patient acceptability, or assist in product identification, or enhance any other attribute of the overall safety and effectiveness of the drug product during storage or use (2).

Excipients can be considered to be the "Cinderellas" of formulation science and drug delivery. They do not treat the disease, nor should they have a pharmacological effect of their own (although they may exert a physiological effect). However, an understanding of the reasons for their presence in the formulation and how they are used is key to the design of robust, reliable medicines that deliver the drug to the patients in the correct amount, at the correct rate, throughout their shelf-life, consistently batch after batch. But whilst excipients can bring tremendous benefits when used correctly, they also have the potential to cause problems when used inappropriately.

Inappropriate use of an excipient can be defined simply as using an excipient in a formulation in a way that ignores certain characteristic properties of the excipient, to the detriment of the formulation performance. It is important to consider all the ways an excipient can interact in a formulation, and then with the physiological fluids after administration of the medicine to the patient. We may include an excipient in a formulation to take advantage of a particular physical or chemical property, but that does not mean that all its other properties are somehow switched off. If not properly considered in the context of the particular formulation, these "other" properties can cause unexpected problems.

In the following paragraphs, we shall discuss the different types of interactions and give examples, and also discuss their possible significance for the performance of the medicinal product.

The use of excipients goes back to centuries. Even before the advent of the capsule and later the tablet, the available botanical drugs were made into powders or mixtures to make them more convenient for the patient, although sometimes not that palatable. Ointments and salves, with similarities to topical formulations that have been used in more recent times, were known in Ancient Greece. However, the scientific basis for the use of certain excipients has emerged only in the last few decades; for example, tablet lubricants—until a few years ago we knew they were needed and when to use them, but not why they functioned as they do.

More recently, we have seen the development of drug delivery systems as a specialized sector in the pharmaceutical industry. This whole concept is based on the interaction of excipients with the API and manufacturing process, and sometimes with other excipients, to produce a formulation of a medicinal product that meets a particular performance specification.

Today we formulate drugs for a variety of reasons, including the following:

1. Convenience: A bottle of tablets or capsules, or a bottle of liquid, is more convenient for the patient than a bag of powder.
2. Accuracy of dose/consistency of dosing: Tablet and capsule machines are simply quick and accurate volumetric sampling devices that allow us to manufacture unit doses far more quickly than we could by hand. It is easier for a patient to measure a volume of liquid accurately than a weight or volume of powder.
3. Improved bioavailability: For some drugs it is necessary to prepare a formulation to achieve the required bioavailability.
4. Taste masking/improvement of palatability: Many drugs have a poor taste, and formulation can overcome this.
5. Reduction in side effects: We can use formulation to reduce the rate of dissolution of a drug and thereby reduce the peaks in the blood level versus time curve such that the incidence of side effects can be reduced.
6. Controlled dissolution/release: By these means, it is possible to optimize the rate of delivery of the drug to improve therapy, to reduce the frequency of dosing, and to aid patient compliance.

Many of these reasons are evident in the products and services offered by the drug delivery sector of the industry.

EXCIPIENT INTERACTIONS

Excipient interactions are a large part of why medicines work (and sometimes why they do not work in development). They can be either beneficial or detrimental, and can be classified simply as

- physical,
- chemical, and
- physiological/biopharmaceutical.

Physical interactions do not involve chemical change. The components retain their molecular structure. Chemical interactions, on the other hand, involve chemical reactions; i.e., a different molecule (or molecules) is (are) created. Physiological interactions are the interactions between the excipient(s) and the body fluids. In reality, they are also physical interactions, but since they are so important, and because they occur after the medicine has been administered to the patient, they will be considered separately for the purposes of this discussion.

Many in the pharmaceutical industry, when they hear the term "excipient interactions," think immediately of excipient compatibility studies. These studies are important in the development of new products, but as we shall discuss, they are only a small part of the overall scope of excipient interactions. The significance of excipient interactions can extend well beyond the development of the particular medicinal product. Excipient interactions can have implications for

- drug stability,
- product manufacture,
- drug release (dissolution; both in vitro and in vivo),
- therapeutic activity, and
- side effect profile.

Many interactions will directly influence the efficacy of the product, and thus potentially the health and/or treatment of the patient. However, it must be reemphasized that excipient interactions are not always detrimental. Sometimes they can be used to our advantage, particularly in the areas of product manufacture and drug delivery systems (see below).

Excipients bring properties to formulations that facilitate the conversion of the API to a medicine. These "functional properties" will depend on the particular formulation. For parenteral products, open wound treatments, and ocular treatments, there are specific additional requirements concerning impurities, microbiological load, and endotoxins. However, excipients intended for nonsterile applications very often function, because they are not single chemical compounds. There are other "functional" or "concomitant" components frequently present, which are necessary to achieve the required performance (functionality) of the excipient in use. These should be considered separately from any impurities, process residues, or foreign substances that may be present. (In some applications, certain components that have traditionally been considered to be "impurities" or "residues," may actually be concomitant components.) It is important to understand that these other components, whatever their source, may also interact with the API or other excipients.

Some excipients are specifically formulated as mixtures to obtain the required performance. Such excipients are often referred to as being coprocessed or compounded. These excipients make beneficial use of excipient–excipient interactions to derive improved functional performance in a particular type of application; examples include

- Cellactose®—a proprietary combination of powdered cellulose and lactose;
- Microcellac®—a proprietary combination of microcrystalline cellulose and lactose;
- Starlac®—a proprietary combination of starch and lactose;
- ProSolv®—a proprietary combination of microcrystalline cellulose and fumed silica;
- Ludipress®—a proprietary combination of lactose, povidone, and crospovidone; and
- Opadry® and Opadry II®—the proprietary formulations of easily dispersed film coating systems.

The presence of additives in the excipient is another issue that can directly influence our understanding of how a particular excipient interacts. The inclusion of additives should not be viewed in a negative light. Most often the additive will contribute to the overall performance of the excipient. However, there is a widely held belief that additives in excipients conforming to a pharmacopeial monograph do not have to be declared. This is not correct. The pharmacopeias are very clear on this, and where additives are allowed they are specifically addressed in the monograph. This misunderstanding can cause problems between vendors and customers, and generally manufacturers and vendors should be encouraged to declare any additives that are present in a particular excipient. Nevertheless, if an unexpected interaction arises, e.g., during product development, it may be prudent to ask your supplier if there are any additives present in the particular excipient that could cause such an interaction. The presence or absence of additives must always be considered when changing the source of supply of a particular excipient.

In the following discussion, examples of different types of excipient interaction are given. Many of these examples are from the area of solid dosage forms because these are the most common types of medicine available today, and the potential for interaction is probably more complex. However, examples from other types of dosage form are included wherever possible.

Physical Interactions

Physical interactions involving excipients are quite common. However, they are also the most difficult to detect because there is often no convenient chemical "handle," as is usually the case with a chemical interaction. Physical interactions are frequently used in pharmaceutical science, for example, to aid processing and to aid or modify drug dissolution (such as oral modified release) or distribution in the body (such as with the use of a parenteral modified release product). Some of these interactions are deliberately invoked to produce a certain effect. Others are unintended, and it is these interactions that usually cause the problems. When considering physical interactions, particularly the ability to predict them in terms of product stability, differential scanning calorimetry (DSC) (see for example Refs. 3–5), or isothermal microcalorimetry (6) may be useful screening tools, especially when used in conjunction with another technique that can confirm that the interaction is physical rather than chemical, e.g., thermogravimetric analysis, high performance liquid chromatography, or thin layer chromatography. Differential thermal analysis has also been used to investigate interactions between pharmaceutical compounds (7) amongst other techniques.

The essential difference between physical and chemical interactions is that in the former the interacting molecules are not chemically modified in any way. Hydrogen bonding may change, but there are no chemical changes that create a different molecule. However, this does not mean that the different components of the interaction can be easily separated; the resultant mixture may be so intimate that separation is not possible. For example, silicified microcrystalline cellulose after processing cannot be separated entirely into its two separate components (fumed silica and microcrystalline cellulose). But on examination, using a number of vibrational spectroscopic methods, it was shown to be an intimate physical mixture and not a new chemical entity (8).

As has been stated earlier, physical interactions can be either beneficial or detrimental to product performance. The distinction often depends on the particular application or context. For example, what may be beneficial for a prolonged release product may be detrimental in an immediate release product, and vice versa. This type of interaction can be between the drug and the excipient(s) or between two or more excipients. Silicified microcrystalline cellulose, mentioned above, is an end product of a beneficial interaction between excipients that can improve performance (functionality) under certain circumstances. In this particular case, it is thought that the fumed silica particles prevent the structural collapse of the microcrystalline cellulose that can occur on drying after wet massing during wet granulation (9). The benefits of the silicification were not seen when attempts were made to prepare the material in situ using conventional pharmaceutical processing (10).

Another example of an excipient–excipient interaction that can be used to our advantage is the one between xanthan gum and locust bean gum (carob gum or ceratonia) in the presence of water. This interaction forms the basis of the identification test for Xanthan Gum NF. The interaction creates a much more viscous gel system than can be created using either component alone. This has been used in the formulation of controlled release oral solid dosage forms in the TimeRx® drug delivery system (11).

In freeze-drying (lyophilization), the integrity of the lyophilized plug remaining in the vial is important for the efficient reconstitution of the solution prior to use. It is important to avoid the collapse of the plug during drying because this can make reconstitution difficult. We make use of the interaction between the components of the plug to prevent collapse. For example, the use of combination of an amorphous disaccharide and an excipient that crystallizes can prevent the collapse of plugs (12).

An example of a physical interaction between an API and an excipient is that between certain primary amine drugs and microcrystalline cellulose. When dissolution is carried out in water a small percentage of the drug may be bound to the microcrystalline cellulose and not released. For high-dose drugs, this may not be a major issue, but for low-dose drugs it can lead to dissolution failures. This has caused problems in the past, but the phenomenon can be remedied by carrying out the dissolution using a weak electrolyte solution for the dissolution medium (e.g., 0.05 M HCl). Under these revised dissolution test conditions, adsorption onto the microcrystalline cellulose is very much reduced and 100% dissolution may be achieved even for low-dose APIs (13).

Another, more general example of a physical interaction usually between a drug and an excipient, but possibly also between two excipients, is in interactive mixing (also known as ordered mixing). In interactive mixing, smaller particles (typically the API) interact with the surface of larger carrier particles (typically the excipient) through physical forces. These forces are sufficient to trap the smaller particles onto the surface of the larger particles, and thus reduce the propensity of the smaller particles to segregate due to percolation through the bed of larger particles. In this way, we obtain a more homogenous powder blend, and eventually a more uniform batch of product. After the medicine, e.g., a tablet, has been administered to the patient, the aqueous environment of the gastrointestinal tract (GIT) either causes the smaller API particles or the carrier particles to dissolve, or causes the surface interactions to change to allow the smaller particles to be released from the larger carrier particles.

An example of interactive mixing between two excipients is the interaction between fumed silica and other components in the formulation. At low concentrations, e.g., 0.05% to 0.1%, the fumed silica is an effective glidant. It appears to function by being adsorbed onto the surface of the other components and thereby disrupting the cohesive forces within the powder bed. However, above 1% the fumed silica may begin to impede the flow, because the available adsorption sites are occupied and the excess material is mixed in with the rest of the components. On its own, fumed silica does not flow well.

A further example of interactive mixing concerns the addition of magnesium stearate to tablet and capsule blends. The magnesium stearate is typically in the form of very fine particles that appear to adhere to the surfaces of the other components on mixing. This was demonstrated by Bolhuis et al. (14) using sodium chloride, and using acetylsalicylic acid by Johansson and Nicklasson (15). The important point to remember is that magnesium stearate is hydrophobic and extended mixing appears to abrade the magnesium stearate particles so that the surface coverage of the blend increases and eventually creates a water repellant barrier at the surface of the blend that may in turn delay dissolution. Lerk et al. (16) and Lerk and Bolhuis (17) were also able to show that this detrimental effect of magnesium stearate could be remedied through an interaction between magnesium stearate and colloidal silicon dioxide.

Another application of interactive mixing is in certain dry powder inhaler devices that use a carrier particle in combination with the smaller API particles. In this case, the carrier particle is often a grade of lactose (18). During administration of the dose, the combination of the velocity of the inhaled air, possibly the change in velocity of the inhaled dose, possibly together with the change in relative humidity (RH), and other phenomena overcomes the interaction between the API and the carrier particles, allowing the API particles to be stripped from the carrier particles. The API particles are small enough to be carried into the deep lung where they are needed for effective therapy, whereas the larger lactose carrier particles lodge higher up the respiratory tract and are cleared from the lungs in the normal manner.

One very common beneficial interaction involving an excipient is the interaction between magnesium stearate and the metal of tablet punches and dies, or the equivalent parts on a powder encapsulation machine. Magnesium stearate is an example of a "boundary" lubricant. As such it has a polar head and a fatty acid tail. It is believed that the polar head of the magnesium stearate is oriented toward the die wall or tablet punch face. In these ways it is able to reduce the ejection force (the force required to eject the tablet from the die after compaction) and prevent sticking to the punch faces. The other boundary lubricants, e.g., calcium stearate and sodium stearyl fumarate, will also function in a similar manner. However, the so-called "liquid film" lubricants function in a very different manner (19).

But as we have already stated, interactions can also be detrimental, and magnesium stearate is recognized within the pharmaceutical industry for causing problems such as reduced tablet "hardness" and dissolution from tablets and capsules. For magnesium stearate it will always be necessary to achieve a balance between its beneficial and detrimental effects during processing (14).

Oral liquid and semisolid formulations containing water as part of the vehicle may be prone to microbial spoilage in the absence of a preservative. In the case of pharmaceutical creams, these are usually oil-in-water emulsions stabilized using a surfactant. Phenolic preservatives, e.g., parabens esters, are inactivated in the presence of nonionic surfactants, and this detrimental interaction can have serious consequences for preservation of the product (20).

Transdermal delivery of certain APIs is now common for the treatment of some medical conditions, and there are several excipients that are promoted as transdermal "penetration enhancers." One of the earlier materials developed was laurocapram (Azone®). There is a detrimental interaction between laurocapram and mineral oil (liquid paraffin) whereby when both are included in the same formulation, the skin penetration–enhancing properties of laurocapram are lost. Such interactions have implications for extemporaneous mixing of different cream and ointment formulations in the pharmacy.

Chemical Interactions

Chemical interactions are almost always detrimental to the product because they usually indicate an incompatibility that gives rise to chemical compounds that would be classified as degradation products under ICH[a] Q3B, thus leading to requirements for quantitation, identification, and ultimately qualification (some form of safety assessment) depending on the level found.

One notable exception to the detrimental nature of most chemical interactions is the beneficial interaction of the effervescent couple whereby sodium bicarbonate reacts with an organic acid, typically citric acid, in the presence of water to generate carbon dioxide, thereby disintegrating the tablet, and forming a solution or suspension of the drug in water depending on the solubility of the API, that can be administered orally. Of course, the presence of bicarbonate and citric acid in the same tablet or granule requires that precautions be taken such as manufacturing and packing at very low RH ($<20\%$) to prevent premature activation of the couple. The packaging also needs to be impermeable to moisture for the same reason.

For chemical interactions involving the API, for example in pharmaceutical dosage forms, we are largely concerned with six main types.

- Primary amines will undergo a Maillard reaction with reducing sugars (21). The glycosidic hydroxyl group of the reducing sugar interacts with the primary amine to form an imine (Schiff's base) that then breaks down to form Amidori compounds. These are intensely colored compounds and are responsible for the yellow-brown coloration characteristic of this type of interaction (e.g., chlorpheniramine and dextrose). This series of reactions appears to be accelerated in the presence of free moisture (e.g., at higher RHs) and catalyzed by magnesium ions, e.g., magnesium stearate.
- Secondary amines may also interact with reducing sugars. However, the reaction cascade does not proceed beyond the formation of the imine, and thus no coloration develops (22).
- Esters (and certain other compounds) may be susceptible to hydrolysis by low or especially high pH, or in the presence of alkaline metal or alkaline earth salts. In the presence of acid, i.e., anion and hydrogen ion, the reaction is at equilibrium. However, in the presence of base and the associated cations, the reaction is driven to completion (e.g., acetyl salicylic acid and the effect of sodium and magnesium salts on the rate and extent of reaction).
- Primary amines may interact with double bonds in a reaction analogous to a Michael addition reaction (e.g., fluvoxamine maleate, where the fluvoxamine primary amine group can interact with the double bond in the maleic acid counterion). Examples of excipients that contain double bonds include sodium stearyl fumarate and sorbitan monooleate.
- Lactone formation because of the close proximity of heteroatoms and an active hydrogen atom in the molecule, e.g., benazepril.
- Certain APIs are susceptible to oxidation, e.g., atorvastatin and cytidine nucleoside analogues. Fumed metal oxides (e.g., fumed silica, fumed titania, and fumed zirconia) can promote such oxidation reactions. These reactions are more complex in some ways, and less easy to predict.

[a]ICH—International Conference on Harmonization of Technical Requirements for Registration of Pharmaceuticals for Human Use.

All of these reactions can involve excipients, either as reactants or as catalysts.

However, these reactions do not always occur. In some instances there may be steric factors in the API molecule that restrict access to the reactive group and the reaction does not occur, or occurs at a much-reduced rate. For almost all chemical interactions, a key component is presence of "free" (unbound) water (23,24). In the absence of a sufficient amount of "free" water, the reactions do not proceed. This is the basis for using very low humidity manufacturing and packaging facilities for the manufacture of effervescent products. The "free" water layer serves to dissolve sufficient of the drug and the excipient, or to form bridges between particles, such that the components/reactants come into sufficiently close contact for the reaction to occur.

EXCIPIENT COMPATIBILITY STUDIES

Excipient compatibility studies are an important part of any preformulation screen for a new API. However, it is important to remember that an excipient compatibility screen can only indicate the excipients to be avoided because of an obvious chemical incompatibility. The results from excipient compatibility studies are not always easy to interpret, particularly if a physical interaction is found. As stated above, physical interactions can be detected using some form of calorimetry in conjunction with, e.g., chromatography, but the interpretation of the significance of the interaction probably requires prior experience of the excipient and its interactions. It is difficult to predict that the molecular structure of the excipient will interact physically with the chemical structure of the API molecule.

Often, chemical interactions are catalyzed by the presence of other components, e.g., the Maillard reaction between primary amines and reducing sugars appears to be catalyzed by magnesium ions, and both sodium and magnesium ions catalyze the hydrolysis of esters. For this reason, it is useful to include potential trial formulations in the excipient compatibility screen so that the effects of two or more excipients on the stability of the API can be assessed. The other major factor to be considered is water, as discussed above, and the expansion of the compatibility screen to include both "dry" (as is) samples and moistened samples is common. This is particularly important if aqueous, wet granulation is being considered, and/or eventual sales in countries included in ICH climatic zones 3 and 4 are likely. In the latter case, the results of such studies will give an early indication of the need for moisture protective packaging for such markets.

Excipient compatibility studies are a form of preliminary stability assessment. It is important that they be executed appropriately. The precise details of the testing will probably be different for each organization carrying out such studies. However, certain general assumptions are implicit in this approach. The underlying principle is the Arrhenius relationship:

$$k = A.e^{\frac{-E_a}{RT}}$$

In simple terms, the reaction rate increases as the temperature increases. Broadly, the reaction rate doubles with a 10°C rise in temperature. The compatibility studies are intended to provide information quickly. Generally, the studies are carried out at elevated temperature, and the resultant mixture examined analytically to determine if a chemical interaction has taken place, or if a physical interaction occurred.

There are two main approaches to excipient compatibility screening: isothermal studies at an elevated temperature and variable temperature studies in which the temperature is steadily increased, as in DSC. Both approaches are valid, but it is important to note, as has been stated above, that excipient compatibility testing is not a definitive test. We cannot state that an interaction will not take place, even though one may not have been found. We can only state which excipients to avoid because there is a very obvious interaction. A typical scheme is given in Figure 1 for a DSC-based excipient compatibility study. (There are other schemes that are used successfully.)

Excipient compatibility and stability studies rely on two underlying assumptions. One is that there is no change in reaction mechanism as temperature increases; the second is that the excipient is also chemically stable under the conditions of test. However, if the reaction mechanism does change with temperature, it is likely the result will show a disproportionately greater breakdown than would be anticipated from lower temperature studies. Thus the risk is that an excipient is rejected that might in reality be perfectly suitable for the formulation. In many cases this is probably an acceptable risk.

The stability of excipients is almost always taken for granted. Obviously, there is the potential for a phase change with certain lower melting excipients, e.g., semisolid materials, however, this is not a chemical phenomenon; although it may enhance the potential for interaction by increasing the effective interface available at which the interaction can take place. However, some materials are not stable under conditions encountered in excipient compatibility screening or accelerated stability testing. A notable example is dibasic calcium phosphate dihydrate. At temperatures as low as 37°C, under certain conditions, the dihydrate can dehydrate to form the anhydrous material with the concomitant loss of water of crystallization (25), and at 25°C, it is a stable solid with a shelf life, when stored correctly, of more than two years.

The objective of the excipient compatibility screening is to quickly find those excipients/processes that should be avoided for the particular API. In order to obtain a result as rapidly as possible we carry out these studies at elevated temperature as discussed above. The question arises as to how long and at what temperature? We need to be able to extrapolate the results to a convenient time frame at 25°C/60% RH for ICH Climatic Zones I and II (or 30°C/65% RH for ICH Climatic Zones III and IV). Based on the approximation from the Arrhenius equation (see above) that the reaction rate doubles for a 10°C rise in temperature, we have standard multipliers that have been widely accepted within the pharmaceutical industry. For example, a study carried out at 40°C for one month would equate to three

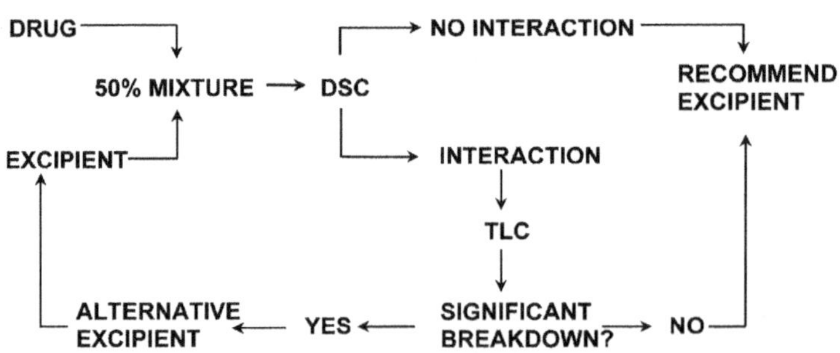

Figure 1 DSC-based excipient compatibility testing program.

months at 25°C, i.e., a multiplier of 3. At 50°C the multiplier is 6 (2 × 3), and at 60°C the multiplier is 12 (2 × 2 × 3). Based on the author's own experience, if a product is stable at 25°C/60% RH for 12 months, this is sufficient stability to get through phase I and probably phase II studies, depending on their duration. Manufacture of further batches may be needed to resupply longer phase II studies. It is rare using modern analytical techniques that a problem that is sufficiently serious to require either reformulation due to physical changes in the formulation, or toxicity assessment because a degradant has exceeded the ICH Q3B threshold, would not be detected before 12 months stability at 25°C. However, again in the author's experience, formulations can appear satisfactory at six months only to show signs of a failing trend by 12 months. On this basis, in the author's opinion, carrying out excipient compatibility studies at 40°C for six or eight weeks is insufficient to meet the needs of the modern formulator, whereas two months at 50°C would suffice. Using the standard multipliers, 12 months stability at 25°C would be equated with one month at 60°C, two months at 50°C, or four months at 40°C. But, there is the added assumption (discussed above) that the reaction mechanism does not change with elevated temperature. It must also be stressed that these times and multipliers are only guides and should not be taken as definitive. The recently issued ICH Q1E: Evaluation of Stability Data document is more conservative in extrapolation of accelerated stability data. Assuming it is adopted by the regulatory agencies, this Guideline gives the requirements for registration applications for "new molecular entities and associated drug products." But this would not prevent the use of other accelerated tests for information purposes within a company.

Water

Excipients both typically contain water and are required to interact with it. The water associated with excipients can exist in various forms. Studies with different materials have shown that water can exist in association with excipients in at least four forms that may be termed "free" water, "bound" water, "structural" water, and water of crystallization. Water associated with a particular excipient may exist in more than one form (26). The type of water will govern how it is implicated in interactions between the excipient and the API or another excipient. The so-called "free" water is the form that is most frequently implicated in excipient interactions. "Bound" water is less easily available for interaction, and structural water is usually the least available one. Water of crystallization can be very tightly bound into the crystal structure; however, there are some comparatively labile hydrates, e.g., dibasic calcium phosphate dihydrate (see above). If water of crystallization remains tightly bound within the crystal structure, it is unlikely to participate in an excipient interaction. However, any material that is in equilibrium with air above 0% RH will have some "free" moisture associated with it. In reality, below about 20% RH, the amount of moisture will probably be insufficient to cause problems. However, if sufficient moisture is present (e.g., at a higher RH), it can facilitate the interaction between components of the formulation.

The dibasic calcium phosphate dihydrate example discussed above is probably an extreme example of the instability of an excipient relating to the release of water. But many excipients exist in a hydrated state, and heating them for the purposes of compatibility studies, or accelerated stability testing, can cause any "free" water, and sometimes other types of water, to be released, which can then influence any potential interaction, or even interact itself with the drug.

The interaction of many excipients with water is an important property because the physiological fluids the medicine will encounter after administration are based on water. For example, we use the interaction between the tablet disintegrant and water to facilitate the breakup of a conventional immediate release tablet or capsule in the stomach, and thereby aid the dissolution of the API.

Certain APIs are administered topically, e.g., topical anti-inflammatory steroids and topical anti-fungal agents. The release of these APIs from their formulation was thought to be maximized when the solution of the drug in the vehicle is just saturated (27). However, Davis and Hadgraft (28) have shown that supersaturation can increase the penetration even further. The interaction of the moisture present in the skin with the topical formulation can be used to change the degree of saturation of the formulation and thereby enhance the release of the API from the formulation into the skin.

Oxygen

Like water, oxygen frequently interacts with pharmaceutical materials, both APIs and excipients. Antioxidants are included in many formulations to inhibit oxidation reactions. Oxidation reactions do not necessarily require molecular oxygen. For example, the oxidation of cytidine analogues to the equivalent uridine analogues can occur in the presence of water even in the absence of oxygen. But oxygen is implicated, for example, in the breakdown of unsaturated fatty acids (rancidity). Oxygen can also react with, e.g., polyethylene glycols (PEGs). These materials are stabilized, during manufacture, by either the addition of antioxidants, or manufacturing under a nitrogen blanket to exclude oxygen. It is important to know which material is stabilized, because a change in source might lead to unexpected stability problems because of the presence or absence of the antioxidant, and thus changes in the potential for interaction.

PHYSIOLOGICAL/BIOPHARMACEUTICAL INTERACTIONS

By this we mean interactions that occur after the medicine has been administered to the patient. For the most part, they are physical interactions. However, the major distinctions are that the interaction is between the medicine (including excipients) and the body fluids, primarily comprising aqueous solutions, and that they have the potential to influence the rate of absorption of the drug. They will vary depending on the route of administration. Because physiological and biopharmaceutical interactions are so important, and they are not specifically linked, for example, to the stability of the medicinal product, and also because they occur after the medicine has been administered to the patient, they have been included as a special category for the purposes of this discussion. The importance and potential impact of biopharmaceutical interactions of excipients has been recognized for some years (see for example Ref. 29).

All excipients interact in a physiological sense when they are administered as part of the medicine. For example, we can get dilution of the vehicle leading to changes in viscosity or precipitation of the drug, disintegration of a tablet or capsule, activation of a controlled release mechanism, etc. This may be stating the obvious to many, but in many instances this interaction with physiological fluids is not actively considered; although it may be understood or assumed to be occurring. However, these physiological interactions are very important for the correct functioning of the product, and it is thus important to understand when other interactions occur and what the physiological impact will be. For example, we consider the physical properties of magnesium stearate as a tablet or capsule lubricant, but, as was discussed in "Physical Interaction" section, it is well documented in the literature that when magnesium stearate is used incorrectly these same physical properties can lead to problems that can impact dissolution, and thus could possibly affect bioavailability.

We include certain excipients in a formulation specifically because they interact with the physiological fluids and the bodily functions in a certain way. For example, as discussed above, we include disintegrants in immediate release tablet and capsule formulations, because we know that when they encounter the aqueous environment of the stomach, they will cause the tablet or capsule to disintegrate and thereby aid dissolution of the API. Another example is the general case of hydrophilic colloid matrices used as prolonged release drug delivery systems. We know that when these materials contact the aqueous environment of the GIT they swell and create a diffusion barrier that slows the rate of dissolution of the dissolved drug.

One physiological interaction that can potentially cause serious problems for the patient is the interaction between enteric coatings and antacids. Enteric coatings on tablets or capsules are intended to allow the formulation to pass though the stomach into the duodenum before dissolution and then rupture of the enteric coating, disintegration of the tablet core, and release of the drug. Certain products may be enteric coated to protect the API from degradation in the stomach, e.g., pro-drugs. Other APIs are enteric coated to protect the stomach from the API, e.g., nonsteroidal anti-inflammatory drugs (NSAIDs). The enteric-coating polymers, e.g., cellulose acetate phthalate and hydroxypropyl cellulose acetate phthalate, rely on their pH-solubility profile for their function; they are soluble at a more basic pH, but insoluble at acid pH. Antacids raise the pH of the stomach contents and thus cause the enteric coating to begin to dissolve in the stomach. The enteric coating thus begins to breakdown allowing the premature release of the API in the stomach. For the pro-drugs, this might mean that more of the drug is degraded than is desirable and the patient would receive a suboptimal dose. For the NSAIDs, the premature breakdown of the enteric coat may cause unwanted side effects, such as gastric bleeding.

A classic biopharmaceutical incompatibility is the interaction between tetracycline antibiotics and calcium and magnesium ions (30). A complex is formed that is not absorbed from the GIT. This is a well-known interaction, and tetracycline antibiotics usually carry a warning against taking them with certain types of food. But magnesium and calcium salts are quite common excipients, in terms of both the range of formulations containing them and the level of inclusion in those formulations. How many of us would think about not including magnesium stearate in a formulation of medicine intended to be an adjunct therapy to treatment with a tetracycline? The point is that the design of a formulation cannot be undertaken in isolation, and the possibility of excipient interactions on final administration to the patient must be considered, not only with the formulation being developed, but also with other medicines administered concomitantly.

Some drugs, such as aspirin, appear to be well absorbed along the length of the lower GIT (the ileum and colon). Certain other drugs, e.g., metoprolol, have a limited absorption window in the GIT. That is to say that they are not absorbed along the whole length of the lower GIT but only to a small segment of it. For these drugs, it is clear that the speed with which the drug passes down the GIT (gastrointestinal motility) will influence absorption of the drug. Certain excipients can increase gastrointestinal motility, i.e., speed up the passage of material down the GIT, and thereby reduce the time available at the site of absorption for drugs such as metoprolol. Excipients that can increase gut motility include the polyols [e.g., sorbitol and xylitol—(31)]. The effect is very much dependent on the amount of the polyol administered at one time. PEG 400 has also been reported to influence the absorption of ranitidine in a similar fashion (32). These are examples of physiological interactions between an API and an excipient.

As our understanding of the mechanisms whereby drugs are absorbed increases, we have come to understand that not only are there mechanisms of drug absorption, but also mechanisms whereby drugs are actively secreted back into the lumen of the GIT. These are known as efflux mechanisms, and a major efflux system concerns p-glycoprotein (there may be others). The significance of p-glycoprotein is that if a drug is a p-glycoprotein substrate (or a substrate for any other efflux mechanism), it may not matter how well absorbed the drug is, the efflux mechanism is likely to pump the drug back out into the GIT. In the past it has been assumed that certain drugs were just not well absorbed, and that may have been the case, but there is now another possible explanation—that they are substrates for an efflux mechanism. There are recent reports in the literature that at least one excipient, α-tocopheryl PEG 1000 succinate, appears to inhibit the p-glycoprotein efflux pathway (33,34). If confirmed, generally, this potentially beneficial biopharmaceutical interaction may have important implications for the oral delivery of certain drugs.

SUMMARY

Excipient interactions are what make the formulations work, or not work in some cases. In one sense, pharmaceutics might be described as the science and investigation of excipient interactions. Excipient interactions may be classified as physical, chemical, or physiological/biopharmaceutical. They can also be beneficial or detrimental. Not all excipient interactions are detrimental, and many interactions between two or more excipients are used to enhance the performance attributes of the medicinal product, or to improve the manufacturing process. However, prediction of excipient interactions and relating them to product stability is complex, particularly for physical interactions. Excipient compatibility studies can provide information on which excipients to avoid because of a probable chemical interaction. Interactions between excipients and active drugs, or between two excipients, can occur during administration of the medicine to the patient. Excipient interactions can also occur throughout the development life cycle. Formulation design relies on excipient interactions, but the formulator must take into account all the known potential interactions of the excipient to realize a robust formulation that will make it to market. It is often as important to understand the limitations of a system, as it is to understand its advantages, and this applies very much to excipients and pharmaceutical formulation.

Water is an essential component of many excipient interactions, whether chemical, physical, or physiological/biopharmaceutical.

Our understanding of the biopharmaceutical and physiological processes that occur during drug absorption is rapidly increasing. It is clear that excipients can influence some of these processes, and the formulation scientist needs to be aware of the potential of these effects. Not all effects are detrimental, but many are. They can mean the difference between success and failure of a development project. As stated in the "Introduction" section, we need to consider three components in order to design a successful formulation: the advantages and limitation of the API, the excipients, and the manufacturing process. We can now add a fourth component—how they all interact? Perhaps this is the key to understanding the science of pharmaceutical formulation.

REFERENCES

1. Blecher L. The formation of the International Pharmaceutical Excipients Council. Pharm Technol 1991; 15(6):54, 56.
2. Blecher L. Excipients–the important components. Pharm Process 1995; 12(1):6–7.
3. Chrzanowski FA, Ulissi LA, Fegely BJ, Newman AC. Preformulation excipient compatibility testing: application of a differential scanning calorimetric method versus a wet granulation simulating, isothermal stress method. Drug Devel Ind Pharm 1986; 12(6): 783–800.
4. Botha SA, Lotter AP. Compatibility study between Naproxen and tablet excipients using differential scanning calorimetry. Drug Dev Ind Pharm 1990; 16(4):673–683.
5. Foda NH. Compatibility study between Fluvoxamine Maleate, mebeverine hydrochloride and tablet excipients using differential scanning calorimetry. Egypt J Pharm Sci 1992; 33(1–2):73–81.
6. Wilson RJ, Beezer AE, Mitchell JC. Determination of thermodynamic and kinetic parameters from isothermal heat conduction microcalorimetry: applications to long term reaction studies. J Phys Chem 1996; 99:7108–7113.
7. Guillory JK, Soon CH, Lach JL. Interactions between pharmaceutical compounds by thermal methods. J Pharm Sci 1969; 58(3):301–308.
8. Tobyn MJ, McCarthy GP, Staniforth JN, Edge S. Physicochemical comparison between microcrystalline cellulose and silicified microcrystalline cellulose. Int J Pharm 1998; 169:183–194.
9. Sherwood BE, Hunter EA, Staniforth JN. Pharmaceutical excipient having improved compressibility. US Patent 5,585,115 (1995).
10. Sherwood BE, Becker JW. A new class of high functionality excipients: silicified microcrystalline cellulose. Pharm Tech 1998; 22(10):78–88.
11. Baichwal AR, Staniforth JN. Directly compressible sustained release excipient. US Patent 4,994,276 (1988).
12. Nail SL, Jiang S, Chongprasert S, Knopp SA. Fundamentals of freeze-drying. Pharm Biotechnol 2002; 14:281–360.
13. Steele DF, Edge S, Tobyn MJ, Moreton RC, Staniforth JN. Adsorption of an amine drug onto microcrystalline cellulose and silicified microcrystalline cellulose samples. Drug Dev Ind Pharm 2003; 29(4):475–487.
14. Bolhuis GK, Lerk CF, Zijlstra HT, De Boer AH. Film formation by magnesium stearate during mixing and its effect on tableting. Pharm Weekly 1975; 110(16):317–325.
15. Johansson ME, Nicklasson N. Investigation of the film formation of magnesium stearate by applying a flow-through dissolution technique. J Pharm Pharmacol 1986; 38:51–54.
16. Lerk CF, Bolhuis GK, Smedema SS. Interaction of lubricants and colloidal silica during mixing with excipients: I. Its effect on tableting. Pharm Acta Helv 1977; 52(3):33–39.
17. Lerk CF, Bolhuis GK. Interaction of lubricants and colloidal silica during mixing with excipients: II. Its effect on wettability and dissolution velocity. Pharm Acta Helv 1977; 52(3):39–44.
18. Timsina MP, Martin GP, Marriott C. Drug delivery to the respiratory tract using dry powder inhalers. Int J Pharm 1994; 101:1–13.
19. Staniforth JN. Use of hydrogenated vegetable oil as a tablet lubricant. Drug Dev Ind Pharm 1987; 13(7):1141–1158.

20. Patel N, Kostenbauder HB. Interaction of preservatives with macromolecules I: binding of parahydroxybenzoic acid esters by polyoxyethylene 20 sorbitan monooleate (Tween 80). J Am Pharm Assoc (Sci) 1958; 47:289–293.
21. Blaug SM, Huang W-T. Interaction of dextroamphetamine sulfate, with dextrates in solution. J Pharm Sci 1973; 62(4):652–655.
22. Wirth DD, Baertschi SW, Johnson RA, et al. Maillard reaction of lactose and fluoxetine hydrochloride, a secondary amine. J Pharm Sci 1998; 87(1):31–39.
23. Ahlneck C, Zografi G. The molecular basis of moisture effects on the physical and chemical stability of drugs in the solid state. Int J Pharm 1990; 62:87–95.
24. Badawy SIF, Gawronski AJ, Alvarez FJ. Application of sorption-desorption moisture transfer modeling to the study of chemical stability of a moisture sensitive drug product in different packaging configurations. Int J Pharm 2001; 223:1–13.
25. Weiner M, Bernstein IL. Adverse Reactions to Drug Formulation Agents: A Handbook of Excipients. New York: Marcel Dekker, 1989:93–94.
26. Maquet J, Théveneau H, Djabourov M, Leblond J, Papon P. State of water in gelatin solutions and gels: An 1H n.m.r. investigation. Polymer 1986; 27(7):1103–1110.
27. Schultze JDR, Waddington WA, Ell PJ, Parsons GE, Coffin ME, Basit AW. Concentration-dependent effects of polyethylene glycol 400 on gastrointestinal transit and drug absorption. Pharm Res 2003; 20(12):1984–1988.
28. Dintaman JM, Silverman JA. Inhibition of P-glycoprotein by D-α-tocopheryl polyethylene glycol 1000 succinate (TGPS). Pharm Res 1999; 16:1550–1556.
29. Monkhouse DC, Lach JL. Drug–excipient interactions. Can J Pharm Sci 1972; 7(2): 29–46.
30. Landín M, Martinez-Pacheco R, Gómez-Amoza JL, Souto C, Concheiro A, Rowe RC. The effect of country of origin on the properties of dicalcium phosphate dihydrate powder. Int J Pharm 1994; 103:9–18.
31. Jain NK, Rosenberg DB, Ulahannan MJ, Pitchumoni CS. Sorbitol intolerance in adults. Am J Gastroenterol 1985; 80(9):678–681.
32. Katz M, Poulsen BJ. Corticoid, vehicle and skin interaction in percutaneous absorption. J Soc Cosmet Chem 1972; 23:565–590.
33. Davis AF, Hadgraft J. Supersaturated solutions as topical drug delivery systems. In: Walters KA, Hadgraft J, eds. Pharmaceutical Skin Penetration Enhancement. New York: Marcel Dekker, 1993:243–268.
34. Yu L, Bridgers A, Polli J, et al. Vitamin E-TPGS increases absorption flux of an HIV protease inhibitor by enhancing its solubility and permeability. Pharm Res 1999; 16: 1812–1817.

9

Improved Excipient Functionality by Coprocessing

Piyush Gupta, Satish K. Nachaegari, and Arvind K. Bansal
Department of Pharmaceutical Technology (Formulations), National Institute of Pharmaceutical Education and Research (NIPER), Punjab, India

INTRODUCTION

Recent decades have seen tremendous strides in the designing of novel dosage forms, but tablets still remain an attractive option for pharmaceutical scientists and clinicians because they offer advantages of accurate unit-dosing, better patient compliance, ease of large-scale manufacturing, and low production cost (1). The formulation of a tablet involves combining the active ingredient, the "drug," with pharmacologically inactive ingredients called "excipients." National Formulary Admission Policy of 1994 (2) defines excipient as "any component other than the active substance(s) intentionally added to the formulation of a dosage form." Excipients aid in the manufacturing and performance of a dosage form, and serve different purposes as diluent, binder, disintegrant, glidant, and lubricant. Thus, excipients can be called as the "functional components" of a formulation (3). The total market for excipients is estimated to be US $2.5 billion with an average annual growth of 7% to 8% in volume and 4% to 5% in value (4). The overall contribution of excipients in dosage form designing can be better appreciated from the fact that more than 70% of the formulations contain excipients at a concentration higher than the drug (5). It is now well established that excipients contribute critically toward processing, stability, safety, and performance of solid dosage forms (Table 1).

It is precisely this increasing appreciation of the excipients' role in solid dosage forms that has triggered their metamorphosis from "inert ingredients" to "functional components" of the formulation (6). Most of the substances lack some important characteristics of an ideal excipient. Coprocessing is a novel concept of altering excipient functionality by retaining the favorable attributes and supplementing with newer ones, by processing the parent excipient with another excipient (7–9). This allows production of high-functionality excipients to the formulator's advantage. The high functionality can be in terms of improved process ability such as flow properties, compressibility, content uniformity, dilution potential, and lubricant sensitivity, or improved performance such as disintegration and dissolution profile. The proceeding

Table 1 Dosage Form Parameters Affected by Excipients, and the Mechanisms Involved

Dosage form parameter	Effect of excipients
Stability	Residual moisture content—adsorbed moisture on excipient surface protects drug from hydrolytic degradation
Process ability	• Surface area, surface free energy, crystal defects, and deformation potential affect compressibility and machineability on high-speed tableting machines with reduced compression dwell times • Particle size distribution and shape affect flow properties, efficiency of dry mixing process, and segregation potential • Compressibility, flow ability, and dilution potential affect the choice of direct compression as a manufacturing process
Performance	Cohesive and adhesive properties, surface free energy, and water uptake behavior affect disintegration and dissolution behavior

sections discuss the intensive efforts to harness greater benefits by developing high-functionality excipients targeted at specific formulation needs.

MANUFACTURING PROBLEMS IN SOLID DOSAGE FORMS

Development of solid dosage forms, more specifically tablets, involves three alternate processing methodologies—wet granulation, dry granulation, and direct compression (DC) (1). All these processes share the following common problems:

- Product weight variation due to poor flow properties
- Content nonuniformity during mixing due to wide differences in density
- Loss of excipient compressibility due to wet granulation and repeated compaction cycles in dry granulation, or excessive usage of lubricants and poorly compressible ingredients in the formulation
- Poor disintegration of product due to excessive usage of binders

Introduction of high-speed tableting machines and a perceptible shift in the processing of solid dosage forms toward DC has altogether led in aggravating these problems. Coprocessing offers a suitable alternative in this regard. By overcoming the above-mentioned limitations of physically mixed excipients, the single-bodied coprocessed product provides ready-to-use excipient with predefined multifunctionality.

SHIFT TOWARD DIRECT COMPRESSION

Popularity of tablets, coupled with an increased understanding of the physics of compression and manufacturing process variables, has matured the manufacturing of tablets as a "science" in its own right (10). Until the 1950s, tablets were primarily produced by the wet granulation process. The availability of new excipients, new grades of existing excipients, and manufacturing machinery—such as positive die feeding and precompression stages—has caused a perceptible shift toward DC process in the manufacturing of tablets. Nearly 41.5% of pharmaceutical manufacturers prefer DC, 41.5% prefer both wet granulation and DC, while 17.2% have nonpreference for DC as a tableting method (11). DC processing methodology directly involves the compression of the powder blend of the drug and suitable excipients to form tablets, without any granulation or drying steps. DC requires fewer processing steps, eliminates the possible detrimental effect of heat and moisture from the process, safeguards drug stability, and requires simplified validation efforts (12). All these advantages translate into huge economical gains.

Tableting process, since being introduced in the early 1840s, has witnessed numerous changes in the form of stringent regulatory requirements for the excipients and product stability. Increasing regulatory pressure on purity, safety, and standardization of the excipients has catalyzed the formation of an international body, the International Pharmaceutical Excipients Council (IPEC) (13). IPEC is a tripartite council with representation from the United States, Europe, and Japan, and has made serious efforts to harmonize requirements for purity and functionality testing of excipients (14).

Simultaneously, development challenges have been posed by high-performance tableting machines that can produce 100,000 to 200,000 tablets/hr (15). Interestingly, such developments, apart from increasing productivity, have negatively affected the manufacturing process by severely limiting the number of materials that can fulfill the performance and regulatory requirements (16). A single drug substance or excipient cannot possess all the desired physicomechanical properties for the development of a robust DC manufacturing process that can be scaled up from the laboratory to the production scale.

Although simple in terms of unit processes involved, the DC process is highly influenced by the powder characteristics (16). The physicomechanical properties of excipients required for a successful DC process are good flow ability, low or no moisture sensitivity, low lubricant sensitivity, good compressibility, and good machineability even in high-speed tableting machinery (17). High compression speeds translate into reduced dwell times of the formulation mix during compression, thus putting a greater demand on their functionality. The majority of the currently available excipients fail to live up to these functionality requirements, thus creating the opportunity for the development of new high-functionality excipients.

DEVELOPMENT OF NEW EXCIPIENTS

The excipients industry to date has been an extension of the food industry (18), which has helped in maintaining a good safety profile and assuring faster regulatory clearance. For the past many years, not a single new chemical excipient has been introduced into the market, postulating the fact that the development of new excipients has been "market-driven" rather than "marketing-driven." The primary reason for this is the relatively high cost involved in the discovery and development of new chemical excipients. However, with the increasing number of new drug moieties of varying physicochemical and stability properties being pushed into the development pipeline, there is a growing pressure on formulators to search for newer excipients to achieve the desired set of functionalities.

Other factors fuelling the search for new excipients are

- the fair appreciation of DC as the process of choice for tablet manufacturing,
- a growing demand for an ideal filler-binder that can substitute two or more excipients,
- the increasing speed capabilities of tablet machinery, requiring excipients with good compressibility and flow properties, even at short dwell times,

- the shortcomings of existing excipients, such as high moisture sensitivity, poor die filling due to agglomeration, and loss of compressibility by microcrystalline cellulose (MCC) upon wet granulation (19)
- the lack of excipients catering the needs of a specific class of patients, such as diabetics, hypertensives, and lactose- and sorbitol-sensitives, and
- the growing performance expectations out of excipients to address issues of disintegration, dissolution, and bioavailability.

SOURCES OF NEW EXCIPIENTS

Although excipients have gained much recognition due to their "functional" properties, the excipients market has remained in stasis in terms of number of newer excipients introduced. This does not imply that excipient manufacturers and formulators are moribund. In fact, both have shown a great deal of ingenuity in developing and utilizing new proprietary combinations of existing excipients to achieve new sets of functionalities (15). The following are the possible three routes by which new excipients can be developed (20).

- New chemical entities as excipients
- New grades of existing excipients
- New combinations of existing excipients

Regulatory expectations of safety and toxicity force the new chemical entities being developed as an excipient to undergo various stages of scrutiny (21), which is a lengthy and costly process. In addition, the excipient is required to undergo a phase of generic development, which shortens the market exclusivity period (9), as shown in Figure 1. The high risk and significant investment involved do not justify the meager returns from the marketing of new excipients. This could be partially overcome if the excipient and pharmaceutical manufacturers jointly develop the drug products, during which a new excipient becomes an integral part of the eventual new drug application (15). This type of arrangement has already been successfully applied in the development of intravenous drug delivery products, wherein CyDex and Pfizer

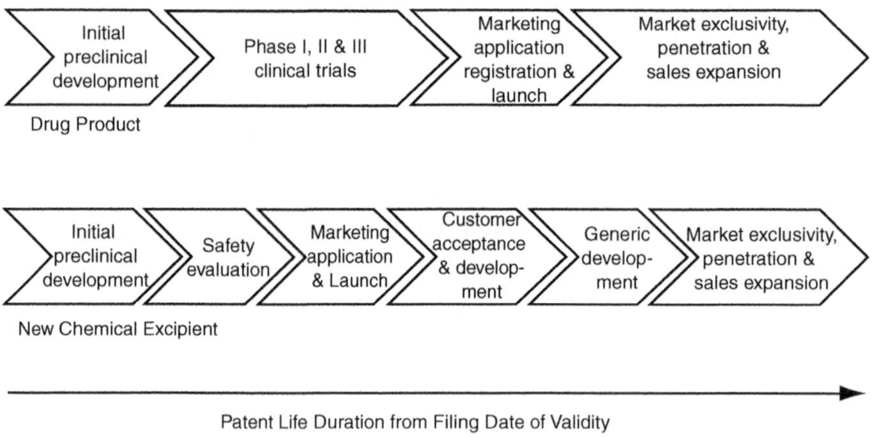

Figure 1 Comparative developmental time lines for a drug product and a new chemical excipient.

worked collaboratively to obtain approval for a solubilizer (22,23). This arrangement, apart from accelerating the development of "tailor-made" innovative excipients, can facilitate their speedy regulatory approval. The latter translates to faster and better return of the investment made on the development of new excipient.

Modification of physicochemical property of the existing excipient has been the most successful strategy in the development of new excipients in the past three decades (24). This model has been successfully adopted for the introduction of excipients with better performance grades, such as pregelatinized starch, croscarmellose sodium, and crospovidone (25). However, the quantum of functionality improvement is limited due to a restricted range of possible modifications.

A majority of the solid dosage forms contain multiple excipients, which opens up a wide window of opportunities by way of combining existing excipients to achieve the desired set of performance characteristics. However, the development of such combinations is a complex process because one excipient may interfere with the existing functionality of another excipient. Over the years, the development of single-bodied excipient combinations at a subparticle level, called coprocessed excipients, has gained huge importance (20). New physical grades of existing excipients and coprocessed excipients are discussed further in the following section, which deals with particle engineering. Particle engineering is a broad concept that involves the manipulation of particle parameters such as shape, size, size distribution, and simultaneous minor changes that occur at the molecular level such as polytypic and polymorphic changes. All these parameters are translated into bulk-level changes, such as flow properties, compressibility, moisture sensitivity, and machineability (26).

PARTICLE ENGINEERING FOR DEVELOPING NEW EXCIPIENTS

The solid state of a substance can be represented by three levels—molecular, particle, and bulk (27). The molecular level comprises the arrangement of individual molecules in the solid state, and includes polymorphs, pseudopolymorphs, and amorphous forms (28). The particle level comprises individual particle properties such as shape, size, size distribution, surface morphology, surface area, and porosity (29). The bulk level is composed of an ensemble of particles, and properties such as cohesive/adhesive strength, flow ability, bulk density, and compressibility (30). These levels are interdependent, with changes in one level getting reflected in the other level (Fig. 2), thus providing a strong scientific framework for the development of new grades and combinations of existing excipients (9).

The fundamental particle properties have a direct bearing on excipient functionalities such as the potential for dilution, disintegration, and lubrication. Hence, the creation of a new excipient must begin with the particle design that is most suited to deliver the desired functionalities (31). Particles with unique characteristics can be created by modulating the conditions like crystallization and drying associated with the preceding molecular level. This is well exemplified by two commonly used excipients, lactose and magnesium stearate, wherein the hydration state has a significant bearing on their functionality (6). However, it is also possible to develop the custom-engineered particles without bringing any changes at the molecular level. Table 2 shows the role of particle engineering, by varying various particle properties, in achieving the desired excipient functionalities (5). Two grades of MCC—Avicel® PH-101 and -102 (FMC BioPolymer, Newark, Delaware, U.S.A.), and spray-dried lactose are examples where such an approach has been successfully applied. However,

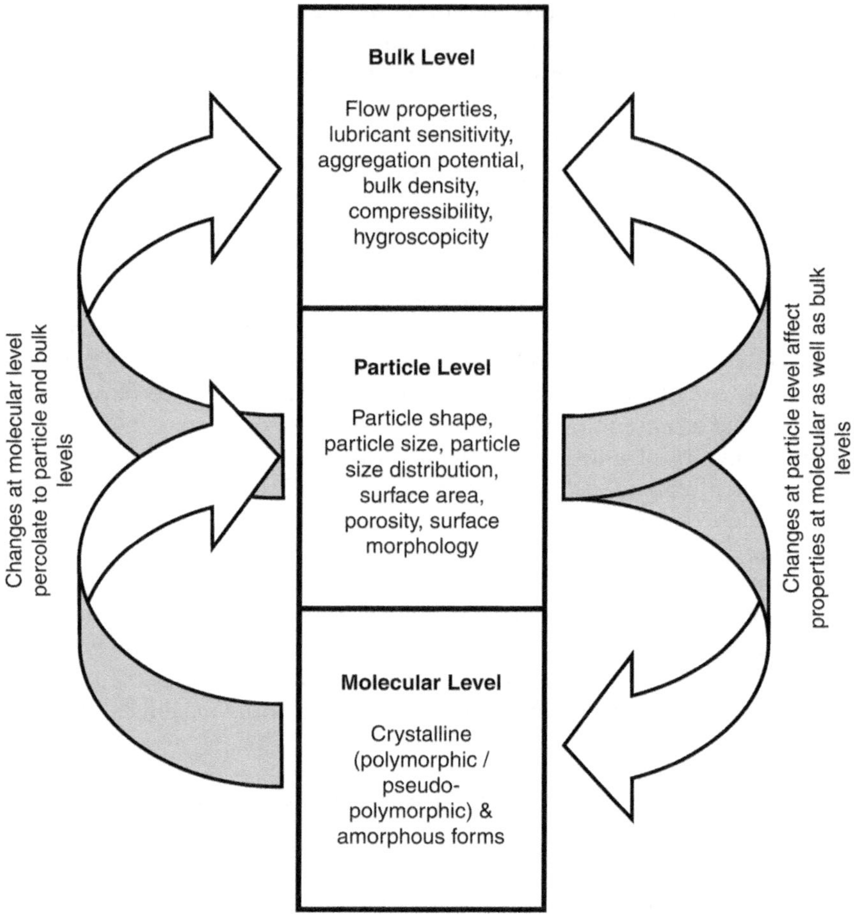

Figure 2 Three levels of solid state.

only a limited functionality improvement is achievable by the particle engineering of a single excipient.

The spectrum of functionality modification can be substantially enlarged by the coprocessing or particle engineering of two or more existing excipients. Coprocessing involves interaction of two or more excipients at the subparticle level, aimed at

Table 2 Particle Properties Influencing Excipient Functionality

Particle property	Excipient functionality affected
Particle size	Flow ability, content uniformity, compressibility, disintegration, dissolution rate
Particle size distribution	Segregation potential
Particle shape	Flow ability, content uniformity, compressibility
Particle porosity	Compressibility, disintegration, dissolution rate
Surface roughness	Flow ability, segregation potential, dilution potential, lubricant sensitivity

providing a synergy of functionality improvements, as well as masking the undesirable properties of the individual excipients (32). The availability of a large number of excipients for coprocessing provides a plethora of opportunities to produce tailor-made "designer excipients" catering to specific functionality requirements.

The preparation of coprocessed excipients involves incorporation of one excipient into the particle structure of another, using processes such as codrying. Figure 3 provides a brief overview of the coprocessing methodology. The coprocessing methodology was initiated in the food industry to improve stability, wettability, solubility, and gelling properties of food ingredients such as coprocessed glucomannan and

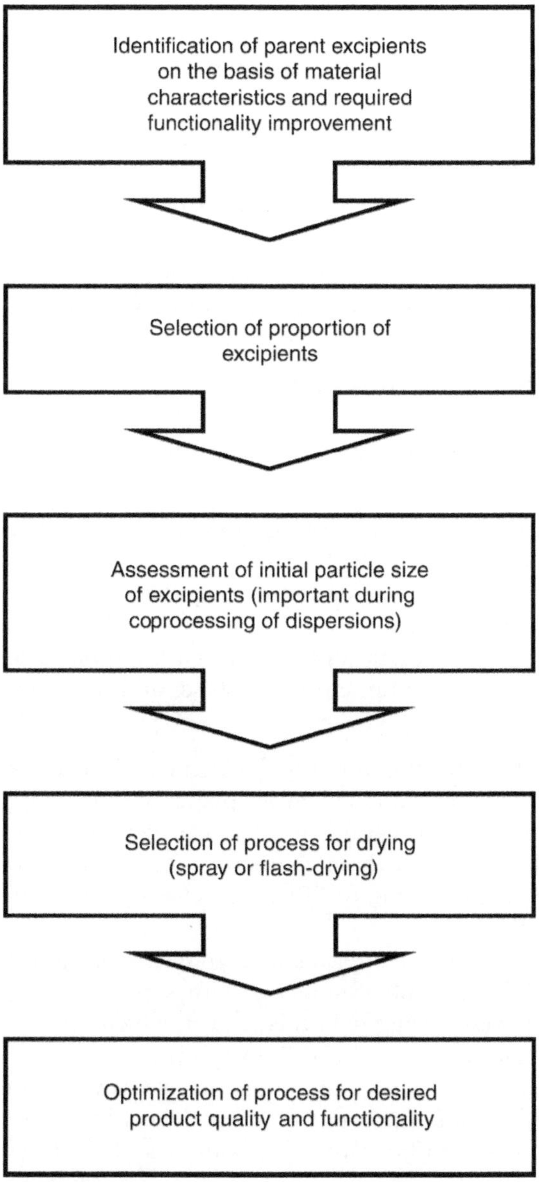

Figure 3 Coprocessing methodology.

galactomannan (33). Coprocessing of excipients in the pharmaceutical industry was introduced around the late 1980s, as exemplified by coprocessed MCC and calcium carbonate (34), followed by Cellactose® (Meggle Corp., Wasserburg, Germany) in 1990, a coprocessed combination of cellulose and lactose, and silicified microcrystalline cellulose (SMCC) in 1996, a coprocessed product of MCC and colloidal silicon dioxide (CSD) (35).

Coprocessing of excipients provides products with superior properties in comparison to their parent excipients, alone or as a physical mixture. Coprocessing is primarily aimed at addressing the issues of flow ability, compressibility, and disintegration potential, and most importantly, the development of filler-binder combinations. The combination of excipients for coprocessing should complement each other to mask the undesirable properties of individual excipients while retaining or improving their desired properties. For instance, a substance used as filler-binder, with a low disintegration property, can be coprocessed with another excipient possessing good wetting properties and high porosity to enhance water uptake, which will aid and hasten the disintegration of the tablets.

ROLE OF MATERIAL CHARACTERISTICS IN COPROCESSING

Material science plays a significant role in altering the physicomechanical characteristics of excipients, especially with regard to their compression and flow behavior.

MATERIAL CHARACTERISTICS AND COMPRESSION

Solid materials, by virtue of their response to applied mechanical force, can be classified under the following three heads (Fig. 4) (36):

Elastic: Any change in shape is completely reversible, and the material returns to its original shape upon release of applied stress.

Plastic: Permanent change in the shape of a material due to applied stress, e.g., MCC, corn starch, and sodium chloride.

Brittle: Rapid propagation of a crack throughout the material on application of stress, e.g., sucrose, mannitol, sodium citrate, lactose, and dicalcium phosphate.

The predisposition of a material to deform in a particular manner depends on its lattice structure, in particular whether weakly bonded lattice planes are inherently present. In definite terms, most of the materials cannot be classified distinctly into individual categories. Pharmaceuticals exhibit all three characteristics, with one of them being the predominant response, thus making it difficult to clearly demarcate the property favorable for compressibility.

Coprocessing offers an interesting tool for altering these physicomechanical properties of excipients. Coprocessing is generally conducted with a plastic and a brittle excipient. Cellactose is an appropriate example in this regard, which involves coprocessing of 75% lactose (a brittle material) with 25% cellulose (a plastic material) (37). Usage of this particular combination prevents the storage of excessive elastic energy during compression, resulting in a small amount of stress relaxation and a reduced tendency for capping and lamination (38). However, examples of the other extreme also exist, e.g., SMCC, which has a large amount of MCC (a plastic material) and a small amount of CSD (a brittle material). These two cases exemplify the

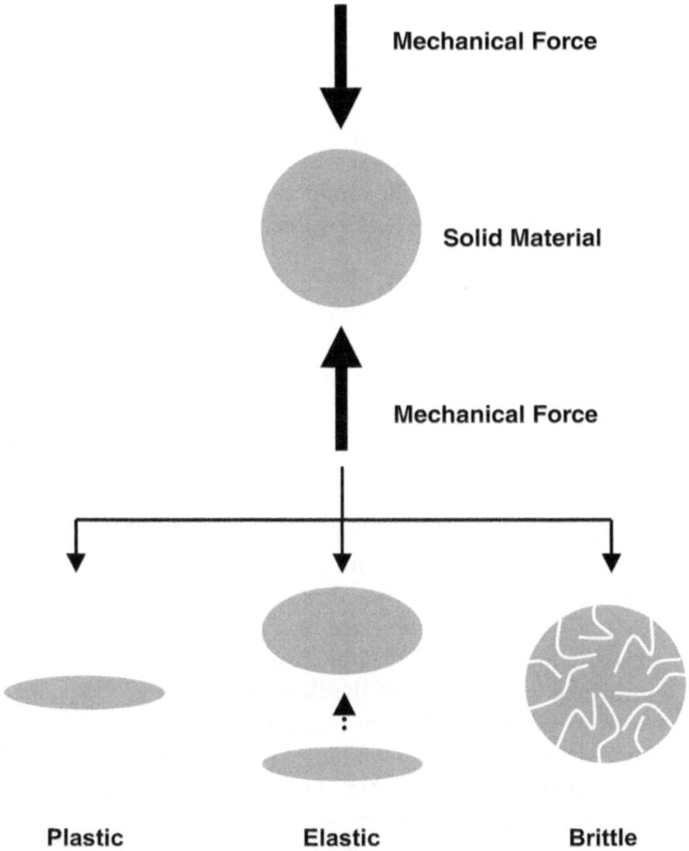

Figure 4 Material classification on the basis of their deformation behavior in the presence of applied stress.

fact that coprocessing is generally performed with a combination of materials possessing plastic deformation and brittle fragmentation characteristics.

MATERIAL CHARACTERISTICS AND FLOW PROPERTIES

Powder flow is typically determined by particle size, particle size distribution, and particle shape (39). Particle size and its distribution have a critical effect on the mixing of powders and the resulting content uniformity of the solid dosage form. Wide differences in particle size result in product segregation during manufacturing. Irregularly shaped particles also contribute to poor flow properties (40,41). Particles having a more regular shape (nearly spherical) are easy to flow and pose minimal hurdles during dosage form production. Coprocessing overcomes all these limitations and provides excipients with predefined attributes.

PROPERTIES OF COPROCESSED EXCIPIENTS

The subject of coprocessing of excipients is multifaceted, with the following characteristic properties.

Absence of Chemical Change

Coprocessing of two excipients results in only a physical change without any chemical alteration. A comprehensive characterization of SMCC with X-ray diffraction, solid-state and C13 nuclear magnetic resonance imaging, and infra-red and Raman spectroscopy confirmed the absence of chemical changes, and indicated a similarity to the physicochemical properties of MCC (19). This reduces the regulatory concerns and encourages the formulators to use coprocessed excipients during the development phase.

Improved Physicomechanical Properties

Coprocessing provides a multitude of improvements in the product's functionality, the most notable of which are discussed below.

Improved Flow Properties

Controlled optimal particle size and size distribution ensures superior flow properties of coprocessed excipients and reduced reliance on addition of glidants. The volumetric flow properties of SMCC were studied in comparison with those of the physical mixture of its parent excipients (42). The particle size range of the two test samples was found to be similar, but the flow of coprocessed excipient was better than that of the physical mixture. A comparison of the flow properties of Cellactose with its parent excipients was also performed (5) by measuring the angle of repose and Hausner ratio, and Cellactose was found to have better flow characteristics than lactose or a physical mixture of cellulose and lactose. The spray-dried coprocessed product had a spherical shape and even surfaces, which resulted in improved flow properties. On similar terms, mechanically coating the 2% CSD over microfine cellulose powder resulted in improving its flow properties (43).

The most common problem manifested due to poor flow property is the variation in fill weight. This problem is much more serious in the case of DC excipients, but coprocessed excipients are devoid of this effect, when compared with the physical mixture of their parent excipients. This is because of the impregnation of one particle into the matrix of another, which reduces the rough particle surfaces and creates a near-optimal size distribution, causing better flow properties. Tablets prepared with M80K, a coprocessed cellulose powder with CSD, showed lesser weight variation than those prepared with Avicel (43).

Fill-weight variation tends to be more prominent with high-speed compression machines. This phenomenon was studied with various machine speeds for SMCC and MCC, and the former showed lesser fill-weight variation than the latter (44).

Improved Compressibility

Coprocessed excipients have been mainly used in DC tableting because of their better flow ability and compressibility, and the excipient formed is a filler-binder. The compressibility of several coprocessed excipients such as Cellactose® (45), SMCC (42,44), and Ludipress® (BASF AG, Ludwigshafen, Germany) (46) have been reported to be superior to the physical mixtures of their constituent excipients. While comparing the compressibility profile of SMCC with MCC in the presence of high compression forces, the former was found to retain the compaction properties, yielding tablets of good hardness. MCC, however, lost its compaction properties. A further utility of SMCC has been reported in the manufacturing of high-dose DC formulations, wherein it reduces the binder requirement by more than half, and results in overall reduction in excipient requirement (47).

Coprocessing of α-lactose monohydrate with cornstarch helped in improving its compressibility, and provided dual benefits of enhanced binding capacity and better disintegration potential, the attributes associated to starch (48). This effect was a result of binding of small starch particles together with α-lactose monohydrate crystals into compound particles.

Although DC seems to be the method of choice for tableting, wet granulation is still widely used in various product manufacturing. Excipients such as MCC lose compressibility upon addition of water, a phenomenon called "quasi-hornification" (49). This property is improved, however, when it is coprocessed into SMCC.

Better Dilution Potential

Dilution potential is the ability of the excipient to retain its functionality even after dilution with another material in a finite proportion. Most drug substances are poorly compressible, and require excipients to achieve better compressibility to retain good compaction even on dilution with them. Cellactose has been shown to possess a higher dilution potential than a physical mixture of its constituent excipients (50).

Reduced Lubricant Sensitivity

Coprocessing endows lesser sensitivity of the product toward loss of their functionality in the presence of lubricants. Most coprocessed products consist of a relatively large amount of brittle material such as α-lactose monohydrate and a smaller amount of plastic material such as cellulose that is fixed between or on the particles of the brittle material (37). The plastic material provides good bonding properties by creating a continuous matrix with a large surface for bonding. The large amount of brittle material provides low lubricant sensitivity by preventing the formation of a coherent lubricant network by forming newly exposed surfaces upon compression, thus breaking up the lubricant network.

Multiple Advantages

Various reports describe improved excipient functionality after coprocessing, with multiple advantages. Roller drying of a solution of anhydrous lactose (95%) and lactitol/sorbitol (5%) resulted in a DC excipient with good tablet strength (51). A free-flowing, compressible powder was obtained by spraying a 4.5% aqueous solution of poly(vinyl pyrrolidone) (PVP) onto a fluid bed of starch and PVP admixture (48:1) (52). Statistical optimization of a coprocessed product of lactose and MCC by various product evaluation parameters such as bulk density, Carr's index, percentage friability, percentage fines, tensile strength, flow rate, and angle of repose resulted in a directly compressible product (with 9:1 composition) with satisfactory flow, compressibility, and friability (12). Coprocessing of lactose monohydrate, PVP, and croscarmellose sodium (79:15:6) by melt agglomeration resulted in a multifunctional DC adjuvant with satisfactory dilution potential, and superior flow ability and compressibility than those of lactose monohydrate (53). Spray drying of rice starch with jet-milled MCC (with volumetric mean diameter of 13.57 μm) in the proportion of 7:3 resulted in spherical composites of a directly compressible excipient with high compressibility, good flow ability, and self disintegration (54).

Other Benefits

Coprocessed excipients offer the following additional advantages:

- Allow the development of tailor-made designer excipients with retention of functional and removal of undesirable properties, which can help in faster product development.
- Provide a single excipient with multiple functionalities, thereby reducing the inventory burden.
- Offer improvement in organoleptic properties, such as those in Avicel® CE-15 (FMC BioPolymer, Newark, Delaware, U.S.A.), a coprocessed excipient of MCC and guar gum, designed for providing chewable tablets with reduced grittiness and tooth packing, minimal chalkiness, better mouth feel, and improved overall palatability.
- Provide more robust tablets at low compression force. Coprocessing of mannitol with sorbitol resulted in interlocked crystals with stronger binding capacity (55). This eased the dispensing of orally dissolving tablet formulations in conventional bottles, eliminating the need for specialized packaging, and thus providing significant cost savings.
- Act as a constant source for development of value-added generic drug products.
- Reduce product cost due to improved functionality (56) and fewer test requirements compared with individual excipients (32).
- Provide intellectual benefits in terms of proprietary combinations, specific for in-house use.

REGULATORY PERSPECTIVE

In the light of the fact that a chemical change is absent during processing, coprocessed excipients can be considered to retain the generally regarded as safe (GRAS) status if the parent excipients are also GRAS certified by the regulatory agencies (20). This reduces the requirement for additional toxicological studies as mandatory for a new chemical entity seeking regulatory approval. IPEC-Americas have submitted a proposal for Excipient Master File, analogous to Drug Master File, to the Food and Drug Administration (57). This document is intended to provide a standard format for submitting excipient safety and manufacturing information to regulatory agencies, and includes provisions for coprocessed excipients also.

The major obstacle to the success of coprocessed excipients in the marketplace is their noninclusion in official monographs. The mixture of excipients was presented as a topic to the National Formulary and was assigned a priority on the basis of its use in marketed dosage forms in which processing provided added functional value to the excipient mixture (35).

Although spray-crystallized dextrose-maltose (EMDEX®, J. Rettenmaier & Sohne GmbH & Co. KG, Germany) and compressible sugars are coprocessed, they are commonly considered as single components and are listed as such in the *United States Pharmacopeia*, while the third edition of the *Handbook of Pharmaceutical Excipients* has listed SMCC as a separate excipient (58).

Table 3 Examples of Marketed Coprocessed Excipients

Coprocessed excipients	Trade name	Manufacturer	Added advantage
Lactose monohydrate (93%), Kollidon® 30 (3.5%), and Kollidon® CL (3.5%)	Ludipress®	BASF AG, Ludwigshafen, Germany	Lower hygroscopicity, good flow ability, tablet hardness independent of machine speed
Lactose monohydrate (96.5%) and Kollidon® 30 (3.5%)	Ludipress LCE	BASF AG, Ludwigshafen, Germany	Lower hygroscopicity, higher tablet hardness
α-Lactose monohydrate (75%) and cellulose powder (25%)	Cellactose® 80	Meggle GmbH & Co. KG, Germany	Highly compressible, good mouth feel, better tableting at low cost
α-Lactose monohydrate (75%) and MCC (25%)	MicroceLac® 100	Meggle GmbH & Co. KG, Germany	Capable of formulating high-dose small tablets with poorly flowable active
α-Lactose monohydrate (85%) and maize starch (15%)	StarLac™	Roquette, Lestrem, France	Good flow, optimized disintegration, excellent tablet hardness
Anhydrous β-lactose (95%) and lactitol (5%)	Pharmatose® DCL14	DMV, Veghel, The Netherlands	High compactibility, superior flowing properties, low lubricant sensitivity
MCC (98%) and colloidal silicon dioxide (2%)	ProSolv HD® 90, ProSolv SMCC® 50, ProSolv SMCC® 90	J. Rettenmaier & Sohne GmbH & Co. KG, Germany	High compactibility, high intrinsic flow, enhanced lubrication efficiency, improved blending properties, reduced sensitivity to wet granulation, better tablet hardness
MCC and guar gum	Avicel® CE-15	FMC BioPolymer, Newark, Delaware, U.S.A.	Less grittiness, reduced tooth packing, minimal chalkiness, creamier mouth feel, improved overall palatability

(Continued)

Table 3 Examples of Marketed Coprocessed Excipients (*Continued*)

Coprocessed excipients	Trade name	Manufacturer	Added advantage
MCC and carboxymethylcellulose sodium	Avicel® RC-581, RC-591, CL-611	FMC BioPolymer, Newark, Delaware, U.S.A.	Viscosity regulator and modifier, thixotropic characteristics, heat and freeze–thaw stable, long shelf-life stability, lengthy hydration times eliminated, stable at pH range 4–11
MCC and calcium sulfate	Celocal®	FMC BioPolymer, Newark, Delaware, U.S.A.	Directly compressible
MCC (65%) and calcium carbonate (35%)	Vitacel® VE-650	FMC BioPolymer, Newark, Delaware, U.S.A.	Direct compression, encapsulation
MCC and carrageenan	LustreClear™	FMC BioPolymer, Newark, Delaware, U.S.A.	Efficient tablet-coating with short hydration time prior to coating and fast drying time
Calcium carbonate (70%) and sorbitol (30%)	Formaxx® $CaCO_3$ 70	Merck KGaA, Darmstadt, Germany	High compressibility, excellent taste masking, free flow, superior content uniformity, controlled particle size distribution
Sucrose (97%) and dextrin (3%)	Di-Pac®	American Sugar Co., New York, U.S.A.	Directly compressible
Carbohydrate system, made from compendial ingredients	Pharmaburst™	SPI Pharma™, Inc., New Castle, U.S.A.	High compactibility, high loading in small tablets, smooth mouth feel, rapid disintegration
Fructose (95%) and Starch (5%)	Advantose™ FS 95 Fructose	SPI Pharma™, Inc., New Castle, U.S.A.	Excellent flow, good compressibility, tablets hold shape well, but are very chewable

Abbreviation: MCC, microcrystalline cellulose.

COMMERCIAL STATUS

Coprocessed excipients are widely available in the market for a vast spectrum of purposes. Table 3 provides a compilation of the marketed coprocessed excipients along with their manufacturers and benefits. Majority of these are produced by a spray-drying process, with very few involving alternate processing technologies. Ludipress is produced by fluidized-bed granulation, while Di-Pac® (American Sugar Co., New York, U.S.) involves minigranulation of sugar crystals glued together with amorphous dextrin.

FUTURE TRENDS

The obvious advantages of solid dosage forms and changing technological requirements will keep alive the search for newer excipients. The newer excipients are required to be compatible not only with the latest technologies and production machineries, but also with the innovative active principles such as those originating from biotechnology. Developments in the field of excipients and manufacturing machinery have helped in establishing traditional inert excipients as functional components. A deeper understanding of their solid-state properties and its impact on excipient functionality is further going to fuel this trend. Functionalities, hitherto unavailable to the formulator, can now be incorporated into the product by judicious choice of high-functionality excipients. Further, a narrow pipeline of new chemical excipients, and an increasing preference for the DC process, creates a significant opportunity for the development of high-functionality excipients. A greater synergy between excipient manufacturers and the pharmaceutical manufacturer in the future is going to help in the development of tailor-made designer excipients complying with safety, performance, and regulatory issues.

CONCLUSIONS

Technological advancements in tablet manufacturing, introduction of high-speed machineries, and a shift in tableting toward DC have catalyzed the search for newer excipients meeting these requirements. Excipients are no more considered as inert ingredients of a formulation, but have a well-defined functional role. Developments in particle engineering have provided wide avenues for designing excipients with predefined functionality requirements. Coprocessed excipients are a result of this arduous innovation only, wherein two excipients are coprocessed to provide products with improved functionality by retaining their favorable and avoiding the unfavorable properties. A better appreciation of this concept can be viewed from the vast number of coprocessed excipients available in the market. The success of these excipients depends on their quality, safety, and functionality. Although the first two parameters have remained constant, significant improvements in functionality provide wide opportunities for the increased use of coprocessed excipients. The advantages of these excipients are numerous, but further scientific exploration is required to understand the mechanisms underlying their performance. The main obstacle in the success of coprocessed excipients is the noninclusion of their monographs in official pharmacopeias, which discourages their use by pharmaceutical manufacturers. With recommendations from IPEC and the continual efforts of excipient manufacturers, these products could find their way into official monographs, either as mixtures or as single-bodied excipients. Once the obstacles are overcome, the use of coprocessed excipients can be expected to increase dramatically.

REFERENCES

1. Banker GS, Anderson NR. Tablets. In: Lachman L, Lieberman HA, Kanig JL, eds. The Theory and Practice of Industrial Pharmacy. Bombay: Varghese Publishing House, 1986:293–345.
2. USP Subcommittee. Excipients. Pharm Forum 1992; 18:4387.
3. Moreton RC. Excipient functionality. Pharm Technol 2004:98, 99, 119.
4. Ermens H. Pharmaceutical excipients—the past and future. Business Briefing: Pharmagenerics 2004:1–3.
5. York P. Crystal engineering and particle design for the powder compaction process. Drug Dev Ind Pharm 1992; 18:677–721.
6. Pifferi G, Santoro P, Pedrani M. Quality and functionality of excipients. Il Farmaco 1999; 54:1–14.
7. Bansal AK, Nachaegari SK. High-functionality excipients for solid dosage forms. Business Briefing: Pharmagenerics 2002:38–44.
8. Bansal AK. Improved excipients by solid-state manipulation. The Industrial Pharmacist 2003:9–12.
9. Nachaegari SK, Bansal AK. Coprocessed excipients for solid dosage forms. Pharm Technol 2004; 28:52–64.
10. Czeisler JL, Perlman KP. Diluents. In: Swarbrick J, Boylan JC, eds. Encyclopedia of Pharmaceutical Technology. New York: Marcel Dekker, Inc., 1998:37–83.
11. Shangraw RF, Demarest DA. A survey of current industrial practices in the formulation and manufacture of tablets and capsules. Pharm Technol 1993; 17:32–44.
12. Gohel MC, Jogani PD. An investigation of the direct-compression characteristics of coprocessed lactose-microcrystalline cellulose using statistical design. Pharm Technol 1999:54–62.
13. IPEC-Americas, "Why IPEC-Americas is needed." http://www.ipecamericas.org/public/whyneeded.html (accessed on 16 January 2003).
14. Blecher L. Pharmaceutical excipients—producers and users strengthen their voice. Pharm Technol 1993; 17:38–39.
15. Restocking the excipient superstore. Pharmaceutical Formulation and Quality. International Pharmaceutical Excipients Council (IPEC), 2000 (February/March).
16. Shangraw RF. Compressed tablets by direct compression. In: Leiberman HA, Lachman L, Schwatz JB, eds. Pharmaceutical Dosage Forms: Tablets. New York: Marcel Dekker, Inc., 1990:195–246.
17. Armstrong NA, Palfrey LP. The effect of machine speed on the consolidation of four directly compressible tablet diluents. J Pharm Pharmacol 1989; 41:149–151.
18. Steinberg M, Blecher L, Mercill A. From inactive ingredients to pharmaceutical excipients. Pharm Technol 2001; 25:62–64.
19. Tobyn MJ, McCarthy GP, Staniforth JN, Edge S. Physicochemical comparison between microcrystalline cellulose and silicified microcrystalline cellulose. Int J Pharm 1998; 169: 183–194.
20. Moreton RC. Tablet excipients to the year 2001: a look into the crystal ball. Drug Dev Ind Pharm 1996; 22:11–23.
21. Robertson MI. Regulatory issues with excipients. Int J Pharm 1999; 187:273–276.
22. CyDex, "Innovative drug delivery technologies." www.captisol.com (accessed on 16 January 2003).
23. CyDex, "Captisol." www.cydexinc.com/CaptisolProductApprovals.pdf (accessed on 16 January 2003).

24. Shangraw RF, Wallace JW, Bowers FM. Morphology and functionality in tablet excipients for direct compression: Part I. Pharm Technol 1981; 5:69–78.
25. Shangraw RF. Emerging trends in the use of pharmaceutical excipients. Pharm Technol 1997; 21:36–42.
26. York P. Molecules, materials science and medicines. Pharm J 2001; 267:437–440.
27. Jones TM. The influence of physical characteristics of excipients on the design and preparation of tablets and capsules. Pharm Ind 1977; 39:469–476.
28. Byrn S, Pfeiffer R, Ganey M, Hoiberg C, Poochikian G. Pharmaceutical solids: a strategic approach to regulatory considerations. Pharm Res 1995; 12:945–954.
29. Brittain HG. Particle-size distribution, Part I. Representations of particle shape, size, and distribution. Pharm Technol 2001:38–45.
30. Standly-Wood N. WS-16, Powder and Bulk Solids. Conference and Exhibition, Chicago, 1989.
31. Reimerdes D, Aufmuth KP. Tableting with co-processed lactose-cellulose excipient. Manuf Chemist 1992; 63:21–24.
32. Reimerdes D. The near future of tablet excipients. Manuf Chemist 1993; 64:14–15.
33. Modliszewski JJ, Ballard AD. Coprocessed galactomannan-glucomannan. US Patent No. 5,498,436, 1996.
34. Mehra DK, West KP, Wiggins JD. Coprocessed microcrystalline cellulose and calcium carbonate composition and its preparation. US Patent No. 4,744,987, 1988.
35. Bolhius GK, Chowhan ZT. Materials for direct compaction. In: Alderborn G, Nystrom C, eds. Pharmaceutical Powder Compaction Technology. New York: Marcel Dekker, Inc., 1996:419–500.
36. Marshall K. Compression and consolidation of powdered solids. In: Lachman L, Lieberman HA, Kanig JL, eds. The Theory and Practice of Industrial Pharmacy. Bombay: Varghese Publishing House, 1986:66–99.
37. Maarschalk KvdV, Bolhuis GK. Improving properties of materials for direct compaction. Pharm Technol 1999; 23:34–46, 96.
38. Casahoursat L, Lemagen G, Larrouture D. The use of stress relaxation trials to characterize tablet capping. Drug Dev Ind Pharm 1988; 14:2179–2199.
39. Howard SA, Lai J-W. Flow properties of solids. In: Swarbrick J, Boylan JC, eds. Encyclopedia of Pharmaceutical Technology. New York: Marcel Dekker, Inc., 1998: 141–169.
40. Heywood H. The evaluation of powders. J Pharm Pharmacol 1963; 15:56T–74T.
41. Ridgway K, Rupp R. The effect of particle shape on powder properties. J Pharm Pharmacol 1969; 21(suppl):30S–39S.
42. Allen JD. Improving DC with SMCC. Manuf Chemist 1996; 67:19–23.
43. Nada AH, Graf E. Evaluation of Vitacel M80K as a new direct compressible vehicle. Eur J Pharm Biopharm 1998; 46:347–353.
44. Sherwood BE, Becker JW. A new class of high functionality excipients: silicified microcrystalline cellulose. Pharm Technol 1998; 22:78–88.
45. Belda PM, Mielck JB. The tableting behavior of Cellactose compared with mixtures of celluloses with lactoses. Eur J Pharm Biopharm 1996; 42:325–330.
46. Schmidt PC, Rubensdorfer CJW. Evaluation of Ludipress as a "multipurpose excipient" for direct compression. Part I: powder characteristics and tableting properties. Drug Dev Ind Pharm 1994; 20:2899–2925.
47. Joshi V. Excipient choice in solid dosage forms. Drug Deliv Technol 2002; 2(6):36–40.
48. Wagner KG, Dressler JA. A corn starch/alpha-lactose monohydrate compound as a directly compressible excipient. Pharm Technol (Europe) March 2003.
49. Staniforth JN, Chatrath M. Towards a new class of high functionality tablet binders. I: Quasi-hornification of microcrystalline cellulose and loss of functionality. Pharm Res 1996; 13:S208.
50. Flores LE, Arellano RL, Esquivel JJD. Study of load capacity of Avicel PH-200 and Cellactose, two direct compression excipients, using experimental design. Drug Dev Ind Pharm 2000; 26:465–469.

51. Meggelaars MM, Biggelaar HAvd, Kussendrager KD. Tableting excipient. US Patent No. 5,534,555, 1996.
52. Menon A, Gillece T, Chakrabarti S. Co-processing method for making a free flowing compressible powder and tablet therefrom. US Patent No. 5,560,927, 1996.
53. Gohel MC, Jogani PD. Functionality testing of a multifunctional directly compressible adjuvant containing lactose, polyvinylpyrrolidone, and croscarmellose sodium. Pharm Technol 2002:64–82.
54. Limwong V, Sutanthavibul N, Kulvanich P. Spherical composite particles of rice starch and microcrystalline cellulose: a new coprocessed excipient for direct compression. AAPS Pharm Sci Tech 2004; 5 (article 30).
55. Joshi AA, Duriez X. Added functionality excipients: an answer to challenging formulations. Pharm Technol (Excipients and Solid Dosage Forms) 2004:12–19.
56. Prosolv. Technical report: tableting binder improves production. New York: Penwest Pharmaceuticals, 2001.
57. DeMerlis CC, Goldring JM. Marketing new excipients: clearing regulatory hurdles. Pharm Technol 2003:102–108.
58. Moreton RC. Cellulose, silicified microcrystalline. In: Kibbe AH ed. Handbook of Pharmaceutical Excipients. Washington, D.C.: American Pharmaceutical Association and London: Pharmaceutical Press, 2000:110–111.

10
A Comparison of Physical and Mechanical Properties of Common Tableting Diluents

Glenn T. Carlson and Bruno C. Hancock
Pfizer, Inc., Groton, Connecticut, U.S.A.

INTRODUCTION

The routine testing of excipients prior to their use in the manufacture of solid oral dosage forms is usually restricted to simple tests that confirm their identity and chemical purity. Whilst most excipient suppliers will guarantee that their products will also pass the tests described in the major pharmacopeias, the results of these testing procedures are likely to tell the end user very little about the functionality of the excipients during normal pharmaceutical manufacturing operations. This is especially true when small differences in primary material properties, such as molecular weight or particle size, can bring about significant changes in ultimate manufacturing performance, such as compressibility.

For the pharmaceutical product development scientist, there is clearly a need for objective information about the practical performance of different excipients and their various grades. In this chapter we set out to bring together the results of some of our ongoing evaluations of the physical and mechanical properties of excipients commonly used for the manufacture of solid oral dosage forms. In this particular article, we have chosen to focus on the fillers that are most commonly used in the manufacture of immediate release tablets: microcrystalline cellulose (MCC), lactose, calcium phosphate, and mannitol (1).

Generally, MCC has good compression properties, imparting strength and robustness to tablet dosage forms. Thus, it is one of the most commonly used ingredients in tablet formulations. The three MCC grades considered in this work, Avicel^{TM} PH102, PH105, and PH302, were chosen because they represented a broad range of mechanical and compaction properties that spanned that of most grades. Lactose is also a first choice excipient for many tablet formulations and, like the other diluents in this manuscript, is available from numerous sources. Grades that were selected for study spanned a wide range of mechanical and physical properties and handling behavior. Dibasic calcium phosphate is a dense inorganic tableting excipient that is available in several different hydration states and grades. The anhydrous as well as the dihydrate forms are used for immediate release tablet formulations. The dihydrate form can undergo dehydration when exposed to elevated temperatures (>60°C) (2), and the granular grades (such as Rhodia's A-TAB^{TM}) are usually intended for direct compression applications. Mannitol is a diluent that is commonly used in direct compression, dry granulated and wet granulated tablet formulations, particularly in chewable tablet formulations because of its sweet taste, its negative heat of solution that imparts coolness in the mouth, and its good "mouth feel." It is not hygroscopic, so it may be used with a moisture-sensitive active pharmaceutical ingredient (API). Wet granulations containing mannitol may be dried relatively easily compared to those with more hygroscopic diluents (3).

The testing procedures used in this work have all been well described in the literature (4) and are focused on understanding the compression behavior of the powder samples and the mechanical properties of the resulting compacts. These methods are summarized in Table 1. For brevity, we have limited our initial studies to single component systems, but recognize that more work is needed in the future to understand the complex behavior of multicomponent mixtures. The current work should provide a sound basis for further work on such systems. It is intended that this treatise will enable pharmaceutical formulation scientists to better understand the similarities and differences between the most common grades and types of excipients, and will facilitate the rational selection of excipients for use in the development of immediate release tablet formulations.

Table 1 Summary of Test Methods

Property	Typical test method	Typical value for excipients
Physical properties		
Particle size distribution	Sieve analysis; laser diffraction	(D4,3) = 25–350 μm
Particle morphology	Scanning electron microscopy; light microscopy	Fibre, agglomerated prisms
True density	Helium pycnometry	1.0–3.0 g/cm^3
Mechanical properties		
Indentation hardness H_0 and H_∞	Pendulum impact and quasistatic indentation	70–600 MPa
Tensile strength (σ_T and σ_{T0})	Diametral compression	0.4–6 MPa
Bonding index ($\times 10^2$)	Calculated from H_0 and σ_T	0.2–3
Brittle fracture index	Calculated from σ_T and σ_{T0}	0.1–0.5
Compaction properties	Compaction simulator or instrumented tablet press	
Tensile strength at 0.85 SF	Values at 0.85 SF or at maximum determined from	1–10 MPa
Compression stress at 0.85 SF	best fit curve in plots of tensile strength vs. SF	100–600 MPa
Maximum tensile strength	or tensile strength vs. compression stress	1.5–13 MPa
Compression stress at max tensile strength		375–700 MPa

Abbreviations: SF, solid fraction; D(4,3) = volume mean diameter.

BACKGROUND

Physical Property Characterization

Typical methods for determining physical properties of powders are summarized in Table 1.

Particle Size Distribution

Particle size is one of the principal determinants of powder behavior such as packing and consolidation, flow ability, compaction, etc., and it is therefore one of the most common and important areas of powder characterization. Typically, one refers to particle size or diameter as the largest dimension of its individual particles. Because a given powder consists of particles of many sizes, it is preferable to measure and describe the entire distribution. While many methods of size determination exist, no one method is perfect (5); two very common methods are sieve analysis and laser diffraction. Sieving is a very simple and inexpensive method, but it provides data at relatively few points within a distribution and is often very operator dependent. Laser diffraction is a very rapid technique and provides a detailed description of the distribution. However, its instrumentation is relatively expensive, the analytical results are subject to the unique and proprietary algorithms of the equipment manufacturer, and they often assume particle sphericity. The particle size distribution shown in Figure 1 was obtained by laser diffraction, where the curves represent frequency and cumulative distributions.

Because particle size is so intimately intertwined with powder performance, it is one of the prime considerations in selecting excipients to develop or improve a formulation. This is particularly important with direct compression formulations where excipient flowability and compaction performance are critical. Typically, excipients for these applications exhibit narrow size distributions with moderate-to-coarse particle size, having a mean size from 100 to 200 μm.

Particle morphology refers to the external features or form of a powder's primary particles. This includes descriptions of shape, including aspect ratio, or crystal habit (plate, needle, lath, equant, etc.). Particles are not always observed as discrete entities. Rather, they are often associated with other particles, sometimes loosely held

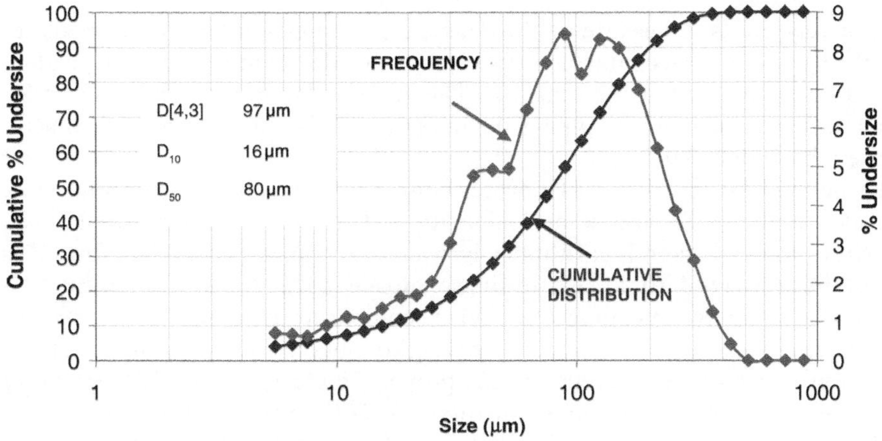

Figure 1 Particle size distribution for D-(−) mannitol.

together (i.e., "agglomerates") and sometimes tightly bound (i.e., "aggregates") such that the clusters of bound particles are not easily separated and may behave as discrete particles. Particle morphology is typically determined by light microscopy, electron microscopy, or other imaging techniques. Morphology can play an important role in determining powder behavior, including particle packing and consolidation, flow ability, compaction (e.g., the propensity for plastic deformation), and segregation. Knowledge of excipient morphology is therefore an important aspect when selecting excipients to develop a formulation.

True Density

Powders are porous materials and their bulk and relative densities can change with consolidation (6). However, a powder's true density is the density of its solid phase only and thus is independent of the state of consolidation. The true density of organic excipients typically ranges from 1.0 to 1.6 g/cm^3 while inorganic excipients (e.g., calcium phosphate) show values greater than 2 g/cm^3. True density is used to determine powder or compact solid fraction (SF) (see below) and it may be a consideration when selecting excipients if segregation is a concern. True density is often determined by gas pycnometry.

Solid Fraction

A powder is a two-phase system consisting of solid particles and gas-filled voids. For a loose powder, the solid portion may occupy less than half the total volume, but for a powder compact, the relative volume of the solid phase is substantially greater compared to that of the gaseous phase. The compaction process brings about particle consolidation where the applied load causes particles to initially rearrange and eventually to fracture and/or deform; the proportion or fraction of the solid phase greatly increases while that of the gaseous phase greatly decreases. A powder or compact's SF is an expression of the relative extent of these two phases. It is also known as the relative density, which is the ratio of the material's "envelope" density, which includes its pores, to its true density, which excludes all pores. It is related to compact "porosity" by Eq. (1).

$$\mathrm{SF} = (1 - \mathrm{Porosity}) \quad (1)$$

Intuitively, SF is a measure of the degree of compression because it increases with applied pressure. Mechanical properties of powder compacts are dependent on SF and best compared at the same SF. The authors' laboratory selected 0.85 SF as its standard because it approximates the midpoint of the range typical of immediate release tablets (0.8–0.9).

The mechanical properties of powder compacts—dynamic indentation hardness, tensile strength (TS), and compression stress (CS)—are dependent on SF, i.e., the properties increase semi-logarithmically over a SF range (4,7). Therefore, it is best to establish a standard SF for testing compacts so that the mechanical properties can be compared meaningfully. When comparing materials compacted to different SFs, a rule-of-thumb can be applied to obtain rough estimates of the mechanical properties at a different SF: the property increases about 10% for every 0.01 increase in SF (personal communication. Dr.G.E. Amioon, Pfizer Inc., Ann Arbor, Michigan, U.S.A.). It must be borne in mind that this is an approximation and that material-to-material differences are likely.

Tableting Indices Characterization

The tableting indices methods, summarized in Table 1, require powder compacts that are prepared under carefully controlled conditions so that they are essentially free of flaws (4,7). These compacts are the samples used for indentation hardness and TS measurements.

Compression Stress

The punch pressure required to form a compact for tableting indices measurements is measured at the end of a long dwell time, typically 1.5 minutes, during which the punches remain in fixed positions and stress relaxation within the compact brings about a decay in the applied load. The reported pressure or CS is calculated from the relaxed force and it is dependent on the compact SF. A sample's CS at a standard SF, such as 0.85, can be interpreted to indicate the ease (i.e., the magnitude of the pressure) of forming compacts under standardized conditions.

Dynamic Indentation Hardness

Dynamic indentation hardness is a compact mechanical property that provides a measure of a material's plasticity or ductility. The sample is deformed by high-speed pendulum impact. Samples with high hardness values show relatively small amounts of deformation (i.e., their indentations are small) and thus are not as ductile as materials that form larger indentations under the same conditions. Thus, during powder compaction, very ductile materials (i.e., having low hardness values) tend to deform to a greater extent than low ductility materials (i.e., having high hardness values). Plastic deformation during powder compaction is thought to be advantageous from a particle bonding perspective because it tends to increase bonding contact area between particles and therefore contributes to particle bonding and tablet formation.

Tensile Strength

A powder compact's TS is the stress required to separate its constituent particles in tensile mode. This is measured for the tableting indices by transverse compression of the square compacts, using narrow platens. Stresses build within the sample until it fails in a tensile mode that is perpendicular to the direction of platen movement. Tablets that are manufactured on a traditional tablet press and that have high TS are considered "hard" and generally robust, and so this is a highly desired attribute for immediate release and other tablet types.

Bonding Index

The bonding index is calculated from the dynamic indentation hardness and TS of powder compacts. It indicates the extent of particle bonding that remains after a tablet has been decompressed. In the tableting process, powder particles in the die first consolidate and then deform and/or fracture during compression, resulting in high bonding contact area and particle bond formation. When the tablet punch retracts, unloading occurs, tablet particles separate, and the bonds formed during the compression phase become stressed in a tensile mode. This often results in reduced bonding contact area and perhaps bond rupture. The particle bonding that remains is represented by the bonding index: high values are significant attributes because they indicate a high probability for forming strong, robust tablets.

Brittle Fracture Index

The brittle fracture index (BFI) describes the propensity for particle bonds to relieve stress by fracturing. It is based on the ratio of the material's TS to its "compromised" TS as measured on powder compacts without and with a macroscopic flaw manufactured into its structure, respectively. The flaw is a circular hole through the compact, which acts as a stress concentrator. The BFI will approach zero for materials for which the regular and compromised TSs are nearly the same (i.e., their regular-to-compromised TS ratio approaches one). This is a highly desirable trait because it indicates a low probability for fracture (lamination or capping) during decompression in a tableting die. BFI values above 0.3—corresponding to a TS ratio greater than 1.6—are often interpreted as "high," signifying that particle bonds have a significant tendency to rupture, thereby forming microscopic voids or structural flaws within the compact during stress unloading. In this situation, the likelihood for crack formation becomes uncomfortably high. The theoretical maximum BFI is one.

Compaction Properties Characterization

Tablet mechanical properties measured on samples prepared on an instrumented tablet press or compaction simulator are an excellent means to characterize excipients under dynamic conditions (8,9). Meaningful data analyses are best achieved if both tablet preparation and tablet property measurements are performed at carefully controlled conditions using standardized procedures. Such testing in the authors'

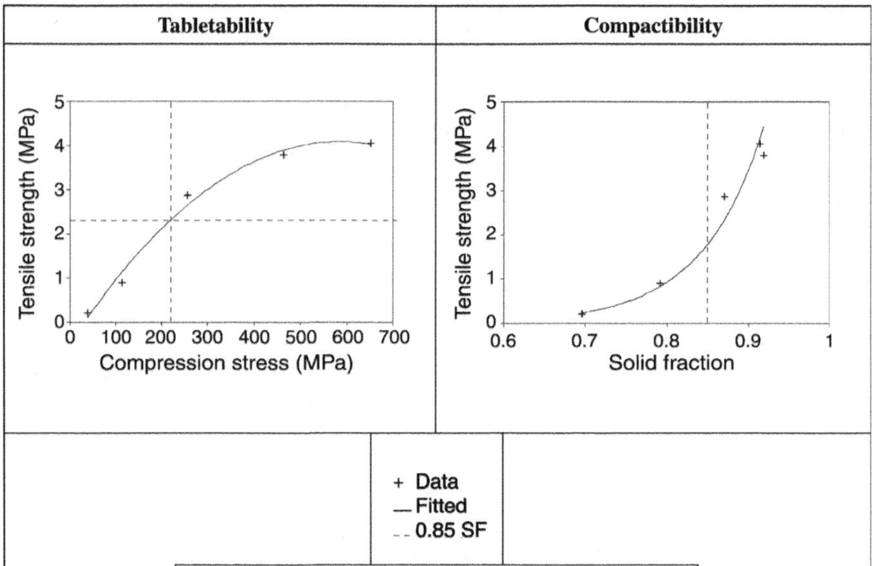

Derived Compaction Properties		
Compression stress at 0.85 SF (MPa)	220	Moderate
Maximum compression stress (MPa)	582	
Tensile strength at 0.85 SF (MPa)	2.32	Moderate

Figure 2 Determination of compaction properties of Mannogem^{TM} EZ. Data points are mean of duplicate trials. Curve is best-fit quadratic equation.

laboratory was performed using the procedures described under "Methods." Tablets prepared over a range of compression forces were collected. Their properties—weight, thickness, diameter, and crushing force—were measured and SF and TS calculated. A "tabletability" profile (10,11) of compression pressure versus TS was plotted and the best quadratic fit was used to predict tablet TS and the corresponding CS at two points on the curve—0.85 SF and at the apex. These standardized properties provide a means of rating and ranking the excipients. A "compactibility profile" graphically depicts the relationship between TS (the dependent variable) and SF (the independent variable) for a series of compacts prepared over a SF range. Examination of a single point in the profile can provide meaningful material comparisons if a standard SF is selected that is representative of tablets in general. Tabletability and compactibility profiles along with summary data for one grade of mannitol are shown in Figure 2, which is an example of data from a standardized compression test.

The above discussions provide background information for the material discussions that follow, where representative grades of MCC, lactose, calcium phosphate dibasic, and mannitol are compared.

EXPERIMENTAL

Materials

MCC, lactose, calcium phosphate dibasic, and mannitol were selected as common tableting diluents and were evaluated as received from their vendors. These materials are summarized in Table 2. Each material is available from several vendors with multiple grades. Three grades within each excipient were selected for their diverse range of physical and mechanical properties (3). These materials typically comprise 5% to 70% of a formulation. The samples were stored at environmentally controlled laboratory conditions of $20 \pm 2°C$ and $40\% \pm 10\%$ relative humidity.

Methods

SEM Images

Photomicrographs of each material were taken with a Jeol JSM-5800 scanning electron microscope (SEM) (Jeol USA Inc., Peabody, Massachusetts, U.S.A.). The photographs were taken at a working distance of 10 mm, with an accelerating voltage of 5 to 10 kV.

True Density

The true densities of the samples were determined with a helium pycnometer (Quantachrome Inc., Boynton Beach, Florida, U.S.A.) operated at $20 \pm 2°C$ according to the manufacturer's recommended methods. Calibration was performed using standard stainless steel spheres, and the mean value of triplicate determinations is reported.

Particle Size Distribution

The particle size distribution of each powder was determined using a Sympatec Helos/Rodos laser diffraction particle size analyzer (Sympatec Inc., Princeton, New Jersey, U.S.A.) with dry powder dispersion capability. The powder dispersion pressure was varied between 0.5 and 2.0 bar (depending on the tendency for agglomeration) with direct feed into the dispersion funnel. The optical concentration was maintained in the range of 5% to 20%. The mean value of duplicate determinations is reported.

Table 2 Common Tablet Diluents

Diluent	Product name	Description	Vendor
Microcrystalline cellulose	Avicel™ PH102	NF, Ph Eur, JP, BP	FMC (Philadelphia, Pennsylvania, U.S.A.)
	Avicel PH105	Fine powder NF, Ph Eur, JP, BP	FMC
	Avicel PH302	High density NF, Ph Eur, JP, BP	FMC
Lactose	Direct Tableting™ lactose	Anhydrous, NF	Quest International (Chicago, Illinois, U.S.A.)
	Lactose-310™	Monohydrate, NF	Foremost (Baraboo, Wisconsin, U.S.A.)
	Lactose-316 Fast Flo™	Monohydrate, spray dried NF	Foremost
Calcium phosphate dibasic	Emcompress™	Dihydrate USP, BP, Ph Eur	JRS Pharma (Rosenberg, Germany)
	CD Anhydrous™	Anhydrous powder, USP	Rhodia (Cranbury, New Jersey, U.S.A.)
	A-TAB™	Anhydrous, granular, unmilled, USP	Rhodia
Mannitol	Mannogem™ 2080	Granular, USP	SPI Pharma (New Castle, Delaware, U.S.A.)
	D-(−) Mannitol	Powder Ph Eur, BP, JP, USP	EMD Chemicals (Gibbstown, New Jersey,U.S.A.)
	Mannogem™ EZ	Spray dried, USP	SPI Pharma

Abbreviations: JP, Japanese Pharmacopeia; NF, National Formulary; BP, British Pharmacopeia; USP, Unitied States Pharmacopeia; Ph Eur, European Pharmacopeia.

$B_{MID}80$, which is a measure of the breadth of the distribution and is also referred to as the span, was calculated according to Eq. (2).

$$B_{MID}80 = (D_{90} - D_{10})/D_{50} \tag{2}$$

Tableting Indices

Samples for mechanical testing were square compacts measuring 1.9 × 1.9 × 1.0 cm, with weights ranging from about 4.5 g to nearly 10 g. They were formed by uniaxial compression (~1 mm/sec compression speed) using a custom-built hydraulic press that permitted controlled, gradual triaxial decompression of the samples. Dwell time for the compression was 1.5 minutes and tri-axial decompression time was two minutes. The punch and die surfaces were sparingly lubricated with magnesium stearate suspended in methanol (~5%). Powder weight was adjusted to produce compacts at 0.85 SF (15% porosity) to directly compare the mechanical properties of the materials. Compacts of dibasic calcium phosphate were prepared at the maximum achievable SF, which was significantly lower than the other excipients due to the limitations of the hydraulic system. Compacts were stored at environmentally controlled laboratory conditions of 20 ± 2°C and 40 ± 10% relative humidity prior to mechanical testing.

Indentation hardness determinations were performed in "dynamic" mode (~1500 mm/sec impact speed) using a pendulum impact device and in "quasistatic" mode (~0.008 mm/sec impact speed) with a custom-built indentation tester. The spherical indenters were of 2.54 cm diameter and 65.6 g mass, and the pendulum length was 92.3 cm with a release angle of 30°. Quasistatic indentation forces were selected to produce indentations of a similar size to the dynamic indentation test (1.5 to 2.0 mm radius). The compact indentations were measured using a white light interferometer (Zygo Corporation, Middlefield, Connecticut, U.S.A.) and the dent depth, dent diameter, apparent radius of curvature, and pendulum initial and rebound heights were used to calculate the indentation hardness of the compacts.

The TS of the compacted samples was determined by transverse compression with a custom-built tensile tester. Tensile failure was observed for all the rectangular compacts when compressed between flat-faced platens at a speed ranging between 0.006 and 0.016 mm/sec. Platen speed was adjusted between materials to maintain a time constant of 15 ± 2 seconds to account for viscoelastic differences; the constant is the time between the sample break point and when the measured force equals F_{BREAK}/e in the force versus time profile, where the denominator is the mathematical *e*. Specially modified punch and die sets permitted the formation of square compacts with a centrally located hole (0.11 cm diameter) that acted as a stress concentrator during tensile testing. This capability permitted the determination of a "compromised" compact TS and thus facilitated an assessment of the defect sensitivity of each compacted material. At least two replicate determinations were performed for each mechanical testing procedure and mean values are reported.

Compaction Properties

Compaction properties of each material were determined with a standardized test performed on a custom-built hydraulic compaction simulator using 8 mm (0.3150 in.) round flat-faced punches. A linear "saw-tooth" upper punch position profile was selected with a punch velocity of 300 mm/sec for both punch extension and retraction. The lower punch position was at a fixed position within the die during the compaction event. The powder weight loaded into the die for each compression was calculated from the equation below so as to form a cylindrical tablet having a thickness-to-diameter ratio of 0.30 at a theoretical SF of 1.0. These dimensions are typical of commercially "elegant" tablets.

$$\text{Powder weight} = 0.3\,(\text{punch diameter}) \times (\text{punch contact area}) \times (\text{true density}) \tag{3}$$

The punch tips and die wall were sparingly lubricated with magnesium stearate from a 5% suspension in methanol. Compression force and tablet SF were adjusted by controlling the minimum separation distance between punch tips during the compaction cycle. Tablets were manufactured in duplicate at several SFs (maximum force < 35 kN) and out-of-die measurements—tablet weight, thickness, diameter, and hardness (crushing strength)—were performed immediately after ejection from the die. Two additional tablets were manufactured at the same compaction conditions for friability testing on the same day.

Tablet SF, TS (in MPa), CS or pressure (in MPa), and friability were calculated from the tablet measurements, the compression forces, and the tooling dimensions. Plots of SF versus TS and TS versus CS were generated; curves were fitted to the data and the properties at 0.85 SF and at the apex of the TS versus CS profile were determined.

RESULTS AND DISCUSSION

The authors determined the mechanical properties of the excipients in this manuscript at or near 0.85 SF except for dicalcium phosphate (DCP). The low SF that could be achieved with DCP (~0.6) demonstrates that its compressibility was less than the compressibility of the other materials and that relevant comparisons to them could not be easily made. When the SF difference between materials is 0.03 or less, property differences due to SF are normally relatively small and comparisons can be made with confidence, but when the SF difference is greater than 0.03, a meaningful material comparison becomes more difficult to achieve. In that situation, material comparisons using only general qualitative statements are appropriate. The latter scenario was the case with DCP, and this has been discussed in the section "Calcium Phosphate Dibasic." In the discussions that follow, excipient mechanical properties were determined at a SF of 0.85 unless otherwise noted.

Microcrystalline Cellulose

Physical Properties

Examination of these materials reveals that many have similar particle morphology—particle shape, aspect ratio, etc.—but they differ significantly in their particle size ranges. SEM images in Figure 3A show Avicel PH102, which has a particle morphology representative of several MCC grades, and Avicel PH105, a specialty grade with a more unique morphology. The particles of Avicel PH102 appear very rough, with many wrinkles and folds, and irregular in shape, with aspect ratios from 1.5 to 3. Other grades with similar morphology include Avicel PH101, PH200, PH301, and PH302, but they differ significantly by particle size. Both SEM images and laser diffraction particle size data indicate that the PH102 grade particle diameters ranged from under 10 μm to about 200 μm, while Avicel PH200, a coarser grade, had about 10% of particles under 50 μm diameter with 300 μm diameter particles being not uncommon. Avicel PH101, a finer grade than PH102, showed about 40% of particles under 50 μm, with the largest approaching 140 μm. Avicel PH302 showed a size range very similar to PH102. The particles of Avicel PH105, a very fine grade, generally show a flake-like morphology with typical length-to-width aspect ratios from 2 to 3 and diameters from less than 5 μm to about 30 μm. Laser diffraction particle size data indicate that about 95% of particles were under 50 μm. Overall, these materials showed narrow particle size distributions ($B_{MID}80 < 2$).

In summary, MCC products are offered in a wide range of particle sizes and they typically show very rough particle morphology. Based on these properties, one would expect a wide range of handling behavior with these materials, from poor flowing and very cohesive to freely flowing, and a range of mechanical properties.

Tableting Indices

Compacts for tableting indices determination for Avicel PH102, PH105, and PH302, as shown in Table 3, were prepared at a SF slightly lower (0.83–0.84) than the standard 0.85 in order to obtain a better measurement of indentation hardness. At SF of 0.85, these materials generally form very hard compacts such that indentations from pendulum impacts were extremely shallow and could not be confidently measured. Compacts prepared at slightly reduced SF were somewhat softer (lower

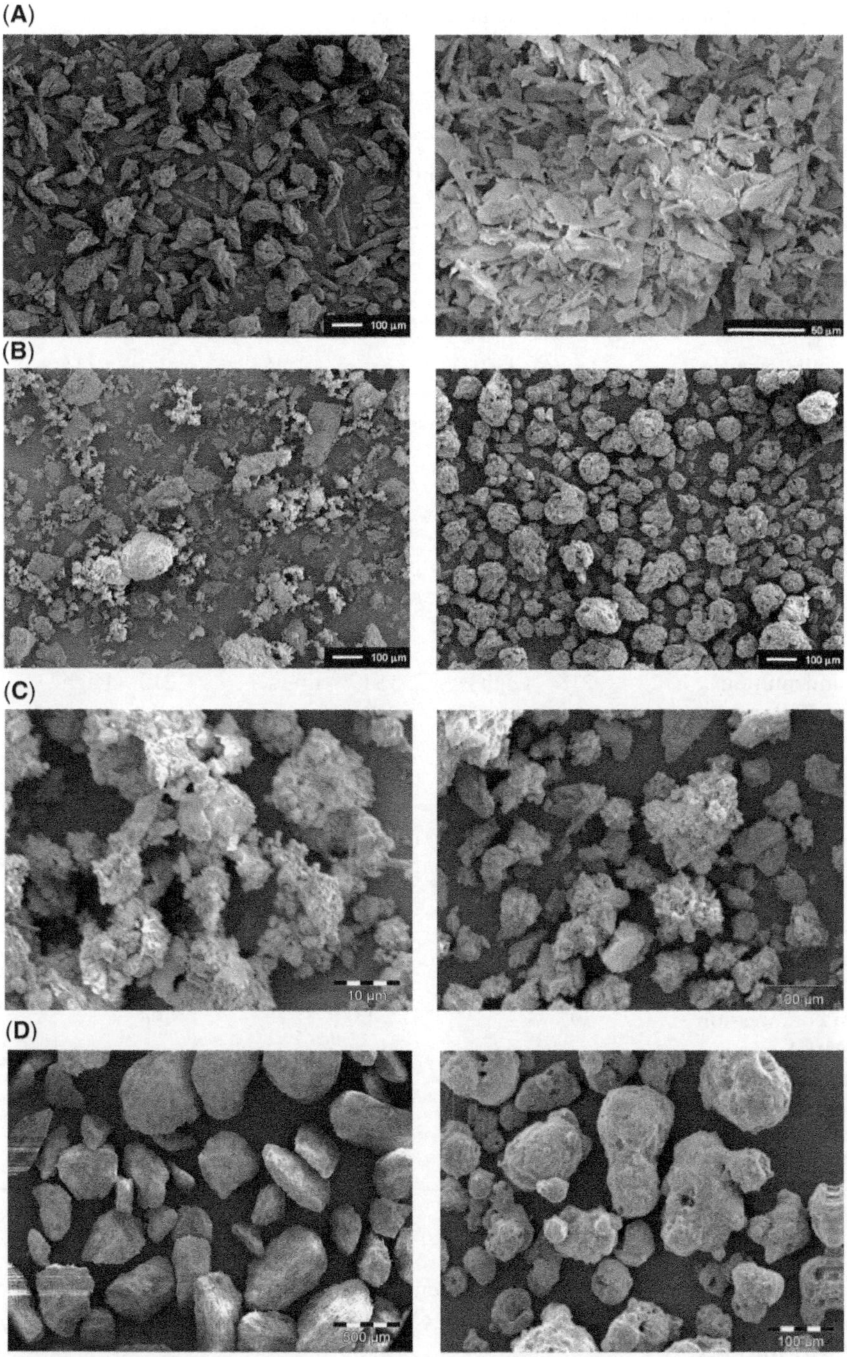

Figure 3 Examples of particle morphology for (**A**) microcrystalline cellulose, (**B**) lactose, (**C**) calcium phosphate dibasic, and (**D**) mannitol. Two images for each material representing different grades.

Table 3 Properties of Microcrystalline Cellulose Samples

	Avicel™ PH102		Avicel PH105		Avicel PH302	
Particle morphology						
Particle aspect ratio	1.5–3		2–3		2–4	
Characteristics	Equant, column		Flake		Equant, lath	
Particle size						
Mean volumetric diameter (μm)	123		21		127	
10th percentile particle diameter (μm)	32		5		27	
50th percentile particle diameter (μm)	112		18		115	
90th percentile particle diameter (μm)	228		39		243	
$B_{MID}80$	1.8		1.9		1.9	
Density						
True density (g/cm^3)	1.58		1.55		1.54	
Tableting indices						
Compression stress (MPa)	81	Moderate	91	Moderate	63	Moderate
SF	0.83		0.84		0.83	
Dynamic indentation hardness (MPa)	218	High	91.7	Low	213	High
Tensile strength (MPa)	6.06	High	11.60	High	3.87	High
Compromised tensile strength (MPa)	4.76		10.20		2.97	
Worst case bonding index ($\times 10^2$)	2.8	High	12.6	High	1.8	High
Brittle fracture index	0.14	Moderate	0.07	Low	0.15	Moderate
Compaction properties						
Compression stress at 0.85 SF (MPa)	175	Moderate			161	Moderate
Maximum compression stress (MPa)	397		443		496	
Tensile strength at 0.85 SF (MPa)	9.7	High			5.7	High
Maximum tensile strength (MPa)	14.3	High	13.6	High	10.0	High

Abbreviation: SF, solid fraction.

dynamic indentation hardness) so that indentations could be measured. Other mechanical property measurements such as CS, TS, and BFI were also slightly reduced at this lower SF.

The range of compression pressures to prepare tableting indices compacts is shown in Table 3; with each considered "moderate" relative to other excipients such as lactose, mannitol, and calcium phosphate dibasic. Avicel PH302 required considerably less pressure than PH102 and PH105, and thus shows greater ease of compression. Their moderate compressibility indicates that a fairly substantial pressure was required to achieve the SF. The rank order of each excipient by the tableting indices mechanical properties is provided in Table 4.

Table 4 Rank Order of Excipient Grades by Property from Lowest to Highest Values

Compression stress (MPa)		Dynamic indentation hardness (MPa)		Tensile strength (MPa)		Bonding index		Brittle fracture index	
D-(−) Mannitol	40	Avicel PH105	92	D-(−) Mannitol	0.46	Mannogem 2080	0.2	Avicel PH105	0.07
Mannogem™ 2080	52	D-(−) Mannitol	145	Mannogem 2080	0.48	A-TAB	0.3	Mannogem 2080	0.10
Avicel™ PH302	63	Mannogem 2080	198	Lactose 310	0.89	D-(−) mannitol	0.3	Emcompress	0.11
Mannogem™ EZ	72	Avicel PH302	213	Emcompress	1.12	Mannogem EZ	0.3	D-(−) Mannitol	0.12
Avicel PH102	81	Emcompress	213	Mannogem EZ	1.17	Direct Tableting lactose	0.4	Avicel PH102	0.14
Avicel PH105	91	Avicel PH102	218	A-TAB	1.32	Lactose 310	0.4	A-TAB	0.14
Direct Tableting™ lactose	92	Lactose 310	232	Direct Tableting lactose	1.56	Fast Flo lactose	0.4	Avicel PH302	0.15
Lactose 310™	98	Mannogem EZ	348	CD Anhydrous	1.60	CD Anhydrous	0.4	Mannogem EZ	0.19
Fast Flo™ lactose	106	Direct Tableting lactose	387	Fast Flo lactose	2.62	Emcompress	0.5	Direct Tableting lactose	0.26
Emcompress™	127	A-TAB	421	Avicel PH302	3.87	Avicel PH302	1.8	Lactose 310	0.30
CD Anhydrous™	158	CD Anhydrous	453	Avicel PH102	6.06	Avicel PH102	2.8	CD Anhydrous	0.42
A-TAB™	163	Fast Flo lactose	666	Avicel PH105	11.60	Avicel PH105	12.6	Fast Flo lactose	0.45

Note: The solid fraction of DCP tableting indices compacts was less than 0.85, particularly for the Emcompress™, CD Anhydrous™ and A-TAB grades™.
Abbreviation: DCP, dicalcium phosphate.

Table 3 shows that Avicel PH102 and PH302 formed powder compacts with similar relatively high dynamic indentation hardness, whereas that of Avicel PH105 was considerably lower, indicating PH105 was the most ductile of the three compacted materials. Relative to other excipients, the values for these materials were fairly low, indicating greater ductility. Ductility is a desirable trait that promotes bonding and strength in tablets.

The three MCC products in Table 3, like most other pharmaceutical grades of MCC, showed high compact TS and within themselves represent a considerable range of strength. They all were regarded as high strength materials compared to other excipients like calcium phosphate dibasic, lactose, and mannitol. Avicel PH105 and PH302 represent the higher and lower ends for values of TS within MCC. The high strength of MCC is a very significant and desirable attribute and is one reason for its popularity; MCC is often used as a binder because it imparts strength to dosage forms, particularly for direct compression blends and dry granulations, rendering them more manufacturable and robust.

Most grades of MCC show high bonding index values and the grades in this discussion are no exception. As with TS, their bonding values generally span the range across most MCC grades with Avicel PH105 at the high extreme and PH302 toward the lower extreme. The bonding index value for Avicel PH102 is fairly representative of many other grades, including PH101 and PH200. The high levels of bonding in general places MCC at the high end of excipient bonding, contributing greatly to its frequent use as a binder and diluent in tablet formulations and promoting tablet manufacturability and robustness.

The BFI values shown by Avicel PH102, PH105, and PH302 in Table 3 were representative of most MCC grades; in general, they were considered low to moderate, indicating a relatively low propensity for relieving stress by fracture, which is an important and desirable attribute. The BFI values of the various grades of MCC are comparable to several other excipients, often depending on their grade (e.g., mannitol and some grades of dibasic calcium phosphate), but also considerably lower than that of other excipients (lactose and other grades of dibasic calcium phosphate). Having a relatively low BFI means that these materials may be useful for overcoming the high brittleness often associated with API and some excipients. This is especially important with direct compression formulations.

Compaction

The high TS observed with MCC tableting indices compacts was similarly observed in tablets prepared by the standardized compaction test at both 0.85 SF and the maximum (at the fitted curve's apex). As shown in Table 3, these values were markedly higher than those observed with the other excipients. Additionally, the corresponding CS at these TS values was moderate relative to the other excipients, confirming the excellent compaction behavior of MCC. Avicel PH105 performed somewhat differently than PH102 and PH302. Tablets at a SF of 0.85 could not be manufactured due to overcompression (manifested as capping). This behavior suggests Avicel PH105 to be more sensitive to tablet press speed than PH102 and PH302, whose compacts exceeded a SF of 0.85 during this high-speed tableting challenge. The maximum strength of PH105, at the fitted curve's apex, was attained at about a SF of 0.80. The fitted curve was extrapolated to determine the predicted TS and CS values at a SF of 0.85, but these values were not realistic (i.e., TS was negative) and thus were not reported in the table.

Lactose

Physical Properties

The SEM images of two grades that are representative of the range of particle morphologies of lactose are shown in Figure 3B. The first image shows Anhydrous Direct Tableting™ Lactose from Quest International, consisting of particles ranging from under 10 μm to well over 200 μm in length. Note that many large particles are roughly rectangular with aspect ratios from 1 to 2. Note also the preponderance of many fine particles under 20 μm diameter and their tendency to agglomerate. The size estimates from this image support the laser diffraction particle size data in Table 5. Thus, it is evident that this material is relatively cohesive and poor flowing. In contrast to this is the material represented by the second SEM image, Lactose-316 Fast Flo™, a grade of lactose monohydrate from Foremost Farms. The image shows that its particles are generally spherical aggregates of smaller particles ranging in diameter from about 30 μm to nearly 200 μm. Note the absence of particles under 20 μm and the lack of association between them. These size estimates also support the laser diffraction data in Table 5. Notice also that the images in Figure 3 and data in Table 5 show the size distribution of Fast Flo™ to be considerably narrower than Anhydrous Direct Tableting and Lactose-310™. These data suggest Fast Flo™-316 is fairly non-cohesive, has a good flow, and is easier to handle relative to the other grades.

Tableting Indices

The compacts of the three grades of lactose required relatively high pressures to achieve 0.85 SF and the tableting indices were about mid-range compared to the common diluents in this discussion. The tableting indices were substantially higher than mannitol; only slightly higher than two of the three MCC grades (Avicel PH102 and PH105) but significantly higher than Avicel PH302; and substantially less than the three DCP products as shown in Tables 3–7. Based on these comparisons, one can describe the ease of compression of these lactose grades as "moderate."

Dynamic indentation hardness values for lactose compacts were typically high (>200 MPa), indicating they showed relatively small deformation (small indentations) from pendulum impact testing, and, thus, they may be described as exhibiting "low ductility." As shown in Table 5, Lactose-316 Fast Flo showed particularly high hardness, > 600 MPa, while Direct Tableting Lactose was also quite hard, > 300 MPa. The softest, most ductile of the three was Lactose-310. Relative to other excipients, these lactose materials were at the high end of the indentation hardness range (i.e., at the low end of the ductility range), showing in general hardness values comparable to DCP, but greater than MCC and mannitol.

Most grades of lactose exhibit low TS (i.e., < 0.8 MPa) or moderate strength (0.8 Mpa $\leq$ TS < 2.0 MPa), as exemplified by Lactose-310 and Direct Tableting Lactose in Table 5. Lactose-316 Fast Flo seems to be an exception, showing high strength. At least two factors may contribute to its higher strength. First, Fast Flo may not be completely crystalline—i.e., it has some amorphous content—because it is processed by spray drying, and second, its particles appear to be porous. Both factors tend to promote plastic deformation of particles and bond formation during compression. Excluding Fast Flo, the TS of most lactose grades are about mid-range compared to excipients in general; they are generally comparable to DCP, stronger than mannitol, and considerably weaker than MCC. High TS as observed in Lactose-316 Fast Flo is a very important attribute because it contributes greatly to particle bonding, tablet strength, and tablet robustness.

Table 5 Properties of Lactose Samples

	Direct Tableting™ lactose		Lactose–310™		Lactose-316 Fast Flo™	
Particle morphology						
Particle aspect ratio	1–2		1.5–2		1–1.5	
Characteristics	Equant, plate		Plate, column		Equant	
Particle size						
Mean volumetric diameter (μm)	153		79		109	
10th percentile particle diameter (μm)	12		9		47	
50th percentile particle diameter (μm)	136		68		104	
90th percentile particle diameter (μm)	324		166		178	
$B_{MID}80$	2.3		2.3		1.3	
Density						
True density (g/cm^3)	1.50		1.54		1.54	
Tableting indices						
Compression stress (MPa)	92	Moderate	98	Moderate	106	High
SF	0.85		0.86		0.85	
Dynamic indentation hardness (MPa)	387	High	232	High	666	High
Tensile strength (MPa)	1.56	Moderate	0.89	Moderate	2.62	High
Compromised tensile strength (MPa)	1.02		0.56		1.38	
Worst case bonding index ($\times 10^2$)	0.4		0.4	Moderate	0.4	Moderate
Brittle fracture index	0.26	Moderate	0.30	High	0.45	High
Compaction properties						
Compression stress at 0.85 SF (MPa)	235	Moderate	157	Moderate	177	Moderate
Maximum compression stress (MPa)	700		700		564	
Tensile strength at 0.85 SF (MPa)	4.1	High	1.2	Low	3.6	High
Maximum tensile strength (MPa)	10.7	High	7.2	High	7.7	High

Abbreviation: SF, solid fraction.

Lactose typically showed similarly moderate bonding index values (0.5 ± 0.2) because of high dynamic indentation hardness values, which are in the bonding index equation's denominator. Lactose-316 Fast Flo was no exception, despite its high strength. Relative to other excipients, lactose bonding index values are about mid-range, slightly greater than mannitol, comparable to DCP, and far lower than MCC.

Lactose typically exhibits relatively high BFI values, approaching or greater than 0.3, as exemplified by the three grades in Table 5. By comparison, MCC, mannitol, and some grades of DCP (e.g., Emcompress™ and A-TAB) show much lower BFI values. Thus, lactose products typically lie at the upper end of the BFI scale for excipients and this trait is one of its main disadvantages, particularly when considered as a component in direct compression formulations. As a general rule in these applications, a high BFI component such as lactose will impart its brittle fracture tendency to the formulation, the degree depending on its percentage in the blend; one or more components with acceptable BFI (i.e., <0.3) should be present at a substantial level to counter its brittle fracture propensity.

Table 6 Properties of Calcium Phosphate Dibasic Samples

	Emcompress™		CD Anhydrous™		A-TAB™	
Particle morphology						
Particle aspect ratio	1		1		1	
Characteristics	Equant		Equant		Equant	
Particle size						
Mean volumetric diameter (μm)	173		13		184	
10th percentile particle diameter (μm)	17		2		50	
50th percentile particle diameter (μm)	183		11		191	
90th percentile particle diameter (μm)	284		26		301	
$B_{MID}80$	1.5		2.2		1.3	
Density						
True density (g/cm^3)	2.26		2.83		2.78	
Tableting indices						
Compression stress (MPa)	127	High	158	High	163	High
SF	0.81		0.67		0.64	
Dynamic indentation hardness (MPa)	213	High	453	High	421	High
Tensile strength (MPa)	1.12	Moderate	1.60	Moderate	1.32	Moderate
Compromised tensile strength (MPa)	0.92		0.87		1.04	
Worst case bonding index ($\times 10^2$)	0.5	Moderate	0.4	Moderate	0.3	Moderate
Brittle fracture index	0.11	Moderate	0.42	High	0.14	Moderate
Compaction properties						
Compression stress at 0.85 SF (MPa)	350	High				
Maximum compression stress (MPa)	700				700	
Tensile strength at 0.85 SF (MPa)	3.3	High				
Maximum tensile strength (MPa)	8.2	High			9.6	High

Abbreviation: SF, solid fraction.

Compaction

Direct Tableting™ lactose outperformed Lactose-310 and 316 Fast Flo™ during compaction testing in terms of TS at 0.85 SF and maximum strength, as shown in Table 5. However, 316 Fast Flo attained these properties at lower CS, particularly the stress at the maximum TS. In fact, only 316 Fast Flo showed an actual apex in its tabletability profile; the fitted curves for Lactose-310 and Direct Tableting™ still had positive slopes at 700 MPa CS, the maximum allowed in this test due to tooling limitations. The TS at that pressure was designated the "maximum." It is apparent that Fast Flo achieved its maximum performance at lower pressures than the other two grades; this could be a criterion for material selection when designing a formulation. Overall, these representative grades of lactose showed a wide array of responses for TS at 0.85 SF, from low (Lactose-310) to high (Direct Tableting™). All lactose grades showed high maximum strength. The compaction properties for lactose indicate that these excipients fall mid-range relative to the other excipients—they were better than mannitol but not as good as MCC.

Calcium Phosphate Dibasic (Dibasic Calcium Phosphate)

Physical Properties

The physical properties of the pharmaceutical grade calcium phosphate dibasic products marketed as Emcompress, CD Anhydrous™ and A-TAB are generally representative of numerous DCP grades. They are high-density powders with true density values greater than 2 g/cm^3, which is typical of inorganic powders and considerably higher than that of organic excipients such as lactose, mannitol, and MCC. The particle morphology of the grades shown in Figure 3C—tightly bound aggregates of smaller particles, often with roughly spherical shape and aspect ratios of two or less—is generally representative of many grades. The various grades often differ significantly by particle size, as demonstrated by the SEM images and by the laser diffraction data in Table 6. The CD Anhydrous™ is typical of many fine powder grades where the largest particles are under 40 μm diameter. On comparison, in the granular grades, e.g., Emcompress or A-TAB, the largest particles typically exceed 200 μm. Because a powder's cohesiveness and thus its handling characteristics are heavily dependent on its physical properties, the wide range of particle sizes available with DCP translates to a broad range of powder handling characteristics, from highly cohesive and poor flowing to freely flowing materials. Their high density relative to organic excipients and API may contribute to segregation, particularly in direct compression blends. This should be a consideration when selecting components for these types of formulations.

Tableting Indices

The testing SF of DCP anhydrous and A-TAB were well below 0.85, so that relevant comparisons to MCC, lactose, and mannitol were not feasible. Emcompress, however, was tested at 0.81 SF so that its properties could be meaningfully compared. Overall, the mechanical properties of Emcompress lay within the ranges of lactose, MCC, and mannitol. It required high CS to form compacts, greater than the other three excipients; its dynamic indentation hardness was moderate to high, being similar to MCC and mannitol but lower than that of lactose; its TS and bonding index were both moderate and comparable to those of lactose and mannitol but substantially lower than those of MCC, and its BFI (rated as moderate) was similar to that of mannitol and MCC but significantly lower than that of lactose.

The mechanical properties of CD Anhydrous and A-TAB "as tested" (i.e., at a low SF) fell within the property ranges of the organic excipients tested at or near SF of 0.85 with the exception of CS. Their properties were rated as follows: for CS, high (the highest for all excipients of interest); for dynamic indentation hardness, high; for TS, moderate; for bonding index, moderate; and for BFI, moderate (A-TAB) and high (CD). Comparison of the mechanical properties to the organic excipients varied by grade and property but their values always stayed within the overall range of the other tablet excipients. This is illustrated in Tables 3–7.

Table 7 Properties of Mannitol Samples

	D-(−) Mannitol		Mannogem™ 2080		Mannogem™ EZ	
Particle morphology						
Particle aspect ratio	2–4		1–2		1–2	
Characteristics	Column, lath		Equant		Equant	
Particle size						
Mean volumetric diameter (μm)	97		457		112	
10th percentile particle diameter (μm)	16		225		51	
50th percentile particle diameter (μm)	80		451		105	
90th percentile particle diameter (μm)	203		709		185	
$B_{MID}80$	2.3		1.1		1.3	
Density						
True density (g/cm^3)	1.44		1.45		1.45	
Tableting indices						
Compression stress (MPa)	40	Moderate	52	Moderate	72	Moderate
Solid fraction	0.86		0.86		0.86	
Dynamic indentation hardness (MPa)	145	Moderate	198	Moderate	348	High
Tensile strength (MPa)	0.46	Low	0.48	Low	1.17	Moderate
Compromised tensile strength (MPa)	0.37		0.40		0.85	
Worst case bonding index ($\times 10^2$)	0.3	Moderate	0.2	Poor	0.3	Moderate
Brittle fracture index	0.12	Moderate	0.10	Moderate	0.19	Moderate
Compaction properties						
Compression stress at 0.85 SF (MPa)	222	Moderate	214	Moderate	220	Moderate
Maximum compression stress (MPa)	365		333		582	
Tensile strength at 0.85 SF (MPa)	1.5	Moderate	1.6	Moderate	2.3	Moderate
Maximum tensile strength (MPa)	2.0	Low	2.2	Low	4.1	Moderate

Abbreviation: SF, solid fraction.

In summary, when tested, the grades of calcium phosphate dibasic discussed above exhibited mechanical properties that were very appropriate for tablet compaction and thus for formulation processing by direct compression, dry granulation, or wet granulation. With this in mind, it is easy to understand the popularity of DCP in pharmaceutical tablet formulations.

Compaction

The compaction data for DCP, at least Emcompress and A-TAB, was very comparable to that of MCC, lactose, and mannitol. Estimates of TS at 0.85 SF and the maximum were higher than that of mannitol, about the same as that of lactose and less than that of MCC. However, the estimated CS values were considerably greater than the three organic excipients. Meaningful data were not obtained for CD Anhydrous. See Table 6 and the following paragraph.

The standard compaction test did not provide complete sets of meaningful data on all the DCP grades, as the empty data fields in Table 6 denote, particularly for estimates at 0.85 SF. This was likely due to the materials' poor compressibility and the near linearity of their TS versus CS profiles. Estimates of maximum TS and the corresponding CS were obtained for Emcompress and A-TAB, but the CS estimates were very high and defaulted to 700 MPa. Property estimates at 0.85 SF are reported in the table for Emcompress, and they seem reasonable because the material compacted in the vicinity of that SF. Estimates for A-TAB at 0.85 SF are not reported because the material compacted at substantially lower SF and extrapolation to 0.85 produced unrealistically high values for both TS and CS. No data were reported for CD Anhydrous because tablets with suitable integrity for measurement were not manufactured owing to the fact that very high die wall friction and capping were encountered at all compression levels.

Mannitol

Physical Properties

Mannitol powder is available in a range of physical forms, including fine powder, granular, and spray-dried forms, and these greatly influence its physical and mechanical properties. The range of particle morphology is illustrated in the SEM images of Figure 3D, where granular and spray-dried forms are shown. The images show the granular material, Mannogem^{TM} 2080, to consist of aggregates that have a boulder-like appearance, with diameters ranging from about 200 μm to well over 500 μm and aspect ratios from 1 to 2. Conversely the spray-dried powder, Mannogem^{TM} EZ, consists of spherical particles or clusters of spheres "glued" together. Their aspect ratios range from 1 to 1.5 but they all have a very rounded appearance. The SEM image shows a lack of very fine particulates, with diameters ranging from 40 μm to about 200 μm. The photomicrographs confirm the laser diffraction particle size data of these materials in Table 7. Mannogem 2080 has a very coarse size distribution, but it is very narrow as indicated by its low $B_{MID}80$ value. Mannogem EZ had a considerably smaller size, but also quite narrow distribution. D-(−) mannitol, the powdered grade, had the smallest and broadest size distribution of these three grades; however, the high aspect ratio of some particles suggests that the laser diffraction technique may not be appropriate because it assumes a spherical particle geometry. Based on these morphological differences, one would expect the particles of these mannitol grades to pack very differently and thus their bulk densities and their handling behavior to be different as well.

Tableting Indices

The compact mechanical properties of the three grades of mannitol are also shown in Table 7. It should be remembered that compacts of each material were prepared at the standard SF, thus allowing for straightforward comparison between them and other excipients. The rank order of each excipient by the tableting indices mechanical properties is provided in Table 4.

The three mannitol products required significantly different pressures to form compacts at equivalent SF; e.g., the powdered material required the lowest pressure, and therefore had the greatest ease of compression, while the spray-dried form required the most pressure. Overall, their ability to be compressed to 0.85 SF ranged from "high" for D-(−) mannitol to "moderate" for Mannogem EZ. These ratings are relative to other excipients. Highly compressible powders require "low" pressure, e.g., less than 40 MPa, to form compacts with high SF (>0.8). In general, moderate-to-high compression ability is desirable.

The mannitol products in this evaluation exhibited large differences in ductility, indicated by their indentation hardness values, as shown in Table 7. Mannogem EZ recorded high hardness, and thus low ductility, relative to the powdered D-(−) mannitol and the granular form, Mannogem 2080, which were considered to be moderately ductile.

As shown in Table 7, compacts of all three mannitol products displayed relatively low TS at the test SF, particularly the powdered and 2080 grades. Only Mannogem EZ formed tablets with moderate strength. Low TS is perhaps one of the main disadvantages of these mannitol grades, because it contributes relatively little to tablet strength and robustness.

All three grades of mannitol showed similar, relatively low bonding index values, indicating low levels of bond survival after tablet decompression. Relatively low bonding is another key disadvantage of mannitol as a tableting material because this trait may result in relatively weak tablets as described above. Relatively poor bonding may have significant tablet manufacturing implications in situations involving unusual tablet shapes—where certain tablet areas may tend to be weak because of unequal force distribution during compaction—or when the maximum tablet hardness is only marginally greater than the required hardness. Thus, the properties of mannitol are best suited to situations where additional high strength and bonding are not required because they are provided by other ingredients such as API, binders, and diluents.

All three grades of mannitol showed similar, relatively low BFI values. This low tendency for brittle fracture represents a very significant advantage of mannitol, particularly with direct compression formulations where material properties, and not powder processing, must be used alone to overcome deficiencies of the API and other ingredients. It should be remembered that low BFI is but one consideration of many when selecting excipients for direct compression formulations.

Compaction

All three mannitol grades exhibited moderate performance during the standardized high speed compaction test performed in the authors' laboratory. Their values for the key comparative parameters—CS and TS at 0.85 SF and the maximum TS—were marginally different, with Mannogem EZ performing better than the other two grades as indicated by higher TS values. Their performance during the standard test was inferior relative to MCC and lactose as well as DCP (Emcompress and A-TAB), confirming that they do not manufacture strong tablets as indicated by the tableting indices.

Table 8 Summary of Excipient Attributes and Deficiencies

Diluent	Product name	Desirable attributes	Nondesirable attributes
Microcrystalline cellulose	Avicel™ PH102	Morphology (compaction) Compressibility Tensile strength Bonding Brittleness	
	Avicel PH105	Morphology (compaction) Compressibility Tensile strength Bonding Brittleness	Particle size (flow ability)
	Avicel PH302	Morphology (compaction) Compressibility Tensile strength Bonding Brittleness	High density (flow ability)
Lactose	Direct Tableting™ lactose	Compressibility	Dynamic indentation hardness
	Lactose-310™	Compressibility	Dynamic indentation hardness BFI
	Lactose-316 Fast Flo™	Morphology (flow ability and compaction) Particle size (flow ability) Compressibility Tensile strength	Dynamic indentation hardness BFI
Calcium phosphate dibasic	Emcompress™	BFI Particle size (flow ability)	Density Compressibility
	CD Anhydrous™	Tensile strength	Density

			Compressibility Dynamic indentation hardness BFI Particle size (flow ability)
	A-TAB™	BFI Particle size (flow ability)	Density Compressibility Dynamic indentation hardness
Mannitol	D-(−) Mannitol	Compressibility Dynamic indentation hardness Brittleness	Particle size (flow ability) Tensile strength
	Mannogem™ 2080	Morphology (flow ability) Particle size (flow ability) Compressibility Dynamic indentation hardness Brittleness	Tensile strength Bonding
	Mannogem™ EZ	Morphology (flow ability) Compressibility Brittleness	Particle size (flow ability) Dynamic indentation hardness

Abbreviation: BFI, brittle fracture index.

SUMMARY

It is apparent that the properties of calcium phosphate dibasic, lactose, mannitol, and MCC are often quite different from one another, and that even within a single excipient there may be a wide range of physical and mechanical properties amongst the various grades. For example, the low BFI and high TS values typically observed with MCC are markedly different from most grades of lactose, where a high BFI and moderate strength are typically observed. Within lactose, 316-Fast Flo exhibits high TS but other grades—as exemplified by Direct Tableting™ and 310—show only moderate strength.

The properties of these excipients often overlap with that of other excipient grades; the TS of Lactose-310 falls between the TS of Mannogem 2080 and Mannogem™ EZ, so that general statements such as "lactose has higher TS than mannitol" are false. This overlap of properties is illustrated in Table 4 where a rank ordering of each excipient grade for each property is presented.

It is reasonable to expect excipients to perform differently during handling or powder processing operations due to their property differences. An excipient with a very small particle size or a very broad size distribution may flow poorly compared to an excipient with a coarser grade or a narrow distribution. Excipients with fibrous or highly irregular particle morphology will likely impede powder flow ability but may promote plastic deformation, TS, and bonding in powder compacts.

Therefore, each material's desirable attributes—its "good" properties—may offer some special functionality to a formulation, which may be used to enhance performance during intermediate processing or final tableting. "Good" mechanical properties include moderate-to-high ductility (moderate-to-low dynamic indentation hardness), high TS, high bonding, and low brittleness. The "goodness" or "badness" of a physical property is dictated by the formulation's need because physical properties often enhance performance in one respect while impeding it in another. The small particle size of Avicel PH105 may greatly enhance a formulation's ability to manufacture strong tablets, while greatly hindering the flow ability of that powder. Table 8 summarizes each material's desirable and nondesirable attributes.

It is therefore logical to select excipients by their properties when designing or optimizing a formulation, and knowledge of excipient properties is an important prerequisite for this process. Selecting excipients with properties that complement the poor qualities of an API or formulation is often an appropriate first step. When designing a formulation, for example, the importance of selecting excipients with complementary properties increases with API loading. Thus, at high loading, a highly brittle API might require a low BFI excipient or a low TS API might require a high-strength ingredient. At low loading the need to counter the API "deficiencies" may be less critical, but appropriate excipients must still be selected to give the formulation suitable properties and performance.

Finally, knowledge of excipient mechanical and physical properties is essential to creating a robust formulation that manufactures tablets that meet specifications in a time- and material-efficient manner. Excipient selection must also take into consideration API stability and biopharmaceutical performance of the dosage form. Uneducated selection of excipients will likely lead to numerous formulating iterations that require much time and material, which are luxuries that product development scientists do not have in the competitive pharmaceutical environment.

ACKNOWLEDGMENTS

Mr. Dauda Ladipo, Ms. Beth Langdon, and Mr. Matthew Mullarney are acknowledged for generating much of the data provided in this manuscript. Ms. Danni Supplee is acknowledged for providing data and for compiling and organizing the data presented in this manuscript. Dr. Cynthia Oksanen, Mr. Ladipo, and Mr. Mullarney are acknowledged for their timely review of this manuscript.

REFERENCES

1. Shangraw RF, Demarest DA. A survey of current industrial practices in the formulation and manufacture of tablets and capsules. Pharma Technol 1993: 32–44.
2. Monograph for Calcium Phosphate, Dibasic Dihydrate from Handbook of pharmaceutical excipients. London: Pharmaceutical Press, 2004.
3. Kibbe AH. Handbook of Pharmaceutical Excipients. Washington, D.C.; London, U.K.: American Pharmaceutical Association and Pharmaceutical Press, 2000.
4. Hiestand EN. Mechanics and physical principles for powders and compacts. West Lafayette, Indiana, U.S.: SSCI, Inc., 2003.
5. Brittain HG. What is the "correct" method to use for particle-size determination? Pharma Technol 2001:96–98.
6. Hancock BC, Colvin JT, Mullarney MP, Zinchuk AV. The relative densities of pharmaceutical powders, blends, dry granulations and immediate-release tablets. Pharma Technol 2003:64–80.
7. Hiestand EN, Smith DP. Indices of tableting performance. Powder Technol 1984; 38:145–159.
8. Roberts RJ, Rowe RC. The compaction of pharmaceutical and other model materials—a pragmatic approach. Chem Eng Sci 1987; 42(4):903–911.
9. Bateman SD, Rubinstein MH, Rowe RC, Roberts RJ, Drew P, Ho AYK. A comparative investigation of compression simulators. Int J Pharma 1989; 49:209–212.
10. Tye CK, Sun C, Amidon GE. Evaluation of the effects of tableting speed on the relationship between compaction pressure, tablet tensile strength and tablet solid fraction. J Pharm Sci 2005; 94(3):465–472.
11. Joiris E, Di Martino P, Bermeron C, Guyot-Hermann AM, Guyot JC. Compression behavior of orthorhombic paracetamol. Pharma Res 1998; 15:1122–1130.

BIBLIOGRAPHY

MCC References

Doelker E. Comparative tabletting properties of sixteen microcrystalline celluloses. Drug Dev Ind Pharma 1987; 13:1847–1875.

Edge S, Fraser Steele D, Chen A, Tobyn MJ, Staniforth JN. The mechanical properties of compacts of microcrystalline cellulose and silicified microcrystalline cellulose. Int J Pharma 2000; 200:67–72.

Lamberson R, Raynor G. Tabletting properties of microcrystalline cellulose. Manufact Chem 1976; 47.

Reier G, Shangraw R. Microcrystalline cellulose in tableting. J Pharma Sci 1966; 55:510.

Williams RO, Sriwongjanya M, Barron MK. Compaction properties of microcrystalline cellulose using tableting indices. Drug Dev Ind Pharma 1997; 23:695–704.

Lactose References

Evdokimov IA, Serov AN. Effect of compaction parameters on mechanical strength of lactose tablets. Izvestiya Vysshikh Uchebnykh Zavedenii Pishchev Tekhnol 2000; (5-6):60–61.

Fell JT. The flow and compaction properties of lactose. Pharmaceutisch Weekblad 1976; 111(28):681–685.

Fell JT. Influence of fines on the flow and compaction properties of lactose. J Pharma Pharmacol 1973; 25(suppl):109P.

Fell JT, Newton JM. The tensile strength of lactose tablets. J Pharma Pharmacol 1968; 20:657–658.

Kazumi D, Kazutoshi K, Kazutaka K, Kotaro I, Hisakazu S, Akinobu O. Effect of particle shape on the compaction and flow properties of powders. Chem Pharma Bull 1989; 37(11):3070–3073.

Lerk CF. Consolidation and compaction of lactose. Drug Dev Ind Pharma 1993; 19(17–18): 2359–2398.

Masaki H, Akinobu O, Fukuji H. Effect of particle size of lactose on the compaction properties of powder and tensile strength of tablets. Yakuzaigaku 1986; 46(1):50–57.

Riepma KA, Lerk CF, De Boer AH, Bolhuis GK, Kussendrager KD. Consolidation and compaction of powder mixtures. I. Binary mixtures of same particle size fractions of different types of crystalline lactose. Int J Pharma 1990; 66(1-3):47–52.

Vromans H, Bolhuis GK, Lerk CF, Kussendrager KD, Bosch H. Studies on tableting properties of lactose. VI. Consolidation and compaction of spray-dried amorphous lactose. Acta Pharmaceutica Suecica 1986; 23(4):231–240.

Whiteman M, Yarwood RJ. Variations in lactose NF from two different sources and their influence on tablet properties. Drug Dev Ind Pharma 1990; 16(11):1815–1827.

Calcium Phosphate References

Carstensen JT, Ertell C. Physical and chemical properties of calcium phosphates for solid state pharmaceutical formulations. Drug Dev Ind Pharma 1990; 16:1121–1133.

Landin M, Rowe RC, York P. Particle size effects on the dehydration of dicalcium phosphate dihydrate powders. Int J Pharma 1994; 104:271–275.

Landin M, Rowe RC, York P. Structural changes during the dehydration of calcium phosphate dihydrate. Eur J Pharma Sci 1994; 2:245–252.

Schmidt PC, Herzog R. Calcium phosphates in pharmaceutical tableting. Pharma World Sci 1993; 15:105–115.

Mannitol References

Roberts RJ, Rowe RC. The compaction of pharmaceutical and other model materials-a pragmatic approach. Chem Eng Sci 1987; 42(4):903–911.

Tomohiro Y, Forbes RT, Peter Y, Yoshiaki K. The improved compaction properties of mannitol after a moisture-induced polymorphic transition. Int J Pharma 2003; 258(1-2):121–131.

Tableting Indices

Hancock BC, Carlson GT, Ladipo DD, Langdon BA, Mullarney MP. The powder flow and compact mechanical properties of two recently developed matrix-forming polymers. J Pharma Pharmacol 2001; 53:1193–1199.

Mullarney MP, Hancock BC, Carlson GT, Ladipo DD, Langdon BA. The powder flow and compact mechanical properties of scrose and three highintensity sweeteners used in chewable tablets. Int J Pharma 2003; 257:227–236

Hancock BC, Carlson GT, Ladipo DD, Langdon BA, Mullarney MP. Comparison of the mechanical properties of the crystalline and amorphous forms of a drug substance. Int J Pharma 2002; 241:73–85

Davar N, Shah R, Pope DG, Ausburger LL. A critical comparison of Hiestand's tabletting indices and traditional measures of compactibility to predict tabletting behavior. Pharma Res 1995; 12:S168.

Hiestand EN. Mechanics and Physical Principles for Powders and Compacts. West Lafayette, Indiana, U.S.: SSCI Inc., 2003.

Hiestandand E, Smith DP. Indices of tabletting performance. Powder Technol 1984; 38:145–159.

Kuppuswamy R, Anderson SR, Hoag SW, Augsburger LL. Practical limitations of tableting indices. Pharma Dev Technol 2001; 6:505–520.

Venkatesh GM, Coleman JAN, Wrzosek TJ, et al. Fractional factorial designs for optimizing experimental conditons for Hiestand's indices of tabletting performance. Powder Technol 1998; 97:151–159.

Williams RO, Sriwongjanya M, Barron MK. Compaction properties of microcrystalline cellulose using tableting indices. Drug Dev Ind Pharma 1997; 23:695–704.

Wurster DE, Majuru S, Oh E. Prediction of the Hiestand bonding indices of binary powder mixtures from single-component bonding indices. Pharma Dev Technol 1999; 4:65–70.

Hancock BC, Carlson GT, Ladipo DD, Langdon BA, Mullarney MP. The powder flow and compact mechanical properties of two recently developed matrix-forming polymers. J Pharma Pharmacol 2001; 53:1193–1199.

Mullarney MP, Hancock BC, Carlson GT, Ladipo DD, Langdon BA. The powder flow and compact mechanical properties of sucrose and three high-intensity sweeteners used in chewable tablets. Int J Pharma. 2003; 257:227–236.

Hancock BC, Carlson GT, Ladipo DD, Langdon BA, Mullarney MP. Comparison of the mechanical properties of the crystalline and amorphous forms of a drug substance. Int J Pharma. 2002;241:73–85

11

Excipients for Oral Liquid Formulations

Meagan Anderson
Aptuit, Kansas City, Missouri, U.S.A.

F. Opawale
Formulation Development, NovaDel Pharma Inc., Flemington, New Jersey, U.S.A.

M. Rao and D. Delmarre
Morton Grove Pharmaceutical, Vernon Hills, Illinois, U.S.A.

Gopal Anyarambhatla
Research and Development, Akorn, Inc., Decatur, Illinois, U.S.A.

INTRODUCTION

Compared to conventional tablets and capsules, oral liquid dosage forms including solutions, syrups, suspensions, elixirs, or concentrates offer unique advantages to many patients. For example, liquids may provide better patient compliance for those with swallowing difficulties and better dosage control versus a fixed tablet dose. However, there are also a number of "challenges" surrounding the formulation and development of these forms. This chapter presents a brief description of some of the typical areas of challenges and opportunities for liquid formulations, the market justification for overcoming these challenges, and some of the typical excipients used to develop solutions, syrups, and suspensions.

IS THE ORAL LIQUIDS MARKET REALLY A "NICHE"?

Liquid dosage forms have typically been targeted for use in geriatric and pediatric patients. In general, these types of patients may have difficulty in swallowing tablets or capsules and have been regarded as a small fraction of the overall population. Therefore, pharmaceutical companies, if they develop oral solutions at all, often develop oral liquid formulations out of necessity rather than responding to a patient need. However, there are potential advantages of oral liquid dosage forms, such as no dissolution time and rapid absorption from the stomach/intestines compared to tablets, which may be an important factor for pain-relieving drugs. Inherent in this benefit is the risk of reaching peak plasma levels too fast, which could be harmful. Finally, as the excipient technology advances, a controlled release profile in liquid dosage forms will likely become readily available.

Even though difficulty in swallowing tablets is regarded as something only the very young or old struggle with, a recent study has shown that difficulty in swallowing tablets also affects the general population. In this study, 48% of noncancer patients and 63% of cancer patients reported swallowing difficulties. Therefore, there may be a large extent of underreported interest in liquid dosage forms (1).

It is estimated that approximately 80% of Food and Drug Administration (FDA)-approved drugs are not labeled for use in infants and children. However, in 1997, the FDA provided an economic incentive by extending patent exclusivity for six months or offering patent protection for conducting pediatric studies. It seems that this initiative has made the justification for investing in a pediatric-friendly dosage form such as liquids an easy one because 73 products have been granted exclusivity and 49 received new labeling between 1997 and 2003 (1).

One of the most anticipated areas of growth for oral liquid dosage forms is in the geriatric population. This is mainly due to the increasing size of this population. According to a report by the U.S. Census Bureau and the National Institute of Aging, the world's population of persons aged 65 and older grows by 800,000 individuals every month. By 2050, the number of people aged 60 and older is predicted to reach almost two billion (2). As a result, the demand for alternative dosage forms such as liquids in this population is likely to spike rapidly. This may be especially true for geriatric patients recovering from a stroke, or who may have lost some physical strength and control of muscle activity.

IMPORTANCE OF EXCIPIENT SELECTION IN THE PROCESS OF ORAL LIQUID FORMULATION DEVELOPMENT

When developing an oral liquid dosage formulation, consideration is first given to the characteristics of the active drug. The major challenges in developing oral liquid dosage forms are (i) the stability of a drug in solution, (ii) the solubility of a drug at the required level, and (iii) an acceptable taste. It is the effective use of excipients, which allows formulators overcome these challenges. Additionally, an excipient's compatibility with a drug in the solid state cannot infer the same compatibility in solution. However, if the mechanism of degradation of the drug is understood, the process of selecting which excipients to use in a solution will be much easier. Finally, some knowledge of the drug's physical and chemical characteristics such as the solubility, pH stability, and pK_a value(s) of reactive functional groups is essential in order to choose the proper excipients effectively.

Ideally, the pH at which the drug is most stable would also be close enough to the solubility for delivering the desired dose in approximately 5 mL. Requiring patients to take more than 10 mL at a time may not be advisable because of lower patient compliance (variability). In this scenario, a simple oral solution or syrup formulation may be developed. However, if the pH at which the drug is most stable is not one at which there is enough solubility, a suspension formulation may be required.

A quick means to identify whether or not a drug may be more suitable for solution or suspension is to overlap the pH-stability profile with the pH-solubility profile. This overlap creates a window, which may suggest which dosage form might be most desirable and subsequently the type of excipients needed. The overlapped figures below demonstrate for aspirin (which is a weak acid) that the pH of greatest stability is also the pH at which there is low solubility (Fig. 1).

The decision to develop a solution versus syrup versus suspension can also be influenced by other factors. The desired release profile of the drug may lead to the

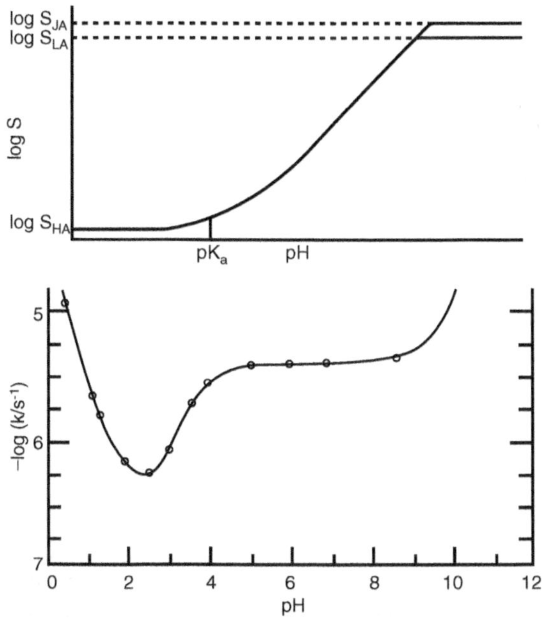

Figure 1 pH stability and solubility curves of aspirin. *Source*: From Ref. 3.

development of a suspension over a solution. In this case, excipients may be used to control or delay release of the Active Pharmaceutical Ingredient (API) from the suspension. Excipients used in an oral concentrate may also be used to protect drugs that are unstable in an acceptable pH range or that are easily hydrolyzed or oxidized by inhibiting the interaction of water with the drug. Additionally, excipients typically used in a syrup or suspension may be able to more effectively mask an extremely bitter tasting drug than a solution formulation. However, there are a variety of compounding and filling challenges for creating suspensions, which often make solutions a more attractive formulation.

Once a dosage form is chosen, this will affect the choice of acceptable excipients for screening. For a list of the excipients, which have been generally regarded as safe (GRAS) see the following Web site: http://www.cfsan.fda.gov/%7Edms/eafus.html. In the following section, excipients, which are commonly used to develop oral liquid dosage forms, are reviewed and summarized by their functionality. The final section addresses the challenges involved in the process of formulation and product development, and various regulatory issues regarding oral liquid dosage forms.

Typically, in preformulation studies, the drug's compatibility with an excipient is studied in a 1:1 mixture, with the excipient under investigation at elevated and/or refrigerated temperatures. When studying the compatibility of a drug and excipient for an oral solution, there are several important parameters that should be closely monitored. Changes in an excipient's viscosity, color, or pH during stability studies can drastically affect the active ingredients in solution. Polymerization or crystallization of certain excipients may also lead to changes in the API stability/concentration/homogeneity.

Once preformulation screening has identified which excipients are able to stabilize the API, a series of prototype formulations can be developed, which will be more reflective of the targeted quantities of excipients and drugs present in the final

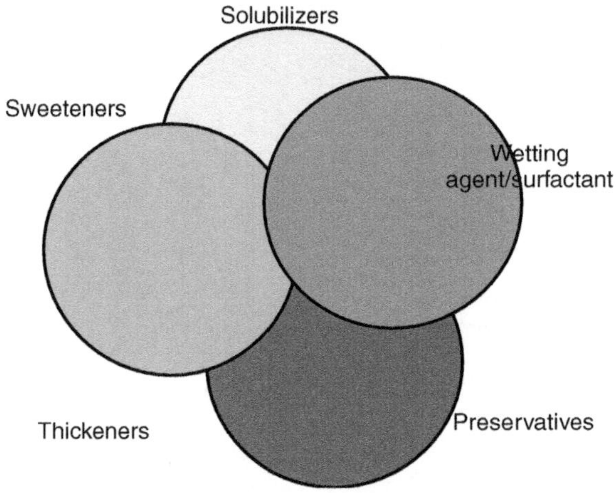

Figure 2 Functional overlap of typical excipients used in oral liquids.

formulation. Many companies use a rapid high-throughput screening approach to determine the optimal combination of excipients for their formulation. In these mixtures, the excipient–excipient (placebo) and excipient–drug incompatibilities should be closely studied for their stability. Additionally, many of the excipients described in the following sections can be used for more than one function. The following figure gives a depiction of the overlapping functionality of some typical excipients (Fig. 2). By understanding this principle, one can develop a formulation that encompasses all the required attributes (sweetening/solubilizing/preservative), using the least number of excipients.

EXCIPIENTS USED IN ORAL LIQUID FORMULATIONS

Solubilizers

In developing a formulation in which the API is dissolved in an aqueous vehicle, the first challenge is to solubilize the drug by breaking the strong hydrogen bonding of water, which causes less polar solutes to be "squeezed out." There are numerous approaches which may be taken to achieve the solubilization of a drug in aqueous solution. For example, the drug's intrinsic water solubility can be modified by the addition of a cosolvent, pH control, complexation, or the use of surfactants. This section focuses solely on the typical cosolvents used in oral liquids.

In aqueous-based solutions, solubilizers are used to modify the polarity of water to allow an increase in the solubility of a nonpolar drug. It is a balance between the forces of entropy that drives the solubilization of a solute and the enthalpic factors that oppose mixing of the solid (typically the API) and liquid phases (delivery vehicle) together. Some typical excipients used as solubilizers in oral liquid dosage forms are propylene glycol (PG), alcohols such as ethanol, sugars such as sorbitol, or polyethylene glycols such as PEG-400. There are many types of derivatives for each of these general groups. The ability of polymers such as polyethylene glycols or polyvinylpyrrolidones to affect solubility is dependent upon the polarity of its monomeric repeating units and end groups. However, it is not only the polarity that is altered upon addition of a solubilizer (cosolvent), but also the density, surface tension, viscosity, boiling point, and specific heat of solution, all of which may be affected in various ways.

Typically, when water-miscible cosolvents are used in combination, the effect is additive (assuming the cosolvents do not interact with each other), and the solubility of the drug is greater than in either of the individual cosolvents alone. However, the partial miscibility of two liquids (cosolvents) may occur if the free energy of a combination of two mutually saturated phases is lower than that of a single phase. Additionally, if the solute is very polar, the addition of a cosolvent may decrease the solubility of the drug. For example, the solubility of phenylalanine is decreased to a greater extent by ethanol than by PG, and least by glycerin. This is because ethanol > PG > glycerin alters (reduces) the polarity of water (3).

Complexation

Another approach to increasing the solubility of a drug in solution is to use a complexing agent such as a cyclodextrin. Currently in the United States, only hydroxypropyl-β-cyclodextrin has been used in an oral liquid formulation. However, many other cyclodextrins are widely used outside the United States in both oral and parenteral formulations. Although these agents are very effective, it is likely that the additional cost of this excipient and the potential approval and licensing challenges have limited the number of products that use cyclodextrins in an oral liquid formulation. Cyclodextrins have various ring sizes, which form complexes with drugs to increase their solubility and/or stability. In addition to various ring sizes, the cyclodextrins have been modified at the sugar hydroxyl groups with nonpolar and polar substituents such as dimethyl, hydroxyalkyl, or glucoside moieties. The degree of substitution can also affect the size and shape of the ring cavity and therefore the complexation of the drug.

The addition of a surfactant or cosolvent to the cyclodextrin–drug complex may have a variety of effects. The complex could be stabilized (an increase in binding coefficient) if the alcohol that surrounds the drug inside the cavity of the cyclodextrin leads to a better "fit." On the other hand, if the alcohol sterically hinders the drug from forming a complex, the solubilizing effect of the cyclodextrin will be decreased. In another scenario, when cyclodextrins are combined with a surfactant, there can be a decrease in the apparent solubility of the drug based on the surfactant being preferentially complexed with the cyclodextrin. Finally, there may be a slight change in the drug's pK_a value when it is complexed. Depending on the drug, this may increase or decrease the overall stability in an oral liquid formulation. This type of change is not likely to be predicted "a priori" but may be observed if an excipient range study is performed.

Sweeteners

A sweetening agent can play a number of important roles in an oral liquid formulation such as enhancing flavor, masking bitter taste, and/or increasing viscosity. The following section describes attributes of each type of sweetener and some potential challenges in their use. To organize the different types of sweeteners used in oral liquid formulations, a distinction was made between the natural and artificial sweeteners.

Natural Sweeteners

Sucrose is the most common sweetener used in oral pharmaceutical formulations. It is produced from sugar cane and sugar beet and is recognized as nontoxic and biodegradable. Its solubility in water at 20°C is 1 part sucrose in 0.5 part water. Sucrose

Table 1 Natural Sweeteners

Excipient	Sweetness compared to sucrose	Solubility (as a ratio of weight or volume in relative amount of water)	Incompatibility	Side effect
Dextrose mono-saccharide	0.75	1:1 in water	Amine, strong alkalis	Nausea vomiting
Fructose mono-saccharide	1.73	1:0.3 in water	Strong acid and alkalis	
Liquid sugar corn syrup	0.74	Miscible with water	Strong oxidizing agent	
Glycerin	0.6	Soluble in water alcohol	Strong oxidizing agent	Laxative at high concentration
Lactose	0.3–0.4	1:1.75 in water		Laxative effect
Maltose dis-accharide	0.32	Miscible with water, glycerin, PG		Flatulence diarrhea
Mannitol	0.5–0.7	1:5.5 in water		Laxative cooling sensation
Sorbitol	0.55	1:0.5 in water	Acid strong oxidizing agent	Laxative cooling sensation
Xylitol	1	1:1.6 in water	Oxidizing agents	Laxative decreases dental plaque and decay cooling sensation

is also the reference (1) by which all other sweeteners are compared. Table 1 summarizes some physical properties of the most commonly used natural sweeteners.

Typically, the concentration of sweeteners in oral solutions or suspensions averages between 30% and 50% of the formulation. In fact, in some cough or cold syrups, the sweetener content is as high as 80%. However, because of a growing population of diabetic patients in the United States, it is advisable to keep the amount of added sweetener (sucrose) as low as possible. Additionally, when natural sweeteners are used, there is an increase risk of microbial contamination and growth in the liquid formulation.

There is also a risk that sweeteners in solution may crystallize with time and/or temperature changes. However, sorbitol is not readily fermented by oral microorganisms and has little effect on dental plaque pH. It is also often used in syrups to prevent crystallization around the cap of bottles.

Artificial Sweeteners

A variety of different artificial sweeteners have been approved for use in oral liquid dosage forms by the FDA. One general characteristic for artificial sweeteners is their very high sweetness compare to sucrose. This also results in a much lower concentration needed in the formulation, which can lower the cost and/or risk of incompatibility with the drug or other excipients. Additionally, a "sugar-free" formulation

Table 2 Artificial Sweeteners

	Sweetness	Solubility	Stability	Daily intake limit[a]
Acesulfame potassium	200	1:3.7 in water	May decompose after long-term storage at 40°C	15 mg/kg of body weight
Aspartame	200	Sparingly soluble in water (1% w/v at pH = 5.2)	More stable at pH = 5.2, known reaction with sugar alcohols	40 mg/kg of body weight
Saccharin (Saccharin sodium)	500, 300	1:290 in water, 1:1.2 in water	High heat and low pH result in significant decomposition	2.5 mg/kg of body weight
Sucralose (modified sucrose)	300–1000	Freely soluble in water and alcohol		15 mg/kg of body weight

[a]Value according to the World Health Organization (WHO).

would be more desirable for diabetic patients. Table 2 represents some physical properties of the most common artificial sweeteners.

Wetting Agents and Surfactants

Wetting agents are routinely used in pharmaceutical formulations, especially in liquid dosage forms. For example, a wetting agent in a solution may be used in medicinal lotions and sprays to remove dirt and debris from wounds so that the preparation will spread out when applied to the surface of the skin and mucous membranes. However, when used in oral liquid dosage forms, these agents are used to create a homogenous dispersion of solid particles in a liquid vehicle. This process can be challenging due to a layer of adsorbed air on the particle's surface. Hence, even particles with a high density may float on the surface of the liquid until the air phase is displaced completely. The use of a wetting agent allows removal of adsorbed air and easy penetration of the liquid vehicle into pores of the particle in a short period of time. For an aqueous vehicle, alcohol, glycerin, and PG are frequently used to facilitate the removal of adsorbed air from the surface of particles. Whereas for a non-aqueous liquid vehicle, mineral oil is commonly used as a wetting agent. Table 3 lists some typical surfactants used in oral liquids based on their ionic character. The selection of a cationic, anionic, or zwitterionic surfactant will depend on a number of factors such as the charge of the API, solution pH, and the type of electrolytes and/or cosolvents in solution.

Typically, hydrophobic API particles are not easily wetted even after the removal of adsorbed air. Hence, it is necessary to reduce the interfacial tension between the particles and the liquid vehicle by using a surface-active agent. Hydrophilic particles, however, do not require the use of such surface-active agents for their solubilization. Structurally, wetting agents comprise branched hydrophobic chains with central hydrophilic groups or short hydrophobic chains with hydrophilic end groups. For example, sodium lauryl sulfate is one of the most commonly used surface-active agents. Such surfactants, when dissolved in water, lower the contact

Table 3 Typical Surfactants Used in Oral Liquid Dosage Forms

	Approved oral dosage form(s)	General structure
Anionic		
Sodium lauryl sulfate	Oral; drops, granules	(see Table 5)
2-Naphthalene sulfonate sodium	Oral; suspension	(see Table 5)
Docusate sodium	Oral; suspension	(see Table 5)
Cationic		$R\text{-}NH_3^+$
Cetylpyridinium chloride	Oral; capsule, soft gelatin	(see Table 5)
Zwitterionics		
Lecithin	Oral; suspension	
Nonionic		
Poloxamer	Oral; suspension (124/338), and solution (188/407)	$HO(C_2H_4O)_a(C3H6O)_b$ $(C_2H_4O)_a$ H
Polysorbate	Oral; suspension (Tween 20,40,60,80), syrup (Tween 40); solution (Tween 80)	(see Figure 3)

angle of water and aid in spreadability of water on the particles surface to displace the air layer at the surface and replace it with the liquid phase. Wetting agents have a hydrophilic–lipophilic balance (HLB) value between 7 and 9, which falls between emulsifying agents, which can have an HLB value between 3 and 6 (for W/O emulsions) and 8 and 18 (for O/W emulsions). While excipients such as detergents have an HLB value between 13 and 16, solubilizing agents have an HLB value from 16 to 18.

The following properties must be considered in the assessment of wetting agents:

- The minimum surface tension that can be attained, regardless of the amount of agent required
- The depression of surface tension achieved with a specified concentration of agent
- The time required for an agent to achieve equilibrium. A good wetting agent permits the depression of surface tension in water by up to 2.5 mN/m in 15 seconds

Careful consideration must be given to the potential changes in activity and bioavailability of the API and/or excipients when a surfactant is used. Dramatic changes in the bactericidal activity of certain excipients take place when they are solubilized by surfactants, and the stability of excipients against oxidation and hydrolysis may be modified by solubilization. Additionally, many nonionic surfactants (at high concentrations) exhibit a characteristic temperature above which the solution becomes cloudy. This cloudiness is due to the formation of very large lamellar micelles, which results from the dehydration of the polyoxyethylene chains. For these types of surfactants, it is essential to consider the risk of exceeding the cloud point. The solubility of some ionic surfactants is dependent on pH. For example, if the ionized form of a compound is surface active [or has a lower critical micellar concentration (CMC) than the ionized form], a change of pH can induce micellization. Furthermore, the formation of micelles invariably alters the dissociation constant of the surfactant. The pK_a of a surfactant with a carboxylic acid moiety is increased by micelle formation; and an amine group is decreased by micelle formation.

The physicochemical characteristics of some typical wetting agents and/or solubilizing agents are listed in Table 4 (4).

Polysorbates (polyoxyethylene sorbitan fatty acid esters, refer to Fig. 3) are a mixture of molecules of varying sizes rather than a uniform mixture of a single chemical entity. All four official polysorbates (numbers 20, 40, 60, and 80), listed in U.S. Pharmacopeia/National Formulary (USP/NF), contain 20 moles of oxyethylene. Polysorbates

Table 4 Physicochemical Characteristics of Wetting/Solubilizing Agents

Agent	Physical state	Solubility[a] Water	Solubility[a] Alcohol	Packaging requirements[b]
Benzalkonium chloride, NF	Gel	VS	VS	TC
Benzethonium chloride	Solid	SOL	SOL	TLR
Cetylpyridinium chloride, USP	Solid	VS	VS	WC
Docusate sodium, USP	Solid	SPSOL	FS	WC
Nonoxynol 9 USP	Liquid	SOL	SOL	TC
Octoxynol	Liquid	MISC	MISC	TC
Poloxamer NF	Solid	FS	FS	TC
Poloxamer 124 NF	Liquid	FS	FS	TC
Poloxamers 188, 237, 338, 407 NF	Solid	FS	FS	TC
Polyoxyl 35 castor oil NF	Liquid	VS	SOL	TC
Polyoxyl 40 hydrogenated castor oil NF	Paste	VS	SOL	TC
Polyoxyl 10 oleyl ether, NF	Semisolid/ Liquid	SOL	SOL	TC
Polyoxyl 20 cetylstearyl ether, NF	Solid	SOL	SOL	TC
Polyoxyl 40 stearate, NF	Solid	SOL	SOL	TC
Polysorbate 20 NF	Liquid	SOL	SOL	TC
Polysorbate 40 NF	Liquid	SOL	SOL	TC
Polysorbate 60 NF	Liquid/gel	SOL	–	TC
Polysorbate 80 NF	Liquid	VS	SOL	TC
Sodium lauryl sulfate, NF	Solid	FS	–	TC
Sorbitan monolaurate NF	Liquid	INSOL	–	TC
Sorbitan monooleate NF	Liquid	INSOL	–	TC
Sorbitan monopalmitate NF	Solid	INSOL	#[c]	WC
Sorbitan monostearate NF	Solid	##[d]	–	WC
Tyloxapol USP	Liquid	MISC	–	TC

[a]Abbreviations of solubility, VS, very soluble, 1 part of solute in less than 1 part of solvent; FS, freely soluble, 1 part of solute in 1 to 10 parts of solvent; SOL, soluble, 1 part of solute in 10 to 30 parts of solvent; SPSOL, sparingly soluble, 1 part of solute in 30 to 100 parts of solvent; SLSOL, slightly soluble, 1 part of solute in 100 to 1000 parts of solvent; VSS, very slightly soluble, 1 part of solute in 1000 to 10,000 parts of solvent; INSOL, practically insoluble or insoluble, 1 part of solute in 10,000 or more parts of solvent; MISC, miscible.

[b]Abbreviations of packaging requirements, TC, tight containers; TLR, tight, light resistant containers; WC, well-closed containers.

[c]Soluble in warm absolute alcohol.

[d]Dispersible in warm water.

Source: From Ref. 1.

are stable to electrolytes and weak acids and bases; however, strong acids and bases lead to saponification. Polysorbates are hygroscopic and should be tested for water content prior to use if necessary. As with other polyoxyethylene surfactants, prolonged storage can lead to the formation of peroxides. Finally, polysorbates may discolor or precipitate with various substances, especially phenols, tannins, or tar-like substances.

Permeation enhancement by excipients has generated some interest, but there is still much research that needs to be done to elucidate the mechanism of these excipients. PEG-400 (and many other excipients such as polyethylene glycol, poloxamers, polysorbates, and vitamin E) is known to inhibit *p*-glycoprotein, which may increase the bioavailability of the API, which was a substrate for this efflux pump. On the other hand, it has been demonstrated that PEG-400 can accelerate small intestinal transit, and thereby reduce the bioavailability of some drugs (e.g., ranitidine) (5).

There are a variety of phenomena that can be observed when surfactants are used in oral liquid dosage forms. It is important to understand the partitioning profile of the API and excipients such as preservatives when using wetting agents or surfactants (Table 5). Above the CMC, there may be significant differences in stability of excipients. For example, the solvolysis of sodium alkyl sulfates by hydrochloric acid was found to be faster above the CMC (6), and increasing the alkyl chain length led to an increase in solvolysis.

Table 5 Structure of Typical Wetting Agents and Surfactants

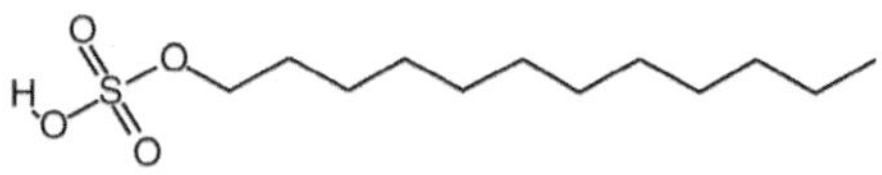	Sodium lauryl sulfate, Sulfuric acid monododecyl ester sodium salt (oral drops)
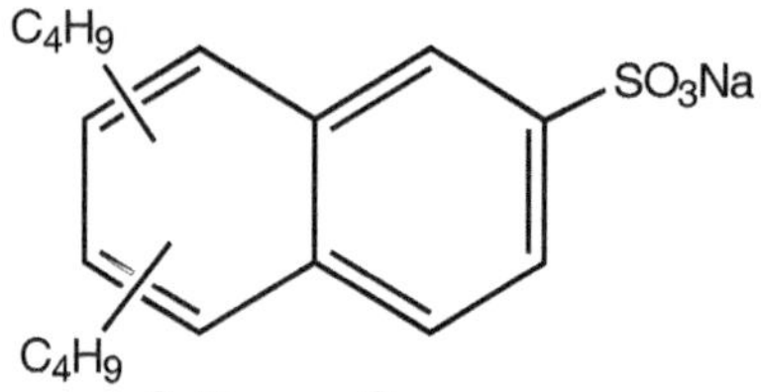	2-Naphthalene sulfonate sodium, Sodium dibutylnaphthalene sulfonate (oral suspension)
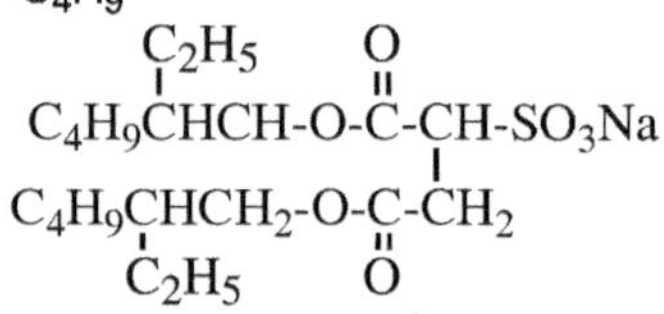	Docusate sodium, sodium di-(2-ethylhexyl) sulfosuccinate, aerosol OT (oral suspensions)
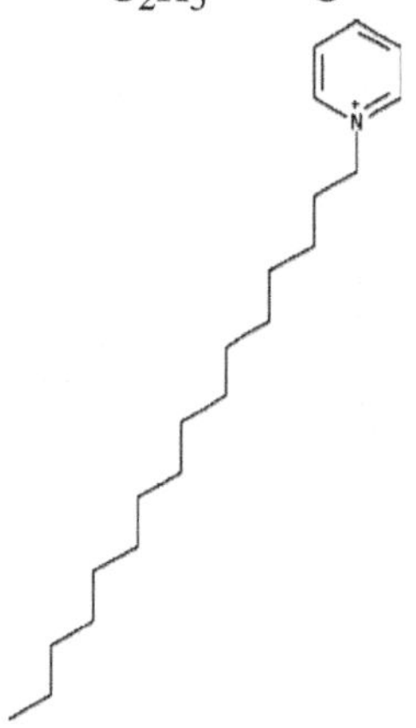	Cetylpyridinium chloride, nitrogenol, Sterogenal

In addition to the concentration of surfactant, the location of the API or excipient in the micelle structure can influence its stability. Surrounding the positive surface of the cationic micelle will be a relatively higher concentration of hydroxyl ions from the surrounding solution. If the drug or excipient is more susceptible to base-catalyzed hydrolysis and exposed to the concentrated hydroxyl area near the surface of the micelle, then the result would likely be more degradation (hydrolysis). However, if it is more stable under alkaline conditions, then there may be less degradation (hydrolysis). Therefore, if a correlation between the location of the drug or excipient in the micelle and its pH-dependent stability can be determined, a formulator may be able optimize the choice of surfactant to prevent degradation.

Lipid-Based Delivery Vehicles

A large number of new drugs being developed are characterized as either Class II or IV according to the biopharmaceutical classification system. To overcome these drugs' low bioavailability and/or solubility, there has been a growing interest in developing novel oral delivery strategies using lipid-based formulations (Table 6). While oral liquid emulsions have been used for many years, self-emulsifying drug delivery systems, which utilize a lipid/surfactant-based vehicle, are becoming a more widely used approach to solubilize water-insoluble drugs. One benefit for this type of formulations is that lipids that keep a hydrophobic drug in solution may facilitate the dissolution and absorption of the drug as the lipid vehicle is metabolized in the GI tract. The erratic bioavailability of some drugs may be overcome by formulation into a microemulsion, which includes oil.

Although the physicochemical stability of these lipid-based formulations may be well characterized, the impact on the physiological factors of drug adsorption and metabolism can vary dramatically. As such, lipid-based formulations may require more extensive BE/BA studies than simple oral liquids such as solutions or syrups.

Another approach in using a lipid-based formulation is to micronize the lipid with the dissolved drug to create a microemulsification. This allows an increased surface area available for the dissolution of the drug from the lipid phase. In these mixtures, a surfactant is usually added to improve the ability of oil to accommodate a hydrophobic drug in solution, and the resulting liquid is almost clear. Also a surfactant can function in the GI tract to help disperse the liquid vehicle on dilution. This allows the drug (dissolved in oil droplets/surfactant) to spread readily along the GI tract.

The manufacture of microemulsions can be an additional challenge due to the difficulty in establishing consistent batch performance using large-scale microfluidizers or high-speed homogenizers.

The following table lists some of the most common oils used in lipid-based drug delivery along with their listing in the various pharmacopeial groups (Table 7).

Table 6 Categories of Lipids

Triacylglycerols	Phospholipids (lecithin)	Lipoproteins
Cerides (waxes)	Glycolipids	Chylomicrons
Sterides	Sulfolipids	
Glycerides (fats, oils)		

Table 7 Pharmacopeial Vegetable Oil

Name	PhEur	USP?NF	J Ph
Almond oil	+	−	−
Castor oil	+	+	+
Coconut oil	+	−	+
Corn oil	+	+	+
Cottonseed oil	+	+	−
Olive oil	+	+	+
Peanut oil	+	−	+
Safflower oil	−	+	−
Sesame oil	+	+	+
Soybean oil	+	−	+
Sunflower oil	+	−	−
Triglycerides, medium chain	+	+	−

Abbreviations: PhEur, European Pharmacopeia; J Ph, Japanese Pharmacopeia.

Although fish oils are used for human consumption, vegetable oils are more typically used in oral liquid dosage formulations.

Phosphatidylcholine

Phosphatidylcholine (PC) is widely used in liposomes, and at physiological pH has a zwitterionic structure, is only slightly able to form salts with divalent cations such as Ca^{2+}, and has a lower transition temperature from liquid to crystalline than other phospholipids.

Phosphatidylethanolamine

Phosphatidylethanolamine (PE) is a good coemulsifier for PC. PE like PC also has a zwitterionic structure; however, it does not form a bilayer membrane in water (Table 8).

Phosphatidic Acid

In water, phosphatidic acid (PA) is 2^{-} negatively charged and may combine with divalent ions to form salts that may precipitate. PA also stabilizes micelles and liposomes because its negative charge prevents fusion with each other.

Table 8 Phospholipids of Soybean Lecithin; Distribution (By % Weight) of Fatty Acids

Fatty acid	PC	PE	PI	PA
Stearic acid	20.5	31.6	47.7	34.0
Palmitic acid	5.5	3.2	8.2	8.1
Oleic acid	10.5	8.7	4.9	11.9
Linoleic acid	58.8	53.2	36.2	44.7
Linolenic acid	4.6	3.2	2.8	1.3

Abbreviations: PC, phosphatidylcholine; PE, phosphatidylethanolamine; PI, phosphatidylinositol; PA, phosphatidic acid.
Source: From Ref. 1.

Phosphatidylinositol. In biological membranes, phosphatidylinositol (PI) is activated by enzymatic digestion to give rise to cell signaling and transport through cell membrane. PI forms salts with divalent ions and creates a negative charge in liposomes.

SUSPENDING AGENTS AND VISCOSITY-MODIFYING AGENTS

One of the most crucial factors involved in formulating a pharmaceutical suspension is the selection of an appropriate suspending agent. Suspending agents function in pharmaceutical systems to impart viscosity, and as such retard particle sedimentation. A number of factors must be considered in the selection of the appropriate agent. These include desired rheological property, suspending ability in the system, chemical compatibility with other excipients, pH stability, length of time to hydrate, batch-to-batch reproducibility, and cost.

Suspending agents can be classified into cellulose derivatives, clays, natural gums, and synthetic gums. In many cases, these excipients are used in combination. Table 9 contains a listing of the suspending agents most commonly used in oral liquid formulations. For each agent, the concentration of use and the respective property such as ionic charge, water dispersibility, pH range, rheological flow behavior, etc. are listed.

pH MODIFIERS AND BUFFERING AGENTS

The pH of an oral liquid formulation is a key point in many regards. Without the ability to control the formulation pH, there may be large changes during storage, based on water loss, or carbon dioxide and oxygen entering the bottle. Therefore, most formulations utilize a buffer to control potential changes in the solution pH. The amount of buffer capacity needed is generally between 0.01 and 0.1 M, and a concentration between 0.05 and 0.5 M is usually sufficient (7). The selection of a suitable buffer should be based on (i) whether the acid–base forms are listed for use in oral liquids, (ii) the stability of the drug and excipients in the buffer, and (iii) the compatibility between the buffer and container. A combination of buffers can also be used to gain a wider range of pH compared to the individual buffer alone. However, not all buffers are suitable for use in oral liquids. For example, a boric acid buffer may be used for optical and IV delivery but not in oral liquids because of its toxicity (7).

Even when there are no ionizable groups present for the excipient or API, the pH of a formulation may also play an important role in the formulation stability. For example, a specific functional group or a particular resonance structure that is stabilized in a specific pH range may facilitate a reaction between the excipient and the drug. Because of additional freedom of movement and rotation in solution, these types of interactions cannot be completely ignored.

However, it may also be possible that the buffer negatively influences the solubility of the drug and other excipients. Buffer salts can either increase or decrease the solubility of organic compounds in water. The effect depends on a combination of the polarity of the solute and of the salt. Nonpolar solutes are solubilized (salted in) by less polar organic salts and are desolubilized (salted out) by polar salts. Conversely, polar solutes are salted in by polar salts and salted out by organic salts. It was shown that for a semipolar solute such as ampicillin, strong electrolytes

Table 9 Suspending Agents, Concentration of Use and Their Properties

Class	Examples	Ionic charge	Concentration of use (%)	Water dispersibility	pH stability	Rheological behavior	Incompatibilities
Cellulose derivatives	Microcrystalline cellulose (and derivatives such as CMC)	Nonionic	1–5	Insoluble in water	5–7	Plastic/ thixotropic	Incompatible with strong oxidizing agents. Small amounts of electrolyte, cationic polymers and surfactants may flocculate MCC
Clays	Magnesium aluminum silicate (Veegum)	Anionic	0.5–2.5	Disperses and hydrates readily. Hot water increases rate of hydration	3–11	Plastic/ thixotropy	Partially flocculated by electrolytes and incompatible with acidic solution < pH 3.5
Natural gums	Sodium alginate	Anionic	1–5	Water-dispersible	4–10 or 4–11.5	Pseudoplastic	Incompatible with heavy-metal ions and sensitive to acids and ethanol in greater than 5%
	Xanthan gum	Anionic	0.3–3	Readily soluble in either hot or cold water	2–13	Plastic or pseudoplastic	Incompatible with cationic surfactants, polymers and preservatives CMC sodium and oxidizing agents
	Carrageenan	Anionic	1–2	Soluble in hot water	4–10	Newtonian/ pseudoplastic	Reactive with cationic materials
Synthetic gums	Carbomer 934	Anionic	0.5–1	Soluble in water	5–11	Plastic	Incompatible with cationic polymers, strong acids and high levels of electrolytes
	Povidone (polyvinylpyrrolidone)	Nonionic	<5	More soluble in hot water than in cold water	5.5–11.5	Newtonian/ pseudoplastic	Incompatible with inorganic salts

(at low concentrations) were able to increase the solubility (salt in), but at high concentrations decreased the solubility (salted out) (8).

The stabilizing effect of buffers that have multiple charged species in solution should also be investigated to determine the potential reaction between excipients and API. For example, buffers that use carbonates, citrate, tartrate, and various phosphate salts may precipitate with calcium ions by forming sparingly soluble salts. However, this precipitation is dependent upon the solution pH. Because phosphate can exist in mono-, di-, and tribasic forms, each calcium salt has its own solubility product, and precipitation will only occur when one of the solubility product is exceeded. Calcium ions may also interact or chelate with various amino acids, and other excipients, which may also lower the effective concentration of calcium that is capable of interacting with phosphate ions. Finally, the activity of phosphate ions may be lowered due to interactions with other solution components.

There are a number of factors that may also affect the solution pH such as temperature, ionic strength, dilution, and the amount and type of cosolvents present. For example, the pH of acetate buffers is known to increase with temperature, whereas the pH of boric acid buffers decreases with temperature. Finally, the drug in solution may itself act as a buffer. If the drug is a weak electrolyte, such as salicylic acid or ephedrine, the addition of base or acid, respectively, will create a system in which the drug can act as a buffer.

PRESERVATIVES

Microbiological contamination presents a significant health hazard in oral liquids. Therefore, the use of preservatives plays an important role in the stability of oral liquid formulations by circumventing the growth of microorganisms during the product's manufacture and shelf life. Although it may be most desirable to develop a "preservative-free" formulation to address the increasing concerns about the biological activity of these compounds, most formulations require some kind of preservative to ensure no microbial growth.

The majority of preservatives are bacteriostatic rather than bacteriocidal, and consist of both acid and nonacid types. Among the acidic types are phenol, chlorocresol, *O*-phenyl phenol, alkyl esters of parahydroxybenzoic acid, benzoic acid, boric acid, and sorbic acid, and their respective salts. Therefore, the pH of solution, and the pK_a of the preservative need to be carefully evaluated prior to selecting a preservative for a formulation. Neutral preservatives include chlorobutanol, benzyl alcohol, and beta-phenylethyl alcohol. Under alkaline conditions, it is generally regarded that most microbial growth is significantly retarded at these pH values, which reduces the need for a preservative.

Choosing an acceptable preservative when developing an oral liquid formulation is primarily limited by the number of approved excipients. As Table 10 demonstrates, there are many preservatives listed in the FDA inactive ingredient guide for dosage forms other than oral liquids; however not many have been commonly used in oral solutions or suspensions.

In addition, the solubility of many preservatives in a mostly aqueous system may not be high enough for effective antimicrobial activity. For example, the parabens often require heating in order to be solubilized. Additionally, it is essential to understand that bacteriostatic agents can partition between organic and aqueous phases in such a way that their activity is significantly reduced. Methyl paraben

Table 10 Typical Preservatives Used in Oral Liquid Dosage Forms

Name	Functional category	Incompatibilities
Alcohol	Antimicrobial preservative, disinfectant, solvent	In acidic conditions may react with oxidizing materials. May react with residual amounts of aldehyde in alkali conditions darkening solution. Incompatible with aluminum containers
Benzyl alcohol	Antimicrobial preservative, disinfectant, solvent	Oxidizing agents, strong acids, plastic containers, and methylcellulose
Bronopol	Antimicrobial preservative; antiseptic (not oral)	Sulfhydryl compounds, sodium thiosulfate, sodium metabisulfite, amine oxide or protein hydrolysate surfactants, aluminum
Chlorbutol	Antimicrobial preservative, plasticizer (not oral)	Plastic containers, rubber stoppers, carboxymethylcellulose, and Sorbate 80
Chlorocresol	Antimicrobial preservative, disinfectant, (not oral)	Calcium chloride, codeine phosphate, diamorphine hydrochloride, papaveretum, and quinine hydrochloride
Butylparaben, Methylparaben, Propylparaben	Antimicrobial preservative	Nonionic surfactants, bentonite, magnesium trisilicate, talc, tragacanth, sodium alginate, essential oils, sorbitol and atropine, yellow iron oxide, and ultramarine blue
Phenol	Antimicrobial preservative; disinfectant (not oral)	Camphor, menthol, thymol, acetaminophen, phenacetin, chloral hydrate, phenazone, ethyl aminobenzoate, methenamine, phenyl salicylate, resorcinol, terpin hydrate, sodium phosphate, or other eutectic formers. Phenol also softens cocoa butter in suppository mixtures
Phenylethanol	Antimicrobial preservative (not oral)	Oxidizing agents, proteins, polysorbates
Sodium benzoate	Antimicrobial preservative; tablet and capsule lubricant	Nonionic surfactants, quaternary compounds, gelatin, ferric salts, calcium salts and salts, of heavy metals, including silver, lead, and mercury

micellization by Tween 80 is a well-known example of this phenomenon (Fig. 3). Preservatives often contain reactive functional groups, which are responsible for their antimicrobial activity but lead to unwanted reactions. Therefore, in addition to the excipient's antimicrobial activity, other parameters should be evaluated during the stability studies such as its compatibility with the API, other excipients, and the container system.

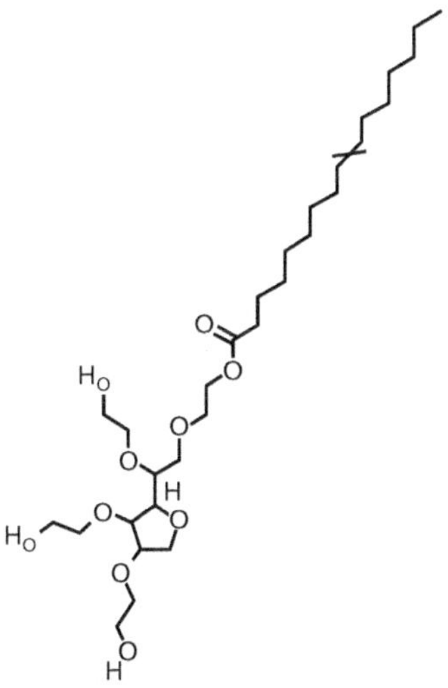

Figure 3 Tween 80

Parabens—(Methyl, Ethyl, Butyl, and Propyl)

Parabens are approved for use in oral solution and suspensions at a concentration of 0.015% to 0.2% w/v. Due to their low solubility, the sodium salts of parabens are often used in aqueous formulations. The parabens are most effective in the pH range of 2 to 6, and their antimicrobial activity decreases with increasing pH. Additionally, they are very unstable at pH 8 or above in solution. Methyl paraben has also demonstrated incompatibility with sorbitol and may show some discoloration in the presence of iron. The absorption of methylparaben by plastics has been reported with the amount absorbed being dependent upon the type of plastic and vehicle. However, no absorption has been reported for low density polyethylene (LDPE) or high density polyethylene (HDPE) containers. Certain coloring agents such as yellow iron oxide, ultramarine blue, and aluminum silicate can extensively absorb ethyl paraben in simple aqueous systems, thus reducing its preservative efficacy.

Parabens have some antimicrobial activity but are most effective against yeasts and molds. Although methyl paraben has the least antimicrobial activity, different combinations of methyl paraben with other long-chain parabens can lead to synergy. As the chain length of the paraben's alkyl moiety is increased, their antimicrobial activity increases.

Benzyl Alcohol

Although benzyl alcohol is listed as an approved antimicrobial under the FDA guide, there are many factors that should be considered before including it in an oral solution formulation. There are numerous reports of adverse reactions to benzyl alcohol following IV and intrathecal administration and it is not recommended for use in premature infants. Benzyl alcohol is incompatible with methylcellulose and is also known to be incompatible with a number of container types. For example, a 2% aqueous solution in a polyethylene container, stored at 20°C, may lose up to 15% of its benzyl alcohol content in 13 weeks. However, it is only slowly sorbed by closures composed of natural rubber or neoprene.

Benzoic Acid

Benzoic acid is an effective antimicrobial in the pH range of 2.5 to 4.4, but may be more recognized for its use as an antifungal agent. When added to a suspension, benzoic acid dissociates, with the benzoate anion adsorbing onto the suspended drug particles. This adsorption alters the charge at the surface of the particles, which may in turn affect the physical stability of the suspension. Benzoic acid is also known to dimerize in many nonpolar solvents. This property, coupled with pH-dependent dissociation in aqueous media, comprises a classic example of the effects of dissociation and molecular association on apparent partitioning behavior. For example, the total concentration of benzoate necessary to provide a bacteriostatic level of benzoic acid in the aqueous phase of an oil-in-water emulsion should be calculated with consideration for its partitioning coefficient into the oil phase.

Potassium Sorbate

Potassium sorbate has both antimicrobial and antifungal properties in formulations below pH 6. Generally, it is used at concentrations of 0.1% to 0.2% in oral formulations (solutions, syrup, and suspensions), and is used much more than sorbic acid because of its higher solubility and stability in water. However, some loss of antimicrobial activity occurs in the presence of nonionic surfactants and some plastics.

Glycerin

At levels greater than 20% v/v, glycerin can be an effective antimicrobial preservative. At this level, the activity of water is low enough to retard the growth of many microbes. However, glycerin may crystallize if stored at low temperatures, and the crystals do not melt until the temperature is raised to 20°C. Additionally, an iron contaminant in glycerin is responsible for the darkening in color of mixtures containing phenols, salicylates, and tannin.

Propylene Glycol

PG, similar to glycerin, is a multifunctional excipient that can be an effective preservative when used at concentrations of 15% to 30% in oral solutions. However, formulations containing 35% PG can cause hemolysis in humans. PG exhibits nonlinear pharmacokinetics and when elimination pathways are saturated, serum levels dramatically increase. Pyruvic and lactic acid are produced from the metabolic degradation of PG and can lead to acidosis. Neonates have a longer PG half-life (16.9 hours) compared with adults (5 hours) and seizures, and respiratory depression has occurred in children who have ingested oral liquid medications containing PG (9). Therefore, special consideration should be placed on the amount of PG in formulations that are intended for infants and children.

ANTIOXIDANTS, CHELATING AGENTS, AND SEQUESTRANTS

The oxidation of an API in an oral liquid formulation can be difficult to control due to the trace amounts of impurities, which may be present from the API or excipient vendor, and oxidation and photolysis have relatively low activation energies (2–12 Kcal/mol) compared to solvolysis, dehydration, and polymorphic transformations (10–56 Kcal/mol) (Table 11) (10).

Table 11 Antioxidants and Sequestrants Used in Oral Liquid Formulations

Antioxidant/ sequestrant	Function(s)	Characteristics and incompatibilities
BHA	Antioxidant (0.01% w/v)	Prevents oxidation of fats and oils and is frequently used in combination with BHT or citric acid. Trace quantities of metals and exposure to light cause discoloration and loss of activity
BHT	Antioxidant (0.01% w/v)	Prevents oxidation of fats and oils and enhance color stability. Heating with catalytic amounts of acids causes rapid decomposition
EDTA	Chelates alkaline earth and heavy metals 0.005–0.01% w/v; antimicrobial activity synergy	Often used in combination with other antimicrobial preservatives, and other antioxidants based on their synergy
Malic acid	Antioxidant, buffering agent, flavoring agent, chelating agent	The powder form has a strongly acid taste and is freely soluble in ethanol and water. However, aqueous solutions are mildly corrosive to carbon steels
Fumaric acid	Acidulant, flavoring agent, chelating agent synergist	Exhibits synergism when used in combination with other true antioxidants. Low aqueous solubility
Tartaric acid	Acidulant, antioxidant, sequestering agent	Soluble in water, glycerin and ethanol. Tartaric acid has a very tart taste
Ascorbic acid	Antioxidant (0.01–0.1% w/v)	Aqueous solutions (especially alkaline) are readily degraded on exposure to air, or light and are catalyzed by copper and iron. Maximum stability occurs at pH 5.4
Citric acid	Buffering agent, antioxidant synergist, chelating agent, flavor enhancer	On storage, sucrose may crystallize from syrups in the presence of citric acid. Dilute aqueous solutions may ferment on standing. Tart acid taste
Alpha tocopherol	Beta, delta, and gamma are considered to be more effective antioxidants	Incompatible with peroxides and metal ions, especially iron, and copper. May be absorbed into plastic. Protect from light, and oxygen during storage
Propyl gallate	Antioxidant (approved for use in oral concentrate), antimicrobial activity	Prevents autoxidation of oils and peroxide formation in ether. Synergistic effects with other antioxidants such as butylated hydroxyanisole

Abbreviations: BHA, butylated hydroxyanisole; BHT, butylated hydroxytoluene; EDTA, edetate calcium disodium.
Source: From Ref. 14.

Oxidation reactions can be prevented by a number of approaches including pH adjustment, the use of chelating agents or antioxidants, or the exclusion of light and oxygen. In most autoxidation reactions, the initiation does not directly involve oxygen but, for the reaction to proceed, oxygen is necessary. It is extremely difficult to prevent the permeation of oxygen into solution once the container is opened.

Because most drugs exist in a reduced form, there may be an increased instability when the solution is consistently introduced into an atmosphere of 20% oxygen. The pH of the solution may effect the oxidation of phenolic and sulfhydryl group containing drugs because it is principally the ionized form of these drugs that participate in the oxidation (11). For example, epinephrine is only slowly oxidized at $pH < 4$ but rapidly degrades under alkaline pH conditions.

Antioxidants can be compounds that can reduce a drug that has been oxidized, or compounds that are more readily oxidized than the agents they are to protect (oxygen scavengers). Many of the lipid-soluble antioxidants act as scavengers. Antioxidants can also act as chain terminators, reacting with free radicals in solution to stop the free-radical propagation cycle. Mixtures of chelating agents and antioxidants are often used because there appears to be a synergistic effect. This occurs because many of the agents act at differing steps in the oxidative process (11).

COLORING AGENTS

Generally, colors are used to match the flavor or color changes of an oral liquid formulation. Pharmaceutical coloring agents are divided into groups that are soluble in water (dyes) and those that are insoluble in water (pigments). The colors approved for clear liquid preparations are limited to the dyes.

For certification as a pharmaceutical coloring agent, the FDA operates a scheme whereby each batch of color that is produced is certified as analytically correct by the FDA prior to issuing a certification number and document, which will permit the sale of the batch in question. Colors requiring certification are described as FD&C (Food, drug, and cosmetic) and D&C (Drug and cosmetics). The requirements for each color are listed in the following section of the CFR Title 21 part 81: "General Specifications and General Restrictions for Provisional Color Additives for Use in Foods, Drugs, and Cosmetics" (see also http://www.cfsan.fda.gov/~dms/cfr81toc.html).

Additionally, a list of the coloring agents that have had their certification cancelled can be found at the following website http://www.cfsan.fda.gov/~lrd/cfr81–30.html.

Based on the reactivity of their functional groups, many coloring agents are poorly compatible or incompatible with citric acid, ascorbic acid, gelatin, glucose, lactose, sodium bicarbonate, and saccharose solutions. Additionally, several groups of dyes have been associated with serious adverse effects. For example, erythrosine (FD&C red #3) was delisted in 1990, based on studies in rats suggesting that it was carcinogenic. However, its use was continued until supplies ran out. Another example is the azo dye tartrazine (FD&C #5), which is known to be potentially dangerous in aspirin-intolerant individuals. Currently, it is estimated that the incidences of cross-reaction to tartrazine may be less than 2.4%. However, patients with aspirin sensitivity may also develop reactions from other dyes such as amaranth, erythrosine, FD&C blue #2, FD&C #1, ponceau, new coccine, sunset yellow, methyl blue, quinone yellow, and FD&C Red #40.

Additionally, the stability of dyes in solution can be dependent upon the specific excipients used in the formulation. For example, FD&C Blue #2 was found to fade more rapidly in the presence of several sugars (sorbitol, mannitol, dextrose, sucrose, and lactose) and that the nonionic surfactant Pluronic F-68 promoted the fading of FD&C Blue #2. The combination of coloring agents can lead to complications during long-term stability if there are physical–chemical interactions between the color and the container, excipient, and/or API. For example, one color may fade with time, leading to an overall color change in the formulation. The impact of this change should be considered before using, especially if the color is intended to mask the color change of the solution with time. To disguise the expected color change of the solution with time, an "aged" sample should be used to select the appropriate color and compared against a "new" sample and placebo to ensure that there is no visible color change.

FLAVORS

There are four general distinctions of flavor: sweet, acid/sour, salty, and bitter (Table 12). Flavor selection is often a patient-driven factor: geriatric and pediatric patients are often very different in their preferences for taste. Flavoring agents as well as coloring excipients are given a type IV drug master file (DMF), which is used by the excipient manufacturers to submit confidential formulation, safety, and manufacturing information about the excipient that may be needed by the FDA in reviewing NDA and ANDA submissions. However, unlike colors, there are no standardized formulas for approved flavors and the individual components of flavors are usually not made available to pharmaceutical manufacturers. Flavor companies are able to use any of the hundreds of GRAS chemicals to prepare a particular flavoring. Based on this variable, the stability of an oral liquid formulation with a particular flavor is a key determinant to formulation development. The addition of a flavoring agent can complicate the analysis of the formulation, namely because flavors are themselves made up of many different compounds. For example, a natural cherry flavor was found to contain more than 70 components, and a natural banana has more than 150. Each component in these flavors may negatively affect the drug's stability and create analytical challenges to separate interfering flavor peaks from the drug's degradation peaks. Other problems relating to flavoring agents may be observed during storage such as adsorption to containers, or partitioning or sorption to suspended materials or into micelles/oil phase of emulsions.

The following table attempts to correlate the taste and odors of some general chemical functional groups, which may be helpful when working with new chemical entity (NCE) (Table 13).

Some spices such as clove and cinnamon can accomplish the desensitizing of taste buds by creating a mild pain reaction through the introduction of heat and numbness. Likewise various sweeteners may provide different sensations in the mouth. Saccharin may give a rapid bitter sensation followed by the sweet flavor

Table 12 Flavors That Typically Mask Each of the Four Types of Tastes

Taste	Flavor
Sweet	Vanilla, grapefruit, bubblegum, and berry
Acid/sour	Lemon-lime, orange, cherry, grapefruit, raspberry, grapefruit
Salty	Nut, butter, butterscotch, spice, maple
Bitter	Anise, coffee, chocolate, mint, grapefruit, cherry, peach, raspberry, orange, lemon-lime

Source: From Ref. 1.

Table 13 Correlation of Chemical Property with Taste

Taste	Chemical property
Sour	H^+
Salty	Simultaneous presence of anions and cations
Bitter	High-molecular-weight salts
Sweet	Polyhydroxyl or polyhydrogenated compounds, alpha amino acids
Sharp, biting	Unsaturation
Odor	
Fruity	Esters, lactones
Pleasant	Ketones
Camphoraceous	Tertiary carbon atom

sensation, whereas sucrose gives a fast sweet sensation that helps to build the full-bodied taste of other flavors. Flavor enhancement using the addition of small amounts of vanilla to the basic flavor is a technique long used in the flavor industry. Vanilla seems to stimulate other flavors to a quicker taste response and intensifies them without altering their basic taste or adding its own vanilla taste. Table 14 describes a few physicochemical characteristics of some widely used flavoring agents.

MANUFACTURING CHALLENGES TO CONSIDER WHEN CHOOSING EXCIPIENTS

Besides an excipient's compatibility and functionality in a given formulation, there are a number of other parameters to be considered. Table 15 lists a number of these parameters, which may be important criteria to include when determining whether or not to include an excipient. For example, from a capital investment perspective, it would not be profitable for a company to use an excipient for a formulation, which costs too much or requires extensive time to prepare. A formulator may choose to avoid including hydroxypropylmethyl cellulose (HPMC) in a product if there are other acceptable options, because homogeneous solutions of HPMC require extra time and can be difficult to produce reproducibly in large tanks. However, evaluating an excipient's advantages and disadvantages can be difficult. In these cases, it may be beneficial to review other closely related excipients to find if any has fewer disadvantages.

From a manufacturing perspective, the excipients used may be very difficult to handle or process. This may lead to any number of problems during compounding, such as foaming, sedimentation, phase separation, particle flocculation, or formation of bubbles. To determine the cause of these problems, careful consideration and planning must be exercised in engineering a plant to ensure reproducible batches that can be efficiently compounded and filled. It is therefore essential that the behavior of each excipient in the formulation be well understood.

Formulations that include alcohol have a number of additional challenges to control, such as evaporation from the tank during manufacturing, evaporation during filling, the effect of headspace on stability, and loss of alcohol during product storage and use.

Other manufacturing challenges related to excipients can involve the filtering process. An oral solution is typically filtered through a 10 μm pore size to remove foreign particles that may have entered the bottle or tank. The pressure drop across

Table 14 Physicochemical Characteristics of Typical Flavoring Agents

Flavoring agent	Dosage in oral liquids	Taste	Physical incompatibilities	Chemical incompatibilities	Solubility
Vanillin	0.01–0.02% w/v in solutions, and syrups	Sweet	Light sensitive, and slowly oxidizes in moist air	Incompatible with acetone. Alkaline solutions turn brown-colored.	1:100 in water, 1:2 in ethanol
Maltol	Suspension 3%, and solution 0.15%	Caramel-like odor, sweet fruity taste in dilute solution	Concentrated solutions in metal containers may discolor on storage	Chelates with aluminum and iron	1:83 in water, 1:28 in propylene glycol
Fructose	Suspension, and solution (up to 25%)	Enhances fruit flavors, and is 20% sweeter than sucrose	Browning may occur when combined with strong acids or alkalis	Aqueous solutions are most stable in pH 3–4	More soluble in alcohol than is sucrose
Menthol	0.003–0.015% in suspensions, and syrups	Cooling effect	Sublimes easily above 25°C; and composition of natural oil may vary with source	Incompatible with thymol, phenol, camphor, and other excipients	Slightly soluble in glycerin; very soluble in alcohol

Table 15 Parameters to Review, Including an Excipient in Formulation

	Yes/Good	No/Bad
GRAS/CDER listed excipient?	X	
Significance in formulation?	X	
Cost and availability of excipient?		X
Ease of handling and processing?		
Robustness and stability?	X	
Cleaning from tank?		X

Abbreviations: GRAS, generally regarded as safe; CDER, center for drug evaluation and research.

a filter can be drastically affected by the type and amounts of excipients used. If this pressure leads to a change in drug concentration or other characteristics such as viscosity or density, the batch may not have good uniformity when filled. Therefore, it can be very useful to determine the potential adsorption of API and/or excipients using various filtering speeds, and pore sizes. Likewise, the size of a screen mesh for a suspension should be determined in a similar manner. Usually, the particle size of the formulation dictates the size of mesh used to break up any agglomeration of particles and/or excipients. However, other parameters such as mixing times should be carefully determined to ensure batch-to-batch uniform drug and excipient concentration. If the mixing times and still times are not controlled, there may be settling, flocculation, or particle size growth of the API and/or excipients. The amount of time spent between stages of compounding, and the temperature changes will also affect the excipients and lead to settling, phase separation, or particle size changes. In summary, understanding the type of rheology of the excipients in the formulation during the filling process will lead to better product uniformity.

The filtering process for oral solutions does not require sterility. However, the FDA does require that the formulation satisfy the USP guidelines for microbial identification and testing procedures, which are set in sections 51 "Antimicrobial effectiveness testing," and 61 "Microbial limit tests." Many excipient manufacturers have begun considerations for transmissible spongiform encephalopathies (TSE) contaminants because of concern regarding the transmission of Creutzfeldt-Jakob Disease. As each manufacturer strives for lower and lower levels of contamination, there is also increasing public demand for reducing the use of preservatives, which creates a dilemma for the manufacturer.

The long-term stability of an oral liquid formulation can also be affected by a number of unexpected reasons. Contamination by solvents used during the tank cleaning or even in the manufacture of excipients or API can be a source of instability of an oral solution. Uncontrolled levels of Class I, II, or III solvents could lead to the rejection of a batch or an excipient vendor. Class III solvents have a permitted daily exposure of 50 mg or less per day. (See the International Conferences on Harmonization, Impurities–Guidelines for Residual Solvents. Q3C, Federal Register 1997; 62(247):67377 and also http://www.fda.gov/cvm/Guidance/guide100.PDF).

The quality of each excipient should be taken as another potential source of instability in an oral solution. For example, a company may rely on the certificate of analysis of an excipient for proof of its purity. However, in solution, the reactivity of any impurity may become especially significant. For example, undetermined trace amounts of peroxides, from excipients such as polyvinyl pyrrolidone, are known to lead to increased oxidative degradation. Likewise, trace amounts of metals such as iron, copper, or lead can be present in excipients which may also lead to instability (i.e., oxidation or complexation) if these levels are not tested and kept at an acceptable level. Not all excipient vendors provide the same level of quality control for every batch of material, and each pharmaceutical manufacturer should therefore evaluate a number of potential vendors and choose the highest-quality material based on their investigation.

Packaging of an oral solution should above all ensure against any leakage (into and out of) the container. As plastic bottles began to replace glass, there have been many examples of increased instability based on the permeability of various types of plastic to atmospheric oxygen, carbon dioxide, and moisture. Additionally, plastic bottles should be evaluated to ensure that the ink or adhesive of the label does not diffuse into the bottle, and likewise prevent the excipients (e.g., cosolvents) and/or API from diffusing out through the plastic bottle. Foil-sealed bottles are often a good means to ensure that no tampering has occurred and that the risk of environmental contamination is reduced. However, it is not uncommon to observe an increase in degradation based on the interaction of either the excipients or API with the foil seal under accelerated conditions. Heavy-metal contamination from the cap foil lining or manufacturing processes may also be a factor influencing the excipient and API compatibility. The USP 661 sets out thorough guidelines for evaluating the performance of plastic [polyethylene terephthalate (PET), LDPE, and HDPE] and glass containers for oral liquids.

However, additional compatibility studies may be necessary to determine any potential incompatibilities between the excipients and/or API and the delivery device closure system.

During the scale-up phase, it may be important to determine the effect of storage in stainless steel tanks to mimic manufacturing conditions. For example, if a drug in combination with excipients is reactive with steel, it may lead to a color change or instability during storage. Finally, the process of cleaning of tanks and swabbing for the residual drug should also ensure that excipients are not overlooked. Small amounts of water and/or an excipient such as sugar may be a source of microbial contamination for the following batch made.

POLYMORPHIC CHANGES IN ORAL LIQUID DOSAGE FORMS

There is an ever-present risk that a polymorphic form that has low solubility may form in solution and precipitate out or have a significant impact on bioavailability. This may be especially true for drugs in suspensions, based on the undissolved crystalline form in the formulation. One example is a theophylline suspension in which the micronized anhydrous crystals were used to create the formulation, and a needle-like crystal formed with time, which was the hydrate form. Another example is ritonavir, which was formulated as an amorphous dispersion and removed from the market in 1998 when an insoluble crystal (new polymorph) was formed during long-term storage. This thermodynamically driven process may be observed if the formulation is studied under refrigerated and/or cyclic temperatures. Excipients can play an important role in the rate of a polymorphic change. For example, a polymorphic transformation of succinylsulfathiazole suspensions was found to be caused by several surfactants, coloring agents, and glycerin. It was also observed that methylcellulose retarded the transformation (12,13).

It is essential to understand how and when the polymorphs of drug substance in oral liquid dosage forms and suspensions can be controlled. One approach to study this phenomenon is to seed the formulation with a small amount of a known polymorphic crystal (other than what is used for the product), which is a common practice to rapidly determine what effect this may have on long-term storage. From these types of studies, the appropriate excipients can be used to preserve the specific polymorphic form desired. However, even when the drug in its crystalline form is studied extensively, there are cases when a previously unknown polymorph may be formed in solution and lead to precipitation (14).

REGULATORY ISSUES OF PHARMACEUTICAL EXCIPIENTS

For an excipient that has not previously been used in an oral dosage form, the 2005 Guidance for Industry Nonclinical Studies for the Safety Evaluation of Pharmaceutical Excipients describes what types of studies are necessary to determine the excipient's safety. For most manufactures, it may be more efficient to use an amount that is already in a currently marketed product. It is important to note that one excipient may have very different limits depending on the type of oral dosage form. For example, benzyl alcohol has been used at up to 5% in oral suspensions, but only 1.5% in oral solutions. Unless the manufacturer can provide evidence that justifies a change, the FDA may not grant approval. However, because most companies are unwilling to spend the necessary time and resources to demonstrate the need for exceeding these limitations, they are not often exceeded.

REFERENCES

1. Sarah Adams. Swallowing a bitter pill. Pharmaceutical Formulation & Quality. Dec/Jan 2003.
2. http://www.census.gov/prod/2001pubs/p95–01–1.pdf.
3. Yalkaowsky SH. Solubility and Solubilization in Aqueous Media. Washington D.C.: American Chemical Society, 1999.
4. Allen LV. Int J Pharmaceut Compounding 2001; 5(4).
5. Basit AW, et al. Pharm Res 2002; 19(9):1368–1374.
6. Nogami H, Kamakubo Y. Chem Pharm Bull 1963; 11:943.
7. Martin A, Swarbrick J, Cammarata A. Physical Pharmacy. 3rd ed. Philadelphia: Lea & Febiger, 1983.
8. Hou JP, Poole JW. Kinetics and mechanisms of degradation of ampicillin in solution. J Pharm Sci 1969; 58:1510–1515.
9. Inactive Ingredients in Pharmaceutical Products: Update (Subject Review). Committee on Drugs Pediatrics, March 30, 2005.
10. Streng WH. The stability of oral liquids. Drug Dev Ind Pharm 1985; 11(9,10):1869–1888.
11. Connors KA, Amidon GL, Stella VJ. Chemical Stability of Pharmaceuticals: A Handbook for Pharmacists. 2nd ed. New York: John Wiley, 1986:Chap 5.
12. Ebian AR, Moustafa MA, Khalil SA, Motawi MM. J Pharm Sci 1975; 64:1481.
13. Moustafa AM, Ebian AR, Khalil SA, Motawi MM. J Pharm Sci 1975; 64:1485.
14. Yu, et al. Adv Drug Deliv Rev 2004; 56:397–414.

12

Use of Nonactive Pharmaceutical Excipients in Oral Drug Formulations: Biopharmaceutical Classification System Considerations

Jane P. F. Bai
ZyxBio, Cleveland, Ohio, U.S.A.

Jian-Hwa Guo
Wyeth Consumer Healthcare, Richmond, Virginia, U.S.A.

Mahesh V. Chaubal
Baxter Healthcare, Round Lake, Illinois, U.S.A.

INTRODUCTION

Pharmaceutical nonactive excipients have long been applied in a variety of pharmaceutical dosage forms to provide a wide range of functional characteristics that facilitate the optimal delivery of a drug to achieve the desired therapeutic effects. Pharmaceutical excipients are inert materials with no adverse effects on the safety and efficacy of therapeutic products. The Food and Drug Administration (FDA) website (1) provides a database listing all the FDA-approved nonactive pharmaceutical excipients. This provides formulation scientists a useful reference for efficient choices of the suitable excipients for the desired formulations of drug.

Formulating a drug in a specific dosage form is a very important step to ensure the adequacy of its bioavailability and therapeutic efficacy. Nonactive pharmaceutical excipients are chemicals with a wide range of molecular sizes, from small molecules to large polymers, and a large variety of individual unique physicochemical characteristics. Therefore, it is important to make sure that drugs and nonactive excipients are compatible and drugs formulated in the designed dosage forms are stable throughout the desired shelf life. Pharmaceutical excipients offer a wide range of properties to influence many characteristics of a pharmaceutical product, thereby achieving the optimal therapeutic efficacy. For instance, in the oral route, the site, duration, and profile of drug release are controlled by the excipients used. Another example is in transdermal products, wherein excipients provide optimal drug release from the patches, enhanced drug permeation into the skin, and adequate adhesion of the patch to the skin.

Dissolution and gastrointestinal permeation are two key factors that affect oral bioavailability of drugs. These two parameters are dictated by the intrinsic physicochemical properties of drug, i.e., its aqueous solubility and lipophilicity. Chemists have long recognized that good drug candidates should have high solubility and high lipophilicity. On understanding the desirable molecular attributes, chemists are making more lipophilic drugs to ensure high membrane permeation. Unfortunately, lipophilicity also translates into high hydrophobicity and poor aqueous solubility. Using pharmaceutical excipients to enhance the dissolution and membrane permeation of drug is a common strategy in optimizing the oral bioavailability of drugs. Certainly, the release of a drug from any dosage form also involves disintegration of the dosage form. Whereas disintegration of dosage form is controlled by pharmaceutical excipients employed, dissolution of drug is largely determined by the aqueous solubility of drug, and may be influenced by the excipients used. Excipients can be used to offer a wide range of impacts on drug delivery, for example, for enhancing or reducing the aqueous solubility of drug through solubilization of drug, reduction of the crystallinity of drug particles, and formation of less-soluble complex with drug, to name a few. Because oral administration is the major target route for pharmaceutical products, this chapter will focus on the use of pharmaceutical, nonactive excipients in oral dosage forms in relation to Biopharmaceutical Classification System (BCS).

BIOPHARMACEUTICAL CLASSIFICATION SYSTEM

To facilitate regulatory submissions for generic drugs as well as for postapproval changes, FDA has published several guidelines regarding the in vitro/in vivo correlations for immediate-release as well as extended-release dosage forms. Dissolution profiles of a drug are used as a "sensitive, reliable, and reproducible surrogate" for ensuring bioequivalence (Guidance for Industry, published by FDA). Therefore, regardless of the types and mechanisms of the impact of pharmaceutical nonactive excipients on drug release, the consistency in the dissolution profiles of drug from pharmaceutical products is the key factor in the successful development of pharmaceutical products. In other words, the pharmaceutical excipients used have to provide this quality of consistency. Furthermore, FDA has also published the BCS to offer general guidance regarding how products could qualify for biowaiver, enabling pharmaceutical companies to avoid some costly in vivo bioavailability and bioequivalence studies (2,3).

The BCS system and its relevance to biowaiver have been extensively discussed in other reviews (4,5). BCS classifies drugs into four classes, as shown in Table 1 (4–6). The candidates for requesting biowaiver are those that are formulated as immediate-release dosage form and have high aqueous solubility and high permeability (Class I). As for Class II drugs, they have low aqueous solubility and high membrane permeability. Class III drugs have high solubility and low membrane permeability, whereas Class IV drugs have low aqueous solubility and low membrane permeability. Examples of drugs in BCS classes I, II, III, and IV are listed in Table 2 (4,6,7). Due to an inherent bias of high-throughput drug discovery technologies toward more lipophilic drugs with lower aqueous solubility, there is a trend for the newer generation of drug candidates to be in class II. Based on the in vitro/in vivo correlations, it is anticipated

Table 1 Biopharmaceutical Classification System

	High solubility	Low solubility
High permeability	Class I	Class II
Low permeability	Class III	Class IV

that class I drugs would have high absorption into the circulation, and class IV drugs are least desirable in drug development because their oral bioavailability is expected to be low and highly variable. Excipients with unique characteristics can be strategically employed to optimize the delivery of drug; however, the challenge is in matching the physicochemical properties of drug with the functional characteristics of excipients to achieve the desired pharmacokinetic profiles of pharmaceutical product.

The delivery profiles of pharmaceutical product could be manipulated via formulations of immediate and sustained–release. Immediate-release formulations are designed to provide a rapid onset of drug action, whereas sustained–release dosage forms are designed to achieve a long-lasting and less-fluctuating plasma level of drug, thereby minimizing efficacy fluctuation, toxicity, and side effects. Whether a drug is a candidate for immediate- or sustained–release formulation depends on its physicochemical characteristics and pharmacokinetic characteristics. In addition to immediate-release and sustained–release, site-specific–release formulations have long been practiced to ensure the highest therapeutic efficacy; the most common practice to achieve site-specific delivery is enteric coatings. Sustained–release formulations release drugs at a constant rate during transit throughout the gastrointestinal tract. Immediate-release formulations release most of the dose shortly after the disintegration of dosage form. The choice of immediate-release versus sustained–release formulations depends on the desired pharmacokinetic profiles, which are determined by the physicochemical and pharmacological characteristics of drug. Class I drugs are anticipated to have less variation in oral absorption because they are readily dissolved in the small intestine, readily released from the dosage forms, and efficiently absorbed across the intestinal epithelium. Therefore, class I is in general suitable for sustained– and immediate-release formulations, with the exception of those that are subject to extensive first-pass metabolisms. Sustained–release of drug below the level of saturating the intestinal and hepatic first-pass enzymes will likely cause more variations and lower bioavailability of drug. For class II drugs, intestinal absorption

Table 2 Some Examples of Drugs in Biopharmaceutical Classification System Classes

Class I	Class II	Class III	Class IV
Verapamil, propranolol, theophylline, caffeine, metoprolol	Amprenavir, carbamazepine, nifedipine, nisoldipine, ketoconazole, nicardipine, mefenamic acid, griseofulvin, naproxen, ketoprofen	Alpha-methyldopa, ranitidine, atenolol, acyclovir, enalaprilate	Furosemide, hydrochlorothiazide

Source: From Refs. 4, 6 and 7.

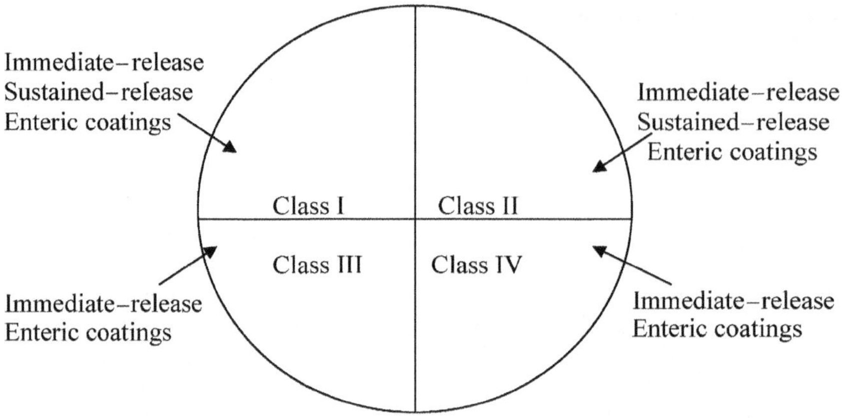

Figure 1 Summary of the suitability of various formulation strategies to drugs in individual Biopharmaceutical Classification System classes.

might be limited by dissolution. Because the absorbed dose fraction of a drug is determined by dose/solubility ratio, oral absorption of class II drugs would be strategically enhanced by excipients chosen to increase their aqueous solubility. For class III drugs, intestinal absorption tends to have higher variations due to their site-dependent intestinal absorption (4,6). Consequently, sustained–release for class III can be very challenging. Class IV drugs are not candidates for sustained–release formulation, due to their poor aqueous solubility and low membrane permeability. The suitability of applying formulation strategies to drugs in the four BCS classes is summarized in Figure 1.

From the formulation perspective, the physicochemical properties of drug and of pharmaceutical-grade excipients are both important for optimizing the oral bioavailability. In addition to serving as inert fillers for oral dosage forms and manipulating drug release, excipients can also serve as carriers to enhance drug dissolution and membrane permeation. In this chapter, we intend to discuss the applications of excipients in controlled release at specific-site and sustained–release formulations, as well as their ability to enhance the dissolution and membrane permeation of drug with reference to the BCS classification.

EXCIPIENTS USED IN SITE-SPECIFIC–RELEASE FORMULATIONS

Site-specific–release formulations usually use an external coating to allow the release of drug in a specific region of the gastrointestinal tract. For example, pH sensitive polymers are used to facilitate release in the small intestine where the pH is much higher than that in the stomach. Similarly, colon-targeting formulations use excipients that are susceptible to specific enzymes present in the colon. Colon delivery is preferably used for local therapy and much less applied than drug delivery through the small intestine, and hence is not covered in this chapter.

Enteric Coatings

Enteric coatings are a commonly used class of excipients for delaying the release of drug from dosage forms until the dosage forms reach the small intestine. Enteric coatings are useful for protection of drugs that are labile to acidic environment in the stomach, from presystemic destruction (8). Enteric coatings are suitable for any drugs in BCS Class I, II, III, and IV as long as their absorption site is the intestine. The pH is the determining factor, controlling the release of drug from enteric coating formulations. Gastric pH is typically less than 2 whereas the small intestine has higher pH, with the pH ranging from 2 to 5 in the duodenum, 6.5 to 7.5 in the jejunum and the ileum, and approximately 7.5 in the colon. Enteric coating materials usually dissolve at pH higher than 5 and readily dissolve at pH 7. Brief introductions of some common enteric coating materials are listed below. Detailed information regarding individual excipients can be found in the *Handbook of Pharmaceutical Excipients* (9).

Shellac (purified lac) dissolves at pH greater than 7 and can be used alone or in combination with other materials. It is soluble in ethanol, propylene glycol, and alkaline solutions.

Cellulose acetate phthalate (CAP) dissolves at pH higher than 6 and is soluble in ketones, ethers, esters, and alcohols. Permeation of water vapor and gastric fluids is a concern but can be overcome by adding other materials such as shellac. Plasticizers that are used with CAP include diethyl phthalate, triacetin, tributyl citrate, and acetylated monoglyceride.

Polyvinyl acetate phthalate (PVAP) dissolves at pH higher than 5 and is soluble in ethanol. Plasticizers that are used with PVAP include triethyl citrate, glyceryl triacetate, and acetyltriethyl citrate.

Hydroxypropyl methylcellulose phthalate (HPMCP) dissolves at pH greater than 5 and has two major grades, HP-55 and HP-55S. HPMCP shares the same problem as CAP and needs shellac to prevent permeation of water vapor and gastric fluids.

Hydroxypropyl methylcellulose acetate succinate (HPMCAS) dissolves at pH higher than 5, and there are three kinds of HPMCAS—AS-LG, AS-MG, and AS-HG. Triethyl citrate is a common plasticizer used with HPMCAS.

EXCIPIENTS USED IN SUSTAINED–RELEASE FORMULATIONS

Sustained–release in the small intestine can be achieved using a film coating or a sustained–release matrix or sustained–release drug–loaded granules. Sustained–release film coatings can be applied to tablets, granules, or beads. Formulations using sustained–release formulations, regardless of a sustained–release film coating or a sustained–release matrix, are suitable for drugs in BCS classes I and II, because both classes require good membrane permeability. One important prerequisite for a drug to be formulated as a sustained–release formulation is that it should not be susceptible to extensive first-pass metabolism. A sustained–release matrix may be a single tablet or multiple small sustained–release tablets housed inside an external coating. For class III and IV drugs, with low membrane permeability, sustained–release formulations may increase the degree of variability in their intestinal absorption, and hence are undesirable.

Sustained–Release Matrix

Polyacrylic acid is a commonly used matrix for sustained–release formulations. These polymers are available as pharmaceutical grade excipients such as Carbomer 910, 934, 934P, 940, 941, 971P, and 974P (8). Carbopol® (Noveon, Cleveland, Ohio, U.S.A.) polymers are polymers of acrylic acid cross-linked with polyalkenyl ethers or divinyl glycol. They are produced from primary polymer particles of about 0.2 to 6 mm average diameter. The flocculated agglomerates cannot be broken down into the ultimate particle when produced. Each primary particle can be viewed as a network structure of polymer chains interconnected by cross-links. Without the cross-links, the primary particle would be a collection of linear polymer chains intertwined but not chemically bonded. In general, based on the dry basis, Carbopol polymers contain 56% to 68% of carboxylic acid. Upon exposure to intestinal fluid, Carbopols swell to form hydrogel-like matrices through which drug molecules could be released in a controlling rate. There are many parameters that can be manipulated to control the rate of release in tablet formulations, including tablet compression processes, Carbopol content, and the ratio between Carbopol and other excipients. Class I drugs, paracetamol and amoxicillin, have been studied in hydrophilic matrix based on Carbopol polymers (10,11).

In addition to Carbopol, there are other materials used to form sustained–release matrix, including methylcellulose, hydroxypropylmethylcellulose, sodium carboxymethylcellulose (CMEC), carnauba wax, and glyceryl palmitostearate (9,12).

Coprecipitates

Another strategy to control drug release is through formation of coprecipitates with pharmaceutical excipients. Ibuprofen is a BCS class I drug with adequate solubility and membrane permeability for complete oral absorption. Coprecipitates of anionic, cationic, or zwitterionic Eudragits (methacrylate polymers and copolymers) with ibuprofen deterred the release rates of ibuprofen (13). Although no significant interactions were observed between ibuprofen and any Eudragit, and the crystalline state of ibuprofen was not altered, the release of ibuprofen was slowed down by the swelling and slow dissolution of Eudragits.

Sustained–Release Film Coatings

Various materials can be used as films for sustained–release of drugs, for various dosage forms including, tablets, drug-loaded beads, and granules. Sustained–release film coatings can form two types of membranes: permeable and semipermeable. The permeable membrane allows the intestinal fluid to enter the dosage forms to dissolve the drug as well as allow the drug to permeate out of the dosage form through the membrane. Permeable membranes are permeable to both intestinal fluid and drug molecules whereas semipermeable membranes are permeable only to the intestinal fluid but impermeable to drug molecules dissolved.

Permeable Membrane

As described in the Fick's law, the factors that determine the rate of drug release from sustained–release permeable membranes include membrane thickness, drug concentration gradient across the permeable membrane, drug solubility in the intestinal fluid, diffusion coefficient of drug molecules through the membrane, the surface area of the dosage form, and the drug particles (12,14).

Materials that form a permeable membrane include fats, bee wax, carnauba wax, cetyl alcohol, cetylsteryl alcohol, zein, acrylic esters, silicone elastomers, and ethylcellulose (14). Aqueous dispersions of water-insoluble polymers are commonly used for sustained–release film coatings. Examples of commercially available aqueous polymer dispersions include Surelease-containing ethylcellulose, Aquacoat-containing ethylcellulose, Eudragit RS 30 D–containing poly(ethylacrylate-methylmethacrylate) triethylammonioethyl methacrylate chloride 1:2:0.1, Eudragit RL 30 D–containing poly(ethylacrylate-methylmethacrylate) triethylammonioethyl methacrylate chloride 1:2:0.2, and Eudragit NE 30 D–containing poly(ethylacrylate-methylmethacrylate) 2:1 (14).

Methacrylic acid copolymer coatings (for example, Eudragit RL and Eudragit RS) are insoluble but permeable throughout the gastrointestinal tract. Plasticizers used with Eudragit to reduce the glass transition temperatures of Eudrugit films include polyethylene glycol (PEG), propylene glycol, diethylphthalate, dibutylphthalate, and triacetin.

Semipermeable Membrane

In osmotic pumps, semipermeable membrane allows water to enter the table matrix through the driving force of osmotic pressure while preventing the permeation of drug molecules across the membranes. Drug molecules are released from an osmotic tablet through the osmotic delivery orifice. Suitable candidates for osmotic tablets include drugs in BCS Class I and II. However, the presence of extensive first-pass metabolism might exclude the application of osmotic pump formulations even for drugs in these two categories. For a drug with good intestinal absorption and extensive first-pass metabolism, saturating the first-pass metabolic enzymes will increase its oral bioavailability whereas a constant delivery rate below the saturating level will most likely result in inadequate and highly variable bioavailability. Excipients used for the semipermeable membranes include polyvinyl alcohol, cellulose acetate, and ethylcellulose (8,9,12,14).

EXCIPIENTS USED TO ENHANCE DISSOLUTION OF BIOPHARMACEUTICAL CLASSIFICATION SYSTEM CLASS II AND IV DRUGS

In addition to controlling the site, the rate, and the duration of drug release, excipients can also be used to improve the dissolution of BCS class II and IV drugs in the gastrointestinal tract. Class II and class IV drugs have poor solubility and consequently incomplete dissolution and less than total release of the dose administered during the transit through the gastrointestinal tract.

The extent of drug release from oral solid formulations is determined by the dissolution rates of drug, which is a function of aqueous solubility and particle sizes as shown in the following equation.

$$\text{Dissolution rate} = \frac{D * S}{h}(C_S - C) \qquad (1)$$

where D is the diffusion coefficient of a drug molecule in the dissolution medium; h, the stagnant layer surrounding the drug particle; S, the total surface area of drug particles; C_s, the saturated solubility of drug in the dissolution medium; and C, the concentration of drug in the dissolution medium. Obviously, there are two parameters for implementing strategies to enhance the dissolution of drug—one being the total surface area of drug particles and the other, solubility in the intestinal fluid. Pharmaceutical excipients can be used to manipulate drug solubility through various mechanisms, from changing the pH in the microscopic environment surrounding drug particles, to affecting the physical state of drug molecules packed with each other, to facilitating wetting by intestinal fluid, and to increasing solubility through emulsifying effects. Implementation of these strategic approaches requires a thorough understanding of the intrinsic characteristics of individual drugs, the desired formulation properties, the physicochemical properties of suitable pharmaceutical excipients, and the stability of drug molecules in the presence of pharmaceutical excipients in oral dosage forms.

Solid Dispersions/Solutions

Solid solutions (solid dispersions) of drugs and pharmaceutical excipients have received wide attention in recent years for enhancing solubility of poorly soluble drugs. There are two commonly used techniques of preparing the drug–excipient solid solutions: solvent method and hot-melt method (15,16). Pharmaceutical excipients can form solid solutions with drug molecules and change the physical state of drug particles from a crystalline state to an amorphous state, thereby facilitating drug dissolution. The amorphous state is not as orderly and tightly packed as the crystalline state, therefore the higher the number of amorphous regions in the solid solution, the quicker the dissolution and release of drug. It is conceivable that intermolecular electrostatic, hydrogen-bonding, or van der Waals interactions between pharmaceutical excipients and drugs will interfere with the orderly packing of drug molecules, resulting in a reduced degree of crystallinity in drug particles. With the aid of X-ray diffraction, it was observed that the relative amount of amorphous versus crystalline areas in drug particles depended on the quantity ratio of drug to excipient in the solid solution, which is determined by individual, unique physicochemical properties of drugs and excipients (15). The excipient:drug ratio is also responsible for the stability of the drug in an amorphous state. As the ratio is decreased, a formulation may have a higher propensity to crystallize over time during storage. Erratic crystallization over storage, and thereby change in performance is a key criterion used in development of solid solution–based formulations.

Griseofulvin, a BCS class II drug (Fig. 2), is a well-known example whose poor aqueous solubility causes low and erratic oral bioavailability. As shown below, griseofulvin has a hydrophobic molecular structure, and is practically insoluble in water. Its oral absorption is highly variable, ranging from 25% to 100%, depending on the crystal size. Ultramicrosize griseofulvin preparations were shown to have 100% oral absorption (12).

Application of pharmaceutical excipients to increase dissolution of griseofulvin from oral solid dosage forms and to increase its oral bioavailability has been well explored. In the solid solutions of griseofulvin with pharmaceutical excipients, such as PEG and HPMCP, amorphous griseofulvin rather than crystalline griseofulvin was found, offering an explanation of higher dissolution rates of griseofulvin resulting from solid solutions than from pure form (17,18). Amorphous griseofulvin dissolved much faster than crystalline griseofulvin. Formation of a eutectic mixture is another mechanism contributing to higher dissolution rates of the drug in the presence of pharmaceutical excipients (succinic acid) (19). Table 3 summarizes the effect of several hydrophilic excipients on the dissolution rate of the drug (17–22).

The principle of forming pharmaceutical excipient–drug solid solutions to enhance the dissolution of drug is applicable to a wide range of drugs regardless of their chemical nature being weakly acidic, weakly basic, or neutral. As shown in Table 4, the dissolution of BCS class II and IV drugs with a wide range of physicochemical properties, including carbamazepine, furosemide, chlorothiazide, nifedipine,

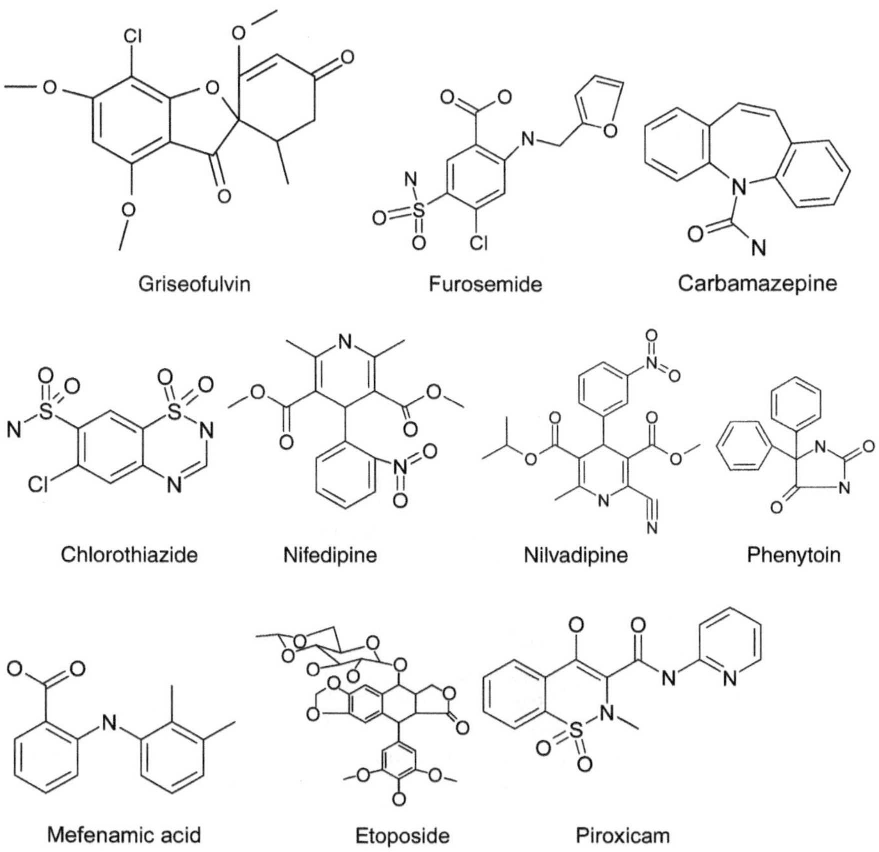

Figure 2 Chemical structures of some of the drugs discussed in this chapter.

nilvadipine, phenytoin, mefenamic acid, etoposide, and piroxicam (Fig. 2), were increased by pharmaceutical excipients (23–36).

With the aid of X-ray diffraction, it was observed that the formation of amorphous dispersion, instead of crystalline dispersion of drug, in the drug–excipient solid solution contributed to quicker dissolution and a higher amount of total drug release in the dissolution test. Furosemide (Fig. 2) was shown to form amorphous dispersions in the solid solution with either polyvinylpyrrolidone (PVP) or crospovidone, as evidenced by X-ray diffraction (26,27). The extent of amorphous dispersions

Table 3 The Effect of Hydrophilic Excipients on the Dissolution Rates of Griseofulvin (a Biopharmaceutical Classification System Class II Drug)

Excipient	Effect	References
Polyvinylpyrrolidone	Y	20
PEG	Y	17
PEG/sodium dodecyl sulphate	Y	21
PEG/talc	Y	22
HPMCP	Y	18
Succinic acid	Y	19

Abbreviations: PEG, polyethylene glycol; HPMCP, hydroxypropylmethylcellulose phthalate; Y, Yes.

Table 4 The Dissolution and Release of Biopharmaceutical Classification System Class II and IV Drugs Are Enhanced by Hydrophilic Excipients

Drug	BCS Class	Excipient	Effect	References
Carbamazepine	II	PEG 4000	Y	23
		PEG 6000	Y	23
		PVA/PVP		24
Furosemide	IV	PEG 6000	Y	25
		PVP		26
		Crospovidone		27
Chlorothiazide	IV	PEG 6000		25
Nifedipine	II	PEG 6000		28
		PVA and nicotinamide		29
		HPMC		29
		CMEC		30
		Chitosan		31
		HPMC and nicotinamde		29
		PVP and nicotinamide		29
Nilvadipine	II	HPMC		30
Phenytoin	II	PEG 6000		32
Mefenamic acid	II	PVP		33
Etoposide	IV	PVP		34
Piroxicam	II	PEG 4000		35
		PVP		36

Abbreviations: PEG, polyethylene glycol; PVP, polyvinylpyrrolidone; PVA, polyvinylalcohol; HPMC, hydroxypropylmethylcellulose; CMEC, carboxymethylethylcellulose, BCS, biopharmaceutical classification system; Y, Yes.

depended on the amount ratio of furosemide–excipient. With the coexistence of crystalline and amorphous dispersions, it was observed that amorphous areas dissolved much more quickly than the crystalline areas. Similar amorphous dispersions were observed in the solid solutions of piroxicam and PVP, of furosemide and crospovidone, and of nifedipine and CMEC (26,27,30).

Another method of using excipients to enhance drug dissolution and release is coprecipitating drug and excipient (37). For a new investigational drug, HO-221, coprecipitates with PVP, HPMCP, or copolyvidone, all had higher bioavailability in dogs than micronized HO-221. HO-221 is insoluble in water (belonging to BCS class II or IV) and has limited oral bioavailability (37). A similar study of another new investigation drug, MFB-1041, for treating fungal infection, showed that solid dispersion strategies enhanced its in vitro dissolution and in vivo oral bioavailability in dogs (38). MFB-1041 has very low solubility in water (1.2 μg/mL). The excipients tested with MFB-1041 included CMEC, HPMCP, and hydroxypropylmethylcellulose. X-ray diffraction also revealed that these excipients induced the formation of amorphous dispersion, thereby facilitating the dissolution of MFB-1041 and improving its oral availability. Another example of using coprecipitates to improve dissolution was the study on water soluble drug (GWX), a research compound belonging to BCS class II with low aqueous solubility and high membrane permeability (39). The excipient used was HPMCP. In this study, the authors mechanistically determined that amorphous state alone was not the determining factor for increased dissolution rates. GWX recrystallized after being dissolved from the amorphous powder alone whereas when used with coprecipitates, with an appropriate amount of HPMCP, there was much less recrystallization of GWX after dissolution. Nonetheless, the pure physical mixture of HPMCP and GWX, unlike the coprecipitates, had no impact on improving the dissolution of GWX. It was suggested by the authors that the excipient might actually have contributed to improved wetting and increased surface area. This study shed additional light on the benefits of using pharmaceutical excipients to improve the dissolution of BCS class II and class IV drugs via the mechanisms of forming coprecipitates or solid solutions. However, in vivo studies were not performed on GWX to confirm whether recrystallization of amorphous materials happens in vivo and whether excipients offer any significant in vivo benefits.

Nalidixic acid is another example of BCS class II drug, with oral bioavailability limited by poor solubility and slow dissolution (40). Compared to drug powder alone, the solid dispersion of nalidixic acid with β-cyclodextrin or PVP or sodium starch glycolate had much faster dissolution. X-ray diffraction studies revealed the formation of amorphous areas and less degree of crystallinity in the solid dispersion of nalidixic acid with excipients.

Solubilization/Emulsification Using Excipients

Various excipients have been used as solubilizers for BCS class II and class IV drugs. Cyclodextrins provide a prime example of the use of excipients as solubilizers, and have been discussed in detail in a separate chapter. Various surfactants have also been used to create emulsion-/microemulsion-type formulations. These have been discussed in a separate chapter as well.

PERMEABILITY-ENHANCING EXCIPIENTS

Drugs that demonstrate poor permeability are classified as BCS class III (if they are freely soluble) or class IV (if they are poorly soluble). Enhancing the bioavailability of such drugs is a major, yet-to-be-resolved challenge for formulation scientists. However, recent progress in the areas of permeation-enhancing excipients has provided significant hope for such drugs.

Poor permeation enhancement of drugs is caused by a number of properties that are intrinsic to the chemical structure of the drug, including presence of strongly charged functional groups, high molecular weight, significant hydrogen-bonding capacity, and high polar surface area. Permeation enhancement can be achieved by the use of excipients that complex with the drugs to reduce their permeation-inhibiting properties, or by serving as receptors for endocytosis of the complex.

For some drugs, permeation is also hampered by efflux of the drug from gastrointestinal epithelial cells, back into the intestinal lumen. Such efflux transporters involve, among others, *P*-glycoprotein (PGP), a family of multidrug resistance–associated proteins. Drugs that are substrates to these proteins are often prone to efflux and hence poor effective permeability. Excipients that serve as efflux pump inhibitors (e.g., PGP pump inhibitors) are useful to reduce the reverse transport of the drugs back into the lumen, thus increasing the drug's effective permeability.

Membrane Transporters

Membrane transporters are excipients that enhance permeability of a drug across the gastrointestinal (GI) epithelium by increasing the flux across the membrane. As illustrated in Figure 3, the enhancement could occur as a result of receptor-mediated

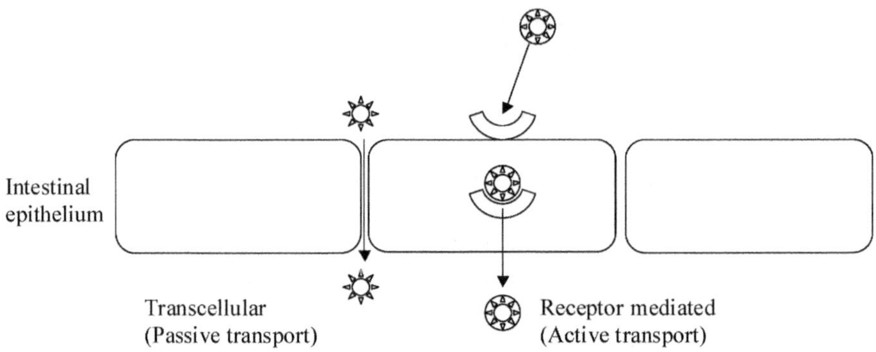

Figure 3 Schematic representation of transcellular and paracellular transport across the gastrointestinal epithelium.

endocytosis (active transporters, transcellular) or by lowering the resistance of the cell membrane (passive transporters, paracellular). The latter property is typically measured via a parameter termed as transepithelial electrical resistance (TEER). Excipients that reduce TEER cause permeation enhancement of the formulation. Various derivatives of chitosan (a natural polysaccharide) have been found to possess permeation enhancement properties, following a reduction in the TEER parameter (41). Naturally, excipients that affect the cell membranes may have toxicity associated with them due to their nonspecific mechanism. For example, palmitoyl carnitine was found to increase the intestinal bioavailability of cefoxitin from 5% to as much as 70% in animal models (42). However, this excipient also caused reversible mucosal damage due to its nonspecific action on multiple cell membranes (43). Increased membrane permeability is also implicated as a reason for certain intestinal inflammatory diseases, such as inflammatory bowel syndrome (44).

Receptor-Mediated Transporters

Receptor-mediated transporters are excipients that serve as substrates to exploit specific receptors present on cell membranes. Examples of various receptors that have been explored for permeation enhancement include bile acids (45), vitamin B_{12} (46), amino acids (47), and folic acid (48). Most of the work in receptor-mediated transporters is conducted via the use of prodrugs. For example, a prodrug of acyclovir conjugated to bile acids was seen to have higher permeability as compared to the original drug, because of receptor-mediated transport of the prodrug via bile acid transporters (49).

Efflux Pump Inhibitors

Certain excipients act as substrates to the efflux proteins, thereby inhibiting their functionality and enhancing the effective permeability of the drug across the gastrointestinal epithelium. A number of surfactants including vitamin E TPGS, Tween 80, and Cremophor EL have been shown to possess efflux pump inhibition properties (50). Sodium lauryl sulfate was found to increase the $CaCo_2$ permeability of seven low-permeable compounds with differing physiochemical properties (51).

Permeation-enhancing excipients have added a significant promise to the concept of oral delivery of macromolecules. Physical complexes of macromolecules and excipients have been shown to be effectively transported in vitro across $CaCo_2$ cell layers, as well as in vivo in humans. For example sodium *N*-[8-(2-hydroxybenzoyl) amino] hcaprylate enables oral heparin absorption via transcellular pathway (52). Other *N*-acylated amino acids have also been successfully used for a number of macromolecules including insulin, human growth hormone, and parathyroid hormone (53).

CONCLUSION

Pharmaceutical excipients are an essential part of pharmaceutical products and play a key role in optimizing the therapeutic delivery and oral efficacy of drug. The FDA-approved pharmaceutical excipients are considered safe and offer a wide range of functionalities for optimizing the therapeutic efficacy of drug and for facilitating high patient compliance. However, caution should be exercised to ensure the stability of the drug in the presence of excipients in the dosage forms. Pharmaceutical excipients make it possible to deliver drugs via various desirable mechanisms including immediate release, site-specific release, and sustained–release. The choice of drug delivery formulations and pharmaceutical excipients is determined by the physicochemical properties and pharmacokinetic characteristics of drug. BCS class I and II drugs are suitable candidates for different release formulations including sustained–release formulations under the prerequisite of no extensive first-pass metabolisms. BCS class III drugs are more problematic for sustained–release dosage forms whereas BCD class IV drugs are least desirable for sustained–release formulations. Pharmaceutical excipients are useful in enhancing the dissolution of drug by forming solid solutions or coprecipitates with drugs. The major contributing mechanisms include changing the physical state of drug from crystals to amorphous solids and improving the wetting of drug particles in intestinal fluid.

REFERENCES

1. www.accessdata.fda.gov/scripts/cder/iig/index.cfm.
2. Guidance for industry. Extended release oral dosage form: development, evaluation, and application of in vitro/in vivo correlations. http://www.fda.gov/cder/guidance/index.htm.
3. Guidance for industry. Waiver of in vivo bioavailability and bioequivalence studies for immediate release solid oral dosage forms based on a biopharmaceutics classification system. CDER/FDA, August 2000.
4. Löbenberg R, Amidon GL. Modern bioavailability, bioequivalence and biopharmaceutics classification system. New scientific approaches to international regulatory system. Eur J Pharm Biopharm 2000; 50:5–12.
5. Blume HH, Schug BS. The biopharmaceutics classification system (BCS): class III drugs-better candidates for BA/BE waiver? Eur J Pharm Sci 1999; 9:117–121.
6. Yu LX, Amidon GL, Polli JE, et al. Biopharmaceutics classification system: the scientific basis for biowaiver extensions. Pharm Res 2002; 19(7):921–925.
7. Rinaki E, Valsami G, Macheras P. Quantitative biopharmaceutics classification system: the central role of dose/solubility ratio. Pharm Res 2003; 20(12): 1917–1925.
8. George A, Agyilirah GA, Banker GS. Polymers for enteric coating applications. In: Tarcha PJ, ed. Polymers for Controlled Drug Delivery. Boca Raton, Florida: CRC Press, 1991.
9. Kibbe AH, ed. Handbook of Pharmaceutical Excipients. Washington, D.C.: American Pharmaceutical Association, 2000.

10. Parojčlć J, Durić Z, Jovanović M, Ibrić S. An investigation into the factors influencing drug release from hydrophilic matrix tablets based on novel carbomer polymers. Drug Deliv 2004; 11:59–65.
11. Tapia-Albarran M, Villafuerte-Roberts L. Assay of amoxicillin sustained release from matrix tablets containing different proportions of Carbopol 971P NF. Int J Pharm 2004; 273:121–127.
12. Remington's Pharmaceutical Sciences. 18th ed. Easton, Pennsylvania: Mack Publishing Company, 1990.
13. Kislaligglu MS, Khan MS, Blount C, Goettsch RW, Bolton S. Physical characterization and dissolution properties of ibuprofen: eudragit copreciptates. J Pharm Sci 1991; 80(8):799–804.
14. Porter CS, Bruno CH. Coating of pharmaceutical solid-dosage forms. In: Lieberman HA, Lachman L, Schwartz JB, eds. Pharmaceutical Dosage Forms: Tablets. Vol. 13. New York: Marcel Dekker, Inc., 1990:77–160.
15. Sekigicjo K, Obi N, Ueda Y. Studies on absorption of eutectic mixtures. II. Absorption of fused conglomerates of chloramphenicol and urea in rabbits. Chem Pharm Bull 1964; 12:134–144.
16. Tachibana T, Nakamura A. A method for preparing an aqueous colloidal dispersion of organic materials by using water-soluble polymers: dispersion of beta-carotene by polyvinylpyrrolidone. Kolloid-Z Polym 1965; 203:130–133.
17. Chiou WL, Riegelman S. Preparation and dissolution characteristics of several fast-release solid dispersions of griseofulvin. J Pharm Sci 1969; 58:1505–1510.
18. Hasegawa A, Kawamura R, Nakagawa H, Sugimoto K. Physical properties of solid dispersions of poorly water-soluble drugs with enteric coating agents. Chem Pharm Bull 1985; 33:3429–3435.
19. Goldberg AH, Gibaldi M, Kanig JL. Increasing dissolution rates and gastrointestinal absorption of drugs via solid solutions and eutectic mixtures III—experimental evaluation of griseofulvin-succinic acid solid solution. J Pharm Sci 1966; 55:487–492.
20. Mayersohn M, Gibaldi M. New method of solid-state dispersion for increasing dissolution rates. J Pharm Sci 1966; 55:1323–1324.
21. Sjökvist E, Nyström C, Aldén M, Caram-Lelham N. Physicochemical aspects of drug release. XIV. The effects of some ionic and non-ionic surfactants on properties of a sparingly soluble drug in solid dispersions. Int J Pharm 1992; 79:123–133.
22. WY Lo, SL Law. Dissolution behavior of griseofulvin solid dispersions using polyethylene glycol, talc, and their combination as dispersion carriers. Drug Dev Ind Pharm 1996; 22:231–236.
23. El-Zein H, Riad L, Elbary AA. Evaluation of carbamazepine dissolution—in vitro and in vivo evaluation. Int J Pharm 1998; 168:169–174.
24. Zingone G, Rubessa D. Release of carbamazepine from solid dispersions with polyvinylpyrrolidone/vinylacetate copolymer (PVP/VA). STP Pharm Sci 1994; 4:122–127.
25. Dubois JL, Ford JL. Similarities in the release rates of different drugs from polyethylene glycol 6000 solid dispersions. J Pharm Pharmacol 1985; 37:494–495.
26. Doherty C, York P. The in-vitro pH-dissolution dependence and in vivo bioavailability of furosemide-PVP solid dispersion. J Pharm Pharmacol 1989; 41:73–78.
27. Shin S, Oh I, Lee Y, Choi H, Choi J. Enhanced dissolution of furosemide by coprecipitating or cogranding with crospovidone. Int J Pharm 1998; 175:17–24.
28. Suzuki H, Sunada H. Comparision of nicotinamide, ethylurea and polyethylene glycol as carriers for nifedipine solid dispersion systems. Chem Pharm Bull 1997; 45:1688–1693.
29. Suzuki H, Sunada H. Influence of water-soluble polymers on the dissolution of nifedipine solid dispersions with combined carriers. Chem Pharm Bull 1998; 46:482–487.
30. Okimoto K, Miyake M, Ibuki E, Yasumura M, Ohnishi N, Nakai T. Dissolution mechanism and rate of solid dispersion particles of nivaldipine with hydroxypropylmethylcellulose. Int J Pharm 1997; 159:85–93.

31. Portero A, Remunanlopez C, Vilajato JL. Effects of chitosan and chitosan glutamate enhancing the dissolution properties of the poorly water soluble drug nifedipine. Int J Pharm 1998; 175:75–84.
32. Jachowicz R. Dissolution rates of partially water-soluble drugs from solid dispersion systems. II. Phenytooin. Int J Pharm 1987; 35:7–12.
33. Ramadan EM, Abd Ed-Gawad AH, Nouh AT. Bioavailaibility and erosive activity of solid dispersions of some non-steroidal anti-inflammatory drugs. Pharm Ind 1987; 49:508–513.
34. Shah JC, Chen JR, Chow D. Preformulation study of etoposide. 2. Increased solubility and dissolution rate by solid-solid dispersions. Int J Pharm 1995; 113:103–111.
35. Fernandez M, Margarit MV, Rodriguez IC, Cerezo A. Dissolution kinetics of piroxicam in solid dispersions with polyethylene glycol-4000. Int J Pharm 1993; 98:29–35.
36. Tantishaiyakul V, Kaewnopparat N, Ingkatawomwong S. Properties of solid dispersions of piroxicam in polyvinylpyrrolidone K-30. Int J Pharm 1996; 143: 59–66.
37. Kondo N, Iwao T, Hira KI, et al. Improved oral absorption of enteric coprecipitates of a poorly soluble drug. J Pharm Sci 1994; 83(4):586–570.
38. Kai T, Akiyama Y, Nomura S, Sato M. Oral absorption improvement of poorly soluble drug using solid dispersion technique. Chem Pharm Bull 1996; 44(3):568–571.
39. Sertsou G, Butler J, Hempenstall J, Rades T. Solvent change co-precipitation with hydroxypropylmethylcellulose phthalate to improve dissolution characteristics of a poorly water-soluble drug. J Pharm Pharmacol 2002; 54:1041–1047.
40. Mallick S, Sahu A, Pal K. Dissolution behavior of nalidixic acid solid dispersions using water-soluble dispersion carriers. Acta Poloniac Pharma-Drug Res 2004; 61(1):21–30.
41. van der Merwe SM, Verhoef JC, Verheijden JH, Kotze AF, Junginger HE. Trimethylated chitosan as polymeric absorption enhancer for improved peroral delivery of peptide drugs. Eur J Pharm Biopharm 2004; 58(2):225–235.
42. Sutton SC, LeCluyse EL, Engle K, Pipkin JD, Fix JA. Enhanced bioavailability of cefoxitin using palmitoylcarnitine, II. Use of directly compressed tablet formulations in the rat and dog. Pharm Res 1993; 10:1516–1520.
43. Lecluyse EL, Sutton SC, Fix JA. In vitro effects of long-chain acylcarnitines on the permeability, transepithelial electrical resistance and morphology of rat colonic mucosa. J Pharmacol Exp Ther 1993; 265:955–962.
44. Hollander D. Intestinal permeability, leaky gut, and intestinal disorders. Curr Gastroenterol Rep 1999; 1(5):410–416.
45. Bahar RJ, Stolz A. Bile acid transport. Gastroenterol Clin North Am 1999; 28(1):27–58.
46. Russell-Jones GJ, Alpers DH. Vitamin B12 transporters. Pharm Biotechnol 1999; 12:493–520.
47. Fagerholm U, Lindahl A, Lennernas H. Regional intestinal permeability in rats of compounds with different physicochemical properties and transport mechanisms. J Pharm Pharmacol 1997; 49(7):687–690.
48. Anderson KE, Eliot LA, Stevenson BR, Rogers JA. Formulation and evaluation of a folic acid receptor-targeted oral vancomycin liposomal dosage form. Pharm Res 2001; 18(3):316–322.
49. Tolle-Sander S, Lentz KA, Maeda DY, Coop A, Polli JW. Increased acyclovir oral bioavailability via a bile acid conjugate. Mol Pharm 2004; 1(1):40–48.
50. Rege BD, Kao JP, Polli JE. Effects of nonionic surfactants on membrane transporters in Caco-2 cell monolayers. Eur J Pharm Sci 2002; 16(4–5):237–246.
51. Rege BD, Yu LX, Hussain AS, Polli JE. Effect of common excipients on Caco-2 transport of low-permeability drugs. J Pharm Sci 2001; 90(11):1776–1786.
52. Malkov D, Wang HZ, Dinh S, Gomez-Orellana I. Pathway of oral absorption of heparin with sodium *N*-[8-(2-hydroxybenzoyl)amino] caprylate. Pharm Res 2002; 19(8):1180–1184.
53. Goldberg M, Gomez-Orellana I. Challenges for the oral delivery of macromolecules. Nat Rev Drug Dis 2003; 2(4):289–295.

13
Excipients for Semisolid Formulations

Prashant Srivastava
Focused Health Solutions, Deerfield, Illinois, U.S.A.

INTRODUCTION

Semisolids constitute a significant portion of pharmaceutical dosage forms. They serve as carriers for drugs that are topically delivered by way of skin, cornea, rectal tissue, nasal mucosa, vagina, buccal tissue, urethral membrane, and external ear lining (1). The primary advantage of topical delivery is the direct accessibility of the drug to affected tissues, with minimal, systemic side effects. In some cases, for systemic delivery, topical application enables delivery of therapeutic agents, avoiding first-pass gastrointestinal tract and hepatic metabolism and allowing maintenance of constant drug levels in the bloodstream. However, it is also generally recognized that the bioavailability of topically applied drugs is very low. The vehicle plays a key role in the appearance, feel, and successful application of a topical drug (2). Excipients, in large part, determine the physical properties of the vehicle as well as its ability to modify the stratum corneum or the mucosa to deliver the drug effectively. For example, it is possible to enhance the bioavailability by employment of an innocuous chemical means to reversibly improve the solubility of the drug in the barrier, e.g., stratum corneum, and facilitate diffusion of the drug through the barrier (3). Excipients, such as fatty acids, alcohols, amines, and amides, are absorbed into the barrier where they alter the overall solvent potential of the barrier. At the same time, the enhancers may disrupt the ordered lipid structure within the barrier, thereby lowering its viscosity. These physicochemical changes will facilitate drug partition from a topically applied formulation into the barrier as well as diffusion of drug molecules through the barrier. Thus, the understanding of excipients and proper selection is critical to successful formulation of semisolid dosage forms to meet the therapeutic needs.

Semisolids dosage forms, as a class, are plastic in behavior, i.e., they retain their shape until acted upon by an outside force, in which case they deform and the deformations are permanent. The common denominator to all semisolid systems which gives them their special rheological character is that they all have a permanent three-dimensional structure. This structure is sufficient when undisturbed to impart solid-like properties but which is easily broken down and realigned under some strain or applied force (4). The semisolid systems used pharmaceutically include semisolid emulsions with fluid internal phases (creams), ointments, pastes, and gels. For the scope of this chapter, we shall also include suppositories in this category because they share similar properties, typically use the same excipients as the semisolids mentioned above, and are typically semisolid during compounding as well as in their application. This chapter will discuss excipient properties, selection criterion, and their impact on dosage form performance for the vast spectrum of semisolids.

CREAMS

The U.S. Pharmacopeia (5) defines creams as "semisolid dosage forms containing one or more drug substances dissolved or dispersed in a suitable base. This term has traditionally been applied to semisolids that possess a relatively fluid consistency formulated as either water-in-oil (e.g., "cold cream") or oil-in-water (e.g., "fluocinolone acetonide cream") emulsions. However, more recently the term has been restricted to products consisting of oil-in-water emulsions, aqueous microcrystalline dispersions of long-chain fatty acids, or alcohols that are water washable and more cosmetically and aesthetically acceptable". Recently Buhse et al. (6) proposed to define a pharmaceutical cream as an emulsion semisolid dosage form that contains less than 20% water and volatiles and/or less than 50% of hydrocarbons, waxes, or polyethylene glycols (PEGs) as the vehicle. Creams are generally used for delivery of active ingredients such as antifungals, antibacterials, and anti-inflammatories across the stratum corneum or the vaginal mucosa for either systemic or local activity. Most generally, all pharmaceutical creams consist of a dispersed oil phase, a continuous water phase, a set of structure-forming excipients, which impart the cream its semisolid properties, a preservative, and a few other excipients (emollients, antioxidants, etc.). Appendix I shows some common creams, their active ingredients and therapeutic class, mode of application, and some commonly used excipients.

Structure-Forming Excipients

Data presented in Appendix I suggests that for most pharmaceutical creams, the oil-in-water emulsion that composes the cream is stabilized not by surfactant mechanical properties or by charge repulsion, but rather by the formation of a gel network consisting primarily of cetyl alcohol, stearyl alcohol, or some combination of the two often referred to as cetostearyl alcohol (7,8). The cetostearyl alcohol is arranged in crystalline bilayers, or lamellae, with surfactant molecules inserted into the layer such that the hydrophilic portion of the surfactant extends into the interlamellar space (9). A diagram of this structure is shown in Figure 1. The hydrophilic portion of the surfactant retards the drainage of water from the interlamellar space (10). This effect produces a gel that can retain large volumes of water within its structure. The oil phase of the emulsion is not necessary to form the gel and is not required for the delivery of a water-soluble drug. However, the oil phase does act as a reservoir for cetostearyl alcohol (7) and contributes to the sensory characteristics of the product such as whiteness and opacity (9). The variation in refractive index caused by the gel network when the "emulsifying wax" (cetostearyl alcohol and surfactant) fraction is high can be seen in both phase contrast and polarized light (11); structures are also clearly visible using electron microscopy (8). The lamellar spacing has been measured using X-ray diffraction, and increases with the fraction of water added to

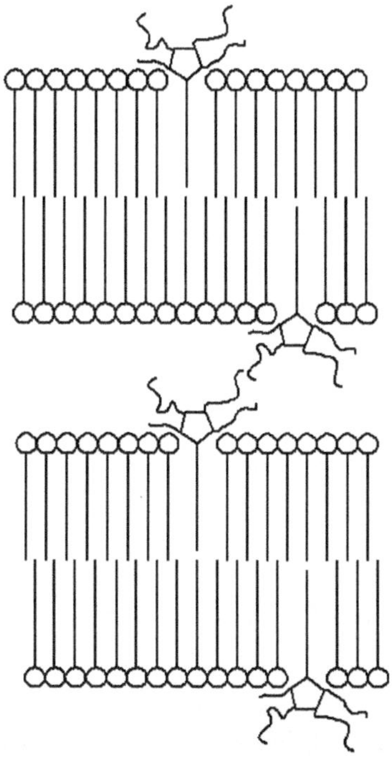

Figure 1 A generalized diagram of the structure of the cetostearyl alcohol gel found in topical and vaginal creams. The bilayers are formed principally of cetostearyl alcohol. The hydrophilic poly(oxyethylene) chains attached to the 5-carbon sorbitan rings in Polysorbate 60 retard water drainage from the interlamellar space and keep the lamellae from collapsing into a dense crystalline structure.

the formulation (12). In gel-forming systems that have not undergone substantial shear, such as creams that have been reheated without additional homogenization, a characteristic particle can be seen in polarized light that exhibits the "maltese cross" birefringence characteristic of concentric crystalline domains (13). This indicates gel formation even when the birefringence of the continuous phase is too low to resolve lamellar structures in the bulk.

Because the cetostearyl alcohol and a hydrophilic surfactant are the primary structure-forming excipients, it is important for pharmaceutical scientists to characterize their interaction as well as any crystalline phase transitions that influence the semisolid nature of the cream, and in turn its physical properties, pharmaceutical elegance, and drug release. Differential Scanning Calorimetry (DSC) has been utilized by researchers, with the intent of observing thermal transitions related to the crystal structure transitions in the gel phase. Significant DSC work using ternary gel systems, composed only of cetostearyl alcohol, a surfactant, and water, was undertaken by Yoon et al. (14). This work found that there are multiple crystalline transitions in such systems, indicating that solid-state transitions occur below 60°C. However, DSC work was also performed earlier by Eccleston (15) on both ternary gels and gel-stabilized emulsions, with the conclusion that the oil phase had a pronounced effect on the transition temperatures and transition enthalpies. This work also found that the composition of the fatty alcohol mixture had a similar effect, and that ageing creams changed the enthalpy, but not the temperature, of at least one transition.

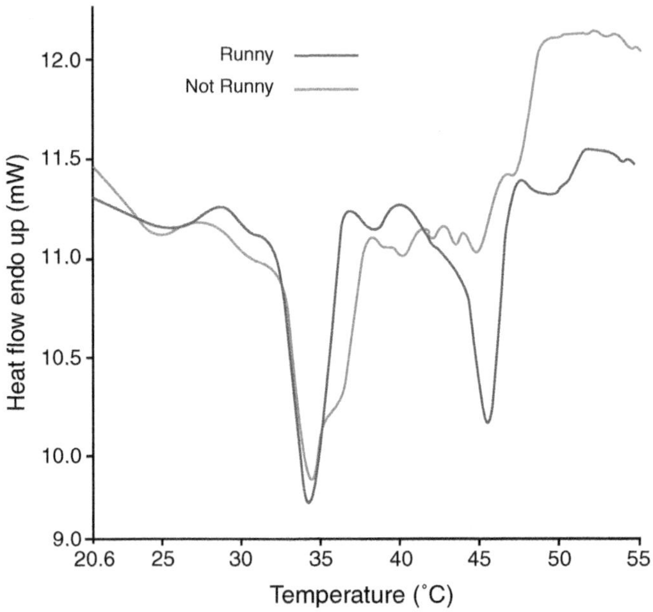

Figure 2 Differential scanning calorimetry results (cooling) for a sample of runny cream and thick cream. Run conditions: sample mass 13.0 mg, cooling rate 10°C/min, sealed aluminum pans.

The polymorphism of cetostearyl alcohol has been suggested as one of the primary mechanisms affecting gel formation by Eccleston (15). Further, the data presented by Eccleston (15), clearly indicates that there is more than one crystalline transition during cooling, and that those transitions are necessary for the formation of a gel. The polymorphic nature of the cetostearyl alcohol has been studied by several investigators (16), Abrahamsson et al. (17), and Ventola et al. (18). Three polymorphs have been identified over the temperature range of interest: these will be referred to here as the alpha, beta, and gamma forms in the order of decreasing crystalline symmetry.[a] The alpha polymorph is a hexagonal phase that always forms from the melt. The crystal structure consists of bilayers of hexadecanol and octadecanol, with their long axes perpendicular to the plane of the bilayer and their hydroxyl groups to the outside of the bilayer. In the alpha form, the hydrocarbon chains are free to rotate about their long axes. At lower temperatures, the alpha form typically transitions to the gamma form, which is monoclinic. In this form, the hydrocarbon chains are at an angle to the plane of the bilayer. It is this form that is the most stable polymorph, in that it has the highest temperature range of stability of the three polymorphs (16,19).

Figure 2 shows the influence of the different phase transitions on the rheological characteristics of a vaginal cream formulation containing cetostearyl alcohol and Polysorbate 60 as the emulsifying wax. Incomplete transition of two observable polymorphs

[a] The names of the three polymorphs vary over the 50-year range spanned by the literature available; the terminology used here is that originally presented in Tanaka et al. (1958), which is the most widely accepted in current literature. Dr. Tanaka's paper also presents a list of synonyms used by other authors for the same crystal forms.

of the emulsifying wax within the cream during compounding resulted in a "runny" cream with milk-like consistency rather than a semisolid product with high-yield stress.

Thermal scanning rheometry (TSR), on the same vaginal cream formulation, were generated by measuring motor torque in a pilot scale reactor during cooling of the product in the compounding process. They were reconfirmed by repeating the measurement during a reheating cycle. Results given below in Figure 3 show a correlation between the phase transitions observed in the DSC and the apparent increase in viscosity of the product.

While these studies indicate that DSC and TSR are powerful tools for examining cetostearyl alcohol creams, it is also clear that pharmaceutical scientists must generate

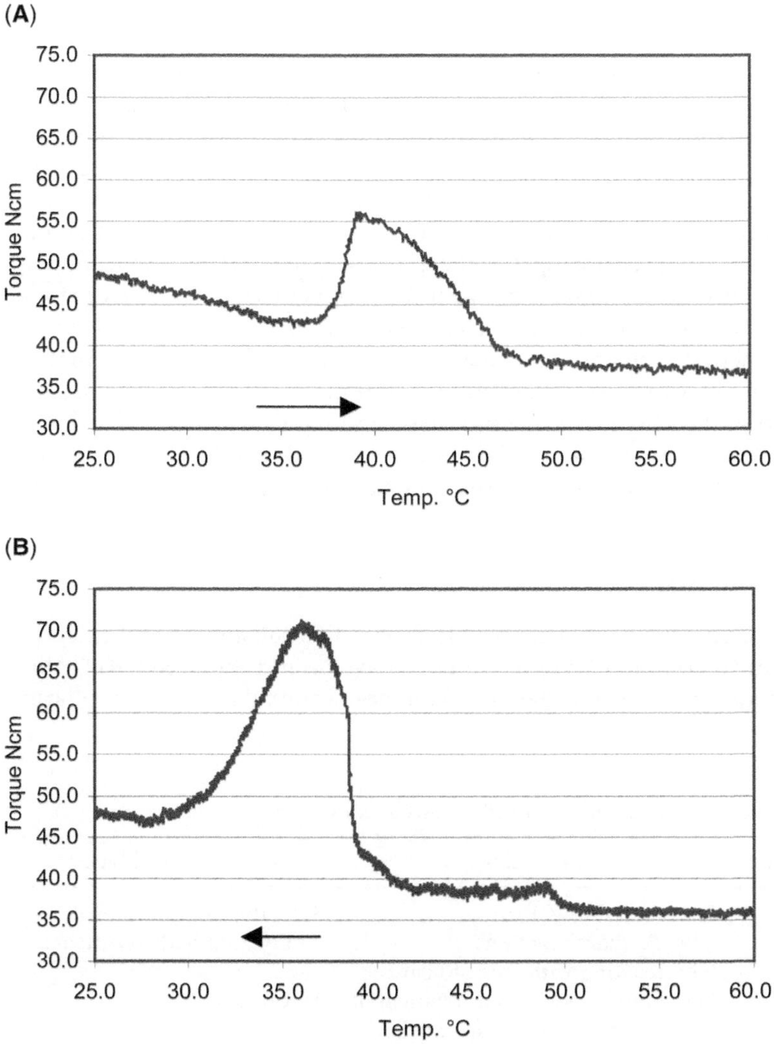

Figure 3 **(A)** Cooling and **(B)** heating thermal scanning rheograms for a vaginal cream formulation. The dashed vertical line indicates the 38°C point in both plots. This is clearly the initiation temperature of the phase transition leading to increased apparent viscosity during both heating and cooling.

this data on their own formulations to account for the influence of their particular oil phase composition as well as source of Cetostearyl alcohol (which could result in a variation in composition). This data can then be utilized in formulation design as well as process design to obtain the desired aesthetic and pharmaceutical properties. For example, for the vaginal cream formulation discussed above where reduced viscosity at 38°C is desired for spreading across the vaginal mucosa at body temperature, a tempering step at this phase transition temperature could be incorporated into the compounding process to ensure a complete transition resulting in a cosmetically elegant semisolid cream that meets patient acceptability goals.

Preservatives

The presence of water in creams requires the use of preservatives to curtail bacterial growth. In addition to preservation against contamination during manufacturing and packaging, most cream formulations are multiple-dose products packaged in tubes and require preservation to combat any organisms that might come in contact with and contaminate the product as a result of reuse during therapy. The following three criteria are considered critical for preservative selection: (i) the preservative system must exhibit the required antimicrobial activity in the proposed formulation over the duration of the product shelf life; (ii) the preservative system must be nontoxic, nonirritant and nonsensitizing for the proposed method of application for the cream; and (iii) it must be compatible with the product (particularly its pH) and package. Commonly used preservatives in cream formulations include benzyl alcohol, propylparabens, methylparabens, chlorocresol, imidazolidinyl urea (Germaben), and sodium benzoate (Appendix I). To provide antimicrobial activity against both gram-positive and gram-negative bacteria, yeasts, and molds often combinations of preservatives are used. Numerous studies and reviews that address the problem of preservative system selection are available to pharmaceutical scientists (20–23).

Other Common Excipients

Antioxidants are often used to reduce oxidation of active substances and excipients in creams. There most common types of oxidants have been recognized (24). Table 1 lists each class of antioxidants and the most common antioxidants used in pharmaceutical creams.

Table 1 Commonly Used Antioxidants in Pharmaceutical Creams

Type of antioxidant	Definition	Most commonly used
True antioxidants	These are thought to block chain reactions by reacting with free radicals	Butylated hydroxyanisole, butylated hydroxytoluene
Reducing agents	These have a lower redox potential than the drug or excipient they are protecting	Ascorbic acid
Antioxidant synergists	These enhance the effect of antioxidants	Edetic acid, sodium edetate

Source: From Ref. 24.

Emollients

Emollients are often added to cream formulations to modify either the characteristics of the pharmaceutical vehicle or the condition of the skin itself to promote penetration of the active ingredient to act either locally or systemically. The stratum corneum, being keratinized tissue, behaves as a semipermeable artificial membrane, and drug molecules can penetrate by passive diffusion. The rate of drug movement depends on the drug concentration in the vehicle, its aqueous solubility, and the oil/water partition coefficient between the stratum corneum and the product's vehicle. Commonly used emollients include glycerin, mineral oil, petrolatum, isopropyl palmitate, and isopropyl myristate.

OINTMENTS

The U.S. Pharmacopeia defines ointments as semisolid preparations intended for external application to the skin or mucous membranes. Most pharmaceutical ointments are hydrocarbon-based semisolids containing dissolved or suspended drugs. These bases, which are known also as "oleaginous ointment bases," are represented by "white petrolatum and white ointment." Only small amounts of an aqueous component can be incorporated into them. Buhse et al. (6) propose to incorporate a water and volatile content less than 20% of water and greater than 50% of hydrocarbons, waxes, or PEGs to distinguish ointments from creams, which have a higher content of water and volatiles. Ointments serve to keep medicaments in prolonged contact with the skin and act as occlusive dressings providing increased and sustained delivery of the active ingredients (25). Hydrocarbon bases are used chiefly for their emollient effects, and are difficult to wash off. They do not "dry out" or change noticeably on aging (5).

Primary structure-forming excipients in most ointments comprise fluid hydrocarbons, possibly mineral oil, and are entrapped in a fine crystalline matrix of long-chain hydrocarbons such as white petrolatum. The mineral oil is incorporated into the petrolatum or waxes by heating together between 60°C and 80°C and mixing in the fluidized state. The system is then cooled with mild stirring until it has solidified. The rate of cooling can be important because rapid cooling tends to impart more structure. The extent and nature of the structure determine the stiffness of the ointment. The drug may be incorporated directly into the congealed system particularly for suspended drugs such as hydrocortisone. Solubilizing excipients such as lanolin, lanolin derivatives, cholesterol or cholesterol esters, or other water-in-oil emulsifiers may be added singularly or in combination with the base to allow aqueous solutions of drugs to be incorporated to obtain higher bioavailability for some hydrophilic drugs (26). Appendix II presents a survey of a wide range of hydrocarbon-based ointments used in various therapeutic applications.

PASTES

Pastes may be defined as a semisolid dosage form that contains a large proportion (i.e., 20–50%) of solids finely dispersed in a fatty vehicle (basically an ointment base) for external application to the skin. The presence of a high concentration of solids makes them much stiffer than ointments. Like ointments, pastes form an unbroken relatively water-impermeable film on the skin surface. Most pastes are opaque due to the high light dispersion by the particulates embedded in the paste matrix. An important criterion to consider in paste formulations is the dispersion of particles such that individual particles are impalpable (i.e., incapable of being individually perceived as particles by touch) (4). Otherwise, pastes will feel gritty upon application. Individual particles are generally regarded as impalpable when their longest dimension is below 20 μm (4). Thus, generally the use of finely micronized insolubles is recommended for the formulation of pastes. Appendix III shows commonly used bases and other excipients in pharmaceutical pastes.

GELS

Gels are semisolid vehicles for drugs aimed at mucosal e.g., ocular, nasal, vaginal, and rectal administration. The gel-forming compound, usually a polymer with a concentration of a few percent, gives a semisolid consistency to the formulation by either physical or chemical cross-linking. This consistency will reduce the drainage rate of the formulation and prolong the residence time at the site of administration. The mucosal surface is covered with a layer of mucus, when administering the dosage form to the mucosal tissue; the polymers in the formulation may interact with the mucus layer. The mucoadhesion in combination with the rheological properties will contribute to an increased contact time and a more intimate contact with the tissue resulting in a more efficient absorption of the drug. In order to fully take advantage of the long residence time of the gels the drug compound must be released at an appropriate rate. Because the gels usually consist of more than 90% water, the small drug molecule will move almost freely in the formulation giving a rapid release of the drug. In order to achieve a sustained release from gels the drug must be incorporated in or interact with a slower diffusing species. Examples of such systems are when including the drug above its solubility giving a suspension of the drug in the gel or incorporation of surface-active molecules in the formulation. The drug can then interact with the gel-forming polymer and/or with aggregates formed by the surfactant.

Among the gelling agents used are: synthetic macromolecules as acrylic acid polymers such as Carbomer 934, cellulose derivatives such as carboxymethyl cellulose or hydroxypropylmethyl cellulose, and natural gums such as xanthan gum. Appendix IV shows the commonly used gelling agents in commercially available pharmaceutical gels.

Acylic Acid-Based Polymeric Gelling Agents

Carbomers[b] and Pemulen® polymeric emulsifiers are acrylic acid polymers cross-linked with polyalkenyl polyethers. Commonly used Carbomers in pharmaceutical applications are Carbomer 934P, Carbomer 940 and Carbomer 941 (1), the difference primarily being in the molecular weight between cross links (*Mc*) which eventually manifests itself in the viscosity and rigidity of the polymer. Aqueous dispersions of carbomers have an approximate pH range of 2.8 to 3.2 depending

---

[b] Carbomers are sold by Noveon Inc. (formerly BF Goodrich) under the brand name Carbopol®.

on polymer concentration. A molecule of these polymers in the dry powder state is tightly coiled, thus limiting its thickening capability. When dispersed in water, the molecule begins to hydrate and uncoil slightly, generating an increase in viscosity. However, to achieve the highest possible performance with the polymer, the molecule must be completely uncoiled.

There are two mechanisms by which the molecule can become completely uncoiled, providing maximum thickening, The most commonly used mechanism is accomplished by neutralizing the polymer with a suitable base such as sodium or potassium hydroxide or amine bases such as Tris® [*tris*(hydroxymethyl) aminomethane]. Neutralization ionizes the carbomer, generating negative charges along the polymer backbone. Repulsions of these like-negative charges cause the molecule to completely uncoil into an extended structure. This reaction is rapid and gives instantaneous thickening (27). Figure 4A and B show the carbomer backbone in its unneutralized and neutralized state respectively. Table 2 provides stoichiometric ratios for most commonly used neutralizers in combination with carbopol® polymers.

The second thickening mechanism involves the use of a hydroxyl donor in addition to the polymer. The combination of a carboxyl donor and one or more hydroxyl donors will result in thickening because of the formation of hydrogen bonds. Some commonly used hydroxyl donors are: polyols (such as glycerin, propylene glycol and PEG), sugar alcohols such as mannitol, nonionic surfactants with five

(**A**)

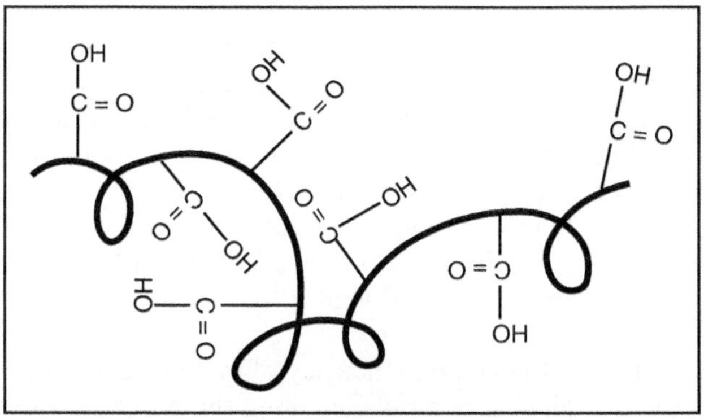

(**B**)

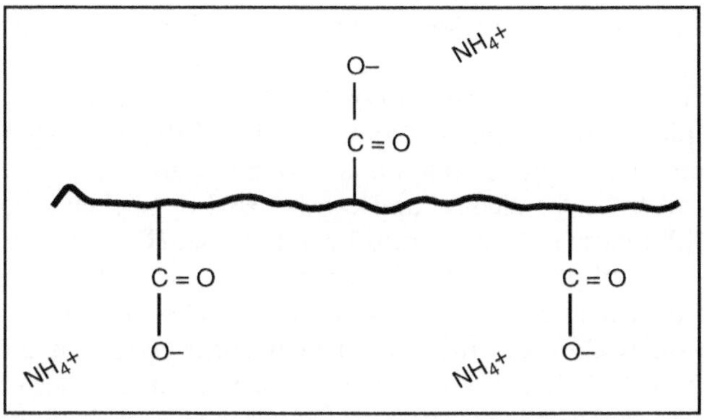

Figure 4 (**A**) Unneutralized and (**B**) neutralized form of the carbomer. *Source*: From Ref. 27.

Table 2 Ten of the Most Common Neutralizers Used, the Manufacturers of These Neutralizers, and the Appropriate Ratio (as Compared to One Part of Carbomer Polymers) to Use to Achieve Exact Neutralization at a pH of 7.0

Trade name	CTFA name	Manufacturer	Neutralization ratio base/carbopol® polymer
NaOH (18%)	Sodium hydroxide		2.3/1.0
Ammonia (28%)	Ammonium hydroxide		0.7/1.0
KOH (18%)	Potassium hydroxide		2.7/1.0
L-Arginine	Arginine	Ajinomoto	4.5/1.0
AMP-95®	Aminomethyl propanol	Angus	0.9/1.0
Neutrol® TE	Tetrahydroxypropyl ethelenediamine	BASF	2.3/1.0
TEA (99%)	Triethanolamine		1.5/1.0
Tris Amino® (40%)	Tromethamine	Angus	3.3/1.0
Ethomeen® C-25	PEG-15 cocamine	Akzo	6.2/1.0
Di-isopropanol amine	Di-isopropanol amine	Dow	1.2/1.0
Tri-isopropanol amine	Tri-isopropanol amine	Dow	1.5/1.0

Abbreviations: CTFA, cosmetic, toiletry and fragrance association; PEG, polyethylene glycol.
Source: From Ref. 29.

or more ethoxy groups, glycol-silane copolymers, polyethylene oxide, and fully hydrolyzed polyvinyl alcohol, among others. These reagents form hydrogen bond with the polymer molecule causing it to uncoil. The hydrogen bonding is not instantaneous—maximum thickening may take from five minutes to three hours. Heating the dispersion hastens thickening, but temperatures above 60°C (140°F) should not be used. The pH of such systems will tend to be acidic (27).

Traditionally, Carbomers are used between concentrations ranging from 0.1% to about 1%. Figure 5 below shows the effect of concentration and pH on the viscosity of an aqueous solution of Carbomer 940 NF neutralized with 10% NaOH solution (28).

Monovalent ions simply reduce the thickening efficiency of systems containing carbomer polymer by reducing the overall charge repulsion along the polymer backbone. By simply adding more carbomer polymer, the loss in thickening efficiency, resulting from the presence of ionic material, can be overcome. Divalent or trivalent ions can, in addition to thinning, also form an insoluble precipitate if present at high enough levels (29).

Different solvent systems are used to satisfy compatibility requirements with active drugs and/or other excipients, and to ensure bioavailability of the drug when applied on skin (30,31). Solvents used include ethanol, isopropanol, and propylene glycol. Carbomers have high tolerance for alcohol and can be used to thicken such hydroalcoholic systems. The differences in fourier transform infra-red (FTIR) spectral responses of water- and alcohol-dominated gels suggest polymer–neutralizer interactions are strongly affected by the polarity of the solvent (32). Thus, the critical factor for successful formulation is choosing the correct neutralizer based on the amount of alcohol that is to be gelled. If the wrong neutralizer is used, the salt of the carbomer will precipitate out because it is no longer soluble in the hydroalcoholic blend. Table 3 gives recommended neutralizers for various alcohol levels.

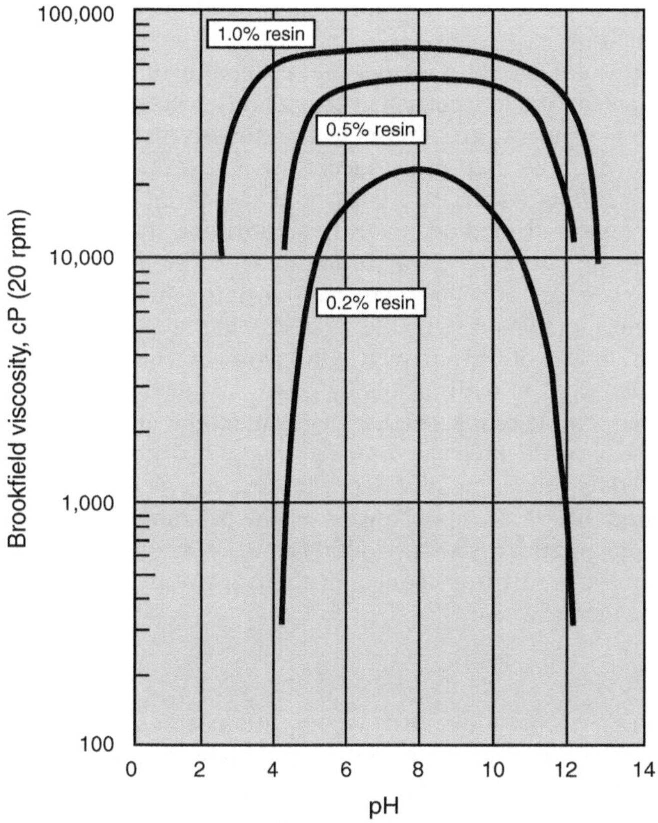

Figure 5 Effect of concentration and pH on the viscosity of an aqueous solution of Carbomer 940NF neutralized with 10% NaOH solution. *Source*: From Ref. 28.

Cellulose-Based Gelling Agents

Commonly used cellulose derivatives include hydroxypropyl cellulose (HPC), carboxymethylcellulose, and hydroxyethyl cellulose (HEC). The choice amongst these cellulose derivatives is primarily based on the type of formulation (aqueous or

Table 3 Recommended Neutralizers for Use in Hydroalcoholic Systems with Carbomers

Up to % alcohol	Neutralizer
20	Sodium hydroxide
30	Potassium hydroxide
60	Triethanolamine
60	Tris Amino
80	AMP-95®
90	Neutral TE
90	Di-isopropanolamine
90	Tri-isopropanolamine
>90	Ethomeen C-25

Source: From Ref. 33.

hydroalcoholic) and compatibility with the physiologically active ingredient and other excipients in the formulation.

HPC is a nonionic water-soluble cellulose ether that is manufactured by reacting alkali cellulose with propylene oxide at elevated temperatures. The polymer is very soluble in water at room temperature, although its solubility decreases with increase in temperature. It is generally used at a concentration of 0.2% in pharmaceutical gels. When used as a gelling agent, care must be taken during addition to ensure a lump-free gel. The preferred method involves preslurrying the powder in a nonsolvent such as glycerin or hot water prior to addition to the main volume of water. Also, whenever possible, HPC should be put into solution before addition of any other soluble ingredients. Other dissolved materials compete for the solvent and slow the solution rate of the polymer. HPC shows excellent solubility in organic solvents and is often used in hydroalcoholic gels.

HEC is a nonionic water-soluble cellulose ether that is manufactured by reacting alkali cellulose with ethylene oxide at elevated temperatures. It dissolves readily in hot or cold water. It is generally present at levels from 1% to 2% in pharmaceutical gels. Solutions of HEC undergo little viscosity change over the pH range of 2 to 12. However, solutions possess the greatest viscosity stability in the pH range of 6.5 to 8.0. The viscosity of solutions of Natrosol changes with temperature, increasing when cooled and decreasing when warmed.

Natural Gelling Agents

Commonly used natural gelling agents are xanthan gum, gellan gum, pectin, and gelatin. Xanthan gum and gellan gum are high molecular weight polysaccharides produced by microbial fermentation. The high viscosity associated with xanthan gum solutions at low shear rates enables products to keep particles suspended or prevent oil droplets from coalescing. Because the viscosity drops when shear is applied, the end-consumer products can be easily scooped, poured, or squeezed from its container. Once the force is removed, the solutions regain their initial viscosity almost immediately. Gellan gum is a gelling agent, effective at extremely low use levels, forming solid gels at concentrations as low as 0.1%.

Pectins are a family of partially methyl esterified polysaccharides produced from citrus peel and sugar beet pulp by extraction and controlled de-esterification. Pectins are classified as high methoxyl (HM) pectin and low methoxyl (LM) pectin. HM pectin requires a minimum amount of soluble solids and a narrow pH range, around 3.0, to form gels; LM pectin requires the presence of a controlled amount of calcium or other divalent cations to form a gel. Apart from adding structure through gelation and viscosity buildup, pectin gels on the skin can provide moisture absorption while being skin friendly.

SUPPOSITORIES

Suppositories are pharmaceutical dosage forms intended for administration of medicine via the rectum, vagina, or urethra that melt, soften, or dissolve in the body cavity. Rectal and vaginal suppositories are most common but urethral suppositories are sometimes used. Suppositories are indicated for administering drugs to infants and small children, severely debilitated patients, those who cannot take medications orally, and those for whom the parenteral route might be unsuitable. Suppositories are used to administer drugs for either systemic or local application. Local applications include the treatment of hemorrhoids, itching, and infections. Systemic application is used for a variety of drugs, including antinauseants, antiasthmatics, analgesics, and hormones.

Suppositories primarily comprise the active ingredient dispersed or dissolved in a base. Suppository bases usually employed are cocoa butter, glycerinated gelatin, hydrogenated vegetable oils, mixtures of PEGs of various molecular weights, and fatty acid esters of PEG (15). The selection of a suppository base is dependent upon a number of physicochemical variables, including the solubility characteristics of the drug. Factors such as the presence of water, hygroscopicity, viscosity, brittleness, density, volume contraction, special problems, and incompatibilities, rate of drug release, pharmacokinetics, and bioequivalence are important. Numerous studies are available in the literature on the influence of the properties of suppository bases using various routes of administration including urethral, rectal, and vaginal administration (34–42).

Suppository Bases

A suppository base should be stable, nonirritating, chemically and physiologically inert, compatible with a variety of drugs, melt or dissolve in rectal fluids, stable during storage, not bind or otherwise interfere with the release or absorption of drug substances, and be aesthetically acceptable. Other desirable characteristics depend upon the drugs to be added. For example, higher melting point bases can be selected for incorporating drugs that generally lower the melting points of the base or when formulating suppositories for use in tropical climates. Lower melting point bases can be used when adding materials that will raise the melting points or when adding large amounts of solids.

For most purposes, it is convenient to classify suppository bases according to their physical characteristics into two main categories and a third miscellaneous group: (i) fatty or oleaginous bases; (ii) water-soluble or water-miscible bases, and (iii) miscellaneous bases, general combinations of lipophilic and hydrophilic substances (43). Appendix V presents a survey of commercial pharmaceutical suppositories and the respective suppository bases.

Fatty or Oleaginous Bases

Fatty bases are perhaps the most frequently employed suppository bases primarily based on cocoa butter. Among the other fatty bases are many hydrogenated fatty acids of vegetable oils such as palm kernel oil and cottonseed oil. Also, fat-based glyceryl esters such as glyceryl monopalmitate and glyceryl monostearate may also be found in fatty suppository bases. The suppository bases in many commercial products employ various combinations of these materials to achieve a base that possesses the desired hardness under conditions of shipment and storage, the desired melting characteristics, and drug release at body temperature.

Cocoa butter NF is defined as the fat obtained from the seed of "Theobroma Cacao" Linné (Family: Sterculiaceae) (44). Cocoa butter softens at 30°C and melts at 34°C. It contains four different forms: alpha, beta, beta prime, and gamma with melting points of 22°C, 34°C to 35°C, 28°C and 18°C, respectively. The beta form is the most stable and is desired for suppositories. The biggest challenge with the polymorphism of cocoa butter is the impact of the manufacturing process on the characteristics of the suppository itself. When cocoa butter is hastily melted at a temperature greatly exceeding the minimum required temperature and then quickly chilled, the result is metastable crystalline form (α crystals), which may not even solidify at room temperature in the molds. The lower melting point polymorphs eventually will convert to the more stable form over time but this process may take several days to several months. Most suppository bases or combinations have the same issues with respect to polymorphism particularly because it relates to sensitivity to heating and recrystallization during compounding. DSC and hot-stage microscopy can be used to understand the various polymorphs that can be formed during cooling of the base and this information can be used to ensure that the compounding process yields product that achieves the desired pharmacokinetic objectives. Table 4 below shows the difference in vitro release rate (at 38°C) observed within the same batch of a vaginal suppository product (using a cocoa butter base) due to different compounding temperatures and the physical differences between them that were identified by hot-stage calorimetry. Tempering the suppositories and recrystallization (with seeding) results demonstrate that these effects are reversible.

Fattibase™ is a preblended suppository base that offers the advantages of a cocoa butter base with few of the drawbacks. It is composed of triglycerides derived from palm, palm kernel, and coconut oils with self-emulsifying glyceryl monostearate and polyoxyl Stearate used as emulsifying and suspending agents. It is stable with a low irritation profile, needs no special storage conditions, is uniform in composition, and has a bland taste and controlled melting range. It exhibits excellent mold release characteristics and does not require mold lubrication. Fattibase™ is a solid, which has a melting point of 35°C to 37°C and a specific gravity of 0.85 to 0.95 at 37°C; it is opaque-white and free of suspended matter.

Wecobee® bases (Stepan Company, Illinois, U.S.A.) are derived from palm kernel and coconut oils, and the incorporation of glyceryl monostearate and propylene glycol monostearate renders them emulsifiable. These bases exhibit most of the desirable features of cocoa butter but few of its shortcomings. Suppocire® and Ovucire bases (Gattefosse SA, St Priest, France) are similar bases consisting mainly of mixtures of C12–C18 triglycerides obtained by esterification or interesterification of common vegetable oils. Special grades may contain a portion of mono- and diglycerides and/or additives such as beeswax, phospholipids, PEG, and sorbitan esters.

Witepsol® bases (Sasol North America Inc., Westlake, Louisiana, U.S.A.) solidify rapidly in the mold, and lubrication is not necessary because the suppositories

Table 4 Influence of Processing Conditions on Drug Release and Morphology of a Typical Suppository Product

Processing condition	Compounded at 47°C	Compounded at 42°C	Compounded at 47°C, tempered and seeded with fresh base at 42°C
Average in vitro drug release at 1 hr (stdev/min/max)	98%	73%	96.8%
Standard deviation ($n = 6$)	4.3%	3.3%	6%
Physical characteristics (microscopy)	Small spherulite crystals	Long needle-like crystals	Long needle-like crystals
Thermal conditioning at 30°C (hot-stage microscopy)	Conversion to long needle-like crystals	No physical transformation	No physical transformation

contract nicely. High melting point Witepsol® bases can be mixed with low melting point Witepsol® bases to provide a wide range of possible melting ranges, i.e., 34°C to 44°C. Because the Witepsol® bases contain emulsifiers, they will absorb limited quantities of water.

Water-Soluble Bases

The use of water-soluble bases may result in some irritation because, as they take up water and dissolve, they may produce slight dehydration of the rectal mucosa. They are widely used, however, and release the drug by dissolving and mixing with the aqueous body fluids. PEG suppository bases and glycerinated gelatin are the most popular in this class.

Glycerinated gelatin suppositories are used most routinely in the preparation of vaginal suppositories, where prolonged localized action of the drug is usually desired. The glycerinated base is slower to soften and mix with physiologic fluids than is cocoa butter and thus provides prolonged release. These suppository bases have a tendency to absorb moisture due to the hygroscopic nature of the glycerin and must be protected from atmospheric moisture to maintain their shape and consistency.

Because PEGs are available in a variety of molecular weight ranges, PEG suppository bases have the advantage of allowing the formulator many degrees of freedom in that the ratios of the low to the high molecular weight individual PEGs can be altered to prepare a base with a specific melting point, or one that will overcome the adverse characteristics of an excess of powder or liquid that must be incorporated into a suppository (45). Depending upon their chain length and molecular weight, PEGs range from being a clear colorless liquid (PEG 300–PEG 600) to a wax-like white solid (PEG 1450, PEG 3350, PEG 8000). Because PEG-based suppositories act by dissolving slowly in the body's fluids, they need not be formulated to melt at body temperature. Several PEG bases may also contain additives to modify their drug-release characteristics. For example, Polybase™ is a preblended suppository base that is a white solid consisting of a homogeneous mixture of PEGs and polysorbate 80.

Miscellaneous Bases

Mixtures of oleaginous and water-soluble bases are included in this category. Some of these bases may be preformed emulsions, generally w/o type, or they are capable of dispersing in aqueous physiologic fluids. Polyoxyl 40 stearate is a common surface-active agent found in a number of commercial bases. Similarly, in recent years some in situ gelling and mucoadhesive liquid suppository bases have been developed that are composed of Poloxamers, sodium alginate, and polycarbophil, which exist as liquid in vitro but gel in vivo, by modulating the gelation temperature of Poloxamer solution (33,46,47).

ACKNOWLEDGMENTS

The author would like to thank Rob Saunders and Jeff Gaudio for generating the experimental data presented in this chapter, and Yuan (Kate) Yuan for her invaluable help in compiling the appendices. The author would also like to thank Pfizer, Inc., for permission to publish experimental data generated at this research laboratory.

REFERENCES

1. Idson B, Lazarus J. Semisolids. In: Lachman L, Lieberman H, Kanig J, eds. The Theory Practice of Industrial Pharmacy. 3rd ed. Philadelphia: Lea and Febiger, 1986.
2. Barry BW. Dermatological formulations: percutaneous absorption. In: Drugs and the Pharmaceutical Sciences: a Series of Textbooks and Monographs. Marcel Dekker, 1983:296–299.
3. Smith WW, Maibach HI. Percutaneous penetration enhancers: the fundamentals. In: Smith EW, Maibach HI, eds. Percutaneous Penetration Enhancers. Boca Raton, Florida: CRC Press, 1995b:1–4.
4. Flynn GL. Topical drug absorption and topical pharmaceutical systems. In: Banker GS, Rhodes CT, eds. Modern Pharmaceutics. New York: Marcel Dekker, 1979.
5. Pharmaceutical Dosage Forms, USP 28-NF 23. General Chapters (1151). The United States Pharmacopeial Convention Inc, 2005.
6. Buhse L, Kolinski R, Westenberger B, et al. Topical drug classification. Int J Pharma 2005; 295(1–2):101–112.
7. Louden JD, Patel HK, Rowe RC. A preliminary examination of the structure of gels and emulsions containing cetostearyl alcohol and cetrimide using laser Raman spectroscopy. Int J Pharm 1985; 25:179–190.
8. Rowe RC, McMahon J. The characterisation of the microstructure of gels and emulsions containing cetostearyl alcohol and cetrimide using electron microscopy—a comparison of techniques. Colloid Surf 1987; 27:367–373.
9. Eccleston GM. Functions of mixed emulsifiers and emulsifying waxes in dermatological lotions and creams. Colloid Surf A 1997:123–124, 169–182.
10. Adamson AW, Gast A. Physical Chemistry of Surfaces. 6th ed. John Wiley and Sons, Inc., 1997.
11. Patel HK, Rowe RC, McMahon J, Stewart RF. A systematic microscopical examination of gels and emulsions containing cetrimide and cetostearyl alcohol. Int J Pharm 1985; 25:13–25.
12. Eccleston GM, Behan-Martin MK, Jones GR, Towns-Andrews E. Synchrotron X-ray investigations into the lamellar gel phase formed in pharmaceutical creams prepared with cetrimide and fatty alcohols. Int J Pharm 2000; 203:127–139.
13. Eccleston GM. Multiple-phase oil-in-water emulsions. J Soc Cosmet Chem 1990; 41: 1–22.
14. Yoon MS, Chung YB, Han K. A study of gel structure in the nonionic surfactant/cetostearyl alcohol/water ternary systems by differential scanning calorimetry. J Disp Sci Tech 1999; 20:1695–1713.
15. Eccleston GM. Phase transitions in ternary systems and oil-in-water emulsions containing cetrimide and fatty alcohols. Int J Pharm 1985; 27:311–323.
16. Kolp DG, Lutton ES. The polymorphism of *n*-hexadecanol and *n*-octadecanol. J Am Chem Soc 1951; 73:5593–5595.
17. Abrahamsson S, Larsson G, von Sydow E. The crystal structure of *n*-hexadecanol. Acta Cryst 1960; 13:770–774.
18. Ventola L, et al. Polymorphism of *n*-alkanols: 1-Heptadecanol, 1-octadecanol, 1-nonadecanol, and 1-eicosanol. Chem Mater 2002; 14:508–517.
19. Tanaka K, Seto T, Hayashida T. Phase transformations of *n*-higher alcohols (I). Bull Inst Chem Res 1958; 35:123–139.
20. Tan E, Shah H, Leister K, et al. Transforming growth factor-α (TGF-α) in a semisolid dosage form: preservative and vehicle selection. Pharma Res 1993; 10(8):1238–1242.

21. Kabara JJ. Food grade chemicals in a systems approach to cosmetic preservation. In: Kabara JJ, ed. Cosmetic and Drug Preservation: Principles and Practice. New York: Marcel Dekker, 1984:345.
22. Sabourin JR. Selecting a preservatice for creams and lotions. Cosmet Toiletries 1986; 101:93–98.
23. Grundy WE. Antimicrobial preservatives in pharmaceuticals. In: Block SS, ed. Disinfection, Sterilization and Preservation. 5th ed. Lippincott Williams and Wilkins, 2000: 757–767.
24. Note for Guidance on Excipients, Antioxidants, and Antimicrobial Preservatives in the Dossier for Marketing Authorization of a Medicinal Product. The European Agency for the Evaluation of Medicinal Products, 2003.
25. Viegas TX, Van Winkle LL, Lehman PA, Franz SF, Franz TJ. Evaluation of creams and ointments as suitable formulations for peldesine. Int J Pharma 2001; 219(1–2):73–80.
26. Velissaratou AS, Papaioannou G. In vitro release of chlorpheniramine maleate from ointment bases. Int J Pharma 1989; 52(1):83–86.
27. Noveon Polymers in Semisolid Products. Bulletin 8. Cleveland, Ohio: Noveon, Inc., 2002.
28. TDS-237. Cleveland, Ohio: Noveon Inc., 2002.
29. Thickening Properties. Bulletin 11. Cleveland, Ohio: Noveon, Inc., 2002.
30. TDS-54. Cleveland, Ohio: Noveon Inc., 2002.
31. Rangarajan M, Zatz JL. Effect of formulation on the topical delivery of alpha-tocopherol. J Cosmet Sci 2003; 54(2):161–174.
32. Rafiee-Tehrani M, Mehramizi A. In vitro release studies of piroxicam from oil-in-water creams and hydroalcoholic gel topical formulations. Drug Dev Ind Pharm 2000; 26(4):409–414.
33. Kim CK, Lee SW, Choi HG, Lee MK, Gao ZG, Kim IS. Trials of in-situ gelling and mucoadhesive acetaminophen liquid suppository in human subjects. Int J Pharm 1998; 174:201–207.
34. Islam MT, Rodríguez-Hornedo N, Ciotti S, Ackermann C. Fourier transform infrared spectroscopy for the analysis of neutralizer-carbomer and surfactant-carbomer interactions in aqueous, hydroalcoholic, and anhydrous gel formulations. AAPS J 2004; 6(4):article 35.
35. Te Wierik GH, Eissens AC, Lerk CF. Preparation, characterization, and pharmaceutical application of lineardextrins. III. Drug release from fatty suppository bases containing amylodextrin. Pharm Res 1994; 11(1):108–110.
36. Tarine Gombkoto Z, Molnar J, Farkasne Gunics G, Regdon G Jr, Selmeczi B. Formulation and in vitro investigation of antibacterial vaginal suppositories. Part 2. In vitro membrane diffusion and microbiologic studies. Acta Pharm Hung 1992; 62(6):302–309.
37. Celebi N, Iscanoglu M, Degim T. The release of naproxen in fatty suppository bases by beta-cyclodextrin complexation. Pharmazie 1991; 46(12):863–865.
38. Schmitt M, Guentert TW. Influence of the hydrophilicity of suppository bases on rectal absorption of carprofen, a lipophilic nonsteroidal anti-inflammatory drug. J Pharm Sci 1990; 79(4):359–363.
39. Adegboye TA, Itiola OA. Formulation effects on the mechanical and release properties of metronidazole suppositories. Afr J Med Med Sci 2003; 32(3):247–251.
40. Takatori T, Shimono N, Higaki K, Kimura T. Evaluation of sustained release suppositories prepared with fatty base including solid fats with high melting points. Int J Pharm 2004; 278(2):275–282.
41. Takatori T, Yamamoto K, Yamaguchi T, Higaki K, Kimura T. Design of controlled-release morphine suppositories containing polyglycerol ester of fatty acid. Biol Pharm Bull 2005; 28(8):1480–1484.
42. Taha EI, Zaghloul AA, Kassem AA, Khan MA. Salbutamol sulfate suppositories: influence of formulation on physical parameters and stability. Pharm Dev Technol 2003; 8(1):21–30.

43. Asikoglu M, Ertan G, Cosar G. The release of isoconazole nitrate from different suppository bases: in-vitro dissolution, physicochemical and microbiological studies. J Pharm Pharmacol 1995; 47(9):713–716.
44. Suppositories and inserts. In: Ansel H, Allen L Jr, Popovich N, eds. Pharmaceutcial Dosage Forms and Drug Delivery Systems. 7th ed. Lippincott Williams and Wilkins, 1999.
45. USP28–NF23. US Pharmacopeia:p2990.
46. Kellaway IW, Marriott C. Correlations between physical and drug release characteristics of polyethylene glycol suppositories. J Pharm Sci 1975; 64(7):1162–1166.
47. Choi HK, Oh YK, Kim CK. In-situ gelling and mucoadhesive liquid suppository containing acetaminophen: enhanced bioavailability. Int J Pharm 1998; 165:23–32.

Appendix I Survey of Pharmaceutical Creams

Indication	Active ingredient	Cream trade name	Structure-forming excipients	Preservative
Antifungal	Ciclopiroxolamine	Loprox	Stearyl alcohol, cetyl alcohol, polysorbate 60	Benzyl alcohol
	Clotrimazole	Canesten	Cetyl stearyl alcohol, polysorbate 60	Benzyl alcohol
	Sulconazole nitrate	Exelderm	Stearyl alcohol, cetyl alcohol, polysorbate 60, sorbitan monostearate, and PEG-100 stearate	None
	Clotrimazole	Lotrimin	Cetearyl alcohol 70/30 (10%), cetyl esters wax NF, Polysorbate 60	Benzyl alcohol
	Clotrimazole	Mycelex 3	Cetyl alcohol, polysorbate 60, PEG 400 stearate	Methylparaben and propylparaben
	Terconazole	Terazol 3	Cetyl alcohol, stearyl alcohol, polysorbate 60 and 80	None
	Clotrimazole	Lotrisone*	Cetearyl alcohol 70/30, ceteareth-30	Benzyl alcohol
	Ciclopiroxolamine	Loprox	Cetyl alcohol, steryl alcohol, polysorbate 60 NF, sorbitan monostearate	Benzyl alcohol NF
	Clotrimazole	Lotrimin	Cetearyl alcohol 70/30 (10%), cetyl esters wax, polysorbate 60, sorbitan monostearate	Benzyl alcohol NF
	Butenafine hydrochloride	Lotrimin ultra	Cetyl alcohol, glyceryl monostearate SE	Benzyl alcohol NF, sodium benzoate
	Butenafine hydrochloride	Mentax	Cetyl alcohol NF, glyceryl monostearate	Benzyl alcohol NF, sodium benzoate
	Naftifine hydrochloride	Naftin	Cetyl alcohol, cetyl esters wax, isopropyl myristate, polysorbate 60, sorbitan monostearate, and stearyl alcohol	Benzyl alcohol
	Ketoconazole	Nizoral	Stearyl and cetyl alcohols, sorbitan monostearate, polysorbate 60, polysorbate 80	None
	Oxiconazole nitrate	Oxistat	Stearyl alcohol, polysorbate 60 NF, cetyl alcohol	Benzoic acid

(*Continued*)

Appendix I Survey of Pharmaceutical Creams (*Continued*)

Indication	Active ingredient	Cream trade name	Structure-forming excipients	Preservative
Hormone replacement	Conjugated estrogens	Premarin	Cetyl esters wax, cetyl alcohol	Benzyl alcohol
	Conjugated estrogens	Estrace	Stearyl alcohol, sodium lauryl sulfate	Methylparaben
Antibacterial	Mupirocin calcium	Bactroban	Cetomacrogol 1000, cetyl alcohol, stearyl alcohol	Benzyl alcohol
	Gentamicin sulfate	Garamycin	Propylene glycol stearate, polysorbate 40	Methylparabens, butylparaben
	Neomycin sulfate; polymyxin b sulfate, pramoxine hcl	Neosporin	Emulsifying wax	Methylparaben
	Silver sulfadiazine	Silvadene	Stearyl alcohol, isopropyl myristate, sorbitan monooleate, polyoxyl 40 stearate	Methylparaben
	Sulfathiazole, sulfacetamide, sulfabenzamide	Sultrin	Cetyl alcohol, diethylaminoethyl stearamide	Methylparabens, propylparabens
	Azelaic acid	Azelex	Cetearyl octanoate, glyceryl stearate, cetearyl alcohol, PEG-5 glyceryl stearate	Benzoic acid
Anti-inflammatory	Betamethasone valerate	Betamethasone valerate	Cetyl alcohol, ethanol (60.4%), polysorbate 60, stearyl alcohol	None
	Hydrocortisone	Cortizone 10 plus	Cetearyl alcohol, cetyl alcohol, sodium lauryl sulfate	Methylparabens, propylparabens
	Triamcinolone acetonide	Aristocort	Emulsifying wax	Benzyl alcohol
	Flurandrenolide	Cordran	Cetyl alcohol, glyceryl monostearate, polyoxyl 40 stearate	Benzyl alcohol
Anti-inflammatory	Fluticasone propionate	Cutivate	Cetostearyl alcohol, ceteth-20, isopropyl myristate	Imidurea

	Amcinonide	Cyclocort	Emulsifying wax	Benzyl alcohol
	Adapalene	Differin	Carbomer 934P, cyclomethicone, edetate disodium, methyl glucose sesquistearate, PEG-20 methyl glucose sesquistearate, phenoxyethanol	Methylparabens, propylparabens
	Pimecrolimus	Elidel	Cetyl alcohol, mono- and di-glycerides, oleyl alcohol, sodium cetostearyl sulphate, stearyl alcohol	Benzyl alcohol
	Hydrocortisone acetate, pramoxine hydrochloride	Proctofoam®-hc	Cetyl alcohol, emulsifying wax, polyoxyethylene-10 stearyl ether	Methylparabens, propylparabens
	Betamethasone dipropionate	Lotrisone	Cetearyl alcohol 70/30, Ceteareth-30	Benzyl alcohol
	Hydrocortisone butyrate	Locoid lipocream cream	Cetostearyl alcohol, Ceteth-20	Propylparabens and butylparabens
	Betamethasone dipropionate	Diprosone	Ceteareth-30; cetearyl alcohol 70/30 (7.2%)	Chlorocresol and propylene glycol
	Alclometasone dipropionate	Aclovate	Cetearyl alcohol, PEG 100 stearate	Chlorocresol
	Hydrocortisone probutate	Pandel	Stearyl alcohol, polysorbate 60, sorbitan monostearat, glyceryl monostearate, PEG-20 stearate, glyceryl stearate SE	Methylparaben, butylparabens
	Fluocinolone acetonide	Tri-luma cream	Cetyl alcohol, stearyl alcohol, PEG 100 stearate	Methylparabens, propylparabens
Antineoplastic antimetabolite	Fluorouracil	Efudex	Stearyl alcohol, polysorbate 60	Methylparabens, propylparabens
Antipruritus	Doxepin hydrochloride	Zonalon	Cetyl alcohol, isopropyl myristate, glyceryl stearate, PEG-100 stearate	Benzyl alcohol
Antiviral	Penciclovir	Denavir	Cetomacrogol 1000 BP, cetostearyl alcohol	None
Burn wounds—anti-infective	Mafenide acetate	Sulfamylon	Cetyl alcohol, stearyl alcohol, cetyl esters wax, polyoxyl 40 stearate, polyoxyl 8 stearate	Methylparaben, propylparaben

(*Continued*)

Appendix I Survey of Pharmaceutical Creams (*Continued*)

Indication	Active ingredient	Cream trade name	Structure-forming excipients	Preservative
Cold sore/fever blister treatment	Docosanol	Abreva™	Sucrose distearate, sucrose stearate	Benzyl alcohol
Depigmenting agent	Monobenzone	Benoquin	Cetyl alcohol NF	Sodium lauryl sulfate NF
Facial hair reduction for women	Eflornithine hydrochloride	Vaniqa	Ceteareth-20, cetearyl alcohol, glyceryl stearate, PEG-100 stearate, stearyl alcohol	Methylparabens, propylparabens
Humectant	Ammonium lactate	Ammonium lactate	Glyceryl stearate, polyxyl 100 stearate, polyoxyl 40 stearate, cetyl alcohol	Methylparabens, propylparabens
Immune response modifier	Imiquimod	Aldara	Cetyl alcohol, stearyl alcohol, polysobate 60, sorbitan monostearate	Benzyl alcohol, methylparabens, propylparabens
Infestation with sarcoptes scabiei (scabies) treatment	Permethrin	Elimite	Carbomer 934 P, polyoxyethylene cetyl ethers	Formaldehyde
Lice treatment	Permethrin	Nix creme rinse	Cetyl alcohol, hydroxyethylcellulose, polyoxyethylene 10 cetyl ether	Methylparabens, propylparabens
Multiple actinic (solar) keratoses	Fluorouracil	Fluoroplex	Emulsifying wax	Benzyl alcohol
Plague psoriasis treatment	Calcipotriene	Dovonex	Cetearyl alcohol, Ceteth-20	Diazolidinyl urea, dichlorobenzyl alcohol

Abbreviations: PEG, polyethylene glycol; NF, National formulary; BP, British Pharmacopeia.

Appendix II Survey of Pharmaceutical Ointments

Indication	Active ingredient	Trade name	Structure-forming excipients
Anesthetic	Lidocaine	Xylocaine	Microcrystalline wax, mineral oil, propylene glycol, white petrolatum
	Lidocaine	Lidocaine	PEG base
Antibacterial	Ciprofloxacin hydrochloride	Ciloxan	White petrolatum, propylene glycol, emulsifying wax
	Bacitracin zinc; polymyxin B sulfate	Polysporin	Propylene glycol, sorbitan sesquioleate, and white petrolatum
	Mupirocin calcium	Bactroban	Mineral oil; white petrolatum; and petrolatum
	Chloramphenicol; desoxyribonuclease; fibrinolysin	Elase-chloromycetin	Mineral oil, white petrolatum
	Gentamicin sulfate	Garamycin	Propylene glycol, sorbitan sesquioleate, and white petrolatum
	Bacitracin; neomycin sulfate, polymyxin B sulfate pramoxine hydrochloride	Neosporin plus pain relief	Liquid paraffin, microcrystalline wax, propylene glycol
Antifungal	Tioconazole	Vagistat-1	White petrolatum and mineral oil
Anti-inflammatory	Hydrocortisone acetate	Anusol	Propylene glycol, USP; propylene glycol stearate (55% monoester); white wax, NF; and white petrolatum, USP
	Triamcinolone acetonide	Aristocort	Mineral oil, petrolatum, propylene glycol
	Prednisolone acetate; sulfacetamide sodium	Blephamide	Liquid petrolatum, polyethylene
	Clobetasol propionate	Cormax	Hexylene glycol; propylene glycol stearate (55% monoester); white wax; white petrolatum
	Fluticasone propionate	Cutivate	Mineral oil; petrolatum; and white petrolatum
	Dexamethasone sodium phosphate	Decadron	Bland, unctuous petrolatum base
	Betamethasone dipropionate	Diprolene	PEG 400, PEG 6000 distearate, PEG 300, PEG 1450, and mineral oil

(*Continued*)

Appendix II Survey of Pharmaceutical Ointments (*Continued*)

Indication	Active ingredient	Trade name	Structure-forming excipients
	Mometasone furoate	Elocon	White petrolatum, mineral oil
	Fluorometholone	FML	White petrolatum, mineral oil
	Fluocinonide	Lidex	Glyceryl monostearate, white petrolatum, propylene carbonate, propylene glycol, white wax
Anti-inflammatory	Hydrocortisone acetate, pramoxine hydrochloride	Pramosone	Mineral oil and PEG
	Clobetasol propionate	Temovate	White petrolatum
	Dexamethasone; tobramycin	Tobradex	Mineral oil USP (liquid petrolatum), emulsifying wax, white petrolatum, propylene glycol
	Desoximetasone	Topicort	White petrolatum and mineral oil
	Halobetasol propionate	Ultravate	White petrolatum
	Hydrocortisone	Cortril	Aluminum stearate, beeswax, pentaerythritol cocoate, petrolatum, propylene glycol
	Amcinonide	Cyclocort	White petrolatum, magnesium aluminum silicate
	Gentamicin sulfate; Prednisolone acetate	Pred-G	White petrolatum; mineral oil; petrolatum
Anti-inflammatory/anti-infective combination	Acyclovir	Zovirax	Mineral oil, paraffin, propylene carbonate, white petrolatum and white wax
Antiviral	Tacrolimus	Protopic	White petrolatum, mineral oil
Treatment of moderate to severe atopic dermatitis	Calcipotriene	Dovonex	White petrolatum, propylene glycol, fatty alcohol citrate fatty acid pentaerythritol ester, beeswax
Treatment of plaque psoriasis			

Abbreviations: PEG, polyethylene glycol; USP, U.S. pharmacopeia; NF, National formulary.

Appendix III Survey of Pharmaceutical Pastes

Indications	Product name	Active ingredient	Structure-forming excipients
Treat ulcers of the mouth called aphthous ulcers or canker sores	Aphthasol	Amlexanox	Gelatin, glyceryl monostearate, mineral oil, pectin, petrolatum, and sodium carboxymethylcellulose
Treatment of inflamed mouth sores and mouth sores caused by injury	Oracort	Triamcinolone acetonide	Gelatin, pectin, and sodium carboxymethylcellulose in a polyethylene and mineral oil gel base
Treatment of diaper rash in babies	Desitin	Zinc oxide	Lanolin and petrolatum
Equine anthelmintic, for the prevention of strongylus vulgaris larval infestation in horses	Strongid C	Pyrantel tartrate	Dehydrated Alfalfa meal, wheat middlings, cane molasses, mineral oil

Appendix IV Survey of Pharmaceutical Gels

Indication	Active ingredient	Trade name	Gelling agents/neutralizer	Aqueous/ hydro-alcoholic
Topical treatment of acne vulgaris	Adapalene	Differin	Carbomer 940/sodium hydroxide	Aqueous
Treatment of lower extremity diabetic neuropathic ulcers	Becaplermin	Regranex	Sodium carboxy-methylcellulose	Aqueous
Topical treatment of acne vulgaris	Benzoyl peroxide	Benzac	Carbomer 940/sodium hydroxide	Aqueous
Topical treatment of inflammatory acne vulgaris	Benzoyl peroxide; Clindamycin phosphate	Duac	Carbomer 940/sodium hydroxide	Aqueous
Anti-inflammatory	Betamethasone dipropionate	Diprolene	Carbomer 940/sodium hydroxide	Aqueous
Topical treatment of interdigital tinea pedis and tinea corporis	Ciclopirox	Loprox	Dimethicone copolyol 190, carbomer 980/sodium hydroxide	Hydroalcoholic
Treatment of acne vulgaris	Clindamycin phosphate	Cleocin T	Carbomer 934 P/sodium hydroxide	Aqueous
Vitamin B_{12} supplement	Cyanocobalamin	Nascobal	Methylcellulose	Aqueous
Indicated for ripening and unfavorable cervix in pregnant women	Dinoprostone	Prepidil	Colloidal silicon dioxide, triacetin	Aqueous
Treatment of acne vulgaris	Erythromycin	Emgel	Hydroxypropyl cellulose	Aqueous
Topical treatment of acne vulgaris	Erythromycin	Erygel	Hydroxypropyl cellulose	Hydroalcoholic
Anti-inflammatory	Fluocinonide	Lidex	Carbomer 940/sodium hydroxide	Aqueous
Antibacterial	Metronidazole	Metrogel-vaginal	Carbomer 934 P, sodium hydroxide	Aqueous
Antifungal	Naftifine hydrochloride	Naftin	Carbomer 934 P/sodium hydroxide	Hydroalcoholic
Antimitotic	Podofilox	Condylox	Hydroxypropyl cellulose	Hydroalcoholic
Progesterone supplementation	Progesterone	Crinone	Carbomer 934P/sodium hydroxide	Aqueous
Anticavity toothpaste for sensitive teeth	Sodium monofluorophosphate	Sensodyne® cool gel	Cellulose gum	Aqueous
Treatment of patients with stable plaque psoriasis of up to 20% body surface area involvement	Tazarotene	Tazorac	Carbomer 934 P/tromethamine	Aqueous
Testosterone replacement therapy	Testosterone	Androgel	Carbomer 980/sodium hydroxide	Hydroalcoholic
Topical treatment of acne vulgaris	Tretinoin	Avita	Hydroxypropyl cellulose, polyolprepolymer-2	Hydroalcoholic

Appendix V Survey of Pharmaceutical Suppositories

Indication	Active ingredient	Product name	Suppository base
Pain reliever, fever reducer	Acetaminophen	Acephen	Hydrogenated vegetable oil, polyethylene glycol 40 stearate, polysorbate 80
Erectile dysfunction	Alprostadil	Muse	Polyethylene glycol 1450
Abort or prevent vascular headache	Caffeine; Ergotamine tartrate	Cafergot	Cocoa butter
Oxytocic for pregnancy termination	Dinoprostone	Prostin E2	Cocoa butter, triglycerides of fatty acids
Anti-inflammatory	Indomethacin	Indocin	Glycerin, polyethylene glycol 3350, polyethylene glycol 8000
Anti-inflammatory	Mesalamine	Canasa	Hard fat NF
Treatment of active ulcerative proctitis	Mesalamine	Rowasa	Hard fat NF
Potent opioid analgesic, relief of moderate to severe pain	Oxymorphone hydrochloride	Numorphan	Polyethylene glycol 1000 and polyethylene glycol 3350
Sedative hypnotics	Pentobarbital sodium	Nembutal	Semisynthetic glycerides
Control of severe nausea and vomiting	Prochlorperazine	Compazine	Glycerin, glyceryl monopalmitate, glyceryl monostearate, hydrogenated cocoanut oil fatty acids and hydrogenated palm kernel oil fatty acids
Perennial and seasonal allergic rhinitis	Promethazine hydrochloride	Phenergan	Ascorbyl palmitate, silicon dioxide, white wax, and cocoa butter
Antifungal	Sulfanilamide	AVC	Polyethylene glycol 400, polysorbate 80, polyethylene glycol 3350, glycerin
Antifungal	Terconazole	Terazol 3	Triglycerides of fatty acids

14
Excipients for Pulmonary Formulations

Hugh Smyth
Pharmaceutical Sciences, College of Pharmacy, University of New Mexico, Albuquerque, New Mexico, U.S.A.

INTRODUCTION

Relative to other types of drug delivery formulations, pulmonary drug delivery systems have used only a small number of excipients. However, the field is rapidly expanding and the need for alternatives has been realized. Two peaks in patent filings have been identified over the last 30 years (1), both related to the findings by Molina and Rowland (2), which implicated chlorofluorocarbons (CFCs) in the depletion of stratospheric ozone. It became clear that the propellants used in metered dose inhalers (MDIs), the most common device used for administering therapeutic aerosols, would need phasing out and replacement by non–ozone depleting alternatives. Although this transition has largely taken place, the number of excipients available for formulators to use in inhalation aerosol products remains low.

Traditionally, inhaled therapeutics have been primarily employed for the treatment and prevention of local respiratory diseases such as asthma and chronic obstructive pulmonary disease (COPD). From this perspective, the treatment of local lung diseases by using an aerosol has key advantages over other routes of administration, including targeting the site of the disease, lowered plasma concentrations, decreased systemic side effects, and rapid onset of action. More recently, the airways have also been viewed with increasing anticipation as a possible route for the administration of a wide range of systemically acting compounds, including protein and peptide therapeutics (3). It is widely anticipated that technologies for aerosol administration of insulin in the treatment of insulin-dependent diabetes mellitus will come into fruition in the near future (4). The expansion of the number of therapeutic compounds and candidate molecules for administration to the lung will undoubtedly lead to a parallel expansion in the number of excipients that will be included in approved and marketed products. Accordingly, the scientific literature and patent databases provide evidence of significant work in this area. However, aerosol administration to the lungs is not without limitations. For formulators and process engineers, pharmaceutical inhalation aerosols have been relatively difficult to develop and manufacture. The following is a nonexhaustive list of some of the foremost challenges faced with these systems:

1. Requirement of tightly controlled particle size range for administration to the lungs (1–5 μm)
2. Avoidance of throat deposition
3. Reproducible lung deposition patterns
4. Delivery of very small doses
5. Drug-blend uniformity
6. Stability of formulations
7. Packaging

Intimately related to these factors is the design of the device, formulation, and the interface with the patient. Much of the discussion below will focus on the implications of excipients on formulation challenges for inhaled aerosol products. This chapter summarizes excipients for pulmonary formulations from several perspectives: (i) excipient selection based on principles of delivery, (ii) physicochemical requirements for excipients, and (iii) specific challenges for formulations faced with aerosol drug delivery systems, including (a) biological aspects, (b) microbiological aspects, (c) analytical issues, and (d) future prospects.

OVERVIEW OF PULMONARY FORMULATIONS AND DELIVERY SYSTEMS

Metered Dose Inhalers

The popularity of pressurized MDIs (pMDIs) grew rapidly after their introduction in the late 1950s. They are currently used by over 25 million Americans for the management of a variety of diseases, such as asthma, COPD, and other lung diseases characterized by obstruction of airflow and shortness of breath. pMDIs contain active drug ingredients dissolved or suspended in a propellant, a mixture of propellants, or a mixture of solvents, propellants, and/or other excipients in pressurized aerosol canisters fitted with metering valves (Fig. 1). The aerosolization of the formulation from the pressurized container is a transient and complex process. When the metering chamber is opened, the pressurized formulation is released and expands rapidly, undergoing preatomization and forming a mixture of gas and liquid before being discharged as a jet through the orifice of the actuator. Aerosolization involves complex fluid dynamic processes that include high shear stress on the formulation and rapid propellant boiling and evaporation. Typically, a MDI product will discharge up

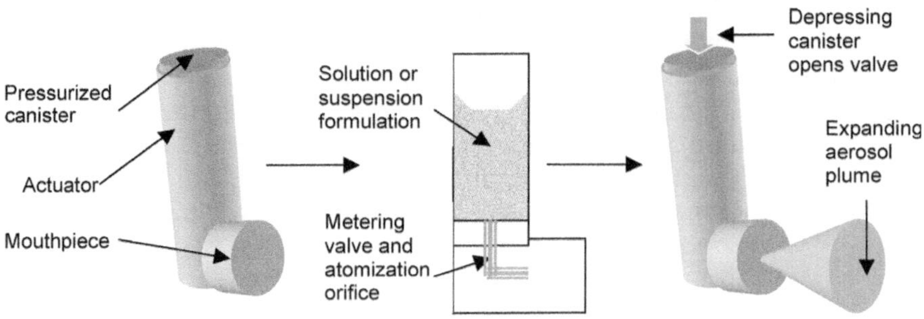

Figure 1 Schematic example of a pressurized metered dose inhaler.

to several hundred metered doses and, depending on the product, each actuation may contain from a few micrograms up to milligrams of the active drug delivered in a volume typically between 25 and 100 μL. Since their introduction, the overall design and function of pMDIs has not evolved significantly. However, environmental concerns realized in the 1970s were translated into global action with the signing of the Montreal Protocol in the mid-1980s, and marked the elimination of CFC propellants that were commonly used in pMDIs. This event has precipitated a large research effort by the industry and the academia to reformulate CFC-based systems with alternative propellant excipients. The implications of this transition are discussed below.

Dry Powder Inhalers

At present, dry powder inhalers (DPIs) are not used as commonly in the United States as are pMDIs. DPIs have been the last pharmaceutical inhalation aerosol system developed. Although the concept of operation is readily envisioned for these devices, the development of an efficient dry powder dispersion device intended for lung delivery has been notoriously difficult. Most of these devices function by using interactive mixtures of fine drug particles (1–5 μm diameter) and carrier excipient particles (usually 75–200 μm). Some evidence suggests that DPI performance is dictated largely by the physicochemical properties of the excipients used (5). However, as will be discussed, the availability of different choices of excipients is very limited, particularly in the United States.

Technical challenges have resulted in a greater variety in the design and function of DPIs relative to MDIs. Figure 2 illustrates the basic components of a passive DPI device and the simplified sequence of events that must occur to achieve aerosolization of a dry powder. Recent efforts over the last decade have focused on design of multidose inhalers that are small and as convenient as the popular pMDI devices (1). Current designs include premetered and device-metered DPIs, both of which are driven by patient inspiration alone. There is significant interest in developing an active DPI system that reduces the burden on the patients inspiratory inhalation effort for dispersing the powder dose (6). Device-metered DPIs have an internal reservoir containing sufficient formulation for multiple doses, which are metered by the device itself during actuation by the patient. Alternatively, blister packaging of the powder doses is another common strategy. The wide array of DPI designs, many with characteristics unique to the design of a specific system, present challenges

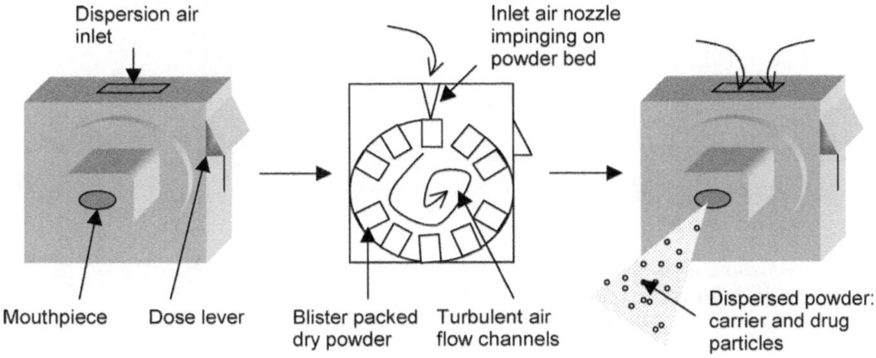

Figure 2 Schematic example of a multidose dry powder inhaler.

in developing a systematic understanding of how dry powder formulations interact with device design and patient variables. Regardless of the DPI design, the most crucial attributes are the reproducibility of the dose and particle size distribution. Maintaining these qualities through the expiration-dating period and ensuring the functionality of the device through its lifetime under patient-use conditions will probably present the most formidable challenge (7).

Nebulizers and Nasal Sprays

Aqueous sprays are used widely in marketed therapeutic inhalation aerosols as well as during product development for evaluation of efficacy and proof-of-concept studies. Furthermore, several patient groups (e.g., pediatric, intensive care, and severe asthmatic patient populations) depend on spray systems in the form of nebulizers for aerosol administration of therapeutic compounds. Three categories of aqueous inhalation aerosol systems include nebulizers, handheld liquid spray systems, and nasal sprays. The potential wide array of inhalation spray drug product designs with unique characteristics present a variety of development challenges, but there are many similarities between these products in terms of formulation and crucial excipient attributes that can be discussed in parallel. Several important output parameters can affect the delivery of the drug substance to the intended biological target for these devices and include the reproducibility of the dose, the spray plume, and the droplet size distribution.

Nebulizers are designed primarily for the atomization of aqueous formulations either as solutions or suspensions, and typically contain additional excipients. These systems are nonpressurized formulations and do not contain propellants. Traditionally, nebulizers operated using one of two basic mechanisms: jet nebulization or ultrasonic nebulization. Jet nebulizers (Fig. 3) function using the Venturi effect to

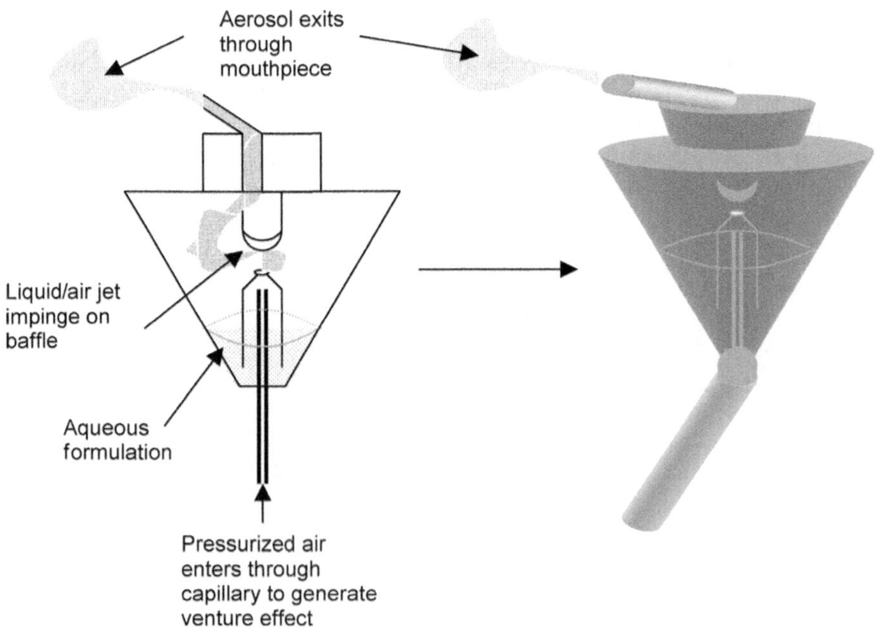

Figure 3 Schematic example of a jet nebulizer.

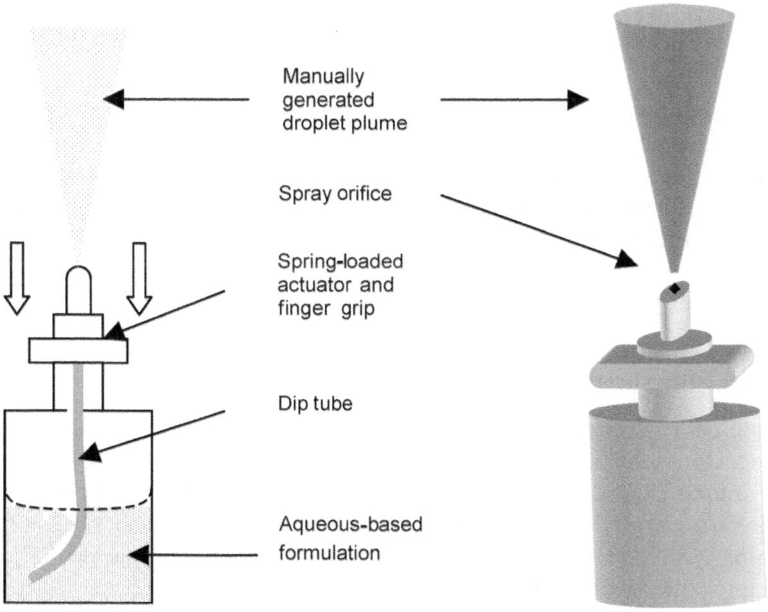

Figure 4 Schematic diagram of nasal spray devices.

draw the aqueous solution through a capillary tube and disperse droplets in air at high velocity (8). Ultrasonic nebulizers employ an oscillating, ultrasonic vibration, conveyed by means of a piezo-electric transducer to a solution creating droplets suitable for inhalation (9). Traditional nebulizers are not portable and require an external compressed air source or connection to a power supply. Recently, handheld aqueous delivery systems have been developed to overcome these problems. The mechanism of aerosol generation from these devices have recently been reviewed (10,11).

Nasal sprays are designed for delivering therapeutic agents to the nasal cavity for local and/or systemic effects (Fig. 4). Nasal spray drug products contain therapeutically active ingredients (drug substances) dissolved or suspended in aqueous solutions that often contain excipients (e.g., preservatives, viscosity modifiers, emulsifiers, and buffering agents) in nonpressurized dispensers that deliver a metered dose spray. In general, nasal sprays require unique design of formulation, container closure system, manufacturing processes, and stability (12). Although the formulation of nasal sprays is similar to that of other aqueous-based drug delivery systems, these aspects should be considered carefully during development because changes to these variables can significantly affect the performance of the product (12).

GENERAL CONSIDERATIONS FOR EXCIPIENT SELECTION FOR PULMONARY DOSAGE FORMS: EXCIPIENT USE DETERMINED VIA PRINCIPLES OF DELIVERY

The preceding section discussed the wide range of principles by which therapeutic aerosols may be generated and delivered to the patient. The excipients used in these formulations are largely determined via mechanisms by which particles and droplets can be successfully aerosolized for efficient targeting of the airways. Due to the variety of mechanisms of operation of pulmonary drug delivery systems, selection of excipients will follow distinct pathways according to the device used. Excipient selection and formulation design should be performed in parallel to device design rather than carried out independently or sequentially. Unfortunately, this has not been common practice, and, as a result, the current understanding of atomization and dispersion in inhalation aerosols is largely empirically based and often results in suboptimal performance of these systems. However, with increasing scientific and commercial interest in these systems, a mechanistic understanding is the research focus of several groups and will contribute significantly to future excipient selection, design, and formulation optimization.

Excipients Used in Pressurized Metered Dose Inhalers

pMDIs have unique characteristics for formulation composition, container design, manufacturing processes, and stability issues compared to other dosage forms. These systems combine unique excipients (i.e., propellants) with specially designed container systems, closures, and dosing components. In terms of regulatory agency control, the container, valve, actuator, formulation, accessories (e.g., spacers), and protective packaging collectively constitute the drug product (12). These factors should, therefore, always be considered together during development decision making. Formulation of pMDIs follows either a solution or suspension pathway. The choice of solution or suspension formulation type depends on the physicochemical properties of the active ingredient and the propellant system. These factors will be discussed in greater detail below. Thus, pMDI systems will typically contain two or three classes of excipients depending on the formulation strategy. The three main classes of excipients are (i) propellants, (ii) surfactants, and (iii) cosolvents. Table 1 provides common examples of the general excipient composition of marketed pMDI products.

Propellants

Propellants generally make up the largest fraction (by mass or volume) of pMDI formulations. Propellants provide the energy for atomization of the formulation and a medium for dispersion or dissolution of the active drug (19). Worldwide concern over the possible deleterious effects of CFCs on stratospheric ozone led, in 1987, to the signing of the Montreal Protocol on Substances that Deplete the Ozone Layer. This protocol committed the signatory nations to cease production of CFCs by 1996 (20). Although specific exemptions were granted for uses of CFCs that were defined essential, the pharmaceutical industry was forced to find alternative propellants for MDIs. The existing MDI propellants, CFC 11, 12, and 14 had no immediate replacements but because several hydrofluoroalkanes (HFAs) shared similar desirable characteristics (nonflammable, non–ozone depleting, chemically stable, and with similar vapor pressures), they were investigated as possible substitutes for CFCs (21). After extensive testing, the propellant tetrafluoroethane (HFA 134a) was demonstrated to have toxicology and safety profiles at least as safe as CFC propellants and has since been incorporated into MDIs approved by regulatory agencies (22).

Surfactants

Surfactants are incorporated into pMDIs for several reasons. In suspensions, surfactants have been used to stabilize the dispersion by reduction of the electrostatic charge of the micronized drug (19,23). Surfactants may help solubilize the drug

Table 1 Examples of Commonly Marketed pMDI Formulations and Their Characteristics

Therapeutic group	Drug	Surfactants/ excipients	Propellant system	Formulation type	Particle size estimate (μm)	References
Bronchodilators						
Maxair	Pirbuterol acetate	Sorbitan trioleate	CFC 11, CFC 12	Suspension	3.3	13
Maxair autohalor	Pirbuterol acetate	Sorbitan trioleate	CFC 11, CFC 12	Suspension	3.1	13
Proventil	Albuterol sulfate	Oleic acid	CFC 11, CFC 12	Suspension		
Proventil HFA	Albuterol sulfate	Oleic acid	HFA 134a, ethanol	Suspension	1.96 2.21	13, 14
Tornalate	Bitolterol mesylate	Ascorbic acid, saccharin, menthol	38% w/w ethanol, CFC 11, CFC 12,	Solution		
Ventolin	Albuterol sulfate	Oleic acid	CFC 11, CFC 12	Suspension	3.33	15
Ventolin HFA	Albuterol sulfate	None	HFA 134a	Suspension		
Corticosteriods						
Aerobid	Flunisolide	Sorbitan trioleate, menthol	CFC11, CFC 12, CFC 114	Suspension	4.14	13
Azmacort	Triamcinolone acetonide		CFC 12, 1% w/w ethanol	Suspension	4.33	16
Beclovent	Beclomethasone dipropionate	Oleic acid	CFC 11, CFC 12	Suspension	4	17
Becotide 100	Beclomethasone dipropionate	Oleic acid	CFC 11, CFC 12	Suspension		
Flovent	Fluticasone propionate		CFC 11, CFC 12	Suspension	2.5	17
QVAR	Beclomethasone dipropionate		HFA 134a, ethanol	Solution	1.0	17
QVAR autohaler	Beclomethasone dipropionate		HFA 134a, ethanol	Solution	1.0	17
Vanceril	Beclomethasone dipropionate		CFC 11, CFC 12	Suspension		
Other anti-inflammatory						
Intal	Cromolyn sodium	Sorbitan trioleate	CFC 11, CFC 12	Suspension	4.65	13
Tilade	Nedocromil sodium	Sorbitan trioleate	CFC 11, CFC 12	Suspension		

Abbreviations: MDI, metered dose inhaler; HFA, hydrofluoroalkane; CFC, chlorofluorocarbon.
Source: From Ref. 18.

and prevent crystal growth during storage in solution formulations (19). In addition, typical pMDI devices need surfactants for valve lubrication over the 100 to 400 doses (19,24). Recently, some have suggested that nonvolatile excipients in HFA systems, such as surfactants, may be useful in modifying particle size where the objective is to produce aerosols that are therapeutically equivalent to CFC predecessors (25). Currently approved surfactants for use in pMDIs include oleic acid, sorbitan trioleate, and soya-derived lecithin (26). Alternatives to these surfactants are being sought due to poor solubility in HFA propellant systems but their use is currently limited by an insufficient toxicological profile with respect to lung delivery.

Cosolvents

Cosolvents will be included in pMDI formulations for several reasons, including (i) increasing the solubility of the active compound (so that sufficient doses can be delivered using small volumes of formulation) and (ii) modulation of the internal package pressure (to influence atomization energy) (20,27). In general, ethanol is universally employed when cosolvents are required. However, some nonconventional cosolvents have been reported in the scientific literature, including diethyl ether (28) and water (29).

Dry Powder Inhalers

Similarly to pMDIs, the formulation of a DPI has a direct effect on the dosing performance and stability of the formulation. The generation of an aerosol from a static dry powder bed involves a complex interaction of gravitational forces, fluid dynamics, and interparticulate forces (electrostatic, van der Waals, capillary, and mechanical interlocking forces). Together, these interactions are responsible for the varying performance of powder formulations and devices (30). The dependence of inhaler performance on powder characteristics is poorly understood. However, it is generally recognized that subtle changes to powder physicochemical properties may have significant effects on aerosolization. These powder properties may affect processes such as blending, dose filling, powder flow, fluidization, and particle dispersion that are critical to DPI functioning.

Thus, DPIs are complex drug products and have several key distinctions from other drug products, which should be kept in mind during formulation and development of these devices. Similarly to pMDIs, the device, any protective packaging, and the formulation together constitute the drug product. This concept is integral to the development of the product because it is accepted that complex and subtle interactions may occur among the drug substance, carrier excipients, and the container/closure system, which significantly affect the safety and effectiveness of the drug product. For example, parameters critical to the performance of DPI systems include particle size distribution, particle morphology, moisture content, and adhesive and cohesive interparticulate forces between the drug substance particles, excipient particles, and device surface. These specific properties will be discussed in greater detail below.

Given the complexity that arises from the multitude of interacting variables associated with DPI systems, there are very few excipients that have been incorporated into DPI formulations. Examples of commonly marketed products are listed in Table 2. Lactose has many benefits including a well-established safety profile, low cost, and wide availability. Physicochemical properties of lactose are also relatively desirable from a DPI formulation standpoint: smooth surfaces, crystalline, and moderate flow properties. However, lactose may not be suitable for some active

Table 2 Selected Dry Powder Inhalers and Some of Their Properties

Inhaler	Company	Energy source	Carrier	Powder supply	Dosing	Doses
Rotahaler®	GSK	Passive	Lactose	Capsule	Single-dose	1
Spinhaler®	Fision/ Aventis	Passive	None	Capsule	Single-dose	1
Inhalator®	Boehringer Ingelheim	Passive	Glucose	Capsule	Multiple unit-dose	6
Diskus®/ Accuhaler®	Glaxo SmithKline	Passive	Lactose	Blister	Multiple unit-dose	60
Aerohaler®	Boehringer Ingelheim	Passive		Capsule	Multiple unit-dose	6
Diskhaler®	Glaxo SmithKline	Passive	Lactose	Blister	Multiple unit-dose	4, 8
Easyhaler®	Orion	Passive	Lactose	Reservoir	Multidose	200
Airmax™	IVAX	Passive		Reservoir	Multidose	
Novolizer®	Sofotec	Passive	Lactose	Reservoir	Multidose	200
Twisthaler®	Schering-Plough	Passive		Reservoir	Multidose	60
Turbuhaler®	AstraZeneca	Passive	None	Reservoir	Multidose	200
Spiros®	Elan Pharma-ceuticals	Impeller	N/A	Blister, cassette	Single-dose, multiple unit-dose	1, 16, or 30
Inhance™	Inhale	Comp-ressed gas	Lactose	Blister	Single-dose	1
Dynamic Powder Dis-perser™	Pfeiffer	Comp-ressed gas	Lactose	Cartridge	Multiple unit-dose	12

Source: Adapted from Ref. 18.

compounds because of its reducing sugar function (31). Also, endotoxin content and concerns over the bovine source of lactose may be considered drawbacks for lactose use (31). Alternatives have included carbohydrates, such as fructose, galactose, sucrose, trehalose, raffinose, melezitose; alditols, such as mannitol and xylitol; maltodextrins, dextrans, and cyclodextrins; amino acids, such as glycine, arginine, lysine, aspartic acid, and glutamic acid; and peptides, such as human serum albumin and gelatin (11). To mask the unpleasant taste of some inhaled drug compounds, particles containing maltodextrin and peppermint oil may be incorporated into dry powder formulations (32). One neglected aspect of the use of sugar carrier particles, particularly in antimicrobial drug therapy, is the potential for microbes to use sugars as substrates for growth.

Nebulizers and Aqueous-Based Systems

Nebulizer systems are universally aqueous in nature and can be either solution or suspension based. Excipients that have been used in nebulizer formulations relate to typical aqueous formulations and the formulation issues common to this type of preparation (i.e., stability and sterility issues). Table 3 provides several examples of nebulizer formulations that are commonly marketed along with the excipients

Table 3 Examples of Commonly Marketed Nebulizer Products and Their Excipients

Therapeutic class	Example of marketed product(s)	Form of supply	Excipients
Inhaled corticosteroids	Pulmicort respules (AstraZeneca)	Inhalation suspension: 0.25 mg/2 mL, 0.5 mg/2 mL. In single-dose envelopes	EDTA
	Budesonide		
Bronchodilators	Tornalate (Elan)	Solution for inhalation: 0.2%. In 10, 30, and 60 mL w/dropper	25% alcohol and propylene glycol
	Bitolterol mesylate		
	Alupent (Boehringer Ingelheim)	Solution for inhalation. In 10 and 30 mL w/dropper	May contain EDTA, benzalkonium chloride
	Metaproterenol Sulfate		
	Xopenex (Sepracor)	Solution for inhalation: 0.31 mg/3 mL (as base)	Preservative-free. Sulfuric acid
	Levalbuterol HCl		
Mast cell stabilizers	Intal (Aventis)	Solution for inhalation: 20 mg/2 mL	None
	Cromolyn sodium		
Mucolytics	Pulmozyme (Genentech)	In 2.5 mL ampoules	Preservative-free. With 0.15 mg/mL calcium chloride dihydrate and 8.77 mg/mL sodium chloride
	Dornase alfa (recombinant human deoxyribonuclease; DNAse)		
	Mucomyst (Apothecon)	Solution, 10% or 20% as salt sodium	May contain EDTA
	Acetylcysteine (*N*-acetylcysteine)		
Anticholinergics	Atrovent (Boehringer Ingelheim)	Solution for inhalation: 0.02% (500 μg/vial). In 25 unit dose vials per foil pouch.	Preservative free
	Ipratropium bromide		
Antiinfectives	TOBI (pathogenesis)	http://www.efactsweb.com/DFC/ Nebulizer solution: 300 mg/5 mL	With sodium chloride, sulfuric acid, and sodium hydroxide
	Tobramycin		

Abbreviation: EDTA, ethylenediaminetetraacetic acid.

included in these formulations. Generally, these excipients function as either preservatives (ethanol, benzalkonium chloride, EDTA, propylene glycol) and/or stabilizers, e.g., EDTA and buffer systems (33). For example, EDTA may inhibit microbial growth in combination with other preservatives, but is primarily added to metaproterenol (all U.S. manufacturers) and some albuterol sterile-filled unit-dose inhalant solutions to chelate metal contaminants and thus prevent solution discoloration (34).

Solution-based systems are common to both nebulizers and nasal formulations. In general, water will form the greatest fraction of the formulation, but, in some cases, cosolvents such as ethanol and propylene glycol may be added for increased stability. Acidifying and alkalizing excipients may also be added to optimize pH from the perspective of the drug stability as well as the physiological effect on the airways. Similarly, iso-osmotic and iso-tonic solutions are preferred.

For nebulizer and other aqueous aerosol products that use suspension systems, excipients are used to influence particle physical and chemical stability (e.g., microcrystalline cellulose for nasal sprays). The suitability of the physicochemical properties of these critical excipients should be thoroughly investigated and documented (12). Far more excipients have been included in formulations designed for nasal administration (Table 4).

PHYSICAL AND CHEMICAL PROPERTIES REQUIRED

Physicochemical Properties of Excipients in Pressurized Metered Dose Inhalers

Propellants

The physicochemical properties of excipients used in pMDIs are different from most dosage forms and are a derivative of the propellant system that constitutes the bulk of the formulation. The transition from CFC-based formulations to HFA-based systems has been lengthened by the historically empirical formulation approach and the dissimilarity of the physicochemical properties of the replacement HFA propellants. Both HFA 134a and HFA 227 show an increased polarity, revealed in increased dipole moments and dielectric constant. The most significant practical change has therefore been a general change in the solvency properties.

Amongst the challenges to reformulation is the increased solubility of water in HFA 134a (35), which may cause either physical or chemical instability (36–38). In addition, the solubility of drug compounds is significantly different in many cases. This has been a particularly challenging problem when the task has been to show equivalence between CFC and the replacement HFA product. For each approach, the solubility characteristics of the drug in the formulation are critical to the stability and performance of the product. In suspension pMDIs, micronized drug particles may float or sediment depending on the relative densities of the drug substance and the liquid phase of the formulation (19). Drug and excipient components have a direct effect on the extent of particle aggregation and suspendability of the micronized drug substance (19,39,40). Adherence of the drug to the walls of the container and/or valve components also influences performance of the formulation. These phenomena, as a function of time, can contribute to inconsistent medication dose delivery and particle size distribution (19). In addition, the formulation composition determines the internal pressure of an inhalation aerosol. This variable is a critical parameter related to features that influence MDI performance, i.e., particle size, evaporation rates, nonvolatile fraction, and plume velocity (20,41).

Table 4 Examples of Nasal Spray Products and Ingredients

Class of nasal product	Examples of marketed products	Active compound(s)	Common excipients
Nasal decongestants: arylalkylamines	Afrin Children's Pump Mist Rhinall Vicks Sinex Ultra Fine Mist Pretz-D	Phenylephrine HCl Ephedrine sulfate	Benzalkonium chloride Thimerosal Chlorobutanol Sodium bisulfite Camphor EDTA Eucalyptol Menthol Tyloxapol Boric acid Sodium borate
Nasal decongestants: imidazolines	12 hr Nasal Afrin 12 hr Duration Vicks Sinex Privine Otrivin Tyzine	Oxymetazoline HCl Naphazoline HCl Xylometazoline HCl Tetrahydrozoline HCl	Benzalkonium chloride Phenylmercuric acetate Glycine Sorbitol Sodium chloride Carboxymethylcellulose Microcrystalline cellulose EDTA Chlorhexidine gluconate Eucalyptol Menthol Camphor

Nasal decongestants: other products	Afrin Saline	Sodium chloride	Glycerin
	Breathe Free	Zinc acetate	Yerba santa
	Mycinaire Saline Mist	Zinc gluconate	Benzalkonium chloride
	NasalEase with Zinc	Ephedrine sulfate	EDTA
			Propylene glycol
			Polyethylene glycol
			Chlorobutanol
			Thimerosal
Nasal steroids	Nasacort	Triamcinolone acetonide	Dichlorodifluoromethane
	Beconase AQ	Beclomethasone dipropionate	Dehydrated alcohol
	Rhinocort Aqua	Budesonide	Dextrose
	Fluinsolide	Flunisolide	Polysorbate 80
	Nasarel		Benzalkonium chloride
			Phenylethyl alcohol
			EDTA
			Butylated hydroxytoluene
			Polyethylene glycol 400
			Sorbitol
Antihistamines	Astelin	Azelstine HCl	Benzalkonium chloride, EDTA
Anticholinergics	Atrovent	Ipratropium bromide	Preservative-free
Endocrine and metabolic agents	Miacalcin	Calcitonin-salmon	Sodium chloride
Antimigraine	Migranal	Dihydroergotamine	Dextrose
		Caffeine	Carbon dioxide
Vaccines	Flumist	Live attenuated virus	None
Mast cell stabilizers	Nasalcrom	Cromolyn sodium	Benzalkonium chloride
			EDTA
Smoking deterrents	Nicotrol NS	Nicotine	Parabens, EDTA

Abbreviation: EDTA, ethylenediaminetetraacetic acid.

Surfactants and Cosolvents

Similarly to the solubility of active drugs, the solubility of surfactants that were used in CFC systems has significantly changed. Surfactant solubility in HFA 134a ranges from 0.005% to 0.02% w/v, much lower than the concentration required to stabilize suspensions (0.1–2.0% w/v) (24,42). The surfactants can be solubilized with the addition of cosolvents such as ethanol. However, it is most likely that cosolvents will be incompatible with suspension formulations because drug solubility will also be promoted and crystal growth will occur.

For suspension systems, electrostatic effects have some influence on suspension stability, but steric forces may be more significant (24,43). Modifying steric forces between drug particles using surfactants has been critical in achieving stable CFC suspensions. Due to the insolubility of traditional surfactants in HFA propellants, the use of alternative surfactants or novel formulation methods may allow similar stability concerns to be addressed. For example, precoating of particles with traditional surfactants for HFA suspensions had some stabilizing effect (42). Due to the commercial interest in developing alternative surfactants, the patent literature has much greater detail of these research efforts (11,22,44,45). For example, Wright described promotion of suspension stability in HFA 227ea, using a variety of polymer and surfactant-based excipients (46). Stefely et al. (45) describe the use of oligolactic acid–based excipients (amphiphiles) in HFA systems. Superior dose uniformity performance was demonstrated in suspension HFA systems over systems without the amphiphilic excipient added. Recently, surfactant complexation methods have been described as an approach to increase solubility and also aid suspension stability (47). Williams used a cogrinding technique to improve the performance characteristics of a triamcinolone acetonide suspension in blends of HFA 134a and 227ea (48). The surfactant, Pluronic F77®, was cogrinded with the drug and suspended in the propellant system. The mass median aerodynamic diameter (MMAD) was decreased and fine particle fraction (FPF) increased at the same time as the physical stability of the suspension was promoted.

Surfactant stabilization is not always suitable or predictable. Many of the interactions of surfactants with drug particles for suspension stabilization are drug specific. Furthermore, surfactants with lowered solubility in HFA systems can be irreversibly precipitated out of solution by competing dipolar molecules such as water (35). Also, surfactant-stabilized suspensions may have suboptimal aerosolization properties (49). Accordingly, some research efforts have been directed toward developing solution-based aerosols. Several alternative formulations for an array of drugs are now marketed, and, in many cases, are solution systems (Table 1). If particle size distributions of the solution system are different from the original suspension system (as is often the case), several strategies can be employed. Droplet evaporation modifiers can be added into the formulation (i.e., surfactants) to achieve equivalence (25). Alternatively, dose modification can be sought, based on the relative differences in the lung-deposition pattern (50).

Lastly, particle engineering as a method to improve suspension stability may be an alternative. Weers et al. and Dellamary et al. describe the use of hollow porous particles to decrease the attractive forces between particles in suspension (43,51). The similarities between the particles and the dispersing medium (the propellant system enters and fills the porous particles) reduces the effective Hamaker constant that corresponds to forces of attraction, and also makes the density difference between the propellant and the particles smaller. The FPF of these aerosols was reported to be around 70%.

Several groups have investigated the effect of surfactants on emitted droplet size. In the early work performed by Polli et al., the surfactant sorbitan trioleate decreased the MMAD of the CFC dexamethasone suspension when added to the formulation (52). A suspension of terbutaline in a CFC system containing sorbitan trioleate surfactant was shown to have little change in emitted particle size when either 2.8 or 14 mg/mL of surfactant was added (53). Interestingly, the surfactant had a significant effect on the obscuration (droplet concentration) of the laser diffraction instrument used to determine particle size. Surfactants may lead to an increase in MMAD due to decreased evaporation rates from aerosol droplets. This may occur because of their tendency to associate at the air–liquid interface (54).

Cosolvents are more commonly used in HFA-based pMDI formulations due to surfactant solubility issues discussed above. This has led to an increased prevalence of solution systems that utilize almost exclusively ethanol as a cosolvent. Propellant systems that have been studied with regard to propellant-driven pMDIs include HFA 134a/ethanol, HFA 227ea/ethanol, and HFA 134a/HFA 227ea mixtures (24,55,56). These three miscible components may allow the formulator to select appropriate densities (for suspension stability), solubility characteristics (for solution and suspension formulations), and also modify the emitted particle size via nonvolatile composition effects (24,56). Density, molar volume, and vapor pressure can be used to assess intermolecular forces within mixtures, and are readily measurable in pMDIs (55). The observed densities of HFA 134a/ethanol and HFA 134a/HFA 227ea mixtures closely matched ideal mixture predictions (55). Vervaet and Byron (24) reported similar results for HFA 134a/ethanol, HFA 227ea/ethanol, and HFA 134a/HFA 227ea mixtures. However, vapor pressure behavior showed positive deviations from Raoult's law with HFA 134a/ethanol and HFA 227ea/ethanol mixtures (24,55,56). Blends of HFA 134a and HFA 227ea did not show any significant deviation from theory (55,57,58). HFA propellants have a higher affinity for the gas–liquid interface than ethanol, which in turn, is surrounded by HFA molecules. In addition, positive deviations from Raoult's law with HFA/ethanol mixtures indicate that the intermolecular forces between the components of the mixture (i.e., between ethanol and HFA molecules) are less than that between molecules of the pure constituents (55,56). The positive deviation in vapor pressure may allow formulators to use higher concentrations of ethanol (for improved solubility) without detrimental effects on droplet size or aerosolization (24).

Physicochemical Properties of Excipients in Dry Powder Inhalers

To generate a dry powder aerosol, the powder in its static state must be fluidized, deaggregated into individual particles, and entrained into the patient's inspiratory airflow. The powder is subject to numerous cohesive and adhesive forces, which must be overcome if it is to be successfully aerosolized in this schema (59). Therefore, fluidization and entrainment requires the input of energy to the static powder bed. The required energy input depends on the physicochemical nature of the powder formulation. Interparticulate forces play an essential role in flowability, mixing, deaggregation, and dispersion. Particle interactions result from "long-range" forces with bonding energies typically less than 40 kJ/mol (i.e., not short-range forces such as covalent bonding) (60). These forces are weaker than chemical bonds but their influence extends over greater distances, and thus they are termed "long-range forces." Interactions are classified as either cohesive or adhesive: cohesion referring to interactions between particles of the same material and adhesion referring to interactions between particles of different materials. These weak forces become significant when the gravitational forces acting on the particles become insignificant. This typically occurs in powder systems where particles are less than approximately 10 μm in diameter (59). Such particulate interactions can be the result of a number of concurrently acting forces that include van der Waals, electrostatic, capillary, and mechanical interlocking forces.

van der Waals forces, also known as dispersion forces (for their role in optical dispersion), are the result of instantaneous differences in the electronic configuration of molecules that give rise to dipolar characteristics. Electrostatic forces can also contribute significantly to particulate interactions. Pharmaceutical powders are usually insulators and therefore are likely to carry charge due to transfer of electrons and ions between particles. Contact between particles can result in the transfer of charge between particles that have different work functions (5). Also, triboelectrification can occur when particles undergo friction forces during movement. Capillary forces arise when water molecules condense on solid–solid interfaces. The force is proportional to the surface tension of the adsorbed liquid layer and may dominate over other forces. Mechanical interlocking occurs because pharmaceutical particulates are rarely uniformly shaped or spherical. Rough surfaces can assist in the interlocking of particles once they have come into contact.

An overview of potential excipient physicochemical factors that influence dry powder dispersion are summarized here.

Particle Size

Increased drug deposition is generally observed with smaller carrier size (61–65). Manufacture method may have significant effects depending on carrier excipient particle size. For example, poor dispersion of nedocromil was obtained using coarse carrier systems, whereas the use of fine carrier particles and high shear mixing techniques physically disrupted the drug–drug contacts and promoted deaggregation (66). Nedocromil sodium powder performance is considered to be dominated by cohesive drug–drug interactions. By decreasing the particle size of the lactose carrier, deaggregation and fine-particle drug dispersion were significantly improved. The carrier's functional effects were achieved by intercalating within the drug self-agglomerates and disrupting the cohesive drug–drug interactions (66).

Shape

Particle shape effects on formulation performance have also been investigated (67–71). Increasing the elongation of lactose increased the fine-particle fraction and dispersibility of salbutamol sulfate after aerosolization of the formulations (61,71). The particle shape of the carrier particles and the carrier surface smoothness were important in determining the extent of dispersion and deaggregation of salbutamol sulfate. Elongated carriers may have an increased duration in the airstream drag forces resulting in increased dispersibility and drug FPF (61,71). Larhrib et al. also looked at lactose elongation ratio effects on the performance of DPI formulations (67). Powder flow was adversely influenced by increasing the elongation ratio, causing poor content uniformity and irreproducible emitted-dose characteristics. However, increasing the elongation ratio of the carrier or drug improved the in vitro deposition profiles of salbutamol sulfate. Thus a balance in particle shape, maximizing dispersion while minimizing poor flow, should be sought.

Surface Roughness

Surface roughness has been shown to influence dispersion and powder performance (68,71–73). In some reports, carriers with smooth surfaces produced higher respirable fractions (71). Low respirable fractions were obtained from carriers with macroscopic surface roughness or smooth surfaces, whereas high respirable fractions were obtained from carriers with microscopic surface roughness, where smaller contact area and reduced drug adhesion occurred at the tiny surface protrusions (72). Macroscopic surface roughness is defined as roughness on the scales equal to or larger than the adhering drug particles. Conversely, microscopic roughness is the surface texture that is on scales smaller than the drug particles. Podczeck investigated the effect of preconditioning lactose to alter surface rugosity. Depending on the initial characteristics of the lactose, preconditioning to smooth the particle surfaces either worsened or improved dispersion properties. Carriers with relatively rough surfaces were reported to lose their capacity to function as a carrier (73). A positive linear trend between surface roughness of lactose and fine particle dispersion performance was reported by Chan et al. (68).

Crystallinity and Polymorphs

Lactose may be obtained in two crystalline forms: α-lactose and β-lactose (in addition to amorphous forms). The alpha form is obtained when water is incorporated into the lattice structure during crystallization (usually by supersaturation below 93.5°C) (5). Alternatively, the beta form does not contain water and exists as a non-hygroscopic and anhydrous form. Amorphous lactose is formed when either the crystallization is rapid or sufficient transient energy is introduced into the crystalline forms (74), i.e., spray drying (75), micronization and milling (76), freeze-drying, and anti-solvent crystallization (77).

Amorphous carrier particles may have significant advantages, such as increased dissolution and altered particle–particle interactions. However, the major disadvantage of amorphous content is the decreased chemical stability (74). The relatively high reactivity of amorphous solids has, therefore, led to formulations that minimize amorphous content in DPI systems. Potential problems with even small amounts of amorphous content has been discussed by Ahlneck and Zografi (78). Water adsorption in powder systems has a high probability of occurring at amorphous regions present in predominantly crystalline material (78). Thus, even small quantities of amorphous content can lead to significant changes in powder properties when exposed to low levels of moisture. However, postproduction treatment can be employed to decrease amorphous content and increase powder stability in DPI formulations. This has been achieved by conditioning the particles through exposure to a controlled environment [e.g., 35–85% relative hymidity (RH) or organic solvent vapor] to induce crystallization (68).

Ternary Blends

Ternary components are fine particles added to the binary drug–carrier mixture to modify the strong interparticulate interactions that often lead to low levels of detachment from the carrier. Ternary components, like carrier particles, must be inert and safe for inhalation. The addition of fine ternary components has increased the FPF of various drug particles. Ternary components examined include magnesium stearate, lactose (discussed below), L-leucine, polyethylene glycol (PEG) 6000, and lecithin (79–84). The mechanism of action of ternary components has not been fully elucidated, but may relate to the saturation of active sites on the carrier, electrostatic shielding, and drug redistribution on the ternary component.

Fine Particle Blends

A subset of ternary blends uses fine excipient particles (i.e., lactose, 5 μm) to enhance powder dispersibility from DPI formulations. In one study, fine lactose was either mixed with the coarse lactose carrier before adding the drug particles or was directly blended with the drug particles without the coarse carrier. Nedocromil sodium undergoes strong particle cohesion, and, in both cases, the drug particles were more easily dispersed from the powder formulation due to reduced cohesion force (85). Recently, a study investigating the effect of removing lactose fines from the carrier system demonstrated that fine particle fraction decreased significantly with reduction in lactose fines (86).

Nebulizers and Aqueous-Based Systems

Aqueous formulations for nebulization generally follow the same principles as those adopted for the development of parenteral products. Excipient selection is based upon achieving pH, osmolarity, and sterility suitable for deposition in the lung. In addition, the stability of formulations, whether solution or suspension based, follows typical guidelines adopted for any aqueous delivery system. Isotonicity is typically achieved by adding buffer salts or sodium chloride (9). Preservatives such as EDTA and benzalkonium chloride have been included in many formulations, but there has been some concern over the possible physiological implication of acute and chronic exposure to these compounds (9). Thus, to manufacture sterile aqueous-based oral inhalation solutions and suspensions, the unit-dose production and packaging in sealed nebules or similar packaging is recommended to prevent microbial contamination. The use of preservatives or stabilizing agents in inhalation spray formulations is discouraged. If these excipients are included in a formulation, their use should be justified by assessment in a clinical setting to ensure the safety and tolerability of the drug product (12).

Ethanol

For inhaled systemically acting drugs, one of the major issues for successful commercialization is the need to improve bioavailability for efficacy, economic, and safety purposes (87). Recently, ethanol has received significant attention for this purpose due to its ability to increase membrane permeability (87). For example, insulin suspensions of up to 9% (w/v) in ethanol were aerosolized with a commercial compressor nebulizer to rats, resulting in increased insulin plasma levels and no detectable acute toxicity (88). In addition, the physicochemical properties of ethanol are thought to have potential benefits in (i) stabilizing tertiary and quaternary protein structure, (ii) preventing microbiological contamination, (iii) delivering a greater drug mass due to absence of solubility limits, (iv) giving rise to high solubility of lipophilic compounds owing to nonpolarity, and (v) permeation enhancing ability (87). However, Edwards et al. also state that several key questions regarding ethanol toxicity, lung epithelia permeability, and molecular weight limitations still need to be addressed (87).

FUTURE CHALLENGES AND OPPORTUNITIES

Analytical Issues

Chemical analysis of excipient materials and trace-level impurities is a challenging task for pharmaceutical chemical analysis. The development of hybrid chromatography techniques such as liquid chromatography–mass spectroscopy, and also combination with magnetic resonance spectroscopy have facilitated better understanding of complex profiles of excipients (89). Specific excipients often require chemical profiling. For example, lecithin used in pMDI formulations may contain various proportions of phosphatidyl choline, triglycerides, fatty acids, and carbohydrates (89,90). Similar analytical issues are present for the detection, identification, and quantification of extractable and leachables in inhaled products. Because inhaled products are packaged and administered from plastic, rubber, and metal materials, the components of these may be extracted into the solvents and excipients of the formulation (91). The Food and Drug Administration in the United States has produced a draft guidance document that states "methods must be developed and validated" for the evaluation of potential extractables (12).

Protein and Peptide Formulations

One of the main drivers for the development of new pulmonary drug delivery systems has been the potential for noninvasive systemic delivery of protein and peptide compounds. The systemic delivery of macromolecules via the airways would overcome the inconvenience and cost associated with current methods of administration (injection), and appears likely given the large surface area of the airways and the thin pulmonary epithelium. Most research has concentrated on pulmonary delivery of insulin for the treatment of diabetes. Recently, one insulin product has completed phase three studies and is now undergoing review by European regulatory agencies for marketing approval.

Specific formulation strategies need to be employed for macromolecule compounds. An excellent review of protein stability in aqueous solutions has been published by Chi et al. (92). In addition to solution stability of proteins and peptides, aerosolization may result in significant surface interfacial destabilization of these compounds if no additional stabilization excipients are added. This is due to the fact that protein molecules are also surface active and adsorb at interfaces. The surface tension forces at interfaces perturb protein structure and often result in aggregation (92). Surfactants inhibit interface-induced aggregation by limiting the extent of protein adsorption (92).

For dry powder formulations of proteins and peptides, the rate of degradation of a particular molecule is a function of the formulation, manufacturing process, packaging, and storage conditions. Protein and peptide drugs can be stabilized by synthetic and formulation techniques. Synthetic techniques include modification of the protein/peptide molecule by cross-linking, adding cation and anion binding sites, and amino acid substitution (93). Formulation techniques that improve protein and peptide stability involve the addition of stabilizing excipients such as glass-forming compounds (94). Protein stabilization by an amorphous sugar has been hypothesized to occur by two possible mechanisms (94). The water substitution hypothesis assumes that in order to maintain higher order protein structure, sugar molecules form hydrogen bonds with dried proteins in place of water molecules. The glassy state theory believes that the high viscosity of an amorphous sugar retards molecular movement, thus preventing physical or chemical degradation. Formulation of proteins in the solid state has been considered more stable than liquid formulations, because many degradation problems observed in solution formulations can be avoided. However, the influence of environmental factors, including temperature, humidity, and light on the physical and chemical stability of protein/peptide powders should be evaluated during formulation development.

Novel Excipients

With an expanding market and new therapeutic entities, inhalation aerosols will likely see a parallel increase in excipients that have not been traditionally used in approved products. The decision to include nontraditional excipients into new products, and bear some risk of potential regulatory delays and product failure due to safety issue, will be facilitated by products that have clear therapeutic benefit and economic value, such as protein and peptide treatments. In pMDIs, for example, the transition to HFA propellants has generated considerable intellectual property covering novel surfactant excipients (47). In addition to basic formulation improvements afforded by these excipients, controlled release and permeation modifiers may also be sought in the future, though these may be far from the marketing stage. DPIs are equally restricted at present for choice of excipients. However, alternative carrier excipients have been investigated (31,95).

SUMMARY

Design and development of inhaled products is the synthesis of several different fields combining the pharmacology of the active substance with formulation and device design activities and the physiology of the airways. Each of these factors is critical to the success of the delivery system and therefore represents a multidisciplinary approach to achieving optimal drug delivery. The properties of the drug and excipients are central to the type of inhaler system chosen. Due to the unique mechanisms by which particles for inhalation are generated, broad expertise on the principles by which different aerosol generators operate is necessary. This understanding is linked to the physicochemical properties of the excipients, and formulators must be aware of subtle changes that give rise to significant differences in aerosolization performance. In addition, a broad understanding of these concepts will facilitate the development of the most appropriate device/formulation for each individual drug compound. In the future, it appears additional excipients will be necessary for inhalation aerosol systems. Design of these will also be directed by a mechanistic understanding of the current systems.

REFERENCES

1. Clark A. Pulmonary delivery technology: recent advances and potential for the new millennium. In: Hickey AJ, ed. Pharmaceutical Inhalation Aerosol Technology. 2d ed. New York: Marcel Dekker, 2004.
2. Molina MJ, Rowland FS. Stratospheric sink for chlorofluoromethane: chlorine atom catalyzed destruction of ozone. Nature 1974; 274:810–812.

3. Patton JS. Deep-lung delivery of therapeutic proteins. Chem Tech 1997; 27(12):34–38.
4. Edwards DA. Delivery of biological agents by aerosols. AICHE J 2002; 48(1):2–6.
5. Zeng XM, Martin AP, Marriott C. Particulate Interactions in Dry Powder Formulations for Inhalation. London: Taylor and Francis, 2001.
6. Crowder TM. Vibration technology for active dry-powder inhalers. Pharma Technol 2004; 52–61.
7. FDA. Guidance for industry metered dose inhaler (MDI) and dry powder inhaler (DPI) drug products. CDERGUID|2180dft.wpd, Nov 1998, http://www.fda.gov/cder/guidance/2180dft.pdf, 1998.
8. Byron PR, Dalby RN, Hickey AJ. Optimized inhalation aerosols. I. The effects of spherical baffle size and position upon the output of several pressurized nonaqueous suspension formulations. Pharma Res 1989; 6(3):225–229.
9. Dalby RN, Hickey AJ, Tiano SL. Medical devices for the delivery of therapeutic aerosols to the lungs. In: Hickey AJ, ed. Inhalation Aerosols. New York, New York: Marcel Dekker, Inc., 1996:441–473.
10. Garcia-Contreras L, Smyth HDC. Dry-powder and liquid spray systems for inhaled delivery of peptide and proteins. Am J Drug Delivery 2005; 3(1):29–45.
11. Crowder TM, Louey MD, Sethuraman VV, Smyth HDC, Hickey AJ. 2001: an odyssey in formulation and design. Pharm Technol 2001; 25(7):99–113.
12. FDA. Guidance for industry: metered dose inhaler (MDI) and dry powder inhaler (DPI) drug products. Draft guidance 1998; www.fda.gov/cder/guidance/index.htm.
13. Tzou TZ. Aerodynamic particle size of metered-dose inhalers determined by the quartz crystal microbalance and the Andersen cascade impactor. Int J Pharm 1999; 186(1):71–79.
14. Cheng YS, Fu CS, Yazzie D, Zhou Y. Respiratory deposition patterns of salbutamol pMDI with CFC and HFA-134a formulations in a human airway replica. J Aerosol Med 2001; 14(2):255–266.
15. Wilkes W, Fink J, Dhand R. Selecting an accessory device with a metered-dose inhaler: variable influence of accessory devices on fine particle dose, throat deposition, and drug delivery with asynchronous actuation from a metered-dose inhaler. J Aerosol Med 2001; 14(3):351–360.
16. Warren SJ, Farr SJ. Formulation of solution metered dose inhalers and comparison with aerosols emitted from conventional suspension systems. Int J Pharma 1995; 124:195–203.
17. Stein SW. Size distribution measurements of metered dose inhalers using Andersen Mark II cascade impactors. Int J Pharma 1999; 186(1):43–52.
18. Smyth HDC, Garcia-Contreras L, Cooney DJ, Garmise RJ, Jones LD, Hickey AJ. Medical and pharmaceutical aerosols. In: Ruzer LS, Harley NH, eds. Aerosols Handbook: Measurement, Dosimetry, and Health Effects. Boca Raton: CRC Press, 2005:265–289.
19. Purewal TS. Formulation of metered dose inhalers. In: Purewal TS, Grant DJW, eds. Metered Dose Inhaler Technology. Buffalo Grove, Illnois: Interpharm, 1998.
20. Smyth HDC. The influence of formulation variables on the performance of alternative propellant-driven metered dose inhalers. Adv Drug Delivery Rev 2003; 55(7): 807–828.
21. Atkins P. Chlorofluorocarbon to hydrofluoroalkane formulations: an industry perspective. J Allergy Clin Immunol 1999; 104(6):S268–S270.
22. Ross DL, Gabrio BJ. Advances in metered dose inhaler technology with the development of a chlorofluorocarbon-free drug delivery system. J Aerosol Med 1999; 12(3):151–160.
23. Johnson KA. Interfacial phenomena and phase behavior in metered-dose inhaler formulations. In: Hickey AJ, ed. Inhalation Aerosols. New York: Marcel Dekker, 1996: 385–415.
24. Vervaet C, Byron PR. Drug-surfactant-propellant interactions in HFA-formulations. Int J Pharm 1999; 186(1):13–30.
25. Brambilla G, Ganderton D, Garzia R, Lewis D, Meakin B, Ventura P. Modulation of aerosol clouds produced by pressurized inhalation aerosols. Int J Pharm 1999; 186(1): 53–61.

26. McDonald KJ, Martin GP. Transition to CFC-free metered dose inhalers—into the new millennium. Int J Pharm 2000; 201(1):89–107.
27. Gupta A, Stein SW, Myrdal PB. Balancing ethanol cosolvent concentration with product performance in 134a-based pressurized metered dose inhalers. J Aerosol Med 2003; 16(2):167–174.
28. Tiwari D, Goldman D, Malick WA, Madan P. Formulation and evaluation of albuterol metered dose inhalers containing tetrafluoroethane (P134a), a non-CFC propellant. Pharma Dev Technol 1998; 3:163–174.
29. DeStefano G, Kelash-Cannavo LJ, inventors; Boehringer Ingelheim Pharmaceuticals, Inc., U.S., assignee. Formulation for a Metered Dose Inhaler Using Hydrofluoroalkanes as Propellants. US Patent 2004184994, 2004.
30. Ashurst I, Malton A, Prime D, Sumby B. Latest advances in the development of dry powder inhalers. Pharm Sci Technol Today 2000; 3(7):246–256.
31. Steckel H, Bolzen N. Alternative sugars as potential carriers for dry powder inhalations. Int J Pharma 2004; 270(1–2):297–306.
32. Buskeyfield LJ, Jay GE, Wright P, inventors; Fisons, PLC, assignee. Pharmaceutical Inhalation Compositions. US Patent US 5,607,662, 4 Mar 1997.
33. Rowe RC, Sheskey PJ, Weller PJ. Handbook of Pharmaceutical Excipients. APhA Publications, 2003.
34. Asmus MJ, Sherman J, Hendeles L. Bronchoconstrictor additives in bronchodilator solutions. J Allergy Clin Immunol 1999; 104(S53–S60):215–229.
35. Blondino FE, Byron PR. Surfactant dissolution and water solubilization in chlorine-free liquefied gas propellants. Drug Dev Ind Pharm 1998; 24(10):935–945.
36. Williams RO III, Hu C. Influence of water on the solubility of two steroid drugs in hydrofluoroalkane (HFA) propellants. Drug Dev Ind Pharm 2001; 27(1):71–79.
37. Williams RO III, Hu C. Investigation of moisture scavengers in pressurized metered-dose inhalers. STP Pharma Sci 2000; 10(3):243–250.
38. Williams RO III, Hu C. Moisture uptake and its influence on pressurized metered-dose inhaler. Pharma Dev Technol 2000; 5(2):153–162.
39. Byron PR, Blondino FE, inventors; The Center for Innovative Technology, USA; Virginia Commonwealth University, assignee. Metered Dose Inhaler Formulations Which Include the Ozone-Friendly Propellant HFC 134a and Suspending, Solubilizing, Wetting, Emulsifying or Lubricating Agent. US Patent 5492688, 1996.
40. Byron PR, Miller NC, Blondino FE, Visich JE, Ward GH. Some aspects of alternative propellant solvency. Paper presented at: RDD IV, 1994.
41. Stein SW, Myrdal PB. A theoretical and experimental analysis of formulation and device parameters affecting solution MDI size distributions. J Pharma Sci 2004; 93(8): 2158–2175.
42. Byron PR. Dosing reproducibility from experimental albuterol suspension metered-dose inhalers. Pharm Res 1994; 11(4):580–584.
43. Weers JG, Tarara TE, Gill H, English BS, Dellamary LA. Homodispersion technology for HFA suspensions: particle engineering to reduce dosing variance. Paper presented at: RDD VII, 2000.
44. Bowman PA, Greenleaf D. Non-CFC metered dose inhalers: the patent landscape. Int J Pharm 1999; 186(1):91–94.
45. Stefely JS, Duan DC, Myrdal PB, Ross DL, Schultz DW, Leach C. Design and utility of a novel class of biocompatible excipients for HFA-based MDIs. Paper presented at: RDD VII, 2000.
46. Wright P. Polymers for the colloidal stabilization of drugs in HFAs. Paper presented at: RDD IV, 1994.
47. Stefely JS. Novel excipients for inhalation drug delivery: expanding the capability of the MDI. Drug Delivery Technol 2002; 2(6):62, 64–69.
48. Williams RO, Repka MA, Barron MK. Application of cogrinding to formulate a model pMDI suspension. Eur J Pharma Biopharma 1999; 48:131–140.

49. Hickey AJ, Dalby RN, Byron PR. Effects of surfactants on aerosol powders in suspension. Implications for airborne particle size. Int J Pharma 1988; 42:267–270.
50. Leach CL, Davidson PJ, Boudreau RJ. Improved airway targeting with the CFC-free HFA-beclomethasone metered-dose inhaler compared with CFC-beclomethasone. Eur Respir J 1998; 12(6):1346–1353.
51. Dellamary LA, Tarara TE, Smith DJ, et al. Hollow porous particles in metered dose inhalers. Pharm Res 2000; 17(2):168–174.
52. Polli GP, Grim WM, Bacher FA, Yunker MH. Influence of formulation on aerosol particle size. J Pharma Sci 1969; 58:484–486.
53. Baeckstroem K, Nilsson PG. Measurements of droplet size distributions from metered dose inhalers with different vapor pressures and contents of surfactant. J Aerosol Sci 1988; 19(7):1097–1100.
54. Dalby RN, Byron PR. Comparison of output particle size distributions from pressurized aerosols formulated as solutions or suspensions. Pharm Res 1988; 5(1):36–39.
55. Tzou TZ. Density, excess molar volume, and vapor pressure of propellant mixtures in metered-dose inhalers: deviation from ideal mixtures. Respir Drug Delivery VI, Int Symp 1998:439–443.
56. Smyth HDC, Mejia-Millan EA, Hickey AJ. The effect of ethanol on solvency vapor pressure, and emitted droplet size of solution metered dose inhalers containing HFA 134a. Respir Drug Delivery VIII 2002; 2:735–738.
57. Williams RO III, Repka M, Liu J. Influence of propellant composition on drug delivery from a pressurized metered-dose inhaler. Drug Dev Ind Pharm 1998; 24(8):763–770.
58. Williams RO III, Liu J, Repka MA. Optimization of drug delivery from a MDI with propellants 134a and 227. Proc Int Symp. Controlled Release Bioact Mater 1997; 24: 745–746.
59. Hickey AJ. Pharmaceutical inhalation aerosol powder dispersion: an unbalancing act. Am Pharma Rev 2003; 6(4):106–110.
60. Israelachvili J. Intermolecular and Surface Forces. 2d ed. London: Academic Press, 1992.
61. Zeng XM, Martin GP, Marriott C, Pritchard J. The effects of carrier size and morphology on the dispersion of salbutamol sulphate after aerosolization at different flow rates. J Pharm Pharmacol 2000; 52(10):1211–1221.
62. Steckel H, Müller BW. In vitro evaluation of dry powder inhalers II: Influence of carrier particle size and concentration on in vitro deposition. Int J Pharma 1997; 154:31–37.
63. Kassem NM, Ho KKL, Ganderton D. The effect of airflow and carrier size on the characteristics of an inspirable cloud. J Pharma Pharmacol 1989; 41:14P.
64. French DL, Edwards DA, Niven RW. The influence of formulation on emission, deaggregation and deposition of dry powder for inhalation. J Aerosol Sci 1996; 27(5): 769–783.
65. Ganderton D. Generation of respirable clouds from coarse powder aggregates. J Biopharma Sci 1992; 3:101–105.
66. Clarke MJ, Tobyn MJ, Staniforth JN. The formulation of powder inhalation systems containing a high mass of nedocromil sodium trihydrate. J Pharm Sci 2001; 90(2): 213–223.
67. Larhrib H, Martin GP, Marriott C, Prime D. The influence of carrier and drug morphology on drug delivery from dry powder formulations. Int J Pharm 2003; 257(1–2): 283–296.
68. Chan LW, Lim LT, Heng PW. Immobilization of fine particles on lactose carrier by precision coating and its effect on the performance of dry powder formulations. J Pharm Sci 2003; 92(5):975–984.
69. Crowder TM, Rosati JA, Schroeter JD, Hickey AJ, Martonen TB. Fundamental effects of particle morphology on lung delivery: predictions of Stokes' law and the particular relevance to dry powder inhaler formulation and development. Pharma Res 2002; 19:239–245.

70. Zeng XM, Martin GP, Marriott C, Pritchard J. Lactose as a carrier in dry powder formulations: the influence of surface characteristics on drug delivery. J Pharma Sci 2001; 90(9):1424–1434.
71. Zeng XM, Martin GP, Marriott C, Pritchard J. The influence of carrier morphology on drug delivery by dry powder inhalers. Int J Pharma 2000; 200:93–106.
72. Kawashima Y, Serigano T, Hino T, Yamamoto H, Takeuchi H. Effect of surface morphology of carrier lactose on dry powder inhalation property of pranlukast hydrate. Int J Pharma 1998; 172:179–188.
73. Podczeck F. The influence of particle size distribution and surface roughness of carrier particles on the in vitro properties of dry powder inhalations. Aerosol Sci Technol 1999; 31(4):301–321.
74. Buckton G. Characterization of small changes in the physical properties of powders of significance for dry powder inhaler formulations. Adv Drug Delivery Rev 1997; 26:17–27.
75. Harjunen P, Lehto VP, Martimo K, et al. Lactose modifications enhance its drug performance in the novel multiple dose Taifun DPI. Eur J Pharm Sci 2002; 16(4–5):313–321.
76. Williams RO III, Brown J, Liu J. Influence of micronization method on the performance of a suspension triamcinolone acetonide pressurized metered-dose inhaler formulation. Pharm Dev Technol 1999; 4(2):167–179.
77. Larhrib H, Martin GP, Prime D, Marriott C. Characterization and deposition studies of engineered lactose crystals with potential for use as a carrier for aerosolized salbutamol sulfate from dry powder inhalers. Eur J Pharma Sci 2003; 19(4):211–221.
78. Ahlneck C, Zografi G. The molecular basis of moisture effects on the physical and chemical stability of drugs in the solid state. Int J Pharm 1990; 62:87–95.
79. Louey MD, Stewart PJ. Particle interactions involved in aerosol dispersion of ternary interactive mixtures. Pharma Res 2002; 19(10):1524–1531.
80. Zeng XM, Martin GP, Tee SK, Ghoush AA, Marriott C. Effects of particle size and adding sequence of fine lactose on the deposition of salbutamol sulphate from a dry powder formulation. Int J Pharma 1999; 182:133–144.
81. Lucas P, Anderson K, Potter UJ, Staniforth JN. Enhancement of small particle size dry powder aerosol formulations using an ultra low-density additive. Pharma Res 1999; 16(10):1643–1647.
82. Zeng XM, Martin GP, Tee SK, Marriott C. The role of fine particle lactose on the dispersion and deaggregation of salbutamol sulphate in an air stream in vitro. Int J Pharma 1998; 176:99–110.
83. Staniforth JN, inventor; Coordinated Drug Development Limited, assignee. Carrier Particles for Use in Dry Powder Inhalers. US Patent US 6,153,224, 28 Nov 2000.
84. Staniforth JN. Improvements in dry powder inhaler performance: surface passivation effects. Paper presented at: Drug Delivery to the Lungs VII, 1996:London.
85. Lucas P, Clarke MJ, Anderson K, Tobyn MJ, Staniforth JN. The role of fine particle excipients in pharmaceutical dry powder aerosols. Paper presented at: Respiratory Drug Delivery VI, 1998.
86. Islam N, Stewart PJ, Larson I, Hartley P. Lactose surface modification by decantation: are drug-fine lactose ratios the key to better dispersion of salmeterol xinafoate from lactose-interactive mixtures? Pharm Res 2004; 21(3):492–499.
87. Edwards DA, Valente AX, Man J, Tsapis N. Recent advances related to the systemic delivery of therapeutic molecules by inhalation. In: Hickey A, ed. Pharmaceutical Inhalation Aerosol Technology. 2d ed. New York: Marcel Dekker, 2004:541–550.
88. Choi WS, Murthy GG, Edwards DA, Langer R, Klibanov AM. Inhalation delivery of proteins from ethanol suspensions. Proc Natl Acad Sci USA 2001; 98(20):11103–11107.
89. Norwood DL, Qiu F. Strategies for the analysis of pharmaceutical excipients and their trace level impurities. Am Pharma Rev 2004; 7(5):92–99.
90. FDA. Metered dose inhaler (MDI) and dry powder inhaler (DPI) drug products chemistry, manufacturing, and controls documentation. CDER/FDA, 1998.

91. Howlett D, Colwell J, Goldsmith S, McCallion ON. Correlation of extractables and leachables from marketed pMDIs. Paper presented at: Respiratory Drug Delivery VIII, 2002.
92. Chi EY, Sampathkumar K, Randolph TW, Carpenter JF. Physical stability of proteins in aqueous solution: mechanism and driving forces in nonnative protein aggregation. Pharma Res 2003; 20(9):1325–1336.
93. Manning MC, Patel K, Borchardt R. Stability of protein pharmaceuticals. Pharm Res 1989; 6:903–918.
94. Immamura K, Iwai M, Ogawa T, Sakiyama T, Nakanishi K. Evaluation of hydration states of protein in freeze-dried amorphous sugar matrix. J Pharm Sci 2001; 90:1955–1963.
95. Smyth HDC, Garmise RJ, Cooney DJ, Zimmerer RO, Pipkin JD, Hickey AJ. Influence of physical form of Captisol® particles on performance as a dry powder aerosol carrier. Paper presented at: Respiratory Drug Delivery IX, Palm Desert, California, 2004.

15
Synergistic Combinations of Penetration Enhancers and Their Discovery by High-Throughput Screening

Pankaj Karande, Amit Jain, and Samir Mitragotri
Department of Chemical Engineering, University of California, Santa Barbara, California, U.S.A.

INTRODUCTION

Interaction of exogenous molecules with human skin has evoked continuous interest in the scientific community. The concept of topical drug administration on human skin has been prevalent since the rise of early human civilization. In the recent past, however, this concept has been extended beyond delivery of medicaments to localized skin areas. Human skin is now viewed as an interface for systemic delivery of therapeutic molecules. This route, the transdermal drug delivery route, offers several advantages compared to conventional modes of drug delivery such as needles and oral drug delivery. Transdermal delivery offers painless and sustained administration of systemic therapeutics while eliminating first-pass hepatic metabolism and gastrointestinal degradation commonly associated with oral administration. The skin, which has evolved as a natural protective barrier of an organism, strictly regulates the transport of molecules in and out of the body. It is not surprising that most candidate therapeutics find it difficult to overcome this natural barrier to transport. Several physical and chemical methods have been studied to perturb or compromise the skin barrier to promote the flux of therapeutics into or across the skin (1–5). The prime requirements of any such technique should be: sufficient barrier disruption for optimal flux and quick reversibility of physiological functions at the site of disruption after termination of the treatment. Iontophoresis (6–15), sonophoresis (16–24), electroporation (25–31), and microneedles (32–37) are some of the physical techniques used to compromise the barrier function of skin. Chemical penetration enhancers (CPEs) (38–47), liposomes (48–52), ethosomes (53–55), transferosomes (56,57), niosomes (58,59), and emulsions (60–63) are some of the chemical methods used for enhancing transdermal drug delivery.

Use of CPEs provides greater design flexibility in tailoring the delivery vehicle and results in a simple Band-Aid®-like delivery device. This method almost completely eliminates the use of any external physical gadgetry and is therefore cost effective and more patient compliant. This chapter provides general comments on the design and testing of CPEs for transdermal as well as topical formulations. Although elaborate discussion will be congruent to transdermal delivery applications, the reader will find the general principles of formulation design applicable to topical as well as other mucosal forms of drug delivery such as nasal and buccal drug delivery.

BACKGROUND

Skin is a complex, multilayered organ designed specifically to inhibit foreign molecules from entering systemic circulation. This barrier resides in the most superficial layer, the stratum corneum (SC). The SC is 15 μm in thickness and is composed of dead cells and lipid bilayers (64,65). An important requirement of a CPE is its ability to cause local changes in the SC, thereby promoting movement of therapeutic molecules across it. Equally important is the pharmacological inactivity of this CPE toward the therapeutic species itself. In the last six decades, extensive effort has been directed toward the search of chemicals that meet these criteria. Unfortunately, most CPEs studied till date are relatively weak. A few CPEs that can achieve significant barrier disruption are not specific in their action toward the dead strata of SC and show activity in the subsequent layers made of live cells and tissue. This nonspecific disruption behavior results in adverse physiological responses such as irritation or adverse systemic responses such as toxicity (46,66–69). In addition to favorable physiological and systemic response, a CPE should also exhibit predictability and reversibility of response, short lag time of response, and finally cosmetic and aesthetic acceptability. No single CPE has been discovered so far that exclusively meets all these criteria. It is not surprising then that the transdermal route has been limited to only 11 low-molecular-weight lipophilic drugs that do not require any significant impetus for transport across skin (3).

The lack of significant impact of CPEs on transdermal delivery vehicles is related to the inherent nonspecific activity of CPEs in the different strata of the skin, as discussed earlier. This limitation may be overcome by utilization of mixtures of CPEs. Research has already shown that binary mixtures of CPEs provide increased permeation enhancement as well as increased safety compared to single enhancers. Such unique chemical combinations, called synergistic combinations of penetration enhancers or SCOPE formulations, offer new opportunities in transdermal drug delivery (46).

The use of skin permeation enhancers in combination for synergistic effects has been studied in the transdermal literature (70). Such "synergistic" methods can be grouped in three categories: (i) combination of two physical methods, e.g., ultrasound and iontophoresis (71–75); (ii) combination of a physical method with a chemical enhancer, e.g., use of ultrasound with sodium lauryl sulfate or isopropyl myristate (76–80); and (iii) combination of two chemicals, e.g., terpenes and propylene glycol (46,81–88). Numerous studies have been published on using combination of two physical methods or use of a physical method in conjunction with a chemical enhancer. Use of a physical method, by itself or in combination with another physical method, increases application cost for delivery purposes as mentioned before. In addition, there are unexplored safety and membrane recovery issues associated with these methods. A few reports have also been published on the use of a mixture of chemical enhancers for enhancing transdermal delivery. Typically, such studies use a chemical enhancer, e.g., terpenes, along with a solvent such as propylene glycol or isopropyl myristate, which increases the thermodynamic activity of the enhancer in the solvent (88,89). This activity may be related to the increased solubilization of the enhancer in the solvent or increased partitioning in the skin. There is, however, no methodical or rational basis for the selection of these enhancers in conjunction. In addition, there is no report on systematic design of such enhancer mixtures to optimize their skin permeation performance. SCOPE formulations on the other hand are based on the peculiar and unique physical chemistry of interaction between the enhancers in the mixture. Also, the discovery of SCOPE formulations is based on a systematic experimentation of enhancer mixtures over a wide range of concentrations and compositions (46).

In the next section, we discuss the practical difficulties and challenges associated with designing such SCOPE formulations.

CHALLENGES IN DESIGNING MULTICOMPONENT CHEMICAL PENETRATION ENHANCER FORMULATIONS

In the last six decades, extensive research in the field of CPE discovery has led to identification of more than 300 potential penetration enhancers. These enhancers can be classified, based on their structure and chemical functionality, into more than 20 different categories. CPEs in different chemical classes show different enhancement and skin irritation. The exact molecular forces responsible for these observed effects remain more or less a mystery. Considerable contributions can be found in the transdermal literature from the works of several prominent investigators, on the mechanism of action of different CPEs on skin (87,90–92). However, a unified model to quantify a priori the exact enhancement and irritation response of any given single enhancer on skin remains absent. In this perspective, the efforts to predict the effect of CPE combinations on skin are severely set back. Coupled with this is an equally challenging problem of modeling the chemistry of enhancer interactions in the formulation phase. Enhancers distributed in various chemical classes interact differently with each other resulting in myriad different species exhibiting polydispersity in concentration, composition, and chemical behavior. Desired responses of lowered irritation and increased permeation enhancement are likely to occur in a very narrow range of chemical compositions of the involved components. Consequently, truly synergistic formulations are expected to exist very rarely. The only feasible way of identifying SCOPE formulations is by brute force testing of all possible combinations of CPEs. Also, because SCOPE formulations are likely to occur in very narrow range of chemical compositions, one needs to map activity of multicomponent formulations at reasonably discrete and yet continuous intervals of possible composition ranges (46). The designing of such multicomponent formulations is discussed next.

DESIGNING MULTICOMPONENT FORMULATIONS

This section outlines the general parameters and rules in the designing of multicomponent formulations. These rules should be generic to the design of multicomponent formulations used in any application. We start with writing down mathematical equations for designing these formulations.

Single-Component Formulations

For designing single-component formulations, we have two degrees of freedom (DOF): choice of component and the chemical potential or concentration of the component in formulation. Let us assume that the candidate pool of possible components is made of "n" different chemicals.

Individual enhancers in this pool may be represented as N_i and the candidate pool as

$$\mathrm{N} \in \{N_i\}, \ (i = 1 - n)$$

Each component may be studied at "c" different chemical potentials or concentrations. If C_i represents the discrete levels of concentrations at which the component is studied in the formulation, then

$$\mathrm{C} \in \{C_i\}, \ (i = 1 - c)$$

The discrete concentrations in the set C may be selected based on different considerations specific to the particular application of the formulation. For example, for transdermal delivery applications, we initially selected concentrations in a narrow range of 0% to 2% wt/vol.

The order of the formulation is represented by "o," which for single-component formulations is simply 1. The total number of single-component formulations that can be designed can be mathematically expressed as:

$$F = {}^{n}C_{o} \sum_{i=1}^{c} i$$

Binary Formulations

For designing binary formulations, we have two DOF in addition to those that define single-component formulations: choice of second component and the composition of at least one component in the mixture. Fixing the composition of the first component automatically fixes the composition of the second component as:

$$X_{\mathrm{A}} = 1 - X_{\mathrm{B}}$$

Where A and B are the two components of the binary formulation. X_{A} and X_{B} are mass fractions of A and B, respectively. The compositions X_{A} and X_{B} can be discretized in intervals between 0 and 1. This discretization may be linear and represented, for example, as

$$X \in \{0, 0.1, 0.2, 0.3, 0.4, 0.5, 0.6, 0.7, 0.8, 0.9, 1.0\},$$

or logarithmic and represented, for example, as:

$$X \in \{0, 0.00001, 0.0001, 0.001, 0.01, 0.1, 1.0\},$$

or can be represented using any other mathematical scheme of choice. This discretization depends partly on the resolution at which the binary combination needs to be studied and partly on the dependence of activity on composition. Any a priori information on dependence of activity on composition of the components in the binary mixture can be used to better design the composition intervals of X_{A} or X_{B}. As we know that SCOPE formulations occur in a very narrow range of chemical compositions, a finer discretization is preferred. Finer discretization also implies increased number of formulations.

Let "*s*" represent the total number of compositions at which a binary combination is studied. For a linear interval with step size of 0.1, *s* is equal to 11. For a logarithmic scheme shown above, *s* is equal to 7.

For a binary formulation, the order of formulation, $o = 2$. The total number of binary formulations of A and B that can be designed can be mathematically expressed as:

$$F = {}^{n}C_{o}\, s \sum_{i=1}^{c} i$$

Figure 1A shows a typical binary formulation made of components A and B studied at four different total concentrations of 0.5%, 1.0%, 1.5%, and 2.0% wt/vol. At each concentration, the composition of A and B is varied from 0 to 1 in steps of 0.1. This results in 44 different formulations. All these formulations can be represented on a two-dimensional (2-D) phase map as shown in Figure 1B, where activity of each formulation is represented as a color, with light gray representing lowest activity and dark gray representing highest activity. This activity may be enhancement, irritation, or any relevant property of choice for a particular application.

The phase map shown in Figure 1B represents the skin permeation enhancement activity of the formulations containing binary mixtures of lauryl sarcosinate and sorbitan monolaurate at different concentrations and compositions. The region of maximum activity lies in a very narrow range of compositions. For such a nonlinear activity–composition behavior, it is very important to probe the binary phase map at as fine a resolution as possible, thus increasing the experimentation volume.

Ternary Formulations

For formulations made of three components, we have six DOF: choice of three enhancer components, one total concentration, and composition of at least two components independent of each other. Fixing the composition of two components automatically fixes the composition of the third component as:

$$X_{\mathrm{C}} = 1 - (X_{\mathrm{A}} - X_{\mathrm{B}})$$

where, A, B, and C represent the three components of the ternary formulation. The compositions X_{A}, X_{B}, and X_{C} can be varied from 0 to 1 in linear or logarithmic intervals as indicated previously.

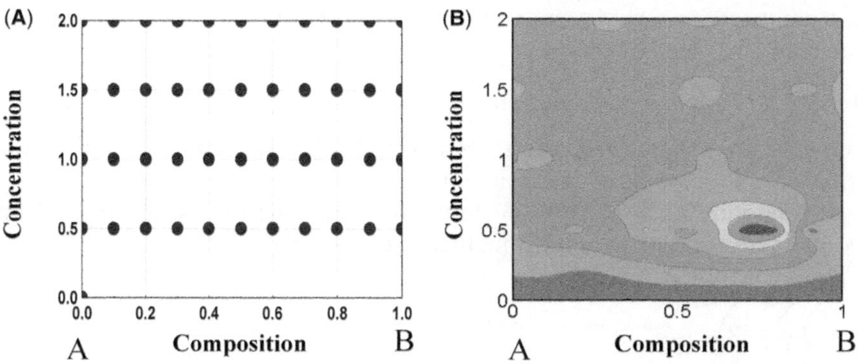

Figure 1 (**A**) Design of binary formulations. (**B**) Activity phase map of a binary combination of chemical penetration enhancers. Dark gray indicates highest skin permeabilization and light gray indicates lowest skin permeabilization.

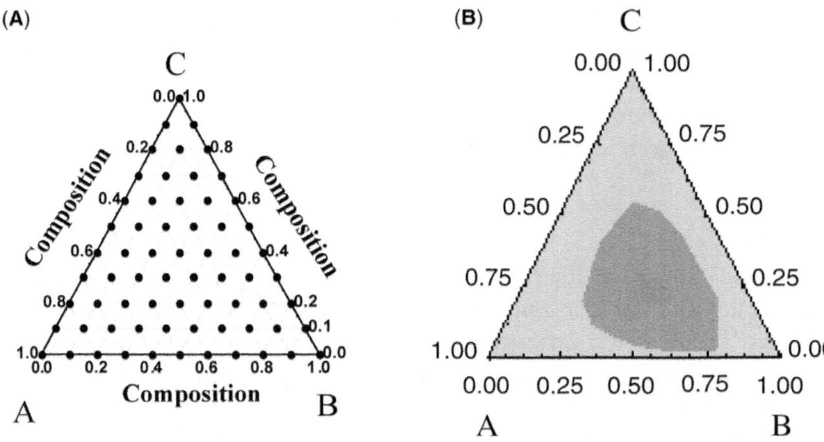

Figure 2 (**A**) Design of a ternary formulation. (**B**) Activity phase map of a ternary combination of chemical penetration enhancers. Dark gray indicates highest skin permeabilization and light gray indicates lowest skin permeabilization.

The order of a ternary formulation is $o=3$. The total number of ternary formulations that can be designed by combining three components A, B, and C can be expressed mathematically as:

$$F = {}^{n}C_{o} \sum_{i=1}^{c} i \sum_{j=1}^{s} j$$

Figure 2A shows a typical ternary formulation with varying compositions of the three components at a given total concentration. Sixty-six ternary formulations can be designed at a fixed total concentration by varying the compositions in step sizes of 0.1. To represent a ternary combination completely at all possible concentrations and compositions would require more than one 2-D phase map. Figure 2B represents the activity (enhancement, irritation, etc.) at each formulation of the combination as a color with light gray representing lowest activity and dark gray representing the highest activity.

As the order of the combination increases, the number of possible formulations that can be designed increases exponentially. Also, it becomes increasingly difficult to represent comprehensively the entire phase behavior on 2-D plots. In general, the DOF increases as

$$\text{DOF} = 2 \times \text{Order of formulation}$$

So, for a quaternary formulation, there are eight DOF: four enhancers, one total concentration, and at least three independent compositions.

The total number of quaternary formulations that one can design from four different components can be expressed mathematically as:

$$F = {}^{n}C_{o} \sum_{i=1}^{c} i \sum_{j=1}^{s} \sum_{k=1}^{j} k$$

where $o=4$

It is impossible to represent a quaternary or higher order formulation as a 2-D phase map.

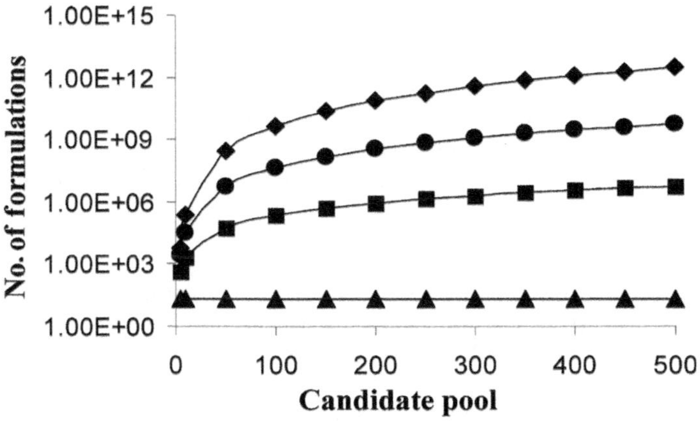

Figure 3 Dependence of experimentation volume on the size of candidate pool. Number of single-component (*triangles*), binary (*square*), ternary (*circles*), and quaternary (*diamonds*) formulations designed from a given candidate pool (*X-axis*).

Writing down such mathematical expressions provides us with the experimentation volume required to completely characterize a test pool of components in multicomponent formulations. It also provides a systematic approach for design of experiments and data interpretation. Using the above mathematical expressions, we can estimate the number of experiments required to characterize a test pool of candidate enhancers for transdermal drug delivery formulations as a function of the size of the test pool.

Figure 3 represents the number of formulations that can be designed from single component, binary, ternary, and quaternary combinations of enhancers as a function of the size of the enhancer pool. The current CPE pool has about 300 enhancers. To completely characterize this candidate pool for discovery of SCOPE formulations would require testing of $O(10^3)$ single-enhancer formulations, $O(10^6)$ binary formulations, $O(10^9)$ ternary formulations, and $O(10^{12})$ quaternary formulations. These numbers are based on use of four total enhancer concentrations and varying compositions linearly in step sizes of 0.1. Conventional tools used to map effect of enhancer formulations on skin permeability typically allow for about 10 experiments a day (93,94). With such low experimental throughputs, screening such large number of formulations will be an impossible task. To screen a million binary formulations arising from a modest pool of 30 enhancers would require approximately 300 years. Also, the current candidate pool of enhancers is by no means comprehensive or complete. As we advance our knowledge of the interaction chemistry between enhancers and skin constituents, more enhancers will be added to this list.

It is then obvious that we need a high-throughput screening system that can test a reasonable fraction of these formulations in a relatively short amount of time.

DESIGNING A HIGH-THROUGHPUT SCREENING ASSAY FOR TESTING TRANSDERMAL FORMULATIONS

Franz diffusion cells (FDC) remain the workhorse of all permeation experiments in transdermal studies. FDCs use the permeation of a solute, assessed by high pressure liquid chromatography (HPLC) or radiation, to evaluate the effect of penetration enhancers on skin permeation. These experiments are cumbersome, have long holdup times, and require manual sampling. A semiautomated version of FDCs has been developed to reduce manual sampling; however, the throughput of FDCs still remain low (10 experiments a day). As a result, FDCs become impractical when used for screening a large library of formulations of $O(10^6)$.

Any high-throughput assay used for screening of transdermal formulations should meet the following requirements:

1. *Ability to screen a large number of formulations*
   To map a reasonable experimental space of $O(10^6)$, formulations would require O(100) years with FDCs. Increasing the throughput by at least two to three orders of magnitude would result in significant improvement in the effort and time spent in the very first stage of formulation development.
2. *Use of a surrogate end point that is quick and easy to obtain*
   Permeation experiments using a radiolabeled, fluorescent, HPLC-detectable, or radio immuno assay/enzyme linked immuno sorbent assay-detectable marker necessitate the need of extensive sample handling and sample analysis. This accentuates the cost of sample analysis and overall time spent in characterizing the efficacy of formulations. Furthermore, current state of the art fluidics systems put a fundamental limit on the number of samples handled in a given time.
3. *Low incubation times to further increase the throughput and hence time efficiency*
   FDC experiments typically use incubation times of 48 to 96 hours, thereby reducing the throughput of permeation experiments. Low incubation times favor high turnover frequencies for assay utilization.
4. *Use of an end point that is not dependent on the physicochemical properties of the model permeant*
   Permeation of a model solute across skin in presence of an enhancer is dependent not only on the inherent capacity of the enhancer to permeabilize skin but also on the physicochemical interactions of the enhancer with the model solute. An end point used to characterize the effect of an enhancer on skin permeability should be able to decouple these two effects. This assures the generality of the results.
5. *Minimal use of test chemicals and efficient utilization of model membrane such as animal skin*
   FDCs typically require 1 to 2 mL of enhancer formulations. This makes it cost prohibitive to include candidates that are expensive in the test libraries. FDCs also generally use about 3 to 4 cm^2 of skin per experiment. This makes screening $O(10^6)$ formulations both cost prohibitive and resource intensive.
6. *Adaptability to automation to reduce human interference*
   Typical FDC setup requires manual sampling with little opportunities for process automation.

In addition to these requirements of the assay tool, the high-throughput screening methodology should also satisfy, if possible, the following experimental constraints:

7. *Use of a common model membrane to represent human skin*
   It is common to find in transdermal literature the use of a variety of different models to represent human skin such as rat skin, pig skin, excised human skin, etc. Whereas human skin is difficult to procure on a large

scale, animal models show deviations in permeability characteristics from human skin. Also, results on one model cannot be directly translated to a different model.

8. *Use of consistent thermodynamic conditions for enhancer formulations* Permeation enhancement efficacy of a CPE is a function of its chemical potential, temperature, pressure, and cosolvent amongst other thermodynamic parameters. These thermodynamic conditions need to be standardized for all the enhancers that are being tested to create direct comparison of their efficacies in increasing skin permeation.

With these general requirements in mind, we have developed a rapid high-throughput assay for screening the skin permeabilization potential of transdermal formulations. This assay is called "in vitro skin impedance guided high-throughput screening" abbreviated as INSIGHT (46).

IN VITRO SKIN IMPEDANCE GUIDED HIGH-THROUGHPUT SCREENING

A schematic of the INSIGHT screening tool is shown in Figure 4A. The INSIGHT screen consists of two polycarbonate or Teflon plates, each 12.7 mm thick (84). The top plate consists of a square matrix of 100 wells (each 3 mm in diameter). The bottom plate contains a symmetric matrix of 100 wells. The wells in the top plate act as the donors, and the wells in the bottom plate act as the receiver chambers in a FDC. Screening of formulations is performed using pigskin as a model. The skin is sandwiched between the donor and receiver plates. The SC is exposed to the test formulations in the donor. The receiver wells are filled with phosphate buffered

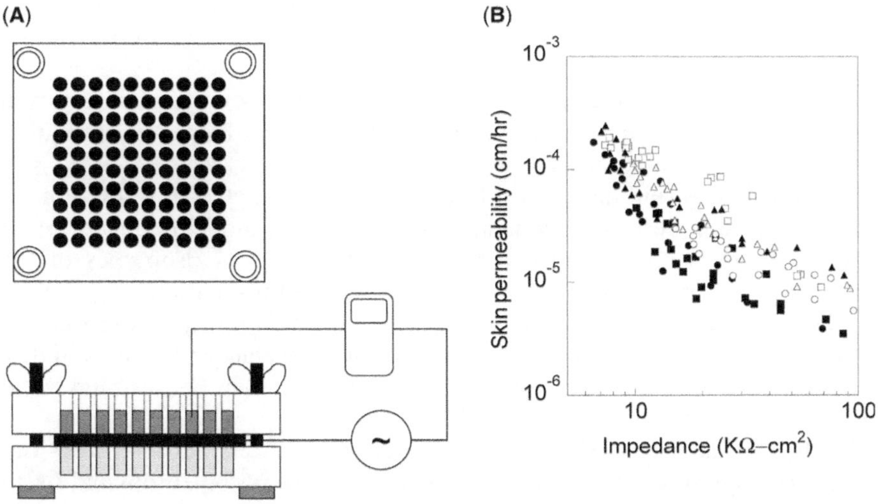

Figure 4 (**A**) Schematic of in vitro skin impedance guided high-throughput screening tool, INSIGHT. (**B**) Skin permeability skin impedance correlation for 1.5% wt/vol. menthol (*open squares*), 1.5% wt/vol. lauric acid (*filled triangles*), 1.5% wt/vol. Brij® 35 (*open circles*), 1.0% wt/vol. Lauryl sarcosinate (*filled circles*), 1.0% wt/vol. sorbitan monolaurate (*open triangles*), and 1.5% wt/vol. benzyldodecyldimethyl ammonium chloride (*filled squares*) in 1:1 PBS:EtOH. *Abbreviation*: PBS:EtOH, phosphate buffered saline:ethanol.

saline (PBS) to keep the skin hydrated during the entire duration of the experiment. The donor and the receiver plates are clamped using four screws. Skin impedance in each well is recorded using two electrodes; a common electrode, which is a hypodermic needle, in the dermis and a second electrode placed sequentially into each donor compartment. An AC signal, 100 mV root mean square at 100 Hz, is applied across the skin with a waveform generator (Agilent 33120A, Palo Alto, California, U.S.A.). Conductivity measurements are performed using a multimeter (Fluke 189, Everett, Washington, U.S.A.) with a resolution of 0.01 μA. Data acquisition parameters (frequency and amplitude of electric potential and number of measurements) were selected to provide maximum resolution and reproducibility in the data. Current measurements are recorded at time 0 and time 24 hours. The conductivity enhancement ratio (ER) at 24 hours is defined as the ratio of current reading at time 24 hours to that at time 0 hours. Each formulation is repeated at least four times on four different skin samples for statistical purposes. It was verified in an independent study that there was no influence of formulations in any particular well on the impedance measurements in the neighboring wells. In other words, lateral diffusion does not contribute or contributes insignificantly to impedance measurements.

With reference to the design criteria highlighted in the previous section, INSIGHT provides the following advantages over FDCs:

1. INSIGHT can assay 1000 to 1500 formulations per day, a 100-fold increase over the screening rate obtained by FDCs (46).
2. INSIGHT uses the fundamental correlation between the electrical and permeability properties of skin. Skin permeability shows a strong correlation with skin impedance, as shown in Figure 4B. Figure 4B shows 150 independent and simultaneous measurements of mannitol skin permeability and skin impedance for six different enhancer formulations. The relationship between skin impedance and permeability to hydrophilic solutes confirms that the former can be used as a surrogate measure for the later. Skin conductance is quick and easy to obtain and does not require additional sample handling and analysis.
3. Skin impedance measured in INSIGHT at 24 hours correlates very well with skin inulin permeability at steady states (Fig. 5). This reduces incubation times, as compared to FDCs.
4. INSIGHT uses skin impedance to evaluate the efficacy of formulations in increasing skin permeability as against the use of a model permeant. As a result, there is no dependence of the physicochemical properties of the model permeant on the measured efficacy of the formulation.
5. INSIGHT uses approximately 80 μL of formulation per experiment, thereby reducing material cost associated with expensive test candidates. Also, INSIGHT uses only, approximately, 0.1 cm^2 skin per test, a 40-fold increase in skin utilization over FDCs.
6. INSIGHT has been designed to easily adapt to process automation. Conductance measurements and sample loading can both be automated using standard robotics.
7. Porcine skin is used as a model membrane in all the screening experiments performed with INSIGHT. Porcine skin closely resembles human skin and serves as an excellent model membrane (95–98).
8. All screening experiments performed with INSIGHT are designed at consistent thermodynamic conditions.

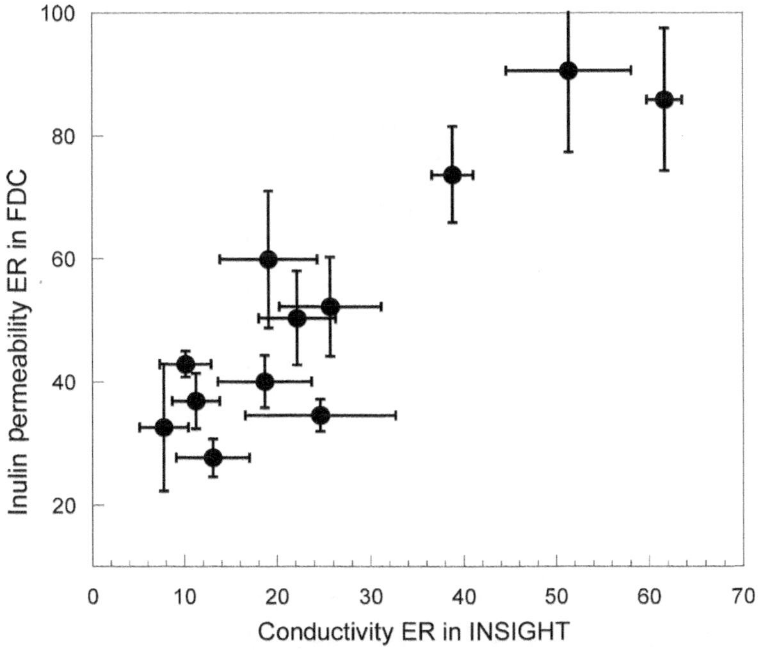

Figure 5 Plot of conductivity enhancement ratios (ERs) in INSIGHT at 24 hours versus permeability ERs in FDC for 12 enhancer formulations. A strong linear correlation indicates the validity of observations in INSIGHT when compared with those from traditional tools such as FDC. *Abbreviations*: INSIGHT, in vitro skin impedance guided high-throughput screening; FDC, Franz diffusion cell.

VALIDATION OF IN VITRO SKIN IMPEDANCE GUIDED HIGH-THROUGHPUT SCREENING WITH FRANZ DIFFUSION CELLS

Conductivity enhancement measurements in INSIGHT were verified against permeability enhancements obtained in FDCs using inulin as a model permeant. Figure 5 shows a plot of conductivity ER from INSIGHT against skin permeability ERs obtained in FDCs for several different enhancers using inulin as a model permeant. Permeability ER is defined as the ratio of skin permeability obtained from an enhancer formulation to the passive skin permeability of inulin. All enhancer formulations were prepared in 1:1 phosphate buffered saline:ethanol. A strong correlation between these values suggests that predictions made by INSIGHT on the potency of enhancers are essentially the same as those obtained in FDCs.

APPLICATIONS OF IN VITRO SKIN IMPEDANCE GUIDED HIGH-THROUGHPUT SCREENING

Further improvements in the INSIGHT screening speed can be obtained by reducing the formulation incubation period. Capabilities of INSIGHT in assessing formulation potency after four-hour incubation are demonstrated in Figure 6, where potency ranking of 438 single and binary formulations, randomly prepared from the enhancer

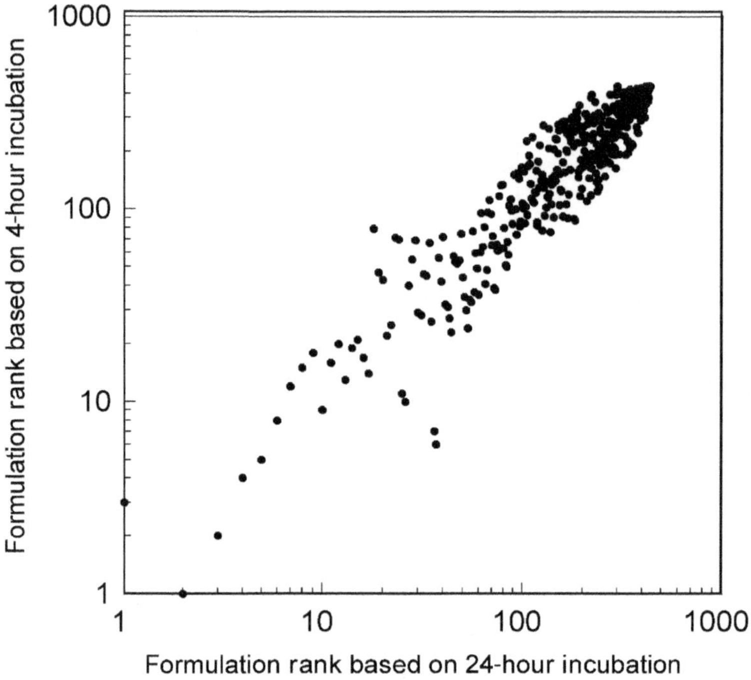

Figure 6 Plot of 24-hour predictions in (INSIGHT) versus 4-hour predictions in INSIGHT on the potency of enhancer formulations. A strong correlation indicates that predictions on potency of formulations can be obtained at significantly lower incubation periods of four hours.

library, based on four-hour screening is compared to that based on 24-hour screening. Rank 1 corresponds to most potent formulation in the library and rank 438 to the weakest formulation. The predictions of potency made in four hours were consistent with those made after a contact time of 24 hours, thus indicating that the efficiency of INSIGHT screening can be further improved. This increased throughput of formulation screening will now make several avenues, previously unexplored, feasible for better design of transdermal formulations.

A few of these are discussed in the next section.

DISCOVERY OF RARE ENHANCER COMBINATIONS

Identifying enhancer combinations that enhance skin permeability to macromolecules is a challenging task. Furthermore, selection of combinations that do so safely is even more challenging. Clearly, potent and safe enhancer combinations exist rarely. Figure 7A shows the distribution of ER values for approximately 5000 formulations obtained by INSIGHT screening. These formulations were generated by combining 32 individual CPEs. Details of the enhancer library and leading hits are reported in Karande et al. and are not discussed here (46). Figure 7A clearly shows that the percent of randomly generated enhancer combinations that exhibit ER above a certain threshold decreases rapidly with increasing threshold. The inset shows a section of the main figure corresponding to high ER values. Less than 0.1% of formulations exhibited more than 60-fold enhancement of skin conductivity. Discovery of such rare formulations by brute force experimentation is contingent on the

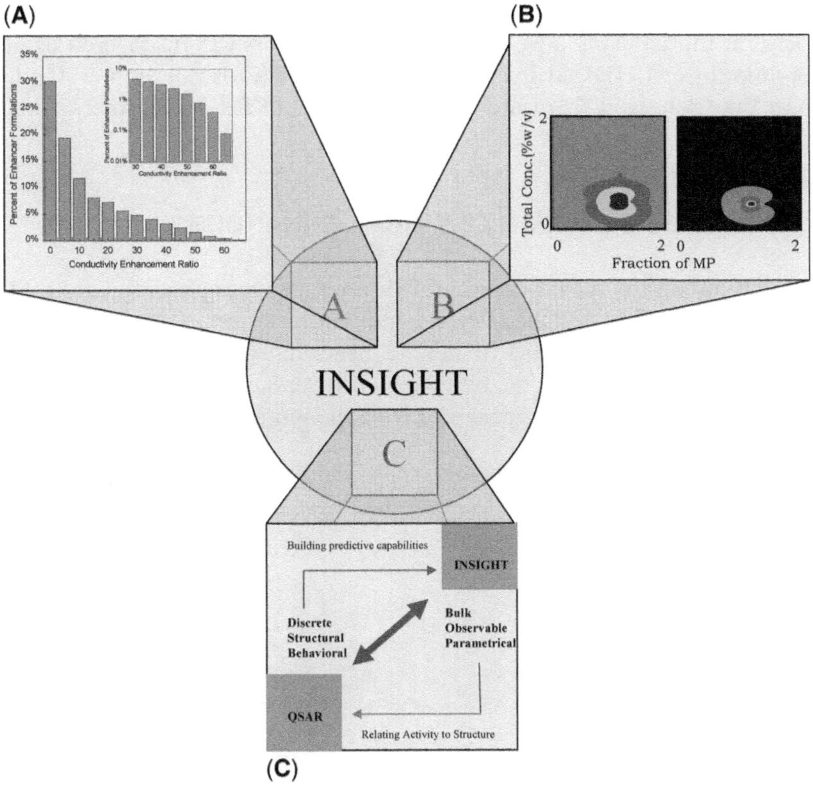

Figure 7 (**A**) Discovery of rare enhancer formulations that are significantly potent in increasing skin permeability. Such formulations are difficult to discover using traditional tools such as Franz diffusion cells due to their low experimental throughput. The success rate of discovering these potent formulations is very small (approximately 0.1%) requiring a tool with high experimental throughput. (**B**) INSIGHT screening is used to quantify the extent of interaction between components of CPEs mixtures in terms of synergy. Regions of high synergy almost always overlap with regions of high potency. (**C**) INSIGHT screening is used to generate a large volume of data on the interaction of CPEs with skin. The information is used to relate chemistry of the enhancer to its potency using quantitative descriptions of structure–activity relation. *Abbreviations*: INSIGHT, in vitro skin impedance guided high-throughput screening; CPEs, chemical penetration enhancers.

throughput of the experimental tool. INSIGHT, the most efficient known tool for screening transdermal formulations, opens up the possibility of discovering such rare formulations.

One of the formulations discovered by INSIGHT, sodium laureth sulfate:phenyl piperazine (SLA:PP), was shown to increase the permeability of macromolecules such as inulin across porcine skin 80- to 100-fold compared to passive skin permeability of inulin. SLA:PP also increased the skin permeability of molecules such as methotrexate, low-molecular-weight heparin, leutenizing hormone–releasing hormone (LHRH), and oligonucleotides 50- to 100-folds. Figure 8A shows the correlation between passive skin permeability and molecular weight of a permeant (*open circles*). Typically, skin permeability decreases with increasing molecular weight, and after about 500 Da, it plateaus off under any relevant therapeutic threshold. However, in the presence of SCOPE formulations, the skin permeability remains independent of molecular weight (*closed circles*). Figure 8B shows the in vivo data on the delivery of LHRH in hairless rat model using a SCOPE formulation, SLA:PP, discovered using INSIGHT. The amount of LHRH delivered using the SCOPE formulation is significantly more than that delivered from a control solution and lies in the therapeutic window of LHRH.

EXPLORING SYNERGIES BETWEEN CHEMICAL ENHANCERS

A number of studies have shown that certain CPEs interact synergistically and offer enhancement higher than that induced by its individual components (46,70,76). Synergies between CPEs not only lead to new transdermal formulations but also potentially offer insight into mechanisms by which CPEs enhance skin permeability. Prediction of synergies from the first principles is challenging. INSIGHT screening

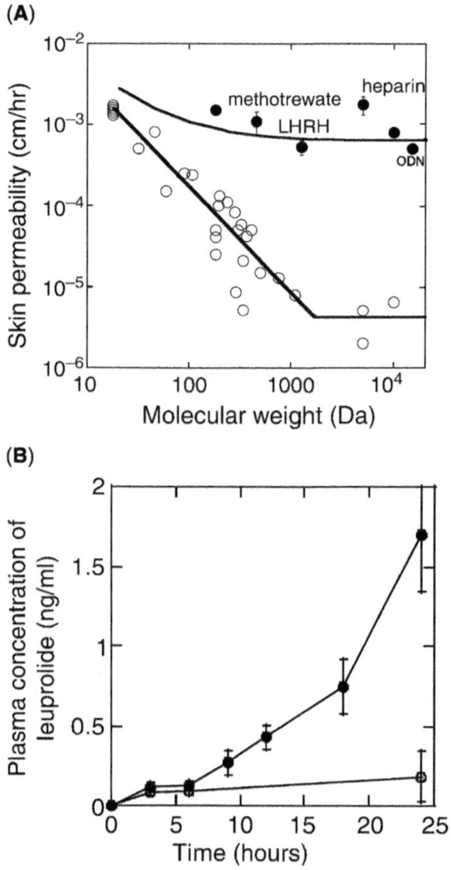

Figure 8 (**A**) In vitro permeability of candidate drug molecules in the presence of synergistic combinations of penetration enhancers (SCOPE) formulations. *Open circles* indicate passive skin permeability and *closed circles* indicate skin permeability in the presence of SCOPE formulations as a function of the molecular weight of the solute. (**B**) In vivo delivery of leuprolide acetate, a synthetic analogue of LHRH in hairless rat model. *Y*-axis shows blood plasma concentration of leuprolide acetate as a function of time for control formulation (*open circles*) and SCOPE formulation (*closed circles*). *Abbreviation*: LHRH, luteinizing hormone-releasing hormone.

offers an effective tool for identifying synergies (positive or negative) between the CPEs. Synergy can be quantified by a parameter, S, indicative of the "extent of interaction" between the two penetration enhancers as follows:

$$S = \frac{\mathrm{ER}_{\mathrm{A+B}}^{X,Y}}{X\,\mathrm{ER}_{\mathrm{A}}^{Y} + (1 - X)\,\mathrm{ER}_{\mathrm{B}}^{Y}}$$

where $\mathrm{ER}_{\mathrm{A+B}}^{X,Y}$ is the ER obtained with a formulation containing two penetration enhancers A and B at a total concentration of Y% wt/vol. and X weight fraction of A; $\mathrm{ER}_{\mathrm{A}}^{Y}$ and $\mathrm{ER}_{\mathrm{B}}^{Y}$ are the ERs obtained with pure components A and B, respectively, at the same total concentration Y.

Figure 7B shows a synergy map and enhancement activity map for combination of methyl pyrrolidone (MP) and S20. The X-axis on the synergy map represents the composition of the formulation (weight fraction of MP), and the Y-axis represents the total concentration at which the MP and S20 are present in the formulation (1:1 EtOH:PBS). The map is color coded, with dark gray representing the highest synergy ($S = 6$) between the enhancers and light gray representing the lowest synergy ($S = 1$). At 0.5% wt/vol. total concentration and 0.6 weight fraction of MP, the combination of MP and S20 is 6.2 times more potent than the weighted average of their individual components. The regions of high synergy overlap with the regions of high potencies, indicating that high permeabilization capacity may be attributed to high synergy.

GENERATING DATABASE FOR STRUCTURE–ACTIVITY CORRELATIONS

Looking beyond searching for potent combinations of enhancers, the sheer volume of information generated via INSIGHT screening on the behavior of a wide variety of penetration enhancers will provide, for the first time, a platform to build further investigation of the fundamental aspects of enhancer–skin interactions. Quantitative descriptions of structure–activity relations for CPEs, which have had limited success in the past, may lead to better outcomes in light of the availability of large volumes of data collected in a consistent manner (99). As exemplified in Figure 7C, this information should help in generating hypotheses relating the chemistry of CPEs with their potency. For working hypotheses, this knowledge can then help refine our selection rules for designing next generation transdermal formulations. Repeating the experiment–hypothesis loop over a vast but limited number of candidate penetration enhancers will provide the missing pieces in solving a vast multivariate problem. Also, this knowledge should significantly reduce the cost and effort of designing therapeutics for use on skin in the future.

ACKNOWLEDGMENTS

The authors acknowledge the technical assistance from Richard Keeler, Drew Lassen, Jeffery O'Neil, Maricella Castaneda, Kaitlin Ergun, Tiffany Coleman, and Arthhur Wojicicki. The authors also thank Chemron corp. for gift samples of sodium laureth sulfate.

REFERENCES

1. Asbill C, El-Kattan A, Michniak B. Enhancement of transdermal drug delivery: chemical and physical approaches. Crit Rev Ther Drug Carrier Syst 2000; 17:621–658.
2. Ting W, Vest C, Sontheimer R. Review of traditional and novel modalities that enhance the permeability of local therapeutics across the stratum corneum. Int J Dermatol 2004; 43:538–547.
3. Prausnitz MR, Mitragotri S, Langer R. Current status and future potential of transdermal drug delivery. Nature Rev Drug Discov 2004; 3:115–124.
4. Barry B. Novel mechanisms and devices to enable successful transdermal drug delivery. Eur J Pharm Sci 2001; 14:101–114.
5. Purdon C, Azzi C, Zhang J, Smith E, Maibach H. Penetration enhancement of transdermal delivery—current permutations and limitations. Crit Rev Ther Drug Carrier Systems 2004; 21:97–132.
6. Harris R. Licht S, ed. Therapeutic Electricity and Ultraviolet Radiation. 2d ed. New Haven: Yale University Press, 1967:156–178.
7. Kasting G, Keister J. Application of electrodiffusion theory for a homogeneous membrane to iontophoretic transport through skin. J Control Release 1989; 8:195–210.
8. Banga A, Chien Y. Iontophoretic delivery of drugs: fundamentals, developments and biomedical applications. J Control Release 1988; 7:1–24.
9. Sloan J, Soltani K. Iontophoresis in dermatology. J Am Acad Dermatol 1986; 15: 671–684.
10. Tyle P. Iontophoretic devices for drug delivery. Pharm Res 1986; 3:318–326.
11. Hill R. The use of microiontophoresis in the study of the descending control of nociceptive transmission. Prog Brain Res 1988; 77:339–347.
12. Behl C, et al. Iontophoretic drug delivery: effects of physicochemical factors on the skin uptake of nonpeptide drugs. J Pharm Sci 1989; 78:355–360.
13. Singh J, Roberts M. Transdermal delivery of drugs by iontophoresis: a review. Drug Delivery 1989; 4:1–12.
14. Sanderson J, deReil S, Dixon R. Iontophoretic delivery of nonpeptide drugs: formulation optimization for maximum skin permeability. J Pharm Sci 1989; 78:361–364.
15. Chien Y, Behl C. Iontophoretic (transdermal) delivery of drugs: overview of historical development. J Pharm Sci 1989; 78:353–354.
16. Tyle P, Agrawala P. Drug delivery by phonophoresis. Pharm Res 1989; 6:355–361.
17. Kleinkort J, Wood F. Phonophoresis with 1 percent versus 10 percent hydrocortisone. Phys Ther 1975; 55:1320–1324.
18. McElnay J, Kennedy T, Harland R. The influence of ultrasound on the percutaneous-absorbtion of Fluocinolone. Int J Pharm 1987; 40:105–110.
19. Benson H, McElnay J, Harland R. Phonophoresis of lignocaine and prilocaine from Elma cream. Int J Pharm 1988; 44:65–69.
20. Fahim M, Brawner T, Hall D. New treatment for herpes-simplex virus type-2 [ultrasound and zinc, urea and tannic acid ointment]. Part II: Female patients. J Med 1980; 11: 143–167.
21. Fahim M, Brawner T, Millikan L, Nickell M, Hall D. New treatment for herpes-simplex virus type-2 [ultrasound and zinc, urea and tannic acid ointment]. Part I: Male patients. J Med 1978; 9:245–264.
22. Lavon A, Kost J. Ultrasound and transdermal drug delivery. Drug Discov Today 2004; 9:670–676.
23. Mitragotri S, Kost J. Low-frequency sonophoresis—a review. Adv Drug Delivery Rev 2004; 56:589–601.
24. Mitragotri S, Kost J. Transdermal delivery of heparin and low-molecular weight heparin using low-frequency ultrasound. Pharm Res 2001; 18:1151–1156.
25. Denet A, Vanbever R, Preat V. Skin electroporation for transdermal and topical delivery. Adv Drug Delivery Rev 2004; 56:659–674.

26. Essa E, Bonner M, Barry B. Electrically assisted skin delivery of liposomal estradiol; phospholipid as damage retardant. J Control Release 2004; 95:535–546.
27. Mori K, et al. Design and feasibility assessment of topically applied drug formulations for electroporation. Chem Pharm Bull 2003; 51:617–619.
28. Denet A, Preat V. Transdermal delivery of timolol by electroporation through human skin. J Control Release 2003; 88:253–262.
29. Sung K, Fang J, Wang J, Hu O. Transdermal delivery of nalbuphine and its prodrugs by electroporation. Eur J Pharm Sci 2003; 18:63–70.
30. Zhang L, Nolan E, Kreitschitz S, Rabussay D. Enhanced delivery of naked DNA to the skin by non-invasive in vivo electroporation. Biochimica et Biophysica Acta (BBA) (General Subjects) 2002; 1572:1–9.
31. Pliquett E, Gusbeth C. Perturbation of human skin due to application of high voltage. Bioelectrochemistry 2000; 51:41–51.
32. Cormier M, et al. Transdermal delivery of desmopressin using a coated microneedle array patch system. J Control Release 2004; 97:503–511.
33. Martanto W, et al. Transdermal delivery of insulin using microneedles in vivo. Pharm Res 2004; 21:947–952.
34. Prausnitz M. Microneedles for transdermal drug delivery. Adv Drug Delivery Rev 2004; 56:581–587.
35. Reed M, Lye W. Microsystems for drug and gene delivery. Proc IEEE 2004; 92:56–75.
36. McAllister D, et al. Microfabricated needles for transdermal delivery of macromolecules and nanoparticles: fabrication methods and transport studies. Proc Natl Acad Sci USA 2003; 24:13755–13760.
37. Henry S, McAllister D, Allen M, Prausnitz M. Microfabricated microneedles: a novel approach to transdermal drug delivery. J Pharm Sci 1998; 87:922–925.
38. Bauerova K, Matusova D, Kassai Z. Chemical enhancers for transdermal drug transport. Eur J Drug Metabol Pharmacokinet 2001; 26:84–94.
39. Fang J, et al. Capsaicin and nonivamide as novel skin permeation enhancers for indomethacin. Eur J Pharm Sci 2001; 12:195–203.
40. Kaplun-Frischoff Y, Touitou E. Testosterone skin permeation enhancement by menthol through formation of eutectic with drug and interaction with skin lipids. J Pharm Sci 1997; 86:1394–1399.
41. Walker R, Smith E. The role of percutaneous penetration enhancers. Adv Drug Delivery Rev 1996; 18:295–301.
42. Kararli T, Kirchhoff C, Penzotti S. Enhancement of transdermal transport of Azidothymidine (AZT) with novel terpene and terpene-like enhancers—in vivo in-vitro correlations. J Control Release 1995; 34:43–51.
43. Hadgraft J. Recent developments in topical and transdermal delivery. Eur J Drug Metabol Pharmacokinet 1996; 21:165–173.
44. Weichers J. Use of chemical penetration enhancers in transdermal drug delivery—possibilities and difficulties. Acta Pharmeceutica Nordica 1992; 4:123–123.
45. Ranade V. Drug delivery systems. Vol. 6. Transdermal drug delivery. J Clin Pharmacol 1991; 31:401–418.
46. Karande P, Jain A, Mitragotri S. Discovery of transdermal penetration enhancers by high throughput screening. Nature Biotechnol 2004; 22:192–197.
47. Finnin B, Morgan T. Transdermal penetration enhancers: applications, limitations, and potential. J Pharm Sci 1999; 88:955–958.
48. Shigeta Y, et al. Skin whitening effect of linoleic acid is enhanced by liposomal formulations. Biol Pharm Bull 2004; 27:591–594.
49. Mezei M. In: Hsieh D, ed. Drug Permeation Enhancement. Vol. 62. New York, Basel, Hong Kong: Marcel Dekker, Inc., 1994:171–198.
50. Ulrich A. Biophysical aspects of using liposomes as delivery vehicles. Biosci Rep 2002; 22:129–150.

51. Raghavachari N, Fahl W. Targeted gene delivery to skin cells in vivo: a comparative study of liposomes and polymers as delivery vehicles. J Pharm Sci 2002; 91:615–622.
52. Honzak L, Sentjurc M, Swartz H. In vivo EPR of topical delivery of a hydrophilic substance encapsulated in multilamellar liposomes applied to the skin of hairless and normal mice. J Control Release 2000; 66:221–228.
53. Godin B, Touitou E. Ethosomes: new prospects in transdermal delivery. Crit Rev Ther Drug Carrier Syst 2003; 20:63–102.
54. Touitou E, Godin B, Weiss C. Enhanced delivery of drugs into and across the skin by ethosomal carriers. Drug Dev Res 2000; 50:406–415.
55. Touitou E, Dayan N, Bergelson L, Godin B, Eliaz M. Ethosomes—novel vesicular carriers for enhanced delivery: characterization and skin penetration. J Control Release 2000; 65:403–418.
56. Cevc G. Lipid vesicles and other colloids as drug carriers on the skin. Adv Drug Delivery Rev 2004; 56:675–711.
57. Cevc G, Schatzlein A, Richardsen H, Vierl U. Overcoming semipermeable barriers, such as the skin, with ultradeformable mixed lipid vesicles, transfersomes, liposomes, or mixed lipid micelles. Langmuir 2003; 19:10753–10763.
58. Uchegbu I, Vyas S. Non-ionic surfactant based vesicles (niosomes) in drug delivery. Int J Pharm 1998; 172:33–70.
59. Cevc G. Transfersomes, liposomes and other lipid suspensions on the skin: permeation enhancement, vesicle penetration, and transdermal drug delivery. Crit Rev Ther Drug Carrier Syst 1996; 13:257–388.
60. Kreilgaard M. Influence of microemulsions on cutaneous drug delivery. Adv Drug Delivery Rev 2002; 54:S77–S98.
61. Marti-Mestres G, Nielloud F. Emulsions in health care applications—an overview. J Disp Sci Technol 2002; 23:419–439.
62. Piemi M, de Luca M, Grossiord J, Seiller M, Marty J. Transdermal delivery of glucose through hairless rat skin in vitro: effect of multiple and simple emulsions. Int J Pharm 1998; 171:207–215.
63. Schwarz J, Weisspapir M, Friedman D. Enhanced transdermal delivery of diazepam by submicron emulsion (SME) creams. Pharm Res 995; 12:687–692.
64. Wertz P, Downing D. In: Hadgraft J, Guy R, eds. New York, Basel, Hong Kong: Marcel Dekker, Inc., 1989:1–17.
65. Friberg S, Kayali I, Margosiak M. In: Osborne D, Amann A, eds. Topical Drug Delivery Formulations. Vol. 42. New York, Basel, Hong Kong: Marcel Dekker, Inc., 1990:29–45.
66. Lynch D, Roberts L. In: Osborne D, Amann A, eds. Topical Drug Delivery Formulations. Vol. 42. New York, Basel, Hong Kong: Marcel Dekker, Inc., 1990:87–108.
67. Drexler H. Skin protection and percutaneous absorption of chemical hazards. Int Arch Occup Environ Health 2003; 76:359–361.
68. Kanikkannan N, Singh M. Skin permeation enhancement effect and skin irritation of saturated fatty alcohols. Int J Pharm 2002; 248:219–228.
69. Sasaki H, Kojima M, Nakamura J, Shibasaki J. Acute toxicity and skin irritation of pyrrolidone derivatives as transdermal penetration enhancers. Chem Pharm Bull 1990; 38:2308–2310.
70. Mitragotri S. Synergistic effect of enhancers for transdermal drug delivery. Pharm Res 2000; 17:1354–1359.
71. Fang J, Hwang T, Huang Y, Tsai Y. Transdermal iontophoresis of sodium nonivamide acetate—V. Combined effect of physical enhancement methods. Int J Pharm 2002; 235:95–105.
72. Le L, Kost J, Mitragotri S. Combined effect of low-frequency ultrasound and iontophoresis: applications for transdermal heparin delivery. Pharm Res 2000; 17:1151–1154.
73. Denet A, Ucakar B, Preat V. Transdermal delivery of timolol and atenolol using electroporation and iontophoresis in combination: a mechanistic approach. Pharm Res 2003; 20:1946–1951.

74. Fang J, Hung C, Fang Y, Chan T. Transdermal iontophoresis of 5-fluorouracil combined with electroporation and laser treatment. Int J Pharm 2004; 270:241–249.
75. Badkar A, Banga A. Electrically enhanced transdermal delivery of a macromolecule. J Pharm Pharmacol 2002; 54:907–912.
76. Tezel A, Sens A, Tuchscherer J, Mitragotri S. Synergistic effect of low-frequency ultrasound and surfactants on skin permeability. J Pharm Sci 2002; 91:91–100.
77. Monti D, Giannelli R, Chetoni P, Burgalassi S. Comparison of the effect of ultrasound and of chemical enhancers on transdermal permeation of caffeine and morphine through hairless mouse skin in vitro. Int J Pharm 2001; 229:131–137.
78. Pillai O, Nair V, Panchagnula R. Transdermal iontophoresis of insulin: IV. Influence of chemical enhancers. Int J Pharm 2004; 269:109–120.
79. de Graaff A, Li G, van Aelst A, Bouwstra J. Combined chemical and electrical enhancement modulates stratum corneum structure. J Control Release 2003; 90:49–58.
80. Zewert T, Pliquett U, Vanbever R, Langer R, Weaver J. Creation of transdermal pathways for macromolecule transport by skin electroporation and a low toxicity, pathway-enlarging molecule. Bioelectrochem Bioenerg 1999; 49:11–20.
81. Barry B. Breaching the skin's barrier to drugs. Nature Biotechnol 2004; 22:165–167.
82. Jain A, Panchagnula R. Combination of penetration enhancers for transdermal drug delivery—studies with imipramine hydrochloride. Pharmazeutische Industrie 2004; 66:478–482.
83. Evrard D, Touitou E, Kolusheva S, Fishov Y, Jelinek R. A new colorimetric assay for studying and rapid screening of membrane penetration enhancers. Pharm Res 2001; 18:943–949.
84. Karande P, Mitragotri S. High throughput screening of transdermal formulations. Pharm Res 2002; 19:655–660.
85. Gwak H, Oh I, Chun I. Transdermal delivery of ondansetron hydrochloride: effects of vehicles and penetration enhancers. Drug Dev Ind Pharm 2004; 30:187–194.
86. Wang Q, Hikima T, Tojo K. Skin penetration enhancement by the synergistic effect of supersaturated dissolution and chemical enhancers. J Chem Eng Japan 2003; 36:92–97.
87. Kanikkannan N, Kandimalla K, Lamba S, Singh M. Structure-activity relationship of chemical penetration enhancers in transdermal drug delivery. Curr Med Chem 2000; 7:593–608.
88. Mollgaard B. In: Walters K, Hadgraft J, eds. Pharmaceutical Skin Penetration Enhancement. Vol. 59. New York, Hong Kong, Basel: Marcel Dekker, Inc., 1993:229–242.
89. Barry B, Williams A. Human skin penetration enhancement: the synergy of propylene glycol with terpenes. Proc Int Symp Control Release Bioactive Mater 1989; 16:33–34.
90. Williams A, Barry B. Penetration enhancers. Adv Drug Delivery Rev 2004; 56:603–618.
91. He N, et al. Mechanistic study of chemical skin permeation enhancers with different polar and lipophilic functional groups. J Pharm Sci 2004; 93:1415–1430.
92. Chantasart D, et al. Mechanistic studies of branched-chain alkanols as skin permeation enhancers. J Pharm Sci 2004; 93:762–779.
93. Bronaugh R. In: Bronaugh R, Maibach H, eds. Percutaneous Absorbtion: Mechanisms-Methodology-Drug Delivery. New York, Basel, Hong Kong: Marcel Dekker, Inc., 1989:423–427.
94. Moody R. Automated in vitro dermal absorption (AIVDA): a new in vitro method for investigating transdermal flux. ALTA (Alternatives to Laboratory Animals) 1997; 25:347–357.
95. Meyer W. Porcine skin as biological model for the human integument. Hautarzt 1996; 47:178–182.
96. Sekkat N, Kalia Y, Guy R. Biophysical study of porcine ear skin in vitro and its comparison to human skin in vivo. J Pharm Sci 2002; 91:2376–2381.
97. Meyer W, Zschemisch N, Godynicki S, Neurand K. Investigations in the structure of the ear skin in the domesticated pig, with special reference to the use for human dermatological

research: 2. Specific histology of the integument. Berliner Munchener Tierarztliche Wochenschrift 2001; 114:100–111.
98. Simon G, Maibach H. The pig as an experimental animal model of percutaneous permeation in man: qualitative and quantitative observations—an overview. Skin Pharmacol Physiol 2000; 13:229–234.
99. Moss G, Dearden J, Patel H, Cronin M. Quantitative structure-permeability relationships (QSPRs) for percutaneous absorption. Toxicol In Vitro 2002; 16:299–317.

16

Excipient Selection and Criteria for Injectable Dosage Forms

Mahesh V. Chaubal, James Kipp, and Barrett Rabinow
Baxter Healthcare, Round Lake, Illinois, U.S.A.

INTRODUCTION

Development of injectable formulations typically requires specific considerations on the type and quality of excipients used. The criteria for excipient selection depend on a number of factors, including the type of drugs and route of injection. Excipient considerations for biopharmaceutical drugs (proteins and vaccines) have been addressed in separate chapters, hence, this chapter will focus on formulation development only for small molecules (molecular weight <1000). The site of injection could be intravenous (into a vein), subcutaneous (SC) (into adipose tissue), intramuscular (IM) (into deep muscle), and, in rare occasions, intradermal (within the skin), intraocular (in the eye), intrathecal (in the spinal fluid), or intratumoral (in the tumor). Injections can be made in a bolus fashion (rapid) or infused over a longer period of time. A number of devices can be used to inject drugs, from conventional needle and syringe to needle-free injectors. The container-closure system may vary from a lyophilized vial to a prefilled syringe or a flexible plastic bag. The drug itself can be formulated as a lyophilized cake, a solution, or a dispersed system (liposomes and suspensions). All these factors play a vital role in selection of the excipients for injectable formulations. Whereas some specific systems, such as liposomes, have specific requirements, the concepts of parenteral science and excipient selection discussed in this chapter are valid for all injectable systems.

Excipients used in injectable formulations have to meet several stringent requirements. A positive identification test uniquely applicable to the excipients is required (e.g., infrared spectrophotometry and chromatography). It is important that manufacturers identify and set appropriate limits for impurities. These limits should be based upon appropriate toxicological data, or the limits described in national compendial requirements. Manufacturing processes should be adequately controlled so that the impurities do not exceed such established specifications. Solvents or catalysts used in the excipient production process should be removed to appropriate levels. If naturally derived, excipients should meet endotoxin levels and may require further testing for bovine spongiform encephalopathy (BSE) /

transmissible spongiform encephalopathy (TSE) considerations. The latter topic is discussed in greater detail in a separate chapter.

IMPACT OF INJECTABLE ROUTE OF ADMINISTRATION UPON SELECTION OF EXCIPIENTS

This section will review how physiological factors at the site of injection impact the design of dosage forms and affect choice of excipients. First, pharmacokinetic factors affecting rates of delivery of drug to the blood will be considered. Then, biocompatibility or safety issues will be addressed. This analysis focuses on the intravascular (IV), IM, and SC routes of administration.

Goals of the injectable route of administration, versus oral route of administration, for example, are several fold (1):

1. Exert direct control over pharmacokinetic parameters, such as t_{max} (fast onset vs. sustained release), C_{max}, tissue concentration, rate of elimination, including supplying nutritional needs and rapid correction of fluid and electrolyte imbalance
2. Guarantee dosage and drug compliance
3. Achieve a delivery effect that is not possible by the oral route, because drug degrades or the patient has nonfunctioning gut, is unconscious, etc.
4. Achieve a desired local effect, e.g., local anesthetic.

Pharmacokinetics

Depending upon the specific goal, different routes of administration are utilized. The choice of route is dictated by the volume of solution, immediate or depot pharmacokinetics desired, tolerability, and convenience. A larger volume (more than 5 mL) requires the IV route as does fast onset time, the need to accurately control blood levels, and high C_{max}. Sustained release with lower peak levels requires the IM or SC spaces, which are also chosen as more convenient ways to access the vascular space. Intravenous injection places drug directly into the blood, from which it is rapidly delivered throughout the body. Injection in an extravascular site, such as the SC or IM sites, requires a preliminary absorption process, across the biological membrane posed by the cells lining the blood capillaries or lymph system. Because of this, there is a delay, which can be advantageous for sustained drug release. The sequence of steps involved to enable drugs to be absorbed into the blood from different dosage forms at different sites of injection is often effectively simplified in a compartmental model, as shown in Figure 1. The unit processes of partitioning, diffusion, and dissolution all involve movement of the drug molecule, and the diffusional pressures associated with drug migration are often described by Fick's Law [Eq.(1)]:

$$dC/dt = (DA/l)(C_o - C_t) \tag{1}$$

where dC/dt is the rate of change of concentration of the drug in the compartment following migration; D, the diffusion coefficient of the drug through the applicable medium; A, the surface area available for diffusion; and l, the thickness of the medium through which the drug is diffusing. $(C_o - C_t)$ is the concentration gradient

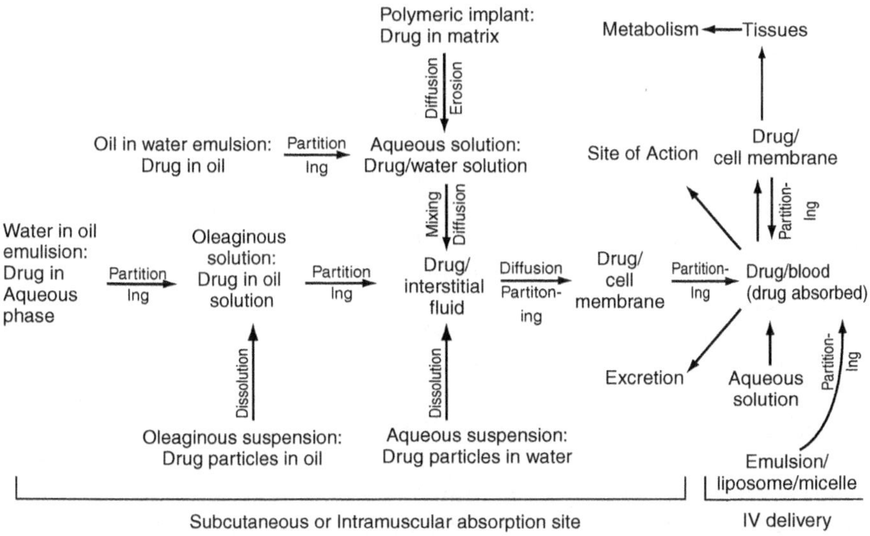

Figure 1 Sequential biodistribution of various formulations of injectable drugs.

driving diffusion, where C_o is the concentration of the drug in the originating compartment, e.g., the depot concentration adjacent to a cell membrane for partitioning, or the saturation solubility of the drug in the diffusion layer adjacent to a solid suspension and C_t is the concentration of drug in the receiving compartment at time t following migration (e.g., the blood level following cellular partitioning, or dissolved concentration in the bulk medium surrounding a drug suspension).

Eq. (1) does not account for capillary forces between cells, the effects of thermal convection, and the binding of drugs to biological molecules (e.g., proteins), in solution and at compartmental interfaces (e.g., membranes). Thus, the factor "DA/l" in Eq. (1) may be considered to be an "effective" permeability coefficient. The rate can be modified by excipients, which alter drug solubility, and therefore concentration, in a depot for example. An oil solution for SC implant may contain a higher concentration of lipid-soluble drug than an aqueous solution. But the effective concentration gradient driving diffusion into the surrounding interstitial space depends upon the drug partitioning at the oil–interstitial fluid boundary. A high octanol–water drug partition coefficient, K_{ow}, will favor the oil phase, leading to lower aqueous drug concentration and slower, more sustained release than would formulation in an aqueous depot. For membrane partitioning, the rate of drug permeation in the absence of active transport mechanisms or ionic transport through channels is also proportional to K_{ow} multiplied by the nonionic drug fraction, because of the interior lipid bilayer component of cell membranes. The pH of the aqueous medium, in the cytosol and extracellular spaces, is therefore important in determining the amount of uncharged species available for partitioning and passive diffusion. Mostly, the nonionized, unbound drug is involved in passive membrane transport. This is an important consideration considering the multitude of ways in which drug may be complexed or ionized both by the biological milieu and by the formulation excipients. Aggregation of the drug solute and precipitation in that milieu are also important factors. For a purely diffusive process, the diffusion coefficient is inversely proportional to viscosity of the medium, an excipient-modifiable variable (2).

Intravenous injection of an aqueous drug solution permits rapid (within minutes) delivery to all organs of the body. The drug may be freely dissolved in the aqueous compartment of the plasma pool or, if hydrophobic, bound to proteins. Intravenous delivery of a lipidic formulation of a drug incorporated in a liposome, emulsion, or micelle, results in drug partitioning from this vehicle into the plasma pool, either bound to carrier proteins (e.g., lipoproteins and albumin) or available as free solute. This is essential for passive drug uptake by the cells of the tissues, which behave as semipermeable membranes, effectively excluding permeants of very high molecular weight or drug entrained in lipid vesicles. Only the freely diffusible drug may therefore enter the cell membrane.

Similarly, drugs injected into the SC or IM space are separated from the blood compartment by the endothelial cells of the capillaries. From the interstitial space, such drug molecules must first diffuse toward and then partition into the endothelial cell membrane. After traversing these cells, the drugs must then partition on the luminal, or blood facing, side of these cells into the blood, which carries them away. By this dilution effect, the blood presents sink conditions, thus maintaining a maximal concentration gradient, $\mathrm{d}C_\mathrm{m}/\mathrm{d}x$, to drive diffusion toward the blood.

It follows that those areas that have a larger diffusing area of blood capillaries will have a larger concentration gradient, and higher rates of diffusion away from the injection sites. Because of this, IM injection leads to faster appearance of drug in the blood than does SC injection, because muscular tissue is more vascularized than is fatty SC tissue. Furthermore, resting blood flow is greater in the deltoid muscle of the arm, intermediate in the thigh, and least in buttock, affecting the relative rates of absorption from IM injections. Lower levels of plasma lidocaine when injected in buttocks compared to plasma lidocaine levels in the deltoid injection were attributed to high affinity of fatty tissue for lidocaine and lower vascularity in this area (3). Access to the capillaries can be affected by prior scarring, affecting tissue vascularity, increased muscular exertion involving these areas, and temperature, as well as disease state, affecting blood flow.

By constricting the vascular bed, such coadministered vasoactive excipients as epinephrine can reduce the rate of uptake from the SC sites (4a). By contrast, the excipient hyaluronidase breaks down the interstitial barrier by lysing hyaluronic acid, a polysaccharide that helps form the intercellular ground substance of connective tissue (4b). This in effect spreads the injected drug solution over a larger area of connective tissue, increasing the absorption surface, and thereby increasing both the volume that can normally be injected SC (Table 1) and the rate of uptake (6).

For an aqueous solution depot in Figure 1, the drug concentration gradient can also be affected by using excipients to modify the osmolality of the depot. This in turn affects movement of water into or out of the depot with respect to the surrounding

Table 1 Physiological Considerations for Injectable Excipients

Route	Intravascular	Subcutaneous	Intramuscular
pH	2–12	2.7–9	2–12
Organic cosolvent	≤70% (for small volume injectables)	≤15%	≤100%
Volume of injection	<3 L	≤2 mL	≤5 mL

Source: From Ref. 5.

tissue. Dilution of the drug in the depot reduces the concentration gradient, thus reducing drug uptake rate, with prolongation of effect.

Besides the concentration gradient, diffusive flow also depends upon the diffusion coefficient of the drug molecule, which is inversely dependent upon the viscosity of the medium. Thus, the diffusion rate of a drug from a SC or IM injection may be slowed, and the duration of action prolonged, by utilizing excipients that increase the viscosity of the aqueous depot. If the same quantity of drug is contained within a smaller volume of excipient vehicle, then the rate of delivery increases and duration of action decreases, as the drug is released. This occurs because the surface area decreases with the square of the depot radius, whereas concentration gradient is inversely proportional to the cube of the radius.

As emphasized earlier, the concentration gradient of the drug in Eq. (1) refers to that of the unbound drug and its ionic distribution, which depends upon its acid–base properties. This can be modified by appropriate choice of excipients to ionize the drug by salt formation, thereby affecting the distribution of ionic versus nonionic species by acid–base equilibrium, using the Henderson–Hasselbach equation. All of the drug will eventually leave the depot and enter the body, but the rate may be reduced if membrane transport is limited by solubility of the neutral species within the membrane.

If simple excipient modification of the aqueous depot, as described above, is inadequate to reduce uptake rate for a desired prolongation of effect, then additional nonaqueous phases may be introduced, adding rate-controlling partitioning (7). These may involve an oleaginous solution of the drug, or incorporation of the drug in the oil phase of an oil-in-water emulsion. In this latter case, the drug must first partition from the oil to the aqueous phase, whereupon it can subsequently mix with and be diluted by the interstitial fluid. Alternatively, the drug may be formulated in the aqueous phase of a water-in-oil emulsion. In this case, the drug must first partition into the oil phase of the depot, and then partition again into the aqueous interstitial fluid, as shown in Figure 1. For still more extended release, the drug may be incorporated into a polymeric implant, requiring either outward diffusion of the drug or erosion of the device, enabling the drug to dissolve in the aqueous phase.

Alternatively, solid suspensions of the drug, either in aqueous or oil phases, have been used, necessitating a dissolution step, to further slow down the release rate. These are typically crystal suspensions of the drug, stabilized by surfactants against agglomeration, so as to control particle size. The rate of dissolution is proportional to the surface area of the dosage form; hence, reducing particle size will increase the dissolution rate. The dissolution rate is also proportional to the concentration gradient across the diffusion layer adjacent to the particles, where the concentration is saturated. Phenytoin is adjusted to a pH of 12 because it is insoluble. Following IM administration, the decrease in pH due to dilution converts the sodium salt to the free acid, which then precipitates in the interstitial space. These crystals have a slow rate of dissolution providing depot release over four to five days. While providing for sustained release, this method is variable and painful (8). The insoluble zinc complex of insulin gives a classic example of slower release provided by the suspension. Depending upon crystal size in the range 5 to 50 μm, the hypoglycemic effects ranged from 4 to 10 hours (9). For IM procaine penicillin G, serum concentrations after one hour postadministration ranged from 2.14 to 1.37 units/mL for particles in the size range of 1 to 2 μm to 150 to 250 μm, respectively (10).

The lipidic vehicles and nanosuspensions described above are also used for increasing the loading of poorly water-soluble drugs, particularly in volume-constrained injections.

Targeting of Drugs

Until now, pharmacokinetics of nontargeted injectable dosage forms has been considered. Another modality for improving drug delivery involves targeting of drugs by the injectable route. This has been accomplished by directed placement of the needle in such applications as regional anesthesia, intra-articular delivery of steroids for arthritis, and intra-arterial therapy for liver cancer. It would be attractive to design dosage forms to automatically target the sites of action. This developing field has utilized IV nanoparticulate carriers to target areas of tumor growth, infection, and inflammation. Associating the active drug in colloidal drug carriers, such as micelles, liposomes, emulsions, and nanoparticles, has demonstrated increased loading, reduced toxicity, achievement of controlled or sustained release, and potentially decreased immunological reactions (11). Following IM or SC delivery, liposomes can target the lymphatic vessels if they are sufficiently small—less than 100 nm. This provides enhanced delivery of drug to lymphatic sanctuaries of tumor cells and viral disease, such as HIV, while minimizing systemic exposure. The rate of drug release from the site of injection is a function of liposome composition and size (12), as well as drug solubility in water and lipids.

Normally, the body removes particulates by the monocyte phagocytic system (MPS). Efficiency of this process is increased for moderately hydrophobic surfaces of contact angle of 50° and decreases on either side of this, as well as for increasing surface charge (13). Clearance of liposomes increases in proportion to their size, with increase in phase transition (tighter bilayers), absence of cholesterol, and dose, due to saturation of the MPS uptake mechanism (14). MPS avoidance is performed by altering the surface by conjugating the phospholipids with polyethylene glycol residues that entrain water molecules, thus preventing opsonizing proteins from initiating the recognition and phagocytosis sequence. Extended circulation time, following IV delivery, then permits such surface-functionalized particles to survive until they target the pores in the vascular endothelium, at areas of tumor growth, infection, or inflammation. Once particles escape through these typically 150 to 400 nm–sized gaps, they are localized at the sites of tumor, in part because of inadequate lymphatic return. This method provides for enhanced delivery of drug to the site of action, while minimizing toxicity associated with systemic exposure to nontarget tissues (15).

Biocompatibility and Toxicity

Injectable formulations must be designed to be compatible with the biological tissues with which they come in contact. Extremes of pH, for example, can induce venous spasm and may hemolyze red cells. SC injections require more safeguards than IM injections, because the tissue residence time of drug formulations is longer for SC than for IM injections, due to reduced vascularity. This confers more potential trauma for SC injections, which therefore must have tighter constraints on pH and amounts of organic solvent injected as well (Table 1). If cosolvents are used, inadequate quantities may lead to precipitation of the drug at the injection site following dilution by body fluids. This will affect bioavailability of drug and cause potential vascular irritation or tissue injury.

The formulations should also be iso-osmotic with surrounding tissues, to avoid causing volume changes of cells. Because some excipients can penetrate the red cell, thus lowering osmotic strength outside, but raising it within the cell, the concept of isotonicity is distinguished from iso-osmoticity. An isotonic solution will not result in volume change of the red cell, potentially leading to hemolysis. The hemolytic potential of certain excipients, such as glycerin and propylene glycol, which penetrate the red cell, can be abrogated by increasing the osmotic strength with a nonmigrating salt such as NaCl (Table 2) or a macromolecular nonpenetrant such as hetastarch or polyethylene glycol.

Biocompatibility of injectable formulations with tissues can be tested by observing microscopic histology of the tissues so exposed, or by using erythrocyte hemolysis as a surrogate for these other tissues. Alternatively, one can measure the level of the cytosolic enzyme creatine phosphokinase that is released from damaged tissues (18).

Besides local toxicity, discussed above, there are numerous other modes of potential adverse interactions involving excipients (19,20). Many of these pose little threat provided the amounts of excipients are constrained to certain levels. Excessive amounts, however, can cause problems, particularly for patients who are intolerant of even modest levels. Commonly used phosphate buffers may cause calcium loss with formation of insoluble calcium phosphates when such buffers are administered in over-ambitious amounts (21). Calcium phosphate precipitation has been noted particularly in nutritional parenteral admixtures for neonates because of the high nutrient requirements. Similarly, renal toxicity has been associated with depletion of zinc and other trace metals caused by large parenteral doses of ethylenediaminetetraacetic acid (EDTA) (22). Excessive absorption of glycine solutions, when used as irrigants during transurethral resections, can cause hyponatremia, hypertension, and confusion (23). The use of preservatives has been associated with cardiac effects in a few patients (24). Premature neonates were found to be at risk for receiving toxic amounts of benzoic acid or benzyl alcohol in bacteriostatic solutions used to flush intravenous catheters (25).

As with drugs themselves, a variety of sensitivity reactions have also been linked to the use of excipients. These range from life-threatening anaphylactic, immunoglobulin E–mediated type reactions to less serious pseudoallergic responses. Because of widespread exposure to thimerosal, the mercury-containing preservative in vaccines, there is now a 10% incidence of allergic sensitivity to mercurials in the United States (26). In sensitive individuals, sulfite antioxidants can cause a range of reactions spanning a broad range of mechanism and severity (27). Because of the lack of suitable substitutes, they are not banned, but sulfite-containing dosage forms typically carry warning labels. All of the colloids, such as human serum albumin,

Table 2 50% Lethal Dose (LD_{50}) Values Expressed as Total Volume Percents of Various Cosolvents, with or Without Various Levels of NaCl, for Lysis of Erythrocytes

	Aqueous NaCl concentration				
Cosolvent	0%	0.9%	1.8%	2.7%	3.6%
Glycerin	3.7	3.3	8.3	12.7	11.9
PG	5.7	6.2	14.7	20.0	19.3
10% EtOH, 40% PG	10.3				
EtOH	21.2	20.5	20.0	20.5	19.7
DMA	37.0	36.6	40.4	39.3	36.9
PEG 400	30.0	29.6	33.5	27.6	23.9
DMI	39.5	17.9	16.6	15.9	9.6

Abbreviations: DMA, dimethylacetamide; DMI, dimethylisosorbide; PEG, polyethylene glycol.
Source: From Refs. 16,17.

dextran, and hydroxyethyl starch, have an approximately 0.01% frequency of shock or cardiorespiratory arrest associated with their usage (28). Residuals of the sterilizing gas, ethylene oxide, remaining in medical devices and dialyzers have been associated with anaphylaxis, due to conjugation reactions with host proteins causing antigenicity (29). Approaches to the problems noted above have ranged from outright banning of certain excipients, to minimizing their use, to inclusion of warning statements, if their retention is deemed essential by an appropriate risk–benefit analysis.

EXCIPIENTS FOR INJECTABLE FORMULATIONS

Cosolvents

Drugs that possess moieties that cannot readily form hydrogen bonds with water may nonetheless be formulated by use of water-miscible cosolvents that can readily interact with water and disrupt hydrogen bonding between water molecules. Water miscibility requires some degree of hydrogen bonding between the cosolvent and water, and yet the ability to interact with the solute. This interaction occurs by the hydrophobic effect in which nonpolar regions of the cosolvent associate with nonpolar moieties of the solute. Thermodynamics, in turn, maximizes mobility and entropy of the surrounding water molecules. Examples of such cosolvents include ethanol, propylene glycol, polyethylene glycol, tetrahydrofurfuryl alcohol, polyethyleneglycol ether, and glycerin. "Melphalan hydrochloride for injection" (Alkeran®, GSK, Australia) exemplifies the application of cosolvency. The drug is supplied as a lyophilized powder with a sterile diluent for reconstitution that is comprised of sodium citrate (0.2 g), propylene glycol (6.0 mL), ethanol (96%, 0.52 mL), and "water for injection," diluted to a total volume of 10 mL. Alkeran for injection is administered intravenously. A representative list of products on the market, utilizing cosolvents, is presented in Table 3. Whereas the number of formulations with

Table 3 A Representative List of Products on the Market Utilizing 100% Cosolvents

Product	Drug	Cosolvents	Route
Sandimmune	Cyclosporin	65% Cremophor® EL, 35% ethanol	IV infusion
VePeSid	Etoposide	30% ethanol, 60% PEG 300, 8% Tween 80, 2% benzyl alcohol	IV infusion
Faslodex	Fulvestrant	Ethanol, benzyl alcohol, benzyl benzoate, castor oil	IM
Taxol	Paclitaxel	51% Cremophor EL, 49% ethanol	IV infusion
Prograf	Tacrolimus	20% Cremophor RH60, 80% ethanol	IV infusion
Eligard	Leuprolide acetate	160 mg NMP	SC, sustained release
Vumon	Teneposide	50% Cremophor EL, 42% ethanol, 6% dimethylacetamide, 2% benzyl alcohol	IV infusion
Valstar	Valrubicin	50% Cremophor EL, 50% ethanol	Intravesical
Viadur	Leuprolide acetate	DMSO	SC, sustained release
Delatestryl	Testosterone enanthate	Sesame oil	IM

Abbreviations: IV, intravascular; IM, intramuscular; SC, subcutaneous; NMP, *N*-methyl pyrrolidone; PEG, polyethylene glycol.

100% nonaqueous solvent is limited, cosolvents represent a fairly common solubilization approach, especially when pH and/or salt formation is not sufficient to dissolve the drug. Challenges associated with use of cosolvents include concerns with tissue irritability (30), potential hemolysis (31), toxicity (32), and uncertainty of drug precipitation (33). Due to these factors, cosolvent formulations are often infused over extended time periods.

For compounds with large hydrophobic groups, and thus high log *P* (e.g., greater than 3), the use of cosolvents in combination with surfactants (e.g., polysorbate 80 and Cremophor® EL, BASF, Ludwigshafen, Deutschland) may be required. Paclitaxel is an example of a molecule possessing a large hydrophobic surface, and high log *P* (3.96). Accordingly, the drug must be formulated in a combination of cosolvent and surfactants. The current formulation for infusion is a clear nonaqueous solution that is intended for dilution with a suitable parenteral fluid prior to intravenous infusion. Each milliliter contains 6 mg paclitaxel, 527 mg of purified Cremophor EL (polyoxyethylated castor oil), and 49.7% (v/v) dehydrated alcohol, US Pharmacopoeia (USP).

Biodegradable Polymers

Biodegradable polymers represent a unique class of excipients that may be used to provide sustained release of the injected therapeutic, thereby reducing frequency of injections. Table 4 provides a list of biodegradable polymer–based formulations that have been approved by Food and Drug Administration (FDA) (34). Whereas the early polymer-based products were based on polylactides and glycolides (based on their known safety and biocompatibility data from approved biodegradable suture products), a number of novel biodegradable polymers have been developed including polyanhydrides, poly(phosphate ester)s, polyesters, polyphosphazenes, and polyamino acids. A polyanhydride-based product, Gliadel®, became the first FDA-approved product for sustained–release of chemotherapeutic agents, when it was approved by FDA in 1996. This product used a copolymer of sebacic anhydride and carboxyphenoxypropane, as shown in Figure 2. The rate of degradation of biodegradable polymers may be variable, depending on the linkages and monomers used; however, in general, anhydride linkages are most labile, whereas amide linkages are most stable. The choice and design of biodegradable polymers for sustained release

Table 4 Representative List of Polymeric Sustained–Release Products Approved by Food and Drug Administration

Product	Drug	Polymer	Administration route
Lupron depot	Leuprolide acetate	PLA	SQ/IM
Zoladex	Goserelin acetate	PLA	SQ
Trelstar depot	Triptorelin pamoate	PLGA	SQ/IM
Gliadel®	Carmustin	Polyanhydride	Intracranial implantation
Nutropin depot	Human growth hormone	PLGA	SQ/IM
Sandostatin LAR	Octreotide	PLGA-glucose	SQ/IM
Risperdal consta	Risperdone	PLGA	SQ

Abbreviations: IM, intramuscular; PLA, polyactic acid; SQ, subcutaneous; PLGA, poly(lactide-co-glycolide).

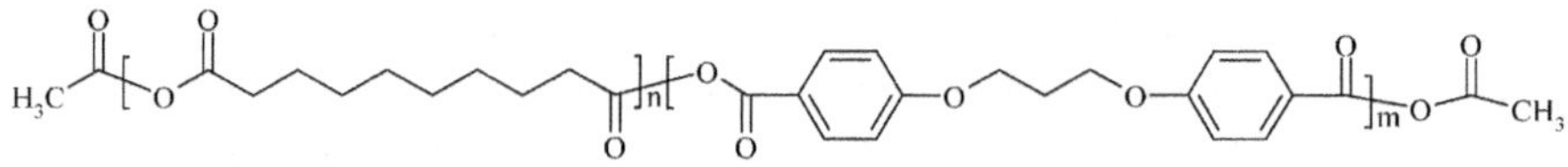

Figure 2 Structure of polifeprosan 20, a polyanhydride-based excipient used for sustained–release of Carmustine, an antineoplastic agent.

requires a thorough understanding of the polymer-degradation pathway, drug-release kinetics, and the polymer safety and biocompatibility. Biocompatibility testing typically involves an understanding of the inflammatory and healing responses of implantable materials, using inflammation, wound healing, and foreign body responses as components of the tissue or cellular host responses to injury (35–38). A number of other processing and formulation factors play a role in determining the rate of drug release from a polymer matrix (39). Some of these factors are listed in Table 5.

Tonicity Agents

Tonicity agents are added to injectable preparations to prevent osmotic shock at the site of injection upon administration, and thereby reduce local irritation. Typical excipients used for tonicity adjustment include saline, glycerin, mannitol, dextrose, and trehalose. Tonicity is a colligative property that depends primarily on the number of dissolved particles in solution. Hence, the amount of tonicity agent to be added depends on the specific formulation. Typically, osmolality of 280 to 320 mOsm is considered iso-osmotic.

Preservative Agents

Preservative agents are added to injectable preparations when multiple dosing is expected out of the same unit. Examples of preservatives used in injectable preparations include phenol, benzyl alcohol, methyl and ethyl parabens, and benzalkonium

Table 5 Typical Factors That Affect Release Kinetics for Parenteral, Polymer Sustained–Release Dosage Forms

Factor	Effect
Type of polymer	Crystallinity and molecular weight affect polymer degradability and consequently erosion-based release
Drug	Hydrophobic drugs typically have slower release rates
Polymer–drug interactions	Increased interactions may lead to slower release
Drug loading	High drug loading may cause increased release rates due to channeling effects
Encapsulation process	Encapsulation process may affect drug distribution in the matrix and subsequently the release rates
Excipients	Excipients present in the matrix may slow down or increase the drug-release rates by changing the microclimate pH, matrix porosity, or polymer–drug interactions
Processing and storage (aging)	Phase changes (e.g., crystallization of drug or carrier) within the matrix and/or chemical degradation of the matrix and drug may affect the release profile

chloride. Preservative effectiveness may depend on the pH and the presence of other excipients in the formulation. For example, a detailed screening may be required for identifying appropriate preservatives for emulsions (40). This is due to the propensity of several preservatives to partition into vesicular structures in disperse formulations such as emulsions and liposomes (41). Preservatives may also cause unanticipated instability to the formulation, especially for labile drugs such as proteins. For instance, a recent study indicated that the presence of benzyl alcohol can induce aggregation within protein solutions, through perturbation of the secondary structure (42). Efficacy of preservatives may be evaluated by in vitro tests involving challenge of the formulation with microorganisms at varying levels (43). USP tests are also available for investigating the effectiveness of preservatives (44).

Buffers

Buffers are added to injectable formulations to maintain pH control during storage. Buffers are especially important when a drug has been solubilized using the pH effect. In such a situation, buffer capacity and strength play a major role in stabilizing the formulation and avoiding drug precipitation during or after injection (45). The buffer and pH itself may induce phlebitis during injection. A detailed study indicated that pH range from 3 to 11 may not induce phlebitis upon bolus injection (46). Naturally this is not a standard rule, because other excipients and the drug itself may also play a role in the phenomenon. Some of the buffers commonly used for injectable formulations are listed in Table 6.

Inclusion Complexation

Solubilizing agents are used as excipients to dissolve poorly soluble drugs, when pH adjustment or salt formation is not sufficient to ensure adequate solubility, and the cosolvent approach is not feasible or desirable. A number of solubilizers have been explored in recent years, due to the growing trend toward poorly soluble drugs. Complexation using soluble carriers such as cyclodextrins is becoming more frequently used for solubilization and screening of poorly soluble compounds. Cyclodextrins solubilize molecules by a host-complexation mechanism. Release of the encapsulated drug occurs upon dilution after injection. The only cyclodextrins that are currently used in approved injectable products are 2-hydropropyl-beta-cyclodextrin (HP-BCD) and sulfobutylether beta-cyclodextrin. HP-BCD is found in itraconazole for injection (Sporanox® IV). Each milliliter of Sporanox® IV contains 10 mg (1%, w/v) of itraconazole, solubilized by hydroxypropyl-beta-cyclodextrin (400 mg, 40% w/v) and 2.5% (v/v) propylene glycol, and the pH adjusted to 4.5, in water for injection. The product is packaged in 25 mL colorless glass ampules, each containing 250 mg of itraconazole, the contents of which are diluted in 50 mL of 0.9% Sodium Chloride Injection, USP

Table 6 Typical Buffers Used in Injectable Formulations

pH range	Buffers
3–6	Acetate, citrate, carbonate
6–8	Phosphate
9–11	Tris(hydroxymethane)aminomethane (rarely used)

(normal saline) prior to infusion. After final dilution, the infusion contains approximately 13% HP-BCD (w/v). Cyclodextrins have been discussed in greater detail in a separate chapter and, hence, will not be covered in further detail in this review.

Surfactants

Surfactants used in injectable formulations are generally nonionic and include Tween 80, Cremophor EL, and Solutol HS. These are amphiphilic molecules that solubilize through formation of micelles. These agents have been discussed in a recent excellent review that lists various excipients that are used in FDA-approved products (47). One drawback with this approach is the need for high amounts of surfactants, and excipient-induced toxicity that can result. For example, Cremophor EL (polyoxyethylated castor oil), used in Taxol® and other FDA-approved drug products, has been associated with several side effects including severe anaphylactoid hypersensitivity reactions, hyperlipidemia, abnormal lipoprotein patterns, aggregation of erythrocytes, and peripheral neuropathy (48). These excipient-induced toxicities limit the dose level and infusion rates for the final formulation, as is the case with Taxol® (49). Concerns with such known solubilizers have resulted in intensified research in development of novel, synthetic polymer–based excipients with higher micellar solubilization capacities (50). One such novel polymeric solubilizer, poly(lactide-co-glycolide)-polyethylene glycol (PLGA-PEG), has been used in the clinic in a Cremophor EL–free formulation of paclitaxel. Because of the elimination of Cremophor EL, the newer formulation did not require premedication with anti-inflammatory agents (e.g., steroids and antihistamines) and could deliver a higher amount of paclitaxel (51a).

EXCIPIENTS FOR DELIVERY OF WATER-INSOLUBLE AGENTS

Figure 3 reveals the multiple strategies available for the formulation of poorly soluble drugs. For the most part, solubility is driven by entropy. In general, compounds that are soluble in water have two important properties—the ability to mix with water, thus resulting in a large positive entropy change, and a reasonably low cohesive energy, as reflected by the vapor pressure of the material. If the drug is a solid, then this cohesion is reflected by the melting point, which must not be prohibitively high. The propensity for the solute to mix with water, as expressed by mole fraction (χ), is reflected by its activity coefficient (γ), whereas the effect of melting is dependent on the enthalpy of fusion (ΔH_{f}) of the solute:

$$\ln\ x = -\frac{\Delta H_{\mathrm{f}}(T_{\mathrm{m}} - T)}{RT} - \ln\ \gamma \tag{2}$$

Compounds that are more polar, and which can better hydrogen-bond with water, require less drastic alterations to the solvent environment to cause dissolution to occur. On the right side of Figure 3, we associate solute polarity with each formulation concept. Drugs that are good hydrogen donors, in the extreme sense, have acidic properties. Likewise, those that are very good acceptors have basic properties. For these compounds, formation of a salt by protonation or deprotonation is a feasible route.

This strategy has limitations when the solubility (S) of the neutral drug molecule is very low. For an acidic drug molecule, an increase of one pH unit above the

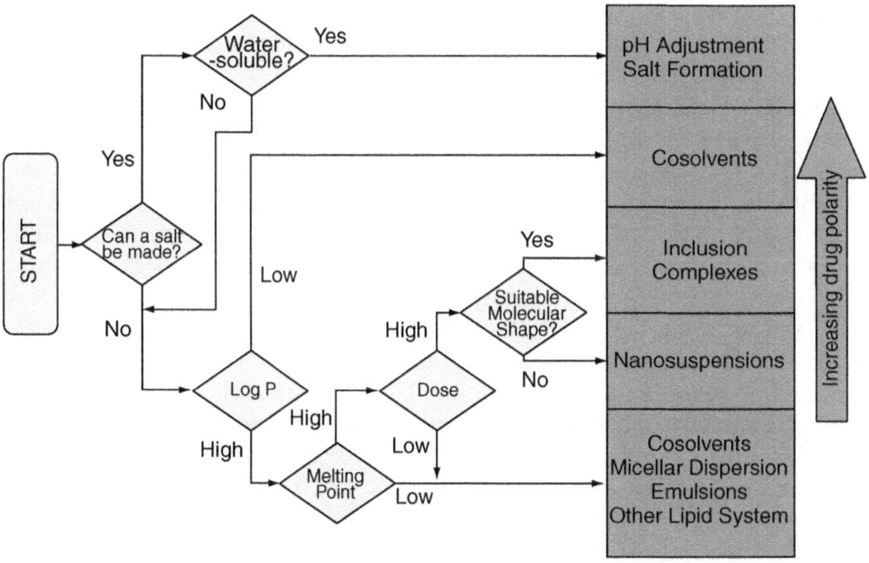

Figure 3 Formulation strategies for delivery of water-insoluble drugs.

pKa results in approximately a tenfold increase in solubility. Likewise, a decrease of one pH unit below the pKa of a basic molecule causes a tenfold solubility increase. Thiopental has a pKa of 7.4 (52a,b) and $\log S(\mathrm{mol/L}) = -3.68$ (51 μg/mL) (51b). The formulation pH of thiopental is approximately 11. Another example is phenytoin, which has a pKa of 8.06 to 8.33 (53), and an intrinsic solubility of $\log S(\mathrm{M}) = -4.26$, or 14 μg/mL (51b). At physiologic pH (7.4), the solubility of these compounds is therefore quite low, and injection of these high pH formulations, when diluted in blood, may cause precipitation near the injection site if introduction of the drug solution is rapid. Extravasation of these formulations may also cause tissue inflammation and even necrosis (54).

Pharmaceutical agents that are too hydrophobic (high log P) may not be formulated by using simple cosolvency as the primary formulation principle. This is because water miscibility requires some degree of hydrogen bonding with water. Hydrophobic compounds, which cannot effectively disrupt hydrogen bonding between water molecules, may also not interact with the water-miscible cosolvent. In such cases, solubility is often approximated as an exponential function of the volume fraction of the cosolvent (55). Adequate enhancement of solubility to reach efficacious therapeutic levels may result in solutions with very high osmolality. This may limit the applicability of cosolvents for the injection of certain drugs. In the polarity hierarchy indicated in the right-hand column in Figure 3, the use of cosolvents, such as ethanol and propylene glycol, requires some degree of solute hydrophilicity (polarity), whereas drugs with high log P (low polarity) may require either dispersed systems such as emulsions or suspensions, formulation using inclusion complexation (e.g., cyclodextrins), or a combination of cosolvency and micellization by surfactants.

Dispersed systems, such as emulsions, have also been employed to achieve high drug loading for parenteral administration. Emulsions generally consist of a vegetable oil (e.g., soybean), a phospholipid surfactant (e.g., lecithin), and glycerol added for isotonicity. The surfactant (emulsifier) is necessary to provide a barrier to agglomeration of the emulsion droplets. Unlike micellar solutions that are thermodynamically stable, emulsions and suspensions exist in a metastable form that is thermodynamically unstable. Stabilization thus hinges upon the ability to kinetically impede coalescence of droplets or particles. Derjaguin–Landau–Verwey–Overbeek (DLVO) theory has been invoked as a mechanism. The DLVO interaction energy for a system of like-charged colloidal particles comprises an attractive van der Waals interaction and a repulsive electrostatic double layer interaction (56).

The anesthetic, propofol, has been formulated as the drug emulsion, Diprivan, and has been a very successful product. Diprivan exemplifies the potential of lipid emulsions in drug delivery. This success is due primarily to the physical properties of the propofol molecule (2,6-di-isopropylphenol). Propofol is a liquid at room temperature, and yet has a high oil solubility (log $P = 3.83$, pH 6–8.5) (57). Because the drug concentration in the continuous aqueous phase affects pain on injection, further benefit is realized by keeping the drug in an oil phase. The pKa of propofol is 11, and therefore not amenable to salt formation. In addition to the active component, the formulation also contains soybean oil (10%, w/v), glycerol (2.25%, w/v), egg lecithin (1.2%, w/v), and disodium edetate (0.005%, w/v). Solution pH is 7 to 8.5. The manufacturer claims that disodium edetate acts as an antimicrobial agent.

The number of solid drugs that can be formulated as emulsions, however, is limited because the degree of water insolubility may influence insolubility of the drug in oil. Drugs that are the least soluble in water interact via hydrophobic forces (London dispersion forces), which are weaker than dipole–dipole interactions, including hydrogen bonding. Lattice energy is largely determined by these polar interactions. Thus, any solid drug with a high melting point must possess some degree of polarity, and these intermolecular forces cannot be overcome by the much weaker dispersion forces present between solute and oil. This limits solubility in oils. The energy of solution is then determined by the energy required to melt the solid, as determined by the heat of fusion and melting point, that is, the first term in Eq. (2). Most drugs that can be formulated in emulsions are generally liquids or low melting solids with high log P values.

The number of other lipid-containing drug carrier systems reported in the pharmaceutical literature is vast. Surprisingly, only a few parenteral formulations have gained acceptance. For example, liposomes have been reported in the research literature for decades but have only gained recent acceptance for amphotericin B and a small handful of antineoplastic agents such as doxorubicin. Liposomes are built from a multimolecular assemblage of phospholipids, which arrange themselves into bilayered structures, or lamellae, with one or several lamellae surrounding an aqueous center. Because the aqueous core comprises a significant fraction of the volume occupied by the unilamellar liposome, most pharmaceutical research has focused on water-soluble drugs entrapped in the aqueous core. Only a few amphiphilic drugs, such as amphotericin B, have been developed in liposomal formulations. Although they share some similar lipid components with emulsions, the ability to load lipophilic drugs into liposomes is significantly more limited. Another disadvantage is that liposomes are generally less physically stable than emulsions, and must often be lyophilized.

Suspensions offer another dispersed platform for delivery of poorly water-soluble agents. As compared with solutions, suspensions afford higher loading. However, as with emulsions, the suspended particles must be stabilized with surfactants to prevent aggregation. A number of steroids have been available for years as suspensions for IM and intra-articular delivery. Examples include DEPO-MEDROL® (Pfizer, Kalamazoo, Michigan, U.S.A.) Sterile Aqueous Suspension. The marketed product contains methylprednisolone acetate, a white, water-insoluble, crystalline powder that melts at about 215°C. Because of its high melting point and moderate log *P* (2.91) [calculated by HyperChem (Hypercube, Gainsville, Florida, U.S.), version 7.51 for Windows], it cannot be formulated in oil as an emulsion. Each milliliter contains active drug (40 or 80 mg/mL), polyethylene glycol 3350 (29 or 28 mg/mL), myristyl gamma-picolinium chloride (0.195 or 0.189 mg/mL), and sodium chloride added to adjust tonicity. Solution pH is within 3.5 to 7.0 [Depo-Medrol® (Pharmacia & Upjohn Company, Kalamazoo, Michigan, U.S.A.), methylprednisolone acetate injectable suspension, USP]. Myristyl gamma-picolinium chloride is a cationic surfactant, added in small quantities as a preservative.

The principles of particle stabilization are even more critical to the stabilization of very small solid particles, less than one micron in diameter (e.g., nanosuspensions), which have much greater surface area. The Ostwald–Freundlich equation,

$$\ln\frac{S}{S_0} = \frac{2v\gamma}{rRT} = \frac{2M\gamma}{\rho rRT} \tag{3}$$

which pertains to spherical particles, defines the effects of particle radius (r), molar volume (v), density (ρ), and interfacial tension (γ) on solubility, S, at temperature T. S_0 is the solubility of a flat, solid sheet ($r \to \infty$), M is the molecular weight of the solid, and R is the ideal gas constant. Reducing the particle size increases drug solubility, all other factors being constant. Long-term stabilization of nanosuspensions is an uphill battle against the thermodynamics of a metastable, dispersed system, and stability rests on the ability to kinetically impede this process. Instability can result from a shift in size distribution to larger particles (Ostwald ripening), irreversible agglomeration, and secondary and polymorphic nucleation. The first phenomenon is a consequence of Eq. (3), in which smaller particles must have higher saturation solubility than large particles. This concentration gradient causes growth of large particles at the expense of dissolving smaller ones. To obtain a stable suspension, the considerable potential energy created by the large interface between the solid and the surrounding medium must be reduced by adding surface-active agents. Surface stabilization may be achieved by using charged amphiphiles that migrate to the solid–liquid interface and provide an electrostatic barrier to particle agglomeration. Nonionic polymeric surfactants such as poloxamer 188, a triblock copolymer of ethylene glycol and propylene glycol, are very effective nonionic stabilizers because of multiple attachments of hydrophobic domains at the particle surface. Entropically, the probability of detachment of all of these hydrophobic moieties is very low at room temperature, thus providing a strong surface affinity (58). Nonionic surfactants may also create a hydration zone, a layer of tightly bound water molecules around each particle. When two particles meet, work is required to dislodge this water layer because of osmotic forces. Other entropic factors are also involved. The hydrophobic domains of the surfactant associate with the particle surface, with pendant hydrophilic domains extending into the aqueous medium. Attraction between particles necessitates the intertwining of these pendant chains leading to a restriction in chain mobility, and hence an unfavorable lowering of entropy (59). The combination of the entropic and enthalpic factors comprises the so-called "steric" stabilization, and may provide an effective barrier to aggregation. A combination of steric and electrostatic stabilization is often required to achieve desired shelf life. Glycol copolymers such as poloxamers or polyethylene glycols suffer, however, from reduced solubility in water at high temperatures, which may lead to particle aggregation. This results from thermally induced cleavage of hydrogen bonds between the hydrated polymer and water, leading to formation of visible polymer aggregates ("cloud point"). The ability to autoclave such formulations is limited if the cloud point lies below the sterilization temperature (121°C). Addition of cloud-point modifiers, usually anionic surfactants such as sodium dodecyl sulfate, may raise the cloud point and enhance stability at high temperature (60). Polysorbates (Tweens), poloxamines, and poloxamers have been used as nonionic surfactants. Bile salts (e.g., sodium cholate) and alkyl sulfonates (e.g., sodium dodecyl sulfate, sodium dioctylsulfosuccinate, and sodium lauryl sulfate) have been effectively used as anionic surfactants.

CONTAINER–EXCIPIENT INTERACTIONS IN INJECTABLE DOSAGE FORMS

The type of container closure system to be used should be kept in mind while selecting and optimizing the excipients, and vice versa. Typically, two issues can exist regarding container closure and formulation compatibility. The first issue may be adsorption of the drug or excipients onto the container surface (this can be the vial surface, rubber stopper, flexible container surface, injection port, etc.). For example, parabens are known to adsorb onto fluoropolymer surfaces. Hence, if the final container requires the use of Teflon-coated stoppers (e.g., FluoroTec®), it may be advisable to choose a nonparaben preservative (61). Another hypothetical example is drug absorption onto the bag, in the case of a ready-to-use formulation. In this situation, a surfactant may be added to coat the surface and minimize drug adsorption. Drug or excipient adsorption is typically characterized by exploring adsorption kinetics, isotherms, and effect of storage temperature. The second issue that can exist with container closures is leachables and extractables, especially for plastic containers or rubber stoppers (62). A formulation can at times actively extract leachables from the container, especially when organic solvents or surface-active excipients are present in the formulation.

Each container material has particular failure modes, which determine the nature of the interaction with excipients. Glass is an excellent barrier to water vapor and oxygen, but may leach aluminum, particularly at low pH or in the presence of trivalent anion chelators, such as citrate, gluconate, or phosphate (63,64). Plastics are permeable not only to gases, but potentially to excipients as well, particularly parabens, water vapor, and oils. This could potentially compromise the sterility of a solution with inadequate content of antimicrobial agent. Permeation through a plastic barrier depends on the composition of the plastic, degree of crystallinity of the plastic, permeation area, thickness of the barrier, partial pressure differential of the permeant across the barrier, and time. Specific additives, primarily plasticizers, can increase the permeation rate greatly. Loss can also occur for a nonvolatile, migratable excipient, from the solution into the container material. This phenomenon depends upon the following factors: (i) the initial amount of excipient present; (ii) the solubility limit of the leachable material in the plastic phase; (iii) the equilibrium partitioning of the leachable component between the container and the solution; and (iv) the rate of migration of the excipient from the solution into the container (65). The equilibrium solution concentration of an excipient remaining after equilibration with an absorbing plastic container material is given by:

$$C_{\mathrm{e}} = (I)/[(W_c \times E_b) + V_s] \tag{4}$$

Where I is the initial amount (g) of excipient in the solution, W_c is the weight of the container (g), V_s is the solution volume (L), and E_b is the equilibrium partitioning constant, the ratio of the concentration of solute in the film to that in water, at equilibrium (66). This can be calculated from the more familiar, and referenced, solvent–solvent partition coefficients. Plastics and rubber stoppers can also leach stabilizers and plasticizers into the contained injection volume. The extent of this can be calculated by considering the same factors described above.

SUMMARY

Injectable formulations involve a number of configurations, all of which are important for excipient selection. In spite of the challenges associated with the stringencies of injectable formulations, the number of excipients available for specific functional needs such as solubilization, stabilization, or sustained release is limited. Excipients are typically chosen from the FDA Inactive ingredients database (for the relevant route of administration). Only a handful of new excipients have been developed specifically for injectable functionalities—Captisol, Solutol HS, and polyanhydrides being three examples. A number of new excipients are now being explored for a variety of applications, as discussed in a separate chapter. It is anticipated that as challenges with injectable formulation development increase (with more challenging molecules such as proteins, DNA-based drugs, and water-insoluble drugs), drug delivery companies will lead the development of novel excipients that fit these needs. The long exercise of developing a new excipient will need convergence of drug delivery, excipient safety, and parenteral science.

REFERENCES

1. Duma RJ, Akers MJ, Turco SJ. Parenteral drug administration: routes, precautions, problems, complications, and drug delivery systems. Chapter 2. In: Avis KE, Lieberman HA, Lachman L, eds. Pharmaceutical Dosage Forms: Parenteral Medications. 1. New York: Marcel Dekker, Inc., 1992.
2. Prescott B, Kauffman G, James WG. Further studies on effect of glycerine on toxicity of isoniazid-streptomycin mixtures in mice. Antibiot Chemother 1958; 8:225.
3. Cohen LS, Rosenthal JE, Horner DW, et al. Plasma levels of lidocaine after intramuscular administration. Am J Cardiol 1948; 29:520.

4a. Feldman S, Brazeau GA. Biopharmaceutic factors influencing drug availability. Chapter 7. In: S Turco, ed. Sterile Dosage Forms. 4th ed. Philadelphia: Lea & Febiger, 1994.

4b. Steiner N, Bruera E. Methods of hydration in palliative care patients. J Palliat Care 1998; 14:6–13.

5. Strickley RG. Solubilizing excipients in oral and injectable formulations. Pharm Res 2004; 21:201–230.
6. Britton RC, Habif DV. Clinical uses of hyaluronidase. Surgery 1953; 33:917–942.
7. Zuidema J, Pieters FA, Duchateau G. Release and absorption rate aspects of intramuscularly injected pharmaceuticals. Int J Pharm 1988; 47:1–12.
8. Kostenbauder HB, Rapp RP, McGovern JP, et al. Bioavailability and single dose pharmacokinetics of intramuscular phenytoin. Clin Pharmacol Ther 1975; 18:449.
9. Schlichtkrull J. Insulin Crystals. Copenhagen: Ejnar Munksgaard Publisher, 1958.
10. Buckwalter FJ, Dickison HL. The effect of vehicle and particle size on the absorption, by the intramuscular route, of procaine penicillin G suspension. J Am Pharm Assoc; Sci Ed 1958; 47:661.

11. Malmsten M. Liposomes. Chapter 4. In: Malmsten M, ed. Surfactants and Polymers in Drug Delivery. New York: Marcel Dekker, Inc., 2002.
12. Jackson AJ. Intramuscular absorption and regional lymphatic uptake of liposome-entrapped insulin. Drug Met Disp 1981; 9:535.
13. Tabata Y, Ikada Y. Phagocytosis of polymer microspheres by macrophages. Adv Polym Sci 1990; 94:107–141.
14. Allen TM, Stuart DD. Liposome pharmacokinetics. Chapter 3. In: Janoff AS, ed. Liposomes. New York: Marcel Dekker, Inc., 1999.
15. Moghimi SM, Hunter AC, Murray JC. Long-circulating and target-specific nanoparticles: theory to practice. Pharmacol Rev 2001; 53:283–318.
16. Reed KW, Yalkowsky SH. Lysis of human red blood cells in the presence of various cosolvents. J Parent Sci Tech 1985 ; 39:64–68.
17. Reed KW, Yalkowsky SH. Lysis of human red blood cells in the presence of various cosolvents. II. The effect of differing NaCl concentrations. Bull Parent Drug Assoc 1986; 40:88–94.
18. Al-Suwayeh SA, Tebbett IR, Wielbo D, Brazeau GA. In vitro-in vivo myotoxicity of intramuscular liposomal formulations. Pharm Res 1996; 13:1384–1388.
19. Akers MJ. Excipient-drug interactions in parenteral formulations. J Pharm Sci 2002; 91:2283–2300.
20. Weiner ML, Kotkoskie LA, eds. Excipient Toxicity and Safety. New York: Marcel Dekker, Inc., 2000.
21. Shackney S, Hasson J. Precipitious fall in serum calcium hypotension and acute renal failure after IV phosphate treatment of hypercalcemia. Arch Intern Med 1968; 122:150.
22. Spencer H. Studies of the effect of chelating agents in man. Ann N Y Acad Sci 1960; 88:435.
23. Thomas D, Hales P. Overhydration during transurethral resection of the prostate using glycine as an irrigating solution. Anesth Intensive Care 1984; 12:366–369.
24. Dowdy EG, Holland WC, Yamanaka I, Kaya K. Cardioactive properties of d-tubocurarine with and without preservatives. Anesthesiology 1971; 34:256–260.
25. Benda GI, Hiller JL, Reynolds JW. Benzyl alcohol toxicity: impact on neurologic handicaps among surviving very low birth weight infants. Pediatrics 1986; 77(4):507–512.
26. Moller H. Merthiolate allergy: a nationwide iatrogenic sensitization. Acta Derm Venereol (Stokh.) 1977; 57:509–517.
27. Gunnison AF, Jaconsen DW. Sulfite sensitivity. A critical review. Crit Rev Toxicol 1987; 17:185–214.
28. Ring J, Messmer K. Incidence and severity of anaphylactoid reactions to colloid volume substitutes. Lancet 1977; 1:466–469.
29. Poothullil J, Shimizu A, Day RP, Dolovich J. Anaphylaxis from the product(s) of ethylene oxide gas. Ann Intern Med 1975; 82:58–60.
30. Gupta PK, Patel JP, Hahn KR. Evaluation of pain and irritation following local administration of parenteral formulations using the rat paw lick model. J Pharm Sci Technol 1994(May-Jun); 48(3):159–166.
31. Mottu F, Stelling MJ, Rufenacht DA, Doelker E. Comparative hemolytic activity of undiluted organic water-miscible solvents for intravenous and intra-arterial injection. PDA J Pharm Sci Technol 2001(Jan-Feb); 55(1):16–23.
32. Mottu F, Laurent A, Rufenacht DA, Doelker E. Organic solvents for pharmaceutical parenterals and embolic liquids: a review of toxicity data. PDA J Pharm Sci Technol 2000(Nov); 54(6):456–469.
33. Yalkowsky SH, Valvani SC, Johnson BW. In vitro method for detecting precipitation of parenteral formulations after injection. J Pharm Sci 1983(Sep); 72(9):1014–1017.
34. Chaubal MV. Polyactides/glycolides—excipients for injectable drug delivery and beyond. Drug Delivery Tech 2002; 2(5).
35. Anderson JM. In vivo biocompatibility of implantable delivery systems and biomaterials. Eur J Pharm Biopharm 1994; 40:1–8.

36. Anderson JM. Mechanisms of inflammation and infection with implanted devices. Cardiovasc Pathol 1993; 23(suppl):33S–41S.
37. Anderson JM. Perspectives on the in vivo responses of biodegradable polymers. In: Hollinger JO, ed. Biomedical Applications of Synthetic Biodegradable Polymers. Boca Raton, FL: CRC Press, 1995:223–233.
38. Anderson JM. Inflammation and the foreign body response. Prob Gen Surg 1994; 11: 147–160.
39. Chaubal MV. Role of excipients in parenteral sustained-release formulations. Drug Deliv Tech 2003; 3(7).
40. Sznitowska M, Janicki S, Dabrowska EA, Gajewska M. Physicochemical screening of antimicrobial agents as potential preservatives for submicron emulsions. Eur J Pharm Sci 2002(Jun); 15(5):489–495.
41. El Jastimi R, Edwards K, Lafleur M. Characterization of permeability and morphological perturbations induced by nisin on phosphatidylcholine membranes. Biophys J 1999(Aug); 77(2):842–852.
42. Zhang Y, Roy S, Jones LS, et al. Mechanism for benzyl alcohol-induced aggregation of recombinant human interleukin-1 receptor antagonist in aqueous solution. J Pharm Sci 2004(Dec); 93(12):3076–3089.
43. Akers MJ, Boand AV, Binkley DA. Preformulation method for parenteral preservative efficacy evaluation. J Pharm Sci 1984(Jul); 73(7):903–905.
44. Sutton SV, Porter D. Development of the antimicrobial effectiveness test as USP chapter 51. PDA J Pharm Sci Technol 2002(Nov-Dec); 56(6):300–311.
45. Alvarez-Nunez FA, Yalkowsky SH. Buffer capacity and precipitation control of pH solubilized phenytoin formulations. Int J Pharm 1999(Aug 5); 185(1):45–49.
46. Simamora P, Pinsuwan S, Alvarez JM, Myrdal PB, Yalkowsky SH. Effect of pH on injection phlebitis. J Pharm Sci 1995(Apr); 84(4):520–522.
47. Strickley RG. Solubilizing excipients in oral and injectable formulations. Pharm Res 2004(Feb); 21(2):201–230.
48. Gelderblom H, Verweij J, Nooter K, Sparreboom A. Cremophor EL: the drawbacks and advantages of vehicle selection for drug formulation. Eur J Cancer 2001(Sep); 37(13): 1590–1598.
49. Price KS, Castells MC. Taxol reactions. Allergy Asthma Proc 2002(May-Jun); 23(3): 205–208.
50. Kwon GS. Polymeric micelles for delivery of poorly water-soluble compounds. Crit Rev Ther Drug Carrier Syst 2003; 20(5):357–403.
51a. Kim TY, Kim DW, Chung JY, et al. Phase I and pharmacokinetic study of Genexol-PM, a cremophor-free, polymeric micelle-formulated paclitaxel, in patients with advanced malignancies. Clin Cancer Res 2004(Jun 1); 10(11):3708–3716.
51b. Yalkowsky SH, He Y. Handbook of Aqueous Solubility Data. Boca Raton, FL: CRC Press, 2003:764.
52a. Thiopental package insert (Pentothal, Abbott—US), Rev 7/84, Rec 5/86.
52b. Wood M, Wood AJ, eds. Drugs and Anesthesia: Pharmacology for Anesthesiologists. Baltimore, MD: Williams & Wilkins; 1983:200 19.
53. Larsen JR, Larsen LS. Clinical features and management of poisoning due to phenytoin. Med Toxicol Adverse Drug Exp 1989; 4(4):229–245.
54. Wheless JW. Pediatric use of intravenous and intramuscular phenytoin: lessons learned. J Child Neurol 1998; 13(Suppl 1):S11–S14; discussion S30–S32.
55. Yalkowsky SH. Solubility and Solubilization in Aqueous Media. Washington, D.C.: American Chemical Society, 1999:184–226.
56. (a) Derjaguin BV, Landau L. Theory of stability of highly charged lyophobic sols and adhesion of highly charged particles in solutions of electrolytes. Acta Physicochimica (USSR) 1941; 4:633; (b) Verwey EJ, Overbeek J, Th G. Theory of the stability of lyophobic colloids. Amsterdam: Elsevier, 1948; (c) Crocker JC, Grier DG. When like charges attract:

the effects of geometrical confinement on long-range colloidal interactions. Phys Rev Lett 1996; 77(9):1897–1900.
57. Diprivan® (propofol for injection) package insert. AstraZeneca 2001, 2004.
58. Alexandridis P, Hatton TA. Poly(ethylene oxide)-poly(propylene oxide)- poly(ethylene oxide) block copolymer surfactants in aqueous solutions and at interfaces: thermodynamics, structure, dynamics, and modeling (review). Colloid Surf A: Physicochem Eng Aspects 1995; 96:1–46.
59. Lee RW, Shaw JM, McShane J, Wood RW. Particle size reduction. In: Liu R, ed. Water-Insoluble Drug Formulation. Interpharm, 2000:478.
60. Lee RW, Shaw JM, McShane J, Wood RW. Particle size reduction. In: Liu R, ed. Water-Insoluble Drug Formulation. Interpharm, 2000:487–488.
61. Bahal SM, Romansky JM. Sorption of parabens by fluoropolymer or fluoropolymer laminated flexible tubing. PDA J Pharm Sci Technol 2001(Nov-Dec); 55(6):417–421.
62. Jenke D, Poss M, Story J, Odufu A, Zietlow D, Tsilipetros T. Development and validation of chromatographic methods for the identification and quantitation of organic compounds leached from a laminated polyolefin material. J Chromatogr Sci 2004(Aug); 42(7):388–395.
63. Hayes P, Martin T, Pybus J. Aluminum content of intravenous solutions, additives and equipment used to prepare parenteral nutrition mixtures. Aust J Hosp Pharm 1992; 22:353.
64. Rabinow B et al. Aluminum in parenteral products: analysis, reduction, and implications for pediatric TPN. J Parent Sci Tech 1989; 43:132.
65. Rabinow BE, Roseman TR. Plastic packaging materials. Chapter 54. In: Remington: The Science and Practice of Pharmacy. 21st ed. Philadelphia, PA: Lippincott Williams & Wilkins, 2005.
66. Jenke DR, Chess EK, Zietlow D, Rabinow BE. Model for estimating the accumulation of solutes leaching from polymeric containers into parenteral solutions. Int J Pharm 1992; 78:115.

17
Excipients for Protein Drugs

Yatin R. Gokarn, Andrew Kosky, Eva Kras, Arnold McAuley, and Richard L. Remmele, Jr.
Department of Pharmaceutics, Amgen Inc., Thousand Oaks, California, U.S.A.

INTRODUCTION

With the advent of recombinant DNA technology, protein-based drugs have become continually and increasingly commonplace in the repertoire of medicines available to medical practitioners for the treatment of a wide range of diseases from cancer to rheumatoid arthritis. Along with the scientific and technical advances in the production of recombinant proteins, the principal reason for the success of protein drugs is their high specificity towards targets and superior safety profiles when compared to their small molecule counterparts.

Among the first recombinant protein drugs to be approved were naturally occurring growth factors, hormones, blood factors, and cytokines. These were followed by monoclonal antibody (mAb)–based therapies. The initial antibody-based drugs were met with limited success because they were of nonhuman origin (e.g., murine) and were often prone to immunological reactions. However, as technologies to express and produce recombinant humanized and fully human monoclonal antibodies have matured, there has been a tremendous resurgence in the success and development of mAb-based drugs. As of the year 2005, there were more than 150 approved protein-based pharmaceuticals on the market, and this number is expected to rise dramatically in the coming years (1).

Although stable formulations for small-molecule drugs can be developed by minimizing chemical degradation pathways, protein formulation development can be far more challenging. Proteins are complex molecules with defined primary, secondary, tertiary, and in some cases quaternary structures, all of which are essential for highly specific biological functions. In addition to chemical degradation, protein drugs are susceptible to the physical degradation process of irreversible aggregation. Protein aggregation is of particular interest because it often results in diminished bioactivity that affects drug potency (2), and can also elicit serious immunological or antigenic reactions in patients (3). Chemical degradation of a protein drug has also been implicated in increasing its immunogenic potential (4). Thus, stable protein formulations require that both physical and chemical degradation pathways of the drug be minimized.

The formulation of a parenterally delivered drug can affect the marketability of the product, specifically by impacting the ease and frequency of administration and the pain occurring on injection. The formulation of a protein drug can also have a critical impact on its safety. For example, the immunogenic reaction in patients taking recombinant human interferon alpha2 (rhIFNα2) has not only been attributed to aggregates of rhIFNα2, but also to mixed aggregates of rhIFNα2 and human serum albumin (HSA), used as an excipient in the formulation (5,6). Another example is that of Eprex, an erythropoietic agent for the treatment of patients with chronic renal failure. A sudden increase in the incidence rate of pure red cell aplasia (PRCA), a debilitating immunogenic reaction to endogenous erythropoietin in patients receiving Eprex treatment, coincided with a formulation change of this product (7–9). The safety and efficacy of a protein drug product is inseparably coupled to its formulation.

Excipients are additives that are included in a formulation, because they either impart or enhance the stability, delivery, and manufacturability of a drug product. Regardless of the reason for their inclusion, excipients are an integral component of a drug product and therefore need to be safe and well tolerated by patients. For protein drugs, the choice of excipients is particularly important because they can affect both efficacy and immunogenicity of the drug. Hence, protein formulations need to be developed with appropriate selection of excipients that afford suitable stability, safety, and marketability.

Protein-based drugs have been formulated mainly as stable liquids or in cases where liquid stability is limiting as lyophilized dosage forms to be reconstituted with a suitable diluent prior to injection. This is because their delivery has been limited primarily to the parenteral routes of intravenous (IV), subcutaneous (SC), or intramuscular (IM) administration. There are a few drugs that have been developed for pulmonary delivery, such as rhDNase (Pulmozyme®) and an inhalable formulation of insulin (e.g., Exubra®). However, even such drugs have been formulated as either liquid or lyophilized or spray-dried powders. This chapter will focus only on excipients that are applicable to liquid and lyophilized protein formulations.

We begin with a brief summary of the degradation pathways of proteins, followed by a discussion on the composition of liquid and lyophilized protein formulations and on various excipients in some detail. An important feature of this chapter is a comprehensive table (Appendix), which details the formulations of approved protein drugs through the year 2005. The table has been compiled with the help of several sources (1,10,11).

DEGRADATION PATHWAYS OF PROTEINS

A protein can undergo a variety of covalent and noncovalent reactions or modifications in solution. Identifying solution conditions that minimize the protein's reactivity with respect to all potential degradation pathways is a significant challenge for the formulation scientist. Protein degradation pathways can be classified into two main categories: (i) physical degradation or noncovalent pathways, and (ii) chemical or covalent degradation pathways. These are discussed in brief below.

Physical Degradation of Proteins

Proteins can degrade via the physical processes of interfacial adsorption and aggregation. Proteins are surface-active molecules, i.e., they tend to adsorb at liquid–solid, liquid–air, and liquid–liquid interfaces. It is well established that proteins fold into their unique three-dimensional structures, which consist of a hydrophobic core and a solvent exposed hydrophilic surface. Sufficient exposure of a protein's core at a hydrophobic surface can result in adsorption as a consequence of agitation, temperature or pH induced stresses. Partial unfolding is not always necessary for surface adsorption. Adsorption can occur through hydrophobic patches on a protein's surface or via electrostatic interactions. For example, no changes to secondary or tertiary structure of recombinant human platelet-activating factor acetylhydrolase were observed upon its adsorption to nano-sized silica particles (12). Adsorption can significantly impact a protein drug's potency and stability. It can cause an appreciable loss in potency for low concentration dosage forms. A second consequence is that unfolding-mediated adsorption at interfaces can often be an initiating step for irreversible aggregation in solution (13,14).

Irreversible aggregation, as the term implies, results from an irreversible reaction(s) between protein molecules that leads to the formation of soluble or insoluble higher order multimers. It is also commonly referred to as "aggregation." This process differs distinctly from the reversible self-assembly processes of self or hetero-association which in general are ordered, obey the law of mass action, and play a central role in signal transduction and execution of biological function. Protein aggregation is not only an important instability encountered during biopharmaceutical drug development but has also been implicated in a variety of diseases such as Parkinson's disease, Alzheimer's disease, and systemic amyloidosis. Aggregation has been studied extensively over the last 60 years and excellent reviews are available on the subject (15–17). Here we have discussed it briefly in context with protein formulation development.

Proteins are inherently unstable molecules, sensitive to pH, ionic strength, and thermal, shear, and interfacial stresses, all of which can lead to aggregation. It is now recognized that aggregation often occurs via the non-native protein conformational changes that can include structural alterations (i.e., misfolding) and or partial unfolding (15,18). The aggregation reaction, represented by the reaction scheme I below, can be viewed as an irreversible, kinetically controlled reaction that is coupled to a thermodynamic equilibrium governing the conformational stability of a protein (19).

$$P \overset{\mathrm{K}}{\longleftrightarrow} P* \overset{k}{\longrightarrow} A \qquad (1)$$

In reaction (1), P is the native form of the protein, P^* its aggregation competent non-native form, and A, the aggregate. Thus, aggregation can be inhibited by modulating solution conditions that favor the native form (P) and also by lowering the kinetic reaction rate constant (k). It should be noted that reaction I applies strictly to aggregation occurring in a bulk solution. Aggregation also occurs at interfaces. It is imperative to understand the predominant pathway(s) of aggregation for a given protein during development. This can guide the choice of excipients included in a given formulation. For example, certain excipients such as polyols and sugars help maintain a protein in its more compact native state. Modulation of the ionic strength can lower the kinetic rate constant, k, in reaction 1. Surfactants can inhibit surface-induced aggregation phenomena. In summary, controlling aggregation requires an understanding of the underlying mechanism(s) of aggregation, properties of excipients intended for use, and the interaction between excipients and the protein drug.

Chemical or Covalent Degradation of Proteins

Proteins are subject to a variety of chemical modification/degradation reactions, viz. deamidation, isomerization, hydrolysis, disulfide scrambling, beta-elimination, and oxidation. The principal hydrolytic mechanisms of degradation include peptide bond hydrolysis, deamidation of asparagine and glutamine, and the isomerization of aspartic acid. A common feature of the hydrolytic degradation pathways is that the most significant formulation variable, with respect to the rates of the reactions, is the solution pH. The hydrolysis of peptide bonds is acid or base catalyzed (20,21). The principal sites of the peptide cleavage reaction are at Asp-Yyy bonds (where Yyy is the residue C-terminal to Asp); however, at extreme pH conditions (pH <2 and >11), all peptide bonds are subject to cleavage in the time scale of a therapeutic protein shelf life. Asparagine and glutamine deamidation are also acid catalyzed below pH 4; the acid catalyzed reaction involves direct hydrolysis of the side-chain amide groups (22). Asparagine deamidation at neutral pH occurs through a succinimidyl intermediate that is base catalyzed and rapid (23). The isomerization (generation of isoaspartic acid) and racemization of aspartic acid residues also occurs through a cyclicimide intermediate. The formation of cyclicimide can be rapid in slightly acidic to neutral pH (pH 4–8) and is subject to racemization that leads to the formation of D-aspartic acid and D-isoaspartic acid residues (20). In addition to the generalized pH effects, buffer salts and other excipients can affect the rates of the hydrolytic reactions. For example, buffer salts have been shown to catalyze deamidation reactions (22–24), whereas additives that reduce the dielectric properties of the solvent have been shown to reduce the deamidation rates (25).

Beta-elimination reactions have been observed in a number of proteins. This reaction occurs primarily at alkaline pH conditions. Abstraction of the hydrogen atom from the alpha-carbon of a cysteine, serine, threonine, phenylalanine, or lysine residue leads to racemization or loss of part of the side chain and the formation of dehydroalanine (26).

Oxidation of methionine, cysteine, histidine, tyrosine, and tryptophan residues are common covalent degradation pathways for proteins. The most commonly observed modification is probably the formation of methionine sulfoxide. Interestingly, auto-oxidation of proteins by molecular oxygen has been shown to be very slow and not attributable to significant protein modification (27). Instead, oxidation is induced following the formation of reactive oxygen species. The reactive oxygen species can originate as products of a transition metal catalyzed reaction, upon exposure of oxygen containing compounds to ultraviolet light, or as contaminants in excipients used in the manufacture or formulation of the final product. Excellent reviews are available on this topic (28). Not surprising, the formulation approaches used to minimize oxidation primarily involve identifying the source of the pro-oxidant and obtaining a purer supply of the excipient.

The stability of proteins toward covalent degradation pathways can often depend on the protein's folded state. In each pathway, solvent accessibility and varying degrees of structural freedom of the peptide backbone and/or side chains around the labile residue are required for reactions to take place. Accordingly, stabilization of the protein's folded state (i.e., its compact structure) that minimizes solvent accessibility can lower the reaction rate of some covalent protein modifications, extending the shelf life of the protein product. Therefore, the selection of formulation excipients depends on their direct and indirect influence on the rates of covalent protein degradation.

COMPONENTS OF LIQUID AND LYOPHILIZED PROTEIN FORMULATIONS

In developing any formulation, excipients need to be selected only when their use is essential in imparting a desired pharmaceutical effect (i.e., stability or delivery).

In fact, it is a regulatory expectation that an appropriate excipient be chosen and its level (amount) in a formulation be demonstrated and justified through formulation screening and development studies (29,30). The science of protein formulation development has become increasingly sophisticated over the past 20 years and its discussion is beyond the scope of this chapter. The interested reader is referred to excellent reviews on this subject for further study (31–34).

While each formulation is unique, there are several general aspects with respect to excipient components in both liquid and lyophilized protein formulations. A comparison of the excipient components in liquid and lyophilized protein formulations is provided in Table 1.

A liquid formulation is usually comprised of a buffering agent, a stabilizer (which may also serve as a tonicity agent), a surfactant, and an anti-oxidant when protein oxidation is significant. Chelating agents are employed when metal ion catalyzed reactions predominate. A preservative may be included when a multi-dose formulation is desired.

A lyophilized formulation is usually comprised of a buffer, a bulking agent, and a stabilizer. The utility of a surfactant may be evaluated and selected in cases where aggregation during the lyophilization step or during reconstitution becomes an issue. An appropriate buffering agent is included to maintain the formulation within stable zones of pH during lyophilization.

Bulking agents are typically used in lyophilized formulations to enhance product elegance and to prevent blowout. Conditions in the formulation are generally designed so that the bulking agent crystallizes out of the frozen amorphous phase (either during freezing or annealing above the T_g) giving the cake structure and bulk. Mannitol and glycine are examples of commonly used bulking agents.

Stabilizers include a class of compounds that can serve as cryoprotectants, lyoprotectants, and glass forming agents (35). Cryoprotectants act to stabilize proteins either during freezing or in the frozen state at low temperatures. Lyoprotectants stabilize proteins in the freeze-dried solid dosage form by preserving the native-like conformational properties of the protein during dehydration stages of freeze drying. Glassy state properties have been classified as "strong" or "fragile," depending on their relaxation properties, as a function of temperature (36). It is important that cryoprotectants, lyoprotectants, and glass-forming agents remain in the same phase with the protein to impart stability. Sugars, polymers, and polyols fall into this category and can sometimes serve all three roles.

EXCIPIENTS

The principal challenge in developing formulations for therapeutic proteins is stabilizing the product against the stresses of manufacturing, shipping, and storage. The role of formulation excipients is to provide stabilization against these stresses. Excipients may also be employed to reduce viscosity of highly concentrated protein formulations to enable their delivery and enhance patient convenience. In general, excipients can be classified on the basis of the mechanisms by which they stabilize proteins against various chemical and physical stresses. Some excipients are used to alleviate the effects of a specific stress or to regulate a particular susceptibility of a specific protein. Other excipients have more general effects on the physical and covalent stabilities of proteins. The excipients described in this chapter are organized either by their chemical type or their functional role in formulations. Brief descriptions of the modes of stabilization

Table 1 Excipient Components of Liquid and Lyophilized Protein Formulations

Excipient component	Function in formulation: Liquid	Function in formulation: Lyophilized
Buffer	Maintain pH of formulation through product shelf life	Maintain pH of formulation during lyophilization and upon reconstitution
Tonicity agent/stabilizer	Provides isotonicity to the formulation such that it is suitable for injection	Stabilizers include cryo and lyoprotectants
	Examples include polyols, salts, and amino acids	Examples include polyols, sugars, and polymers
	Help maintain the protein in a more compact state (polyols)	Cryoprotectants protect proteins from freezing stresses
	Minimize electrostatic, solution protein–protein interactions (salts)	Lyoprotectants stabilize proteins in the freeze-dried state
Bulking agent	Not applicable	Used to enhance product elegance and to prevent blowout
		Provides structural strength to the lyo cake
		Examples include mannitol and glycine
Surfactant	Prevent/control aggregation, particle formation and surface adsorption of drug	Employed if aggregation during the lyophilization process is an issue
	Examples include polysorbate 20 and 80	May serve to reduce reconstitution times
		Examples include polysorbate 20 and 80
Antioxidant	Control protein oxidation	Usually not employed, molecular reactions in the lyo cake are greatly retarded
Metal ions/chelating agent	A specific metal ion is included in a liquid formulation only as a co-factor	May be included if a specific metal ion is included only as a co-factor
	Divalent cations such as zinc and magnesium are utilized in suspension formulations	Chelating agents are generally not needed in lyo formulations
		Chelating agents are used to inhibit heavy metal ion catalyzed reactions
Preservative	For multi-dose formulations only	For multi-dose formulations only
	Provides protection against microbial growth in formulation	Provides protection against microbial growth in formulation
	E.g., benzyl alcohol	Is usually included in the reconstitution diluent (e.g., BWFI)

Table 2 Commonly Used Buffering Agents and Their pK_a Values

Buffer	pK_a	Example drug product
Acetate	4.8	Neupogen®, Neulasta®
Succinate	$pK_{a1}=4.8$, $pK_{a2}=5.5$	Actimmune®
Citrate	$pK_{a1}=3.1$, $pK_{a2}=4.8$, $pK_{a3}=6.4$	Humira®
Histidine (imidazole)	6.0	Xolair®
Phosphate	$pK_{a1}=2.15$, $pK_{a2}=7.2$, $pK_{a3}=12.3$	Enbrel® (liquid formulation)
Tris	8.1	Leukine®

are provided when discussing each excipient type. Detailed descriptions of the mechanisms of protein stabilization may be found elsewhere (37–39).

Buffers

The stability of a protein drug is usually observed to be maximal in a narrow pH range. This pH range of optimal stability needs to be identified early during preformulation studies. Several approaches such as accelerated stability studies and calorimetric screening studies have been demonstrated to be useful in this endeavor (40). Once a formulation is finalized, the drug product must be manufactured and maintained within a predefined specification throughout its shelf life. Hence, buffering agents are almost always employed to control pH in the formulation.

Organic acids, phosphates, and Tris have been employed routinely as buffers in protein formulations (Table 2).

The buffer capacity of the buffering species is maximal at a pH equal to the pK_a and decreases as pH increases or decreases away from this value. Ninety percent of the buffering capacity exists within one pH unit of its pK_a. Buffer capacity also increases proportionally with increasing buffer concentration.

Several factors need to be considered when choosing a buffer. First and foremost, the buffer species and its concentration need to be defined based on its pK_a and the desired formulation pH. Second, equally important, ensuring that the buffer is compatible with the protein drug and with other formulation excipients, and does not catalyze any degradation reactions. Recently, polyanionic carboxylate buffers such as citrate and succinate have been shown to form covalent adducts with the side-chain residues of proteins [Unpublished results, personal communication with Nicole Piros (piros.nicole@gene.com) and Thomas Patapoff (patapoff.thomas@gene.com), Genentech Inc., One DNA way, South San Francisco, California, U.S.]. A third important aspect to be considered is the sensation of stinging and irritation the buffer may induce. For example, citrate is known to cause stinging upon injection (41). The potential for stinging and irritation is greater for drugs that are administered via the SC or IM routes, where the drug solution remains at the site for a relatively longer period of time than when administered by the IV route, where the formulation gets diluted rapidly into the blood upon administration. For formulations that are administered by direct IV infusion, the total amount of buffer (and any other formulation component) needs to be monitored. One has to be particularly careful about potassium ions administered in the form of the potassium phosphate buffer, which can induce cardiovascular effects in a patient (42).

Buffers for lyophilized formulations need additional consideration. Some buffers like sodium phosphate can crystallize out of the protein amorphous phase during freezing resulting in rather large shifts in pH. Other common buffers such as acetate and imidazole should be avoided since they may sublime or evaporate during the lyophilization process, thereby shifting the pH of formulation during lyophilization or after reconstitution.

Salts

Salts are often added to protein parenterals to increase the ionic strength of the formulation, which can be important for protein solubility, physical stability, and isotonicity. Salts can affect the physical stability of proteins in a variety of ways. Ions can stabilize the native state of proteins by binding to charged residues on the protein's surface. Alternatively, they can stabilize the denatured state by binding to the peptide groups along the protein backbone (–CONH–). Salts can also stabilize the protein native conformation by shielding repulsive electrostatic interactions between residues within a protein molecule. Electrolytes in protein formulations can also shield attractive electrostatic interactions between protein molecules that can lead to protein aggregation and insolubility.

The effect of salt on the stability and solubility of proteins varies significantly with the characteristics of the ionic species. The Hofmeister series originated in the 1880s as a way to rank order electrolytes based on their ability to precipitate proteins (43). In this report, the Hofmeister series is used to illustrate a scale of protein stabilization effects by ionic and nonionic co-solutes. In Table 3, co-solutes are ordered with respect to their general effects on solution state proteins, from stabilizing (kosmotropic) to destabilizing (chaotropic). In general, the differences in effects across the anions are far greater than that observed between the cations, and, for both types, the effects are most apparent at higher concentrations than are acceptable in parenteral formulations. High concentrations of kosmotropes (e.g., >1 M ammonium sulfate) are commonly used to precipitate proteins from solution by a process called "salting-out" where the kosmotrope is preferentially excluded from the protein surface reducing the solubility of the protein in its native (folded) conformation. Removal or dilution of the salt will return the protein to solution. The term "salting-in" refers to the use of

Table 3 The Hofmeister Series of Salts

Cosolute			Stabilization scales	
Anion	Cation	Other		
F^-	$(CH_3)_4N^+$	Glycerol/sorbitol	Stabilizing (salting-out)	Kosmotropic
PO_4^-	$(CH_3)_2NH^+$	Sucrose/trehalose	↑	↑
SO_4^-	NH_4^+	TMAO		
$CHCOO^-$	K^+			
Cl^-	Na^+			
Br^-	Cs^+			
I^-	Li^+		↓	↓
	Mg^{2+}	Guanidine	Destabilizing (salting-in)	Chaotropic
	Ca^{2+}	Arginine		
	Ba^{2+}	Urea		

destabilizing ions (e.g., like guanidine and chloride) that increase the solubility of proteins by solvating the peptide bonds of the protein backbone. Increasing concentrations of the chaotrope will favor the denatured (unfolded) state conformation of the protein as the solubility of the peptide chain increases. The relative effectiveness of ions to "salt-in" and "salt-out" defines their position in the Hofmeister series.

In order to maintain isotonicity in a parenteral formulation, salt concentrations are generally limited to less than 150 mM for monovalent ion combinations. In this concentration range, the mechanism of salt stabilization is probably due to screening of electrostatic repulsive intramolecular forces or attractive intermolecular forces (Debye-Huckel screening) (15,38). Interestingly, chaotropic salts have been shown to be more effective at stabilizing the protein structure than similar concentrations of kosmotropes by this mechanism. The chaotropic anions are believed to bind more strongly than the kosmotropic ions (43,44). With respect to covalent protein degradation, differential effects of ionic strength on this mechanism are expected based upon Debye-Huckel theory. Accordingly, published reports of protein stabilization by sodium chloride are accompanied by those where sodium chloride accelerated covalent degradation (45,46). The mechanisms by which salts affect protein stability are protein specific and may vary significantly as a function of solution pH. An example where an excipient can be useful in enabling the delivery of a protein drug is that of some high concentration antibody formulations. Recently, salts have been shown to be effective in reducing the viscosity of such formulations (47).

Amino Acids

Amino acids have found versatile use in protein formulations as buffers, bulking agents, stabilizers, and antioxidants. Histidine and glutamic acid are employed to buffer protein formulations in the pH range of 5.5 to 6.5 and 4.0 to 5.5, respectively. The imidazole group of histidine has a $pK_a = 6.0$ and the carboxyl group of the glutamic acid side chain has a pK_a of 4.3 which makes them suitable for buffering agents in their respective pH ranges. Acetate, the most commonly used buffer in the acidic pH range of 4.0 to 5.5, sublimates during lyophilization and hence should be avoided in freeze-dried formulations. Glutamic acid is particularly useful in such cases (e.g., Stemgen®). Histidine is commonly found in marketed protein formulations (e.g., Xolair®, Herceptin®, Recombinate®). It provides a good alternative to citrate, a buffer known to sting upon injection. Interestingly, histidine has also been reported to have a stabilizing effect on ABX-IL8 (an IgG2 antibody) with respect to aggregation when used at high concentrations in both liquid and lyophilized presentations (48). Histidine (up to 60 mM) was also observed to reduce the viscosity of a high concentration formulation of this antibody. However, in the same study, the authors observed increased aggregation and discoloration in histidine-containing formulations during freeze-thaw studies of the antibody in stainless steel containers. They attributed this to an effect of iron ions leached from corrosion of steel containers. Another note of caution with histidine is that it undergoes photo-oxidation in the presence of metal ions (49). The use of methionine as an antioxidant in formulations appears promising; it has been observed to be effective against a number of oxidative stresses (50).

The amino acids glycine, proline, serine, and alanine have been shown to stabilize proteins by the mechanism of preferential exclusion (51). Glycine is also a commonly used bulking agent in lyophilized formulations (e.g., Neumega®, Genotropin®, and Humatrope®). It crystallizes out of the frozen amorphous phase, giving the cake structure and bulk. Arginine has been shown to be an effective agent in inhibiting aggregation and has been used in both liquid and lyophilized formulations (e.g., Activase®, Avonex®, and Enbrel® liquid). Furthermore, the enhanced efficiency of refolding of certain proteins in the presence of arginine has been attributed to its suppression of the competing aggregation reaction during refolding (52).

Polyols

Polyols encompass a class of excipients that includes sugars (e.g., mannitol, sucrose, and sorbitol), and other polyhydric alcohols (e.g., glycerol and propylene glycol). We have included the polymer polyethylene glycol (PEG) in this category for ease of discussion. Polyols are commonly used as stabilizing excipients and/or isotonicity agents in both liquid and lyophilized parenteral protein formulations. With respect to the Hofmeister series, the polyols are kosmotropic and are preferentially excluded from the protein surface. Polyols can protect proteins from both physical and chemical degradation pathways. Preferentially excluded co-solvents increase the effective surface tension of the solvent at the protein interface, whereby the most energetically favorable protein conformations are those with the smallest surface areas (53).

Mannitol is a popular bulking agent in lyophilized formulations because it crystallizes out of the amorphous protein phase during freeze drying lending structural stability to the cake (e.g., s®, Enbrel – Lyo, and Betaseron®). It is generally used in combination with a cryo and/or lyoprotectant like sucrose. Because of the propensity of mannitol to crystallize under frozen conditions, sorbitol, and sucrose are the preferred tonicity agents/stabilizers in liquid formulations to protect the product against freeze–thaw stresses encountered during transport or when freezing bulk prior to manufacturing. Sorbitol and sucrose are far more resistant to crystallization and therefore less likely to phase separate from the protein. It is interesting to note that while mannitol has been used in tonicifying amounts in several marketed liquid formulations such as Actimmune®, Forteo®, and Rebif®, the product labels of these drugs carry a "Do Not Freeze" warning. The use of reducing sugars (containing free aldehyde or ketone groups) such as glucose and lactose should be avoided because they can react and glycate surface lysine and arginine residues of proteins via the Maillard reaction of aldehydes and primary amines (54,55). Sucrose can hydrolyze to fructose and glucose under acidic conditions (56), and consequently may cause glycation.

The PEG could stabilize proteins by two different temperature-dependent mechanisms. At lower temperatures, it is preferentially excluded from the protein surface but has been shown to interact with the unfolded form of the protein at higher temperatures, given its amphipathic nature (57). Thus, at lower temperatures, it may protect proteins via the mechanism of preferential exclusion, but at higher temperatures possibly by reducing the number of productive collisions between unfolded molecules. PEG is also a cryoprotectant and has been employed in Recombinate, a lyophilized formulation of recombinant Antihemophilic Factor, which utilizes PEG 3350 at a concentration of 1.5 mg/mL. The low-molecular weight liquid PEGs (PEG 300–600) can be contaminated with peroxides and cause protein oxidation. If used, the peroxide content in the raw material must be minimized and controlled throughout its shelf life. The same holds true for polysorbates (discussed below).

Surfactants

Protein molecules have a high propensity to interact with surfaces, making them susceptible to adsorption and denaturation at air–liquid, vial–liquid, and liquid–liquid (silicone oil) interfaces. This degradation pathway has been observed to be inversely dependent on protein concentration and results in either the formation of soluble and insoluble protein aggregates or the loss of protein from solution via adsorption to surfaces (13). In addition to container surface adsorption, surface-induced degradation is exacerbated with physical agitation, as would be experienced during shipping and handling of the product.

Surfactants are commonly used in protein formulations to prevent surface-induced degradation. Surfactants are amphipathic molecules with the capability of out-competing proteins for interfacial positions. Hydrophobic portions of the surfactant molecules occupy interfacial positions (e.g., air/liquid), while hydrophilic portions of the molecules remain oriented toward the bulk solvent. At sufficient concentrations (typically around the detergent's critical micellar concentration), a surface layer of surfactant molecules serves to prevent protein molecules from adsorbing at the interface. Thereby, surface-induced degradation is minimized. The most commonly used surfactants are fatty acid esters of sorbitan polyethoxylates, i.e., polysorbate 20 and polysorbate 80 (e.g., Avonex, Neupogen®, and Neulasta®). The two differ only in the length of the aliphatic chain that imparts hydrophobic character to the molecules, C-12 and C-18, respectively. Accordingly, polysorabte-80 is more surface-active and has a lower critical micellar concentration than polysorbate-20. The surfactant poloxamer 188 has also been used in several marketed liquid products such Gonal-F®, Norditropin®, and Ovidrel®.

Detergents can also affect the thermodynamic conformational stability of proteins. Here again, the effects of a given excipient will be protein specific. For example, polysorbates have been shown to reduce the stability of some proteins and increase the stability of others (39,58). Detergent destabilization of proteins can be rationalized in terms of the hydrophobic tails of the detergent molecules that can engage in specific binding with partially or wholly unfolded proteins. These types of interactions could cause a shift in the conformational equilibrium toward the more expanded protein states (i.e., increasing the exposure of hydrophobic portions of the protein molecule in complement to binding polysorbate). Alternatively, if the protein native state exhibits some solvent-exposed hydrophobic surface, detergent binding to the native state may stabilize that conformation.

Another aspect of polysorbates is that they are inherently susceptible to oxidative degradation. Often, as raw materials, they contain sufficient quantities of peroxides to cause oxidation of protein residue side chains, especially methionine (59). The potential for oxidative damage arising from the addition of stabilizer emphasizes the point that the lowest effective concentrations of excipients should be used in formulations. For surfactants, the effective concentration for a given protein will depend on the mechanism of stabilization. It has been postulated that if the mechanism of surfactant stabilization is related to preventing surface-denaturation, the effective concentration will be around the detergent's critical micellar concentration. Conversely, if the mechanism of stabilization is associated with specific protein–detergent interactions, the effective surfactant concentration will be related to the protein concentration and the stoichiometry of the interaction (39).

Antioxidants

Oxidation of protein residues arises from a number of different sources. Beyond the addition of specific antioxidants, the prevention of oxidative protein damage involves the careful control of a number of factors throughout the manufacturing process and storage of the product, such as atmospheric oxygen, temperature, light exposure, and chemical contamination. The most commonly used pharmaceutical antioxidants are reducing agents, oxygen/free-radical scavengers, or chelating agents (50). Antioxidants in therapeutic protein formulations must be water soluble and remain active throughout the product shelf life. Reducing agents and oxygen/free-radical scavengers work by ablating active oxygen species in solution. Chelating agents such as EDTA can be effective at binding trace metal contaminants that promote free-radical formation. For example, EDTA was utilized in the liquid formulation of acidic fibroblast growth factor to inhibit the metal ion–catalyzed oxidation of cysteine residues (60). EDTA has been used in marketed products like Kineret® and Ontak®.

In addition to evaluating the effectiveness of various excipients in preventing protein oxidation, formulation scientists must be aware of the potential for the antioxidants themselves to induce other covalent or physical changes to the protein. A number of such cases have been reported in the literature. Reducing agents (like glutathione) can cause disruption of intramolecular disulfide linkages, which can lead to disulfide shuffling (59). In the presence of transition metal ions, ascorbic acid and EDTA have been shown to promote methionine oxidation in a number of proteins and peptides (59,61,62). Sodium thiosulfate has been reported to reduce the levels of light- and temperature-induced methionine-oxidation in rhuMab HER2; however, the formation of a thiosulfate-protein adduct was also reported in this study (50). Selection of an appropriate antioxidant is made according to the specific stresses and sensitivities of the protein.

Metal Ions

In general, transition metal ions are undesired in protein formulations because they can catalyze physical and chemical degradation reactions in proteins. However, specific metal ions are included in formulations when they are cofactors to proteins and in suspension formulations of proteins where they form coordination complexes (e.g., zinc suspension of insulin). Recently, the use of magnesium ions (10–120 mM) has been proposed to inhibit the isomerization of aspartic acid to isoaspartic acid (63).

Two examples where metal ions confer stability or increased activity in proteins are human deoxyribonuclease (rhDNase, Pulmozyme®), and Factor VIII. In the case of rhDNase, Ca^{2+} ions (up to 100 mM) increased the stability of the enzyme through a specific binding site (64). In fact, the removal of calcium ions from the solution with EGTA caused an increase in deamidation and aggregation. However, this effect was observed only with Ca^{+2} ions; other divalent cations, Mg^{2+}, Mn^{2+}, and Zn^{2+}, were observed to destabilize rhDNase. Similar effects were observed in Factor VIII. Ca^{2+} and Sr^{2+} ions stabilized the protein, whereas others such as Mg^{2+}, Mn^{2+} and Zn^{2+}, Cu^{2+}, and Fe^{2+} destabilized the enzyme (65). In a separate study with Factor VIII, a significant increase in the aggregation rate was observed in the presence of Al^{3+} ions (66). The authors note that other excipients like buffer salts are often contaminated with Al^{3+} ions and illustrate the need to use excipients of appropriate quality in formulated products.

Preservatives

Preservatives are necessary when developing multidose parenteral formulations that involve more than one extraction from the same container. Their primary function is to inhibit microbial growth and ensure product sterility throughout the shelf life or duration of use of the drug product. Commonly used preservatives include benzyl alcohol, phenol, and *m*-cresol. Although preservatives have a long history of use with small-molecule parenterals, the development of protein formulations that include preservatives can be challenging. Preservatives almost always have a destabilizing effect (aggregation) on proteins, and this has become a major factor in limiting their use in multidose protein formulations (67). To date, most protein drugs have been formulated for single-use only. However, when multidose formulations are possible, they have the added advantage of enabling patient convenience and increased marketability. A good example is that of human growth hormone (hGH) where the development of preserved formulations has led to commercialization of more convenient, multidose injection pen presentations. At least four such pen devices containing preserved formulations of hGH are currently available on the market. Norditropin® (liquid, Novo Nordisk), Nutropin AQ® (liquid, Genentech), and Genotropin (lyophilized—dual chamber cartridge, Pharmacia & Upjohn) contain phenol, whereas Somatrope® (Eli Lilly) is formulated with *m*-cresol.

Several aspects need to be considered during the formulation development of preserved dosage forms. The effective preservative concentration in the drug product must be optimized. This requires testing a given preservative in the dosage form with concentration ranges that confer antimicrobial effectiveness without compromising protein stability. For example, three preservatives were successfully screened in the development of a liquid formulation for interleukin-1 receptor (Type I), using differential scanning calorimetry. The preservatives were ranked based on their impact on stability at concentrations commonly used in marketed products (68).

As might be expected, development of liquid formulations containing preservatives are more challenging than lyophilized formulations. Freeze-dried products can be lyophilized without the preservative and reconstituted with a preservative-containing diluent at the time of use. This shortens the time for which a preservative is in contact with the protein, significantly minimizing the associated stability risks. With liquid formulations, preservative effectiveness and stability have to be maintained over the entire product shelf life (approximately 18–24 months). An important point to note is that preservative effectiveness has to be demonstrated in the final formulation containing the active drug and all excipient components.

Some preservatives can cause injection site reactions, which is another factor that needs consideration when choosing a preservative. In clinical trials that focused on the evaluation of preservatives and buffers in Norditropin®, pain perception was observed to be lower in formulations containing phenol and benzyl alcohol as compared to a formulation containing *m*-cresol (69). Interestingly, among the commonly used preservatives, benzyl alcohol possesses anesthetic properties (70).

REFERENCES

1. Walsh G. Biopharmaceuticals: recent approvals and likely directions. Trends Biotechnol 2005; 23(11):553–557.
2. Robbins DC, Hirshman M, Wardzala LJ, Horton ES. High-molecular-weight aggregates of therapeutic insulin. In vitro measurements of receptor binding and bioactivity. Diabetes 1988; 37(1):56–59.
3. Patten PA, Schellekens H. The immunogenicity of biopharmaceuticals: lessons learned and consequences for protein drug development. In: Brown F, Mire-Sluis AR, eds. Immunogenicity of Therapeutic Biological Products. Basel, Switzerland: S. Karger, 2003.
4. Hermeling S, Crommelin DJA, Schellekens H, Jiskoot W. Structure-immunogenicity relationships of therapeutic proteins. Pharm Res 2004; 21:897–903.
5. Schellekens H. Bioequivalence and immunogenicity of biopharmaceuticals. Nat Rev Drug Discov 2002; 1(6):457–462.

6. Hermeling S, Aranha L, Damen JM, et al. Structural characterization and immunogenicity in wild-type and immune tolerant mice of degraded recombinant human interferon alpha2b. Pharm Res 2005; 22(12):1997–2006.
7. Hasselbeck A. Epoetins: differences and their relevance to immunogenicity. Curr Med Res Opin 2003; 19(5):430–432.
8. Hermeling S, Schellekens H, Crommelin DJ, Jiskoot W. Micelle-associated protein in epoetin formulations: a risk factor for immunogenicity?. Pharm Res 2003; 20(12): 1903–1907.
9. Schellekens H, Crommelin DJ. Equivocal role of micelles in Eprex adverse events. Nat Biotechnol 2003; 21(11):1265.
10. PDR Electronic Library™ Online: © 2002–2006. Thomson PDR.
11. Spada S, Walsh G, ed. Directory of Approved Biopharmaceutical Products. Boca Raton, Florida: CRC Press LLC, 2005:33431.
12. Chi EY, Weickmann J, Carpenter JF, Manning MC, Randolph TW. Heterogeneous nucleation-controlled particulate formation of recombinant human platelet-activating factor acetylhydrolase in pharmaceutical formulation. J Pharm Sci 2005; 94(2):256–274.
13. Treuheit MJ, Kosky AA, Brems DN. Inverse relationship of protein concentration and aggregation. Pharm Res 2002; 19(4):511–516.
14. Jones LS, Kaufmann A, Middaugh CR. Silicone oil induced aggregation of proteins. J Pharm Sci 2005; 94(4):918–927.
15. Chi EY, Krishnan S, Randolph TW, Carpenter JF. Physical stability of proteins in aqueous solution: mechanism and driving forces in nonnative protein aggregation. Pharm Res 2003; 20(9):1325–1336.
16. Wang W. Protein aggregation and its inhibition in biopharmaceutics. Int J Pharm 2005; 289(1–2):1–30.
17. Gregersen N, Bolund L, Bross P. Protein misfolding, aggregation, and degradation in disease. Mol Biotechnol 2005; 31(2):141–150.
18. Dobson CM. Protein folding and misfolding. Nature 2003; 426(6968):884–890.
19. Lumry R, Eyring H. Conformation changes of proteins. J Phys Chem 1954; 58:110–120.
20. Oliyai C, Borchardt RT. Chemical pathways of peptide degradation. IV. Pathways, kinetics, and mechanism of degradation of an aspartyl residue in a model hexapeptide. Pharm Res 1993; 10(1):95–102.
21. Oliyai C, Borchardt RT. Chemical pathways of peptide degradation. VI. Effect of the primary sequence on the pathways of degradation of aspartyl residues in model hexapeptides. Pharm Res 1994; 11(5):751–758.
22. Patel K, Borchardt RT. Chemical pathways of peptide degradation. II. Kinetics of deamidation of an asparaginyl residue in a model hexapeptide. Pharm Res 1990; 7(7): 703–711.
23. Bhatt NP, Patel K, Borchardt RT. Chemical pathways of peptide degradation. I. Deamidation of adrenocorticotropic hormone. Pharm Res 1990; 7(6):593–599.
24. Patel K, Borchardt RT. Chemical pathways of peptide degradation. III. Effect of primary sequence on the pathways of deamidation of asparaginyl residues in hexapeptides. Pharm Res 1990; 7(8):787–793.
25. Brennan TV, Clarke S. Spontaneous degradation of polypeptides at aspartyl and asparaginyl residues: effects of the solvent dielectric. Protein Sci 1993; 2(3):331–338.
26. Costantino HR, Langer R, Klibanov AM. Moisture induced aggregation of lyophilized insulin. Pharm Res 1994; 11:21–29.
27. Schoneich C, Zhao F, Wilson GS, Borchardt RT. Iron-thiolate induced oxidation of methionine to methionine sulfoxide in small model peptides. Intramolecular catalysis by histidine. Biochim Biophys Acta 1993; 1158(3):307–322.
28. Goolcharran C, Khossravi M, Borchardt RT. Chemical pathways of peptide and protein Degradation. In: Frokjaer S, Hovgaard L, eds. Pharmaceutical Formulation Development of Peptides and Proteins. (Pharmaceutical Science.) U.K.: Taylor and Francis, 1999.

29. CPMP/CVMP/QWP/115/95 Note for Guidance on Inclusion of Antioxidants and Antimicrobial Preservatives in Medicinal Products, 1997.
30. CPMP/QWP/155/96 Note for guidance on development pharmaceutics (CPMP adopted Jan. 98).
31. Chang BS, Hershenson S. Practical approaches to protein formulation development. Pharm Biotechnol 2002; 13:1–25.
32. Shire SJ, Shahrokh Z, Liu J. Challenges in the development of high protein concentration formulations. J Pharm Sci 2004; 93(6):1390–1402.
33. Wang W. Instability, stabilization, and formulation of liquid protein pharmaceuticals. Int J Pharm 1999; 185(2):129–188.
34. Carpenter JF, Chang BS, Garzon-Rodriguez W, Randolph TW. Rational design of stable lyophilized protein formulations: theory and practice. Pharm Biotechnol 2002; 13:109–133.
35. Wang W. Lyophilization and development of solid protein pharmaceuticals. Int J Pharm 2000; 203(1–2):1–60.
36. Ediger MD, Angell CA, Nagel SR. Supercooled liquids and glasses. J Phys Chem 100(31):13200–13212.
37. Arakawa T, Kita Y, Carpenter JF. Protein-solvent interactions in pharmaceutical formulations. Pharm Res 1991; 8(3):285–291.
38. Kendrick BS, Li T, Chang BS. Physical stabilization of proteins in aqueous solution. In: Carpenter JF, Manning MC, eds. Rational Design of Stable Protein Formulations: Theory and Practice. Pharmaceut Biotechnol 2002; 13:61–84.
39. Randolph TW, Jones LS. Surfactant-protein interactions. Pharm Biotechnol 2002; 13: 159–175.
40. Remmele RL Jr, Bhat SD, Phan DH, Gombotz WR. Minimization of recombinant human Flt3 ligand aggregation at the T_m plateau: a matter of thermal reversibility. Biochemistry 1999; 38(16):5241–5247.
41. Laursen T, Hansen B, Fisker S. Pain perception after subcutaneous injections of media containing different buffers. Basic Clin Pharmacol Toxicol 2006; 98(2):218–221.
42. Hollander-Rodriguez JC, Calvert JF Jr. Hyperkalemia. Am Fam Physician 2006; 73(2): 283–290.
43. Cacace MG, Landau EM, Ramsden JJ. The Hofmeister series: salt and solvent effects on interfacial phenomena. Quart Rev Biophys 1997; 30(3):241–277.
44. Yamasaki M, Yano H, Aoki K. Differential scanning calorimetric studies on bovine serum albumin: II. Effects of neutral salts and urea. Int J Biol Macromol 1991; 13(6):322–328.
45. Robinson NE, Robinson AB. Molecular Clocks. In: Deamidation of Asparaginyl and Glutaminyl Residues in Peptides and Proteins. Cave Junction, Oregon: Althouse Press, 2004.
46. Callahan WJ, Narhi LO, Kosky AA, Treuheit MJ. Sodium chloride enhances the storage and conformational stability of BDNF and PEG-BDNF. Pharm Res 2001; 18(3):261–266.
47. Liu J, Nguyen MD, Andya JD, Shire SJ. Reversible self-association increases the viscosity of a concentrated monoclonal antibody in aqueous solution. J Pharm Sci 2005; 94(9):1928–19240 [Erratum in: J Pharm Sci 2006; 95(1):234–235.].
48. Chen B, Bautista R, Yu K, Zapata GA, Mulkerrin MG, Chamow SM. Influence of histidine on the stability and physical properties of a fully human antibody in aqueous and solid forms. Pharm Res 2003; 20(12):1952–1960.
49. Tomita M, Irie M, Ukita T. Sensitized photooxidation of histidine and its derivatives. Products and mechanism of the reaction. Biochemistry 1969; 8(12):5149–5160.
50. Lam XM, Yang JY, Cleland JL. Antioxidants for prevention of methionine oxidation in recombinant monoclonal antibody HER2. J Pharm Sci 1997; 86(11):1250–1255.
51. Arakawa T, Timasheff SN. Preferential interactions of proteins with solvent components in aqueous amino acid solutions. Arch Biochem Biophys 1983; 24(1):169–177.

52. Tsumoto K, Umetsu M, Kumagai I, Ejima D, Philo JS, Arakawa T. Role of arginine in protein refolding, solubilization, and purification. Biotechnol Prog 2004; 20(5): 1301–1308.
53. Lin TY, Timasheff SN. On the role of surface tension in the stabilization of globular proteins. Protein Sci 1996; 5(2):372–381.
54. Chevalier F, Chobert JM, Dalgalarrondo M, Choiset Y, Haertle T. Maillard glycation of beta-lactoglobulin induces conformation changes. Nahrung 2002; 46(2):58–63.
55. Humeny A, Kislinger T, Becker CM, Pischetsrieder M. Qualitative determination of specific protein glycation products by matrix-assisted laser desorption/ionization mass spectrometry Peptide mapping. J Agric Food Chem 2002; 50(7):2153–2160.
56. Kautz CF, Robinson AL. The hydrolysis of sucrose by hydrochloric acid in the presence of alkali and alkaline earth chlorides. JACS 1928; 50(4):1022–1030.
57. Lee LL, Lee JC. Thermal stability of proteins in the presence of poly(ethylene glycols). Biochemistry 1987; 26(24):7813–7819.
58. Bam NB, Cleland JL, Yang J, et al. Tween protects recombinant human growth hormone against agitation-induced damage via hydrophobic interactions. J Pharm Sci 1998; 87(12):1554–1559.
59. Akers MJ, Defelippis MR. Peptides and proteins as parenteral solutions. In: Frokjaer S, Hovgaard L, eds. Pharmaceutical Formulation Development of Peptides and Proteins. (Pharmaceutical Science.) U.K.: Taylor and Francis, 1999.
60. Tsai PK, Volkin DB, Dabora JM, et al. Formulation design of acidic fibroblast growth factor. Pharm Res 1993; 10(5):649–659.
61. Fransson JR. Oxidation of human insulin-like growth factor I in formulation studies. 3. Factorial experiments of the effects of ferric ions, EDTA, and visible light on methionine oxidation and covalent aggregation in aqueous solution. J Pharm Sci 1997; 86(9):4046–4050.
62. Yin J, Chu JW, Ricci MS, Brems DN, Wang DI, Trout BL. Effects of antioxidants on the hydrogen peroxide-mediated oxidation of methionine residues in granulocyte colony-stimulating factor and human parathyroid hormone fragment 13–34. Pharm Res 2004; 21(12):2377–2383.
63. Kaisheva EA, Gupta S, Flores-Nate A. Stable liquid pharmaceutical formulation of antibodies that are prone to isomerization. PCT Int Appl WO 2004039337 A2 20040513 (2004).
64. Chen B, Costantino HR, Liu J, Hsu CC, Shire SJ. Influence of calcium ions on the structure and stability of recombinant human deoxyribonuclease I in the aqueous and lyophilized states. J Pharm Sci 1999; 88(4):477–482.
65. Fatouros A, Osterberg T, Mikaelsson M. Recombinant factor VIII SQ—influence of oxygen, metal ions, pH and ionic strength on its stability in aqueous solution. Int J Pharm 1997; 155:121–131.
66. Derrick TS, Kashi RS, Durrani M, Jhingan A, Middaugh CR. Effect of metal cations on the conformation and inactivation of recombinant human factor VIII. J Pharm Sci 2004; 93(10):2549–2557.
67. Roy S, Jung R, Kerwin BA, Randolph TW, Carpenter JF. Effects of benzyl alcohol on aggregation of recombinant human interleukin-1-receptor antagonist in reconstituted lyophilized formulations. J Pharm Sci 2005; 94(2):382–396.
68. Remmele RL Jr, Nightlinger NS, Srinivasan S, Gombotz WR. Interleukin-1 receptor (IL-1R) liquid formulation development using differential scanning calorimetry. Pharm Res 1998; 15(2):200–208.
69. Kappelgaard AM, Bojesen A, Skydsgaard K, Sjogren I, Laursen T. Liquid growth hormone: preservatives and buffers. Horm Res 2004; 62(suppl 3):98–103.
70. Minogue SC, Sun DA. Bacteriostatic saline containing benzyl alcohol decreases the pain associated with the injection of propofol. Anesth Analg 2005; 100(3):683–686.
71. Bam NB, Cleland JL, Randolph TW. Molten globule intermediate of recombinant human growth hormone: stabilization with surfactants. Biotechnol Prog 1996; 12(6):801–809.

Appendix Formulations of Approved Protein Drugs

Marketed name (generic name) [molecule type]	Manufacturer	Formulation type (dosage forms)	Strengths or concentration (pH of formulation)	Excipients	Route of administration
Actimmune® (interferon–gamma-1b) [*Escherichia coli*–expressed recombinant protein]	InterMune	Liquid (single-dose vial)	100 µg in 0.5 mL; 0.2 mg/mL	Per 0.5 mL: 20 mg mannitol; 0.36 mg sodium succinate; 0.05 mg polysorbate 20	SC
Activase® (Alteplase) [CHO-expressed glycoprotein-recombinant human tissue plasminogen activator]	Genentech	Lyophilized (50 mg single-dose vial; 100 mg single dose vial)	50 and 100 mg doses to be reconstituted with 50 or 100 mL of WFI, respectively	Per 50 mg vial: 1.7 g L-arginine; 0.5 g phosphoric acid; $\leq$ 4 mg polysorbate 80 (under vacuum) Per 100 mg vial: 3.5 g L-arginine; 1 g phosphoric acid; $\leq$11 mg polysorbate 80 (without vacuum)	IV
Adagen® (pegademase bovine) [monomethoxy polyethylene glycol-adenosine deaminase]	Enzon pharmaceuticals	Liquid (single use vial)	250 units/mL (pH 7.2–7.4)	Isotonic solution	IM
Aldurazyme® (Laronidase) [CHO-expressed glycoprotein]	BioMarin Pharmaceuticals	Liquid (single-dose vial)	2.9 mg per 5 mL; 0.58 mg/mL (pH = 5.5)	Per 5 mL: 43.9 mg sodium chloride; 63.5 mg sodium phosphate monobasic monohydrate; 10.7 mg sodium phosphate dibasic heptahydrate; 0.05 mg polysorbate 80	IV (diluted in 0.9% NaCl containing 1% HSA prior to administration)

(*Continued*)

Appendix Formulations of Approved Protein Drugs (*Continued*)

Marketed name (generic name) [molecule type]	Manufacturer	Formulation type (dosage forms)	Strengths or concentration (pH of formulation)	Excipients	Route of administration
Amevive® (Alefacept) [CHO-expressed dimeric fusion protein]	Biogen-Idec	Lyophilized (15 mg single-dose vial; 7.5 mg single-dose vial)	7.5 and 15 mg doses, both to be reconstituted with 0.5 mL of WFI (pH = 6.9)	Per 0.5 mL: 12.5 mg sucrose; 5 mg glycine; 3.6 mg sodium citrate dihydrate; 0.12 mg/mL citric acid monohydrate	IV (7.5 mg vial); IM (15 mg vial)
Avonex® (interferon beta-1a) [CHO-expressed glycoprotein-recombinant human interferon beta-1a]	Biogen-Idec	Lyophilized (30 μg single-dose vial)	Lyo: 30 μg vial to be reconstituted with 1.0 mL WFI (pH = 7.3)	Lyo per 1 mL: 15 mg albumin (human); 5.8 mg sodium chloride; 5.7 mg dibasic sodium phosphate; 1.2 mg monobasic sodium phosphate	IM (30 μg vial); prefilled syringe (30 μg)
		Liquid: 30 μg prefilled syringe	Liquid: 30 μg in 0.5 mL; 0.06 mg/mL (pH = 4.8)	Prefilled syringe per 0.5 mL: 0.79 mg sodium acetate trihydrate; 0.25 mg glacial acetic acid; 15.8 mg arginine-hydrochloride; 0.025 mg/mL polysorbate 20	
BeneFIX® (Nonacog alfa) [CHO-expressed glycoprotein-recombinant human coagulation factor IX]	Wyeth	Lyophilized (250, 500, and 1000 IU single-dose vials)	The 250, 500, and 1000 IU are to be reconstituted with WFI	After reconstitution of the lyophilized drug product, the concentrations of excipients in the 500 and 1000 IU dosage strengths are 10 mM L-histidine, 1% sucrose, 260 mM glycine, 0.005% polysorbate 80 The concentrations of excipients after reconstitution in the 250 IU dosage strength are half those of the other two dosage strengths	IV

Beromun® (Tasonermin) [recombinant nonglycosylated cytokine]	Boehringer Ingelheim	Lyophilized (1 mg single-dose vial)	To be reconstituted with normal saline to a final concentration 0.2 mg/mL	Sodium di-hydrogen phosphate di-hydrate; disodium hydrogen phosphate dodecahydrate; sodium chloride; HSA. The solvent comprises sodium chloride and WFI	ILP (Isolated limb perfusion)
Beatseron (interferon-beta-1b) [*E. coli*–expressed recombinant human interferon beta]	Berlex, Inc.	Lyophilized (0.3 mg single-dose vial)	To be reconstituted with 1.2 mL of sodium chloride, 0.54% solution	Per vial: 15 mg albumin (human); 15 mg mannitol	SC
BEXXAR® (Tositumomab)[murine IgG_{2a} lambda monoclonal Ab]	Glaxo-SmithKline	Liquid (35 mg single-dose vial; 225 mg single-dose vial)	225 mg in 16.1 mL; 35 mg vial in 2.5 mL; 14 mg/mL (pH = 7.2)	Per vial: 10% (w/v) maltose; 145 mM sodium chloride; 10 mM phosphate	IV infusion
Tev-Tropin (somatropin) [*E. coli*–expressed polypeptide-recombinant hGH]	Teva Pharmaceuticals	Lyophilized (5 mg single-dose vial)	To be reconstituted with 5 mL bacteriostatic 0.9% sodium chloride for injection, USP, with normal 0.9% benzyl alcohol as a preservative saline (pH in the range of 7.0–9.0)	Per vial: 30 mg mannitol	SC
Bioclate or Recombinate® (recombinant) [CHO-expressed glycoprotein-recombinant antihemophilic factor]	Bioclate® and Baxter Healthcare	Lyophilized (250, 500, and 1000 IU single-dose bottles)	To be reconstituted with sterile WFI	Per bottle: 12.5 mg/mL albumin (human); 0.20 mg/mL calcium; 1.5 mg/mL PEG (3350); 180 mEq/L sodium; 55 mM histidine; 1.5 µg/AHF IU polysorbate-80	IV injection

(*Continued*)

Appendix Formulations of Approved Protein Drugs (*Continued*)

Marketed name (generic name) [molecule type]	Manufacturer	Formulation type (dosage forms)	Strengths or concentration (pH of formulation)	Excipients	Route of administration
CEREZYME® (imiglucerase) [CHO-expressed monomeric glycoprotein analogue of the human enzyme (beta)-glucocerebrosidase]	Genzyme	Lyophilized [200 unit single-dose vials; 400 unit single-dose vial (an enzyme unit (U) is defined as the amount of enzyme that catalyzes the hydrolysis of 1 mM of the synthetic substrate para-nitrophenyl-(beta)-D-glucopyranoside (pNP-Glc) per min at 37°C)]	200 Unit vial; 212 units reconstituted with 5.1 mL sterile water to a final volume of 5.3 mL of reconstituted product (40 U/mL) (pH = 6.1) 400 Unit vial; 424 units reconstituted with 10.2 mL sterile water to a final volume of 10.6 mL of reconstituted product (40 U/mL) (pH = 6.1)	Per 200 units: 170 mg mannitol; 70 mg sodium citrates; 52 mg trisodium citrate; 18 mg disodium hydrogen citrate; 0.53 mg polysorbate 80	IV infusion
Enbrel® (Etanercept) [CHO-expressed dimeric fusion protein]	Amgen	Lyo: 25 mg multiple-dose vial	Lyo: to be reconstituted with 1 mL of BWFI containing 0.9% benzyl alcohol (pH = 7.4 ± 0.3)	Lyo per vial: 40 mg mannitol; 10 mg sucrose; 1.2 mg tromethamine	SC
		Liquid: 1 mL single-dose prefilled syringe	Liquid: 0.98 mL fill (50 mg/mL) (pH = 6.3 ± 0.2)	Liquid per 1 mL: 10 mg sucrose; 5.8 mg sodium chloride; 5.3 mg L-arginine hydrochloride; 2.6 mg sodium phosphate mono-basic monohydrate; 0.9 mg sodium phosphate dibasic anhydrous	

Eprex® (Epoetin alpha)[CHO-expressed recombinant human protein]	JANSSEN CILAG	Liquid: 1 mL multi-dose vial; 0.3, 0.4, 0.5, 0.6, 0.8, and 1.0 mL single dose syringe(s)	Vials: 20,000 IU per 1.0 mL	Vials: HSA containing formulation; 0.25% albumin (human); sodium chloride; sodium citrate; citric acid; 0.9% benzyl alcohol as a preservative; water for injection	SC and IV
			Pre filled syringe: 1000 IU per 0.5 mL; 2000 IU per 0.5 mL; 3000 IU per 0.3 mL; 4000 IU per 0.4 mL; 5000 IU per 0.5 mL; 6000 IU per 0.6 mL; 8000 IU per 0.8 mL; 10,000 IU per 1.0 mL; 20,000 IU per 0.5 mL; 40,000 IU per 1.0 mL	Pre filled syringe: polysorbate-80 containing (HSA- free) formulation; glycine; polysorbate 80; sodium chloride; sodium phosphate monobasic dihydrate; sodium phosphate dibasic dihydrate; water for injection	
EPOGEN®/procrit (epoetin alfa) [mammalian cell–expressed glycoprotein]	Amgen	Liquid: 1 mL single dose vials	2000, 3000, 4000, 10,000, 40,000 Units/mL (pH 6.9 ± 0.3)	Per 1 mL: 2.5 mg albumin (human); 5.8 mg sodium citrate; 5.8 mg sodium chloride; 0.06 mg citric acid in WFI, USP	IV or SC
				Per 1 mL: 2.5 mg albumin (human); 1.2 mg sodium phosphate monobasic monohydrate; 1.8 mg sodium phosphate dibasic anhydrate; 0.7 mg sodium citrate; 5.8 mg sodium chloride; 6.8 μg citric acid in WFI, USP	

(*Continued*)

Appendix Formulations of Approved Protein Drugs (*Continued*)

Marketed name (generic name) [molecule type]	Manufacturer	Formulation type (dosage forms)	Strengths or concentration (pH of formulation)	Excipients	Route of administration
		1 mL multidose vials	20,000 Units/mL (pH = 6.1 ± 0.3)	Per 1 mL: 2.5 mg albumin (human); 1.3 mg sodium citrate; 8.2 mg sodium chloride; 0.11 mg citric acid; 1% benzyl alcohol as preservative in WFI, USP	
		2 mL multidose vials	10,000 Units/mL (pH = 6.1 ± 0.3)	Per 1 mL: 2.5 mg albumin (human); 1.3 mg sodium citrate; 8.2 mg sodium chloride; 0.11 mg citric acid; 1% benzyl alcohol as preservative in WFI, USP	
Exubera® [Coli expressed recombinant human insulin]	Pfizer	Powder; unit dose blister	1 and 3 mg dose	Sodium citrate (dehydrate), mannitol, glycine, and sodium hydroxide	Inhalation
Fabrazyme® (agalsidase beta) [CHO-expressed recombinant human alpha-galactosidase]	Genzyme	Lyophilized: 35 mg single-dose vial; 5 mg single-dose vial	To be reconstituted with either 7.2 mL or 1.1 mL of sterile WFI, giving 35 mg in 7 mL or 5.5 mg in 1 mL	Per 35 mg vial: 222 mg mannitol; 20.4 mg sodium phosphate monobasic monohydrate; 59.2 mg sodium phosphate dibasic heptahydrate	IV infusion
				Per 5 mg vial: 33.0 mg mannitol; 3.0 mg sodium phosphate monobasic monohydrate; 8.8 mg sodium phosphate dibasic heptahydrate	
Fasturtec/Elitek ELITEK® (Rasburicase) [recombinant enzyme]	Sanofi-aventis	Lyophilized; 1.5 mg single-dose vial	To be reconstituted with 1 mL sterile WFI containing 1.0 mg Poloxamer 188 (0.15 mg/mL)	Per vial: 10.6 mg mannitol; 15.9 mg/mL L-alanine; 12.6–14.3 mg of dibasic sodium phosphate	IV

Forteo® (teriparatide) [*E. coli*–expressed recombinant human parathyroid hormone]	Eli Lilly	Liquid: prefilled pen delivery device	3.3 mL fill (250 μg/mL) (pH 4)	Per pen: 0.41 mg/mL glacial acetic acid; 0.10 mg/mL sodium acetate (anhydrous); 45.4 mg/mL mannitol; 3.0 mg *m*-cresol	SC
Genotropin® (somatropin) [*E. coli*–expressed polypeptide hormone]	Pharmacia & Upjohn	Lyophilized: 1.5 mg two-chamber cartridge (without preservative); preassembled in a GENO TROPI N INTRA-MIX® growth hormone reconstitution device	1.5 mg in 1.13 mL WFI (1.3 mg/mL)	The front chamber contains: 27.6 mg glycine; 0.3 mg sodium dihydrogen phosphate anhydrous; 0.3 mg disodium phosphate anhydrous The rear chamber contains: 1.13 mL dileunt	SC
Genotropin®		Lyophilized: 5.8 mg two-chamber cartridge (with preservative); preassembled in a GENO-TROPIN INTRA-MIX® growth hormone reconstitution device	5.8 mg in 1.14 mL WFI (5 mg/mL)	The front chamber contains: 2.2 mg glycine; 1.8 mg mannitol; 0.32 mg sodium dihydrogen phosphate anhydrous; 0.31 mg disodium phosphate anhydrous The rear chamber contains: 0.3% *m*-cresol (as a preservative); 45 mg mannitol	SC
Genotropin		Lyophilized: 13.8 mg two-chamber cartridge (with preservative);	13.8 mg in 1.13 mL WFI (12 mg/mL)	The front chamber contains: 2.3 mg glycine; 14.0 mg mannitol; 0.47 mg sodium dihydrogen phosphate	SC

(*Continued*)

Appendix Formulations of Approved Protein Drugs (*Continued*)

Marketed name (generic name) [molecule type]	Manufacturer	Formulation type (dosage forms)	Strengths or concentration (pH of formulation)	Excipients	Route of administration
		preassembled in a GENOTROPIN INTRA-MIX growth hormone reconstitution device		anhydrous; 0.46 mg disodium phosphate anhydrous The rear chamber contains: 0.3% *m*-cresol (as a preservative); 32 mg mannitol	
Genotropin		Lyophilized: Genotropin MINIQUICK (single-use syringe device containing a two-chamber cartridge)	0.22–2.2 mg in 0.275 mL WFI (pH = 6.7); individual doses of 0.2–2.0 mg in 0.2 mg increments	The front chamber contains: 0.23 mg glycine; 1.14 mg mannitol; 0.05 mg sodium dihydrogen phosphate; 0.027 mg disodium phosphate anhydrous The rear chamber contains: 12.6 mg mannitol	SC
GlucaGen® (glucagon) [expressed in a *Saccharomyces cerevisiae* vector-recombinant hormone]	Novo Nordisk Pharmaceuticals	Lyophilized: 1 mg single-dose vial	To be reconstituted in 1 mL of sterile WFI (pH = 2.5–3.5)	Per 1 mL: 1 mg hydrochloride; 107 mg lactose monohydrate Diluent syringe contains 12 mg/mL glycerin	SC, IM, or IV injection
Glucagon® (glucagon, rDNA origin) [*E. coli*–expressed single-chain polypeptide hormone]	Eli Lilly	Lyophilized: 1 mg single-dose vial	To be reconstituted in 1 mL of sterile diluent (soluble at a pH of less than 3 or more than 9.5)	Per vial: 49 mg of lactose; diluent syringe contains 12 mg/mL of glycerin WFI, and hydrochloric acid	IV or IM
Gonal-F® (follitropin alfa) [CHO-expressed glycoproteins of	Serono Pharma	Liquid: RFF pen-disposable, prefilled drug	Delivers: 22 µg in 0.5 mL; 33 µg in 0.75 mL, or 66 µg in 1.5 mL;	Per pen: 60 mg/mL sucrose; 3.0 mg/mL *m*-cresol; 1.1 mg/mL di-sodium hydrogen phosphate	SC

recombinant human FSH-two noncovalently linked, nonidentical]		delivery system; 30, 41, or 75 µg	*O*-phosphoric acid and/or sodium hydroxide may be used for pH adjustment	dihydrate; 0.45 mg/mL sodium dihydrogen phosphate monohydrate; 0.1 mg/mL methionine; 0.1 mg/mL Poloxamer 188	
Kogenate® FS (octocog alfa) [BHK cell–expressed recombinant antihemophilic factor]	Bayer Healthcare	Lyophilized: 250, 500, and 1000 IU single-dose vials	To be reconstituted with sterile WFI	Per vial: 0.9–1.3% or 28 mg sucrose; 21–25 mg/mL glycine; 18–23 mM histidine; 2–3 mM calcium chloride; 27–36 mEq/L sodium; 32–40 mEq/L chloride; 96 µg/mL polysorbate 80 (NMT); 20 µg/1000 IU imidazole (NMT); 5 µg/1000 IU tri-*n*-butyl phosphate (NMT); 0.6 µg/1000 IU copper (NMT)	IV
Herceptin® (trastuzumab) [CHO-expressed humanized recombinant IgG1 monoclonal Ab]	Genentech	Lyophilized: single-dose vial	Each 440 mg vial to be reconstituted with 20 mL bacteriostatic WFI (pH of approximately 6)	Per 440 mg vial: 400 mg (alpha)-trehalose dehydrate; 9.9 mg L-histidine HCl; 6.4 mg L-histidine; 1.8 mg polysorbate 20	IV
HUMALOG MIX75/25^{TMa} (mixture of insulin lispro protamine suspension and 25% insulin lispro injection) [*E. coli*–expressed recombinant protein]	Eli Lilly	Liquid suspension and liquid solution in 10 mL vials	100 Units/mL (pH of 7.0–7.8)	Per 1 mL: 0.28 mg protamine sulfate; 16 mg glycerin; 3.78 mg dibasic sodium phosphate; 1.76 mg *m*-cresol; zinc oxide content adjusted to provide 0.025 mg zinc ion and 0.715 mg phenol	SC?
Humatrope® (somatropin) [*E. coli*–expressed recombinant polypeptide hormone]	Eli Lilly	Lyophilized: 5 mg single-dose vials	To be reconstituted with 1.5–5 mL of diluent provided (pH of approxi-	25 mg mannitol; 5 mg glycine; 1.13 mg dibasic sodium phosphate The diluent contains: WFI with	SC or IM injection

(*Continued*)

Appendix Formulations of Approved Protein Drugs (*Continued*)

Marketed name (generic name) [molecule type]	Manufacturer	Formulation type (dosage forms)	Strengths or concentration (pH of formulation)	Excipients	Route of administration
			mately 7.5)	0.3% *m*-cresol as a preservative and 1.7% glycerin	
Humatrope		Lyophilized: 6 mg cartridge	To be reconstituted with 3 mL diluent	18 mg mannitol; 6 mg glycine; 1.36 mg dibasic sodium phosphate The diluent contains: 3 mL WFI; 0.3% *m*-cresol as a preservative; 1.7% glycerin	SC or IM injection
Humatrope		Lyophilized: 12 mg cartridge	To be reconstituted with 3 mL diluent	36 mg mannitol; 12 mg glycine; 2.72 mg dibasic sodium phosphate The diluent contains: 3 mL WFI; 0.3% *m*-cresol as a preservative; 0.29% glycerin	SC or IM injection
Humatrope		Lyophilized: 24 mg cartridge	To be reconstituted with 3 mL diluent	72 mg mannitol; 24 mg glycine; 5.43 mg dibasic sodium phosphate The diluent contains: 3 mL WFI; 0.3% *m*-cresol as a preservative; 0.29% glycerin	SC or IM injection
Humira® (adalimumab) [mammalian cell expression of recombinant human IgG1 monoclonal Ab]	Abbott	Liquid: 1 mL prefilled syringe	40 mg per 0.8 mL (pH of about 5.2)	Each 0.8 mL contains: 4.93 mg sodium chloride; 0.69 mg monobasic sodium phosphate dehydrate; 1.22 mg dibasic sodium phosphate dehydrate;	SC

				0.24 mg sodium citrate; 1.04 mg citric acid monohydrate; 9.6 mg mannitol; 0.8 mg polysorbate 80	
HUMULIN® R REGULAR U-500 (concentrated insulin human injection, USP) [*E. coli*–expressed biosynthetic human insulin]	Eli Lilly	20 mL vials of zinc-insulin crystals dissolved in a clear fluid	500 Units/mL; 100 units/mL (U-100)	16 mg/mL glycerin, 2.5 mg/mL *m*-cresol as a preservative; 0.085 mg/mL (0.017 mg/100 units) zinc-oxide (sodium hydroxide and/or hydrochloric acid may be added during manufacture to adjust the pH)	SC
HUMULIN® N NPH [*E. coli*–expressed human insulin isophane suspension]	Eli Lilly		100 Units/mL (U-100)	Crystalline suspension of human insulin with protamine and zinc	SC
HUMULIN N pen [*E. coli*–expressed) NPH human insulin isophane suspension]		3 mL disposable insulin delivery device	100 Units/mL (U-100)	Crystalline suspension of human insulin with protamine and zinc	SC
INFERGEN® (interferon alfacon-1) [*E. coli*–expressed recombinant protein]	(InterMune) CFP Farmaceutica (European Union) Amgen (U.S.)	Liquid: 9 μg single-dose vials or 15 μg single-dose vials	9 μg in 0.3 mL; 15 μg in 0.5 mL (pH 7.0 ± 0.2)	Per vial: 5.9 mg/mL sodium chloride; 3.8 mg/mL sodium phosphate	SC
KINERET® (Anakinra) [*E. coli* expressed recombinant,	Amgen	Liquid 100 mg prefilled single dose syringe	100 mg per 0.67 mL (pH 6.5)	Per Syringe: 1.29 mg sodium citrate 5.48 mg sodium chloride 0.12 mg disodium EDTA	SC

(*Continued*)

Appendix Formulations of Approved Protein Drugs (*Continued*)

Marketed name (generic name) [molecule type]	Manufacturer	Formulation type (dosage forms)	Strengths or concentration (pH of formulation)	Excipients	Route of administration
nonglycosylated form of the human interleukin-1 receptor antagonist]				0.70 mg polysorbate 80	
Kogenate® FS (Antihemophilic factor) [BHK cells–expressed recombinant human protein]	Bayer Corporation	Lyophilized		Sucrose (0.9–1.3%); glycine (21–25 mg/mL); histidine (18–23 mM); calcium chloride (2–3 mM)	IV
LANTUS® (insulin glargine) [*E. coli*–expressed recombinant human insulin analog]	Sanofi-Aventis	Liquid: 10 cc vials or 3 mL cartridge (cartridge only used with the OptiClik™ (insulin delivery device)	Each vial or cartridge system contains 3.6378 mg/mL of drug (pH = 4).	Per vial: 0.03 mg/mL zinc; 2.7 mg/mL *m*-cresol; 20 mg/mL glycerol 85%	SC
LEUKINE® (Sargramostim) [recombinant human granulocyte-macrophage colony-stimulating factor (rhu GM-CSF]	Berlex	Lyo: 250 mg single-dose vials	Lyo: To be reconstituted with either sterile WFI or sterile bacteriostatic WFI	Lyo per 1 mL: 40 mg mannitol; 10 mg sucrose; 1.2 mg tromethamine	SC or IV infusion
		Liquid: 500 µg multidose vial	Liquid: 500 µg vial in 1 mL	Liquid per 1 mL: 40 mg mannitol; 10 mg sucrose; 1.2 mg tromethamine; 1.1% benzyl alcohol	
CAMPATH® (Alemtuzumab)	Berlex	Liquid: single-use ampoule	30 mg in 3 mL of solution (pH 6.8–7.4)	Per ampoule: 24.0 mg sodium chloride; 3.5 mg dibasic sodium	IV infusion

[CHO-expressed humanized recombinant IgG1 monoclonal Ab]				phosphate; 0.6 mg potassium chloride; 0.6 mg monobasic potassium phosphate; 0.3 mg polysorbate 80; 0.056 mg disodium edetate	
RITUXAN® (Rituximab) [CHO-expressed humanized IgG1 monoclonal Ab]	Biogen Idec and Genente-ch	Liquid: 100 or 500 mg single vial	100 mg in 10 mL or 500 mg in 50 mL (pH = 6.5)	Per vial: 9.0 mg/mL sodium chloride; 7.35 mg/mL sodium citrate dehydrate; 0.7 mg/mL polysorbate 80	IV infusion
TNKase® (Tenecteplase) [CHO-expressed recombinant glycoprotein]	Genentech	Lyophilized: 50 mg single-dose vial	To be reconstituted with 10 mL of sterile WFI (pH of approximately 7.3)	Per vial: 0.55 g L-arginine; 0.17 g phosphoric acid; 0.43 mg polysorbate 20	IV bolus
MYLOTARG® (Gemtuzumab ozogamicin) [Mammalian-expressed recombinant humanized IgG4 conjugated with a cytotoxic antitumor antibiotic]	Wyeth-Ayerst	Lyophilized: 5 mg single-dose amber vial	To be reconstituted with sterile WFI	Per vial: dextran 40, sucrose, sodium chloride, and monobasic and dibasic sodium phosphate	IV infusion
NATRECOR® (Nesiritide) [*E. coli*–expressed recombinant human B-type natriuretic peptide (hBNP]	Scios Inc.	Lyophilized: 1.5 mg single-dose vials	To be reconstituted in 5 mL diluent of choice from an IV bag. The following preservative-free diluents are recommended for reconstitution: 5% dextrose injection (D5W), USP; 0.9% sodium chloride injection, USP; 5%	Per vial: 20.0 mg mannitol; 2.1 mg citric acid monohydrate; 2.94 mg sodium citrate dihydrate	IV infusion

(Continued)

Appendix Formulations of Approved Protein Drugs (*Continued*)

Marketed name (generic name) [molecule type]	Manufacturer	Formulation type (dosage forms)	Strengths or concentration (pH of formulation)	Excipients	Route of administration
			dextrose and 0.45% sodium chloride injection, USP; or 5% dextrose and 0.2% sodium chloride injection, USP		
Aranesp® (Darbepoetin alfa) [CHO-expressed recombinant human erythropoiesis stimulating–protein]	Amgen	Liquid: single-dose vials or prefilled syringes; doses available: 25, 40, 60, 100, 150, 200, 300, or 500 μg; vials and syringes at each Aranesp dosage can be obtained as albumin containing or albumin free (polysorbate containing)	Vial or syringe: 25, 40, 60, 100, 150, 200, 300, or 500 μg in 1.0 mL; each vial at various strengths contains either albumin or polysorbate	Single-dose vials and prefilled syringes are available in two formulations that contain excipients as follows: Per 1 mL: polysorbate solution 0.05 mg/mL polysorbate 80; 2.12 mg/mL sodium phosphate monobasic monohydrate; 0.66 mg/mL sodium phosphate dibasic anhydrous; 8.18 mg/mL sodium chloride (pH 6.2 ± 0.2) Albumin solution contains: 2.5 mg/mL albumin (human); 2.23 mg/mL sodium phosphate monobasic monohydrate; 0.53 mg/mL sodium phosphate dibasic anhydrous; 8.18 mg/mL	IV or SC

				sodium chloride in (pH 6.0 ± 0.3)	
NEULASTA® (Pegfilgrastim) [covalent conjugate of recombinant methionyl human G-CSF and 20 kD PEG]	Amgen	Liquid: single-dose syringe	10 mg in 0.6 mL (pH 4.0)	Per syringe: 0.35 mg acetate; 30.0 mg sorbitol; 0.02 mg polysorbate 20; 0.02 mg sodium	SC
NEUMEGA® (Oprelvekin) [*E. coli*–expressed recombinant thrombotic growth factor]	Wyeth-Ayerst	Lyophilized: 5 mg single-dose vials	To be reconstituted with 1 mL of sterile WFI	Per vial: 23 mg glycine; 1.6 mg dibasic sodium phosphate heptahydrate; 0.55 mg monobasic sodium phosphate monohydrate	SC
NEUPOGEN® (Filgrastim) [*E. coli*–expressed human granulocyte colony-stimulating factor (G-CSF)]	Amgen	Liquid: single-dose vial 300 or 480 μg or prefilled syringe 300 or 480 μg	Vials: 300 μg in 1 mL and 480 μg in 1.6 mL	Per 300 μg vial: 0.59 mg acetate; 50 mg sorbitol; 0.004% Tween® 80; 0.035 mg sodium	IV infusion or SC
			Syringes: 300 μg in 0.6 mL and 480 μg in 0.8 mL	Per 480 μg vial: 0.94 mg acetate; 80 mg sorbitol; 0.004% Tween 80; 0.056 mg sodium	
				Per 300 μg syringe: 0.295 mg acetate; 25 mg sorbitol; 0.004% Tween 80; 0.0175 mg sodium	
				Per 480 μg vial: 0.472 mg acetate; 40 mg sorbitol; 0.004% Tween 80; 0.028 mg sodium	
NORDITROPIN® CARTRIDGES (somatropin) [*E. coli*–expressed polypeptide hormone]	Novo Nordisk	Liquid: 5, 10, or 15 mg single-dose cartridges	Each per 5 mL	Per 5 mg cartridges: 1 mg histidine; 4.5 mg Poloxamer 188; 4.5 mg/mL phenol; 60 mg mannitol	SC
				Per 10 mg cartridges: 1 mg	

(*Continued*)

Appendix Formulations of Approved Protein Drugs (*Continued*)

Marketed name (generic name) [molecule type]	Manufacturer	Formulation type (dosage forms)	Strengths or concentration (pH of formulation)	Excipients	Route of administration
				histidine; 4.5 mg Poloxamer 188; 4.5 mg/mL phenol; 60 mg mannitol Per 15 mg cartridges: 1.7 mg histidine; 4.5 mg Poloxamer 188; 4.5 mg/mL phenol; 58 mg mannitol	
[b]NOVOLIN® 70/30 (human insulin) [Recombinant DNA origin]	Novo Nordisk	10 mL vial (U-100)	100 Units/mL	Mixture of 70% NPH, human insulin isophane suspension and 30% regular, human insulin injection; cloudy or milky suspension of human insulin with protamine and zinc	SC
Novolog® injection (insulin aspart); [human regular (rDNA origin)(B28 asp regular human insulin analog)]	Novo Nordisk	Liquid	100 Units/mL (pH of 7.2–7.6)	Per each 100 U/mL; 16 mg/mL glycerin; 1.50 mg/mL phenol; 1.72 mg/mL *m*-cresol; 19.6 μg/mL zinc; 1.25 mg/mL disodium hydrogen phosphate dehydrate; 0.58 mg/mL sodium chloride	SC
NOVOLOG® MIX 70/30 (70% insulin aspart protamine suspension and 30% insulin aspart injection) [*S. cerevisiae* yeast expresse-recombinant	Novo Nordisk	Liquid: 100 Units/mL suspension; 10 mL vials; 3 mL PenFill cartridges; 3 mL FlexPen prefilled syringe	100 Units/mL human insulin analog suspension containing 70% insulin aspart protamine crystals and 30% soluble insulin	36.4 mg/mL mannitol; 1.50 mg/mL phenol; 1.72 mg/mL *m*-cresol; 19.6 μg/mL zinc; 1.25 mg/mL disodium hydrogen phosphate dihydrate; 0.58 mg/mL sodium chloride;	SC

DNA origin]			aspart (pH of 7.20–7.44)	0.33 mg/mL protamine sulfate (hydrochloric acid or sodium hydroxide may be added to adjust pH)	
NOVOSEVEN® (Eptacog alfa) [BHK-expressed recombinant glycoprotein, coagulation factor VIIa (recombinant) glycoprotein]	Novo Nordisk	Lyophilized: 1.2, 2.4, or 4.8 mg single-dose vial	To be reconstituted with sterile WFI (pH of approximately 5.5)	Per vial: 1.2 mg vial: 5.84 mg sodium chloride; 2.94 mg calcium chloride; 2.64 mg glycylglycine; 0.14 mg polysorbate 80; 60.0 mg mannitol 2.4 mg vial: 11.68 mg sodium chloride; 5.88 mg calcium chloride; 5.28 mg glycylglycine; 0.28 mg polysorbate 80; 120.0 mg mannitol 4.8 mg vial: 23.36 mg sodium chloride; 11.76 mg calcium chloride; 10.56 mg glycylglycine; 0.56 mg polysorbate 80; 240.0 mg mannitol	IV bolus
NUTROPIN® AQ (Somatropin) [*E. coli*–expressed recombinant hGH)]	Genentech	Liquid: 10 mg vial or 10 mg pen cartridge	Vial and pen cartridge both contain 10 mg in 2 mL (pH of approximately 6.0)	Per vial or pen cartridge: 17.4 sodium chloride; 5 mg phenol; 4 mg polysorbate 20; 10 mg sodium citrate	SC
Oncaspar (pegaspargase) [*E. coli*–derived protein]	Enzon, Inc.	Liquid: 5 mL single-dose vials	0.5 mL per vial containing 750 IU/mL Oncaspar in a clear, colorless, phosphate buffered saline solution, pH 7.3. Each vial contains 3750	Per mL: monobasic sodium phosphate, USP 1.20 mg ± 5% dibasic sodium phosphate, USP 5.58 mg ± 5% sodium chloride, USP 8.50 mg ± 5% WFI, USP qs to 1.0 mL	IM or IV route of administration

(*Continued*)

Appendix Formulations of Approved Protein Drugs (*Continued*)

Marketed name (generic name) [molecule type]	Manufacturer	Formulation type (dosage forms)	Strengths or concentration (pH of formulation)	Excipients	Route of administration
			IU of Oncaspar. Each milliliter of Oncaspar contains: PEG-L-asparaginase 750 IU ± 20%		
ONTAK® (Denileukin diftitox) [*E. coli*–expressed recombinant cytotoxic protein]	Seragen, Inc. (Ligand Pharmaceutical)	Frozen liquid: 300 μg single-dose vial	300 μg in 2 mL (pH of 6.9 to 7.2)	Per vial: 20 mM citric acid; 0.05 mM EDTA; < 1% polysorbate 20	IV
ORTHOCLONE OKT® (Muromonab-CD3) [murine monoclonal Ab]	Ortho Biotech	Liquid: 5 mg ampoule	5 mg in 5 mL (pH 7.0 ± 0.5)	Per 5 mL: L2.25 mg monobasic sodium phosphate; 9.0 mg dibasic sodium phosphate; 43 mg sodium chloride; 1.0 mg polysorbate 80	IV
OVIDREL® (choriogonadotropin alfa) [CHO-expressed glycoprotein]	Serono	Liquid: 250 μg prefilled syringe	257.5 μg in 0.515 mL (pH is 6.5–7.5)	Per 0.515 mL: 28.1 mg mannitol; 505 μg 85% *O*-phosphoric acid; 103 μg L-methionine; 51.5 μg Poloxamer 188	SC
PEGASYS® (peginterferon alfa-2a) [covalent conjugate of recombinant alfa-2a interferon and 40 kD PEG]	Roche Laboratories	Liquid: single-dose vial, prefilled single-dose syringe	Vial: 180 μg in 1.0 mL; prefilled syringe: 180 μg in 0.5 mL (pH = 6.0 ± 0.5)	Vial per 1 mL: 8.0 mg sodium chloride; 0.05 mg polysorbate 80; 10.0 mg benzyl alcohol; 2.62 mg sodium acetate trihydrate; 0.05 mg acetic acid	SC

				Prefilled syringe per 0.5 mL: 4.0 mg sodium chloride; 0.025 mg polysorbate 80; 5.0 mg benzyl alcohol; 1.3085 mg sodium acetate trihydrate; 0.0231 mg acetic acid	
PROLEUKIN® (aldesleukin) [*E. coli*–produced human recombinant interleukin-2]	Chiron	Lyophilized: 1.1 mg single dose	To be reconstituted with 1.2 mL of sterile WFI (pH range 7.2–7.8)	Per 1.0 mL: 50 mg mannitol; 0.18 mg sodium dodecyl sulfate; 0.17 mg sodium phosphate; 0.89 mg dibasic sodium phosphate	IV
PULMOZYME® (Dornase alfa) [(CHO-expressed recombinant glycoprotein-human deoxyribonuclease I (rhDNase)]	Genentech	Liquid: single-dose ampoule	2.5 mg in 2.5 mL (pH of the solution is 6.3)	Per ampoule: 0.15 mg/mL calcium chloride dehydrate and 8.77 mg/mL sodium chloride	Administered by inhalation
Retavase® (Reteplase) [nonglycosylated deletion of mutein of plasminogen tissue activator]	Centocor, Inc.	Lyophilized: 18.1 mg single-use vial	To be reconstituted with 10 mL of sterile WFI (pH 6.0 ± 0.3)	Per vial: 8.32 mg tranexamic acid; 136.24 mg dipotassium hydrogen phosphate; 51.27 mg phosphoric acid; 364 mg sucrose; 5.2 mg polysorbate 80	IV
REBETRON® combination therapy containing: REBETOL® (Ribavirin) and INTRON® A (interferon alfa-2b) [*E. coli*–expressed recombinant protein]	Schering	REBETOL: 200 mg capsules	Interferon alfa-2b vials	Each 200 mg ribavirin capsule contains: capsule shell of gelatin and titanium dioxide	Oral administration
		Ribavirin white, crystalline powder	Single use (0.012 mg) 0.024 mg/mL (3 million IU/0.5 mL)	Inactive ingredients: microcrystalline cellulose, lactose monohydrate, croscarmellose sodium, magnesium stearate, pharmaceutical ink, shellac, anhydrous ethyl alcohol,	

(Continued)

Appendix Formulations of Approved Protein Drugs (*Continued*)

Marketed name (generic name) [molecule type]	Manufacturer	Formulation type (dosage forms)	Strengths or concentration (pH of formulation)	Excipients	Route of administration
				isopropyl alcohol, *n*-butyl alcohol, propylene glycol, ammonium hydroxide, FD&C Blue #2 aluminum lake	
		Interferon alfa-2b vials	Multidose 18 million IU (0.088 mg) 0.024 mg/mL (22.8 million IU/3.8 mL) or (3 million IU/0.5 mL)	INTRON A: 7.5 mg/mL sodium chloride; 1.8 mg/mL sodium phosphate dibasic; 1.3 mg/mL sodium phosphate monobasic; 0.1 mg/mL edetate disodium, 0.1 mg/mL polysorbate 80; 1.5 mg/mL *m*-cresol as a preservative	
		0.012 mg (3 million IU) single-use vial; 0.088 mg (18 million IU) multidose vial; 0.058 mg (18 million IU) multidose pen	Pen multidose 18 million IU (0.087 mg) 0.058 mg/mL (22.5 million IU/1.5 mL) or (3 million IU/0.2 mL)		
REBIF® (interferon beta-1a) SC injection [CHO-expressed recombinant glycoprotein]	Pfizer	Liquid: 22 or 44 μg single dose prefilled syringe	22 μg in 0.5 mL or 44 μg in 0.5 mL	Per 0.5 mL: 2 or 4 mg albumin (human) 27.3 mg mannitol; 0.4 mg sodium acetate	SC
REFACTO®	Wyeth-Ayest	Lyophilized: single-	To be reconstituted with	Sodium chloride, sucrose,	IV

(Antihemophilic factor) [CHO-expressed glycoprotein]		use vials 250, 500, 1000, or 2000 IU	4 mL of sterile 0.9% sodium chloride	L-histidine, calcium chloride, polysorbate 80 (amount not indicated)	
REFLUDAN® (Lepirudin) [recombinant hirudin polypeptide expressed in yeast cells]	Berlex	Lyophilized: 50 mg single-dose vial	To be reconstituted in either sterile WFI or 0.9% sodium chloride for injection (pH approximately 7)	Per vial: 40 mg mannitol	IV injection or infusion
REMICADE® (Infliximab) [recombinant chimeric IgG1(kgr) monoclonal Ab)]	Centocor	Lyophilized: 100 mg single-dose vial	To be reconstituted with sterile WFI (pH approximately 7.2)	Per 10 mL: 500 mg sucrose; 0.5 mg polysorbate 80; 2.2 mg monobasic sodium phosphate monohydrate; 6.1 mg dibasic sodium phosphate dihydrate	IV infusion
REOPRO® (Abciximab) [mammalian cell culture–expressed Fab fragment of a chimeric human-murine monoclonal Ab]	Lilly	Liquid: 10 mg single dose vial	10 mg in 5 mL at 2 mg/mL (pH 7.2)	0.01 M sodium phosphate; 0.15 M sodium chloride; 0.001% polysorbate 80 in WFI	IV
ROFERON®-A (Interferon alfa-2a) [*E. coli*–expressed recombinant protein]	Roche Laboratories	Liquid: 11.1 μg single-dose prefilled syringe; 22.2 μg single-dose prefilled syringe; 33.3 μg single-dose prefilled syringe	0.5 mL	Per 0.5 mL: 3.605 mg sodium chloride; 0.1 mg polysorbate 80; 5 mg benzyl alcohol (preservative); 0.385 mg ammonium acetate	SC
SIMULECT® (Basiliximab) [CHO-expressed glycoprotein, chimeric monoclonal Ab IgG1]	Novartis	Lyophilized: 10 mg single-dose vial; 20 mg single-dose vial	10 mg vial to be reconstituted with 2.5 mL sterile WFI 4.0 mg/mL	Per 10 mg vial: 3.61 mg monobasic potassium phosphate; 0.50 mg disodium hydrogen phosphate (anhydrous); 0.80 mg sodium chloride; 10 mg sucrose; 40 mg mannitol; 20 mg glycine	Central or peripheral IV

(Continued)

Appendix Formulations of Approved Protein Drugs (*Continued*)

Marketed name (generic name) [molecule type]	Manufacturer	Formulation type (dosage forms)	Strengths or concentration (pH of formulation)	Excipients	Route of administration
			20 mg vial to be reconstituted with 5 mL of sterile WFI 4.0 mg/mL	Per 20 mg vial: 7.21 mg monobasic potassium phosphate; 0.99 mg disodium hydrogen phosphate (anhydrous); 1.61 mg sodium chloride; 20 mg sucrose; 80 mg mannitol; 40 mg glycine	
SOMAVERT® (Pegvisomant) [recombinant protein covalently attached to 4 to 6 PEG (5 Kd)]	Pharmacia & Upjohn	Lyophilized: 21, 32, 43 mg single-dose vials	To be reconstituted with 1 mL of sterile WFI	Per 1 mL: 1.36 mg glycine; 36.0 mg mannitol; 1.04 mg sodium phosphate dibasic anhydrous; 0.36 mg sodium phosphate monobasic monohydrate	SC
SYNAGIS® (Palivizumab) [humanized recombinant monoclonal Ab IgG1]	MedImmune	Lyo: 50 and 100 mg single-dose vials	Lyophilized: 50 mg vial to be reconstituted with 0.6 mL of sterile WFI; 100 mg vial to be reconstituted with 1.0 mL of sterile WFI (pH = 6.0)	Lyophilized: Per 1 mL of 100 mg vial: 67.5 mg mannitol; 8.7 mg histidine; 0.3 mg glycine Per 0.6 mL of 50 mg vial: 40.5 mg mannitol; 5.2 mg histidine; 0.2 mg glycine	IM
		Liquid: 50 and 100 mg single dose vials	Liquid: 50 mg in 0.7 mL; 100 mg vial in 1.2 mL (pH = 6.0)	Liquid: Per 1.2 mL of 100 mg vial: 4.7 mg of histidine; 0.1 mg of glycine Per 0.7 mL 50 mg vial: 2.7 mg of histidine; 0.08 mg of glycine	

Stemgen® (Ancestim, stem cell factor) [*E. coli*–expressed recombinant human protein]	Amgen	Lyophilized: 1.875 mg vial	Lyophilized: 50 mg vial to be reconstituted with 0.6 mL of sterile WFI	Each vial: 4.5% mannitol; 0.5% sucrose; 5 mM glutamic acid; 10 mM histidine	SC
THYROGEN® (thyrotropin alfa) [CHO-expressed glycoprotein]	Genzyme	Lyophilized: 1.1 mg single-dose vial	To be reconstituted with 1.2 mL Sterile WFI (pH approximately 7.0)	Per 1 mL: 30 mg mannitol; 4.25 mg sodium phosphate; 2 mg sodium chloride	IM
INTRON® A (Interferon alfa-2b) [*E. coli*–produced recombinant protein]	Schering	Lyo: 0.038 mg single-dose vial; 0.069 mg single-dose vial; 0.192 mg single-dose vial	Lyo: to be reconstituted with 1 mL of sterile WFI	Lyo per 1 mL: 20 mg glycine; 2.3 mg sodium phosphate dibasic; 0.55 mg sodium phosphate monobasic; 1.0 mg human albumin	Lyo: IM, SC, IV; only 0.038 mg vial—IL, IM, SC, IV
		Liquid: 0.038 mg single-dose vial; 0.088 mg multidose vials; 0.123 mg multidose vials; 0.058 mg multidose pens; 0.096 mg multidose pens; 0.192 mg multidose pens	Liquid: each single-dose liquid vial contains 0.038 mg in 1 mL Each multidose liquid vial contains: 0.088 mg per 3.8 mL or 0.123 mg per 3.2 mL Each multidose pen contains: 0.087 mg per 1.5 mL; 0.144 mg 1.5 mL; 0.288 mg 1.5mL	Liquid per 1 mL of vial or multidose pen: 7.5 mg sodium chloride; 1.8 mg sodium phosphate dibasic; 1.3 mg sodium phosphate monobasic; 0.1 mg edetate disodium; 0.1 mg polysorbate 80; 1.5 mg *m*-cresol as a preservative	Liquid: SC, IL; multidose pens: SC
PEG-INTRON® (Peginterferon alfa-2b) [covalent conjugate of *E. coli*–produced recombinant	Schering	Lyophilized: 50, 80, 120, 150 μg single-dose vials and 50, 80, 120, 150 μg	Each vial is to be reconstituted with 0.7 mL of sterile WFI to yield. Each Redipen	Per vial: 1.11 mg dibasic sodium phosphate anhydrous; 1.11 mg monobasic sodium phosphate dehydrate; 59.2 mg sucrose;	SC

(*Continued*)

Appendix Formulations of Approved Protein Drugs (*Continued*)

Marketed name (generic name) [molecule type]	Manufacturer	Formulation type (dosage forms)	Strengths or concentration (pH of formulation)	Excipients	Route of administration
alfa-2b interferon with monomethoxy PEG]		single-dose Redipen® dual-chamber glass cartridge	cartridge to be reconstituted with the supplied WFI present in the cartridge chamber to yield 0.5 mL solution	0.074 mg polysorbate 80 Per Redipen: 1.013 mg dibasic sodium phosphate anhydrous; 1.013 mg monobasic sodium phosphate dehydrate; 54 mg sucrose; 0.0675 mg polysorbate 80	
XIGRIS® (Drotrecogin alfa—activated) [recombinant form of human activated protein C, glycoprotein]	Lilly	Lyophilized: 5 mg single-dose vial; 20 mg single-dose vial	5 mg vials to be reconstituted with 2.5 mL of WFI 2 mg/mL; 20 mg vials to be reconstituted with 10 mL of WFI 2 mg/mL	Per 5 mg vial: 40.3 mg sodium chloride; 10.9 mg sodium citrate, and 31.8 mg sucrose Per 20 mg vial: 158.1 mg sodium chloride; 42.9 mg sodium citrate; and 124.9 mg sucrose	IV infusion
XOLAIR® (Omalizumab) [CHO-expressed IgG1 kappa]	Genentech	Lyophilized: 75 mg single-dose vial; 150 mg single-dose vial	75 mg vial: 129.6 mg to be reconstituted with 0.9 mL WFI delivers 75 mg of Omalizumab in 0.6 mL 150 mg vial: 202.5 mg to be reconstituted with 1.4 mL WFI delivers 150 mg of Omalizumab in 1.2 mL	75 mg vial contains: 93.1 mg sucrose; 1.8 mg L-histidine hydrochloride monohydrate; 1.2 mg L-histidine; 0.3 mg polysorbate 20 150 mg vial contains: 145.5 mg sucrose; 2.8 mg L-histidine hydrochloride monohydrate; 1.8 mg L-histidine; 0.5 mg polysorbate 20	SC

ZENAPAX® (Daclizumab) [recombinant humanized IgG1 monoclonal Ab]	Roche Laboratories/Genentech	Liquid: 25 mg single-dose vial	25 mg/5 mL (pH 6.9)	Per vial: 3.6 mg sodium phosphate monobasic monohydrate; 11 mg sodium phosphate dibasic heptahydrate; 4.6 mg sodium chloride; 0.2 mg polysorbate 80	IV
ZEVALIN® (Ibritumomab tiuxetan) [immunoconjugate between CHO-expressed IgG1 kappa and linker-chelator tiuxetan]	Biogen Idec	Liquid: 3.2 mg single-dose vial	3.2 mg in 2 mL of 0.9% NaCl	Packaged as a kit containing four vials: Active ingredient Zevalin packaged in 2 mL of 0.9% NaCl 50 mM sodium acetate buffer vial containing: 13.6 mg sodium acetate trihydrate in 2 mL of water Formulation buffer vial containing: 750 mg if human albumin; 76 mg of sodium chloride; 21 mg of sodium phosphate dibasic heptahydrate; 4 mg pentetic acid; 2 mg potassium phosphate monobasic; 2 mg of potassium chloride; 10 mL of water (pH 7.1)	IV infusion

[a]Humulin (R, L, N, U, 50/50, and 70/30). Human insulin manufactured by Eli Lilly and Company has the trademark Humulin and is available in six formulations—Regular (R), NPH (N), Lente (L), Ultralente® (U), 50% human insulin isophane suspension (NPH)/50% human insulin injection (buffered regular) (50/50), and 70% human insulin isophane suspension (NPH)/30% human insulin injection (buffered regular) (70/30).

[b]Novo Nordisk Novolin 70/30 human insulin 10 mL vials, Novolin N human insulin 10 mL vials, Novolin R human insulin 10 mL vials. Delivery Systems Novolin 70/30 InnoLet, Novolin N InnoLet, Novolin R InnoLet, Novolin R PenFill 1.5 mL Cartridges, Novolin R PenFill 3 mL Cartridges, Novolin N PenFill 3 mL Cartridges, Novolin 70/30 PenFill 3 mL Cartridges.

Abbreviations: Ab, antibody; SC, subcutaneous; IV, intravenous; CHO; HSA, human serum albumin; SC, subcutaneous; USP, United States Pharmacopeia; AHF; IU, international unit; NMT, not more than; Ig, immunoglobulin; BHK, baby hamster kidney; PEG, polyethylene glycol; hGH, human growth hormone; IM, intramuscular; WFI, water for injection.

18
Excipients Used in Vaccines

Manmohan Singh and Derek O'Hagan
Vaccine Delivery Group, Chiron Vaccines, Chiron Corporation, Emeryville, California, U.S.A.

INTRODUCTION

Vaccines available on the market contain various types of antigens, adjuvants, and additives, which in combination provide maximum protection against several infectious diseases. These vaccines might contain live or killed viruses, purified viral proteins, inactivated bacterial toxins, and polysaccharides or purified subunit recombinant proteins. The diverse nature of these antigens requires different excipients to be used to stabilize them for use within their designated shelf life. Because the total vaccine doses used each year globally average several million in number, the selection of an excipient for a vaccine formulation is a serious task having huge implications toward safety, stability, and storage. Also, in recent years, certain preservatives used in some vaccines such as thimerosal have been highlighted for their possible contributions to unwarranted reactogenicity in infants and adults.

In this chapter on vaccine excipients and their use in vaccine formulations, we will describe several categories of parenteral excipients and highlight the ones used in existing marketed vaccine formulations. We will also focus on some emerging vaccine formulations that might become vaccine products in the future. Like any other pharmaceutical excipient intended for human use, the excipients used in vaccines must comply with rigorous standards of quality, purity, availability, and compatibility. Parenteral excipients are further evaluated to meet higher purity and safety standards because these are injected into the human body. Because most commonly used vaccines are administered parenterally, excipients must comply with strict guidelines set forth by the U.S. Food and Drug Administration (FDA) for any parenteral dosage form.

The following sections will detail each component within a vaccine formulation and describe its use, source, and limitations.

ADJUVANTS

The word "adjuvant" is derived from the Greek word "adjure" meaning "to enhance." Clearly, adjuvants are meant for enhancing the immune response to antigens when combined with the vaccines (1,2). Aluminum salts are the most commonly used adjuvants in marketed vaccines and are a part of more than 75% of all available vaccine formulations (3). These may include aluminum hydroxide, aluminum phosphate, and potassium aluminum sulfate. These aluminum salts are often referred to as "alum" in general use. The vaccines are prepared by either adsorption of antigens on aluminum gels or precipitation of antigens in a solution of alum. The aluminum salts are used in the range of 0.2 to 1.0 mg/dose depending upon the vaccine [higher doses for combination vaccines such as diphtheria-tetanus-acellular pertusis vaccine (DTaP) and lower doses for single antigen vaccines such as hepatitis B surface antigen (HBsAg)].

The use of alum in vaccine formulations dates back to early 1940s when it was first used for diphtheria and tetanus toxoid (TT) vaccines (4). Early studies suggested that aluminum salts led to a depot effect (reduced elimination of the antigen from the site of injection). But subsequent studies questioned the importance of the depot effect and showed that alum salts enhanced uptake by antigen-presenting cells or induced production of cytokines or complements. Alum has a good safety record, but comparative studies show that it is a weak adjuvant for antibody induction to subunit vaccines and a poor adjuvant for cell-mediated immunity. Moreover, alum induces the induction of immunoglobulin-E antibodies and has been associated with some adverse effects in human subjects. Over the last six decades, the safety of alum has been well established. A recent review article evaluated the evidence of adverse events and concluded that aluminum salts did not cause any serious and long-lasting adverse events (5). But as the number of available vaccines on alum continues to increase, there is some concern regarding aluminum-induced toxicity, and FDA now seeks evidence on the need of using alum for all new vaccine formulations. Research on other vaccine delivery systems such as emulsions, liposomes, and microparticles may reduce the amount of alum being used in next generation vaccines. Other less commonly used adjuvants are calcium phosphate and L-tyrosine. A ragweed allergy vaccine uses L-tyrosine as an adjuvant and is marketed in Europe (Tyrosin TU, Allergy Therapeutics, Worthing, West Sussex, U.K.).

The most important issue in adjuvant development is safety because safety concerns have restricted the development of adjuvants since alum was first introduced more than 50 years ago. Many experimental adjuvants have demonstrated high potency but are too toxic for routine clinical use. For standard prophylactic immunization in healthy individuals, only adjuvants that induce minimal side effects will prove acceptable. Additional issues that are important in adjuvant development include biodegradability, stability, ease of manufacture, cost, and applicability to a wide range of vaccines. The main supplier of alum adjuvants is Brenntag Biosector (Frederi Kssund, Germany). Although a diverse range of substances have been shown to exert an adjuvant effect, the mechanisms of action are often poorly understood.

Immunostimulatory Adjuvants

Immunostimulatory adjuvants exert their effects predominantly at the cytokine level or through the activation of costimulatory signals. The type of response required for optimal protection depends on the pathogen. One class of immunostimulatory adjuvants is derived from the lipopolysaccharide of gram-negative bacteria. The most extensively evaluated member of this family, monophosphoryl lipid A (MPL), is obtained from *Salmonella minnesota*. MPL has been shown to induce the synthesis and release of cytokines, which promote the generation of specific immune responses.

MPL has been evaluated in the clinic for both infectious disease vaccines and cancer vaccines, and has shown adjuvant activity, with tolerable side effects (6,7). MPL has also been combined with alum and evaluated in the clinic, showing good tolerability in a limited number of volunteers, and is now part of an allergy vaccine (Pollinex Quattro, Allergy Therapeutics, West Sussex, U.K.). A second group of immunostimulatory adjuvants are the triterpenoid glycosides derived from *Quillaja saponaria*. The fraction QS21 was isolated by Kensil, who defined the structural moieties responsible for adjuvant activity (8,9). Although the mechanism of action of QS21 remains poorly defined, it is a potent adjuvant for cytotoxic T lymphocyte (CTL) responses and induces Th1 cytokine responses [interleukin-2 (IL-2) and interferon-γ (IFN-γ)]. QS21 is yet to be used in licensed products.

It has recently been established that bacterial DNA, but not vertebrate DNA, has direct immunostimulatory effects on leukocytes in vitro. The immunostimulatory effect is due to unmethylated CpG dinucleotides, which are underrepresented and "methylated" in vertebrate DNA. CpG DNA and synthetic oligonucleotides from a variety of sources have shown significant promise as new adjuvants (10–12). CpG induces a strong Th1 response, mainly by stimulating cytokine induction and through the expression of costimulatory molecules on antigen-presenting cells. CpG is currently in clinical trials and may become part of a licensed product in the future.

Emulsions and Lipids

Another new adjuvant used in a flu vaccine in Europe is called as MF59, which is a submicron oil in water emulsion containing squalene as the oil phase. This adjuvant emulsion is safe and nontoxic for use in humans and has been tested in several million subjects (13–16). The vaccine formulation contains MF59 (FLUAD)^{TM}, which is a licensed product in Europe and has been shown to be safe and well tolerated in patients over the last seven years.

Liposomes are phospholipid vesicles that have been evaluated both as adjuvants and as vehicles for antigens and adjuvants (17). A liposomal hepatitis A (Hep A) vaccine (virosomes) has been extensively evaluated in the clinic and is currently licensed for a Hep A vaccine (18). Alternative adjuvants that have been used in a few products include L-tyrosine (allergy vaccine) and MPL (cancer treatment). The various adjuvants (mainly alum salts) used in vaccine formulations and their quantities per dose are listed in Table 1.

PRESERVATIVES

Another important component of most vaccine formulations is a suitable preservative. The three most commonly used preservatives in available vaccines are phenol, 2-phenoxyethanol, and ethyl mercurithiosalicylate (thimerosal). Thimerosal, in particular, is used in multidose vials as an antimicrobial preservative. Concerns about the presence of mercury in thimerosal (25 μg/dose) has led to FDA stopping the use of this preservative in all vaccines by an amendment to the FDA Modernization Act of 1997. By 2001, thimerosal was removed from most childhood vaccines as a precautionary measure. The sources of all of the preservatives for vaccines are the same suppliers that supply preservatives for the parenteral dosage forms—(J. T. Baker, Aldrich, Spectrum, etc. from U.S.A.). Table 2 lists some of the preservative concentrations in common vaccines.

Table 1 A Representative List of Adjuvants Used in Commercial Vaccine Formulations

Adjuvant	Vaccine	Amount (mg)
Aluminum hydroxide	DTaP	0.625
	Hep B	0.25
	Hep A	0.25
	DTaP–IPV	0.85
Aluminum phosphate	DTaP	0.33
	Pneumococcal conjugate	0.125
Aluminum sulfate	Hep B	0.5
	Hib-Hep B	0.225
MF59	Influenza	21 mg of squalene in 0.5 mL emulsion

Abbreviations: DTaP, diphtheria-tetanus-acellular pertusis vaccine; Hep B, hepatitis B; Hep A, hepatitis A; IPV, inactivated polio vaccine; Hib, *Haemophilus influenzae* type B.

ADDITIVES

Additives are used in vaccines to stabilize, buffer, and prevent adherence to the glass and aggregation. The most commonly used additives fall into three main classes—sugars (lactose and sucrose), amino acids (glutamic acid, glycine, and histidine), and proteins (gelatin and albumin). The use of sugars and amino acids has no associated concerns, but the use of proteins has raised some issues regarding immediate-type hypersensitivity reactions with gelatin and possible contamination of infectious agents and bovine spongiform encephalopathy (mad cow disease) with albumin. Therefore, the selection and choice of a suitable additive, apart from its ability to stabilize, also depends on the source of the additive. Thus human serum albumin is derived from blood of screened donors only, and the use of gelatin and its source is closely monitored. The common suppliers are Aldrich, Penta manufacturing, Croda Foods, etc. all from U.S.A. Table 3 lists the most commonly used additives in marketed vaccines and their concentrations.

Table 2 A Representative List of Preservatives Used in Commercial Vaccine Formulations

Preservative	Vaccine	Amount (mg)
Phenol	*Salmonella typhi*	1.25
	Pneumococcal polysaccharide	1.25
2-Phenoxyethanol	DTaP	2.5
	Hep A	2.5
	IPV	2.5
	Hep A–Hep B	5.0
	DTaP–IPV–HBV	2.5
Thimerosal	Influenza	0.025
	Japanese encephalitis	0.007

Abbreviations: DTaP, diphtheria-tetanus-acellular pertusis vaccine; Hep A, hepatitis A; Hep B, hepatitis B; IPV, inactivated polio vaccine; HBV, hepatitis B virus.

Table 3 A Representative List of Additives Used in Commercial Vaccine Formulations

Additive	Vaccine	Amount (mg)
Lactose	Meningococcal polysaccharide	5.0
Sucrose	Hib-TT	42.5
	MMR	1.9
Sorbitol	MMR	14.5
Human albumin	MMR	0.3
Gelatin	DTaP	0.0015
	MMR	14.5
	Influenza	0.025
	Varicella	12.5

Abbreviations: Hib-TT, *Haemophilus influenzae* type B-tetanus toxoid, MMR, mumps, measles, rubella; DTaP, diphtheria-tetanus-acellular pertusis vaccine.

SALTS

The most commonly used salts in vaccine formulations are sodium chloride, sodium phosphate, succinic acid, and sodium borate. The concentrations of the salts used in any given formulation are based on isotonicity, pH, and other stabilizers being used in the formulations. A typical range is from 5 to 20 mM salt concentration. These concentrations are also selected to reduce pain on injection and to accord rapid normalization with physiological fluid. Surfactants used in MF59 emulsion include Tween 80 and sorbitan trioleate.

RESIDUALS FROM THE MANUFACTURING PROCESS

A certain amount of reagents used in the manufacturing processes of the antigens do end up in the final formulation, and their amounts are clearly defined and regulated. Some of the most common forms of residuals that are present are inactivating agents (formaldehyde), antibiotics (neomycin sulfate, streptomycin, and amphotericin B), and cellular contents (yeast proteins). The residual content for formaldehyde is required to be below 0.1 mg/unit dose. The amount of yeast proteins in HBsAg vaccines is about 1 mg/mL. The common suppliers are (Dupont, Aqulon, U.S.A.), (Hoechst, Germany), etc. Table 4 lists some of the vaccines and their formaldehyde and neomycin content.

EXCIPIENTS USED TO IMPROVE STABILITY OF VACCINES

Certain additives, salts, and bulking agents may be added primarily to improve vaccine stability upon storage (19,20). These excipients such as mannitol, glycine, and trehalose have a direct impact on the stability of the polypeptide or conjugate and are investigated for this purpose.

EXCIPIENTS USED IN VACCINE FORMULATIONS CURRENTLY IN CLINICAL TRIALS

Several novel vaccine formulations are currently in clinical trials. Some of these will eventually end up as products in the future. Several new polymers are being

Table 4 Residual Contents of Vaccine Formulations

Residual	Vaccine	Amount (mg)
Formaldehyde	Polio	0.1
	Hib-HBsAg	0.0002
	Hep A	0.05
	DTaP	0.1
	Japanese encephalitis	0.1
Neomycin	MMR	0.025
	Rabies	< 0.15
	Measles	0.025
Egg protein	Influenza	0.001

Abbreviations: Hib-HBsAg, *Haemophilus influenzae* type B-hepatitis B surface antigen; Hep A, hepatitis A; DTaP, diphtheria-tetanus-acellular pertusis vaccine; MMR, mumps, measles, rubella.

evaluated as a component of a vaccine formulation. Most common among these are poly(lactide-co-glycolide) (PLG), chitosan, hyaluronic acid polyesters, and hydrogels. Other new excipients include novel lipids and modified cholesterols. These new components will have to prove their safety and tolerability in hundreds of patients before they can be approved for mass immunization, predominantly in young children. Table 5 lists some of these new excipients for next generation vaccines.

ANALYTICAL ASSAYS AND QUALITY CONTROL OF EXCIPIENTS FOR VACCINE FORMULATIONS

Any excipient for a vaccine formulation is treated as a component of a parenteral formulation and must adhere to strict FDA requirements of compliance and regulation of these materials. Most of the excipients commonly used in vaccine formulations (except adjuvants) are also used in many other parenteral formulations and thus have a long safety and tolerability profile. Many components like the adjuvants have to have established guidelines for purity, monomer ratios, and concentrations. Ideally, purity greater than 98% is considered as the minimal criterion. The selection of concentration is based on extensive preclinical evaluation to show minimum reactogenicity and enhanced immunogenicity. Typical assays used for quantifying the excipients are based on reversed phase–high performance liquid chromatography,

Table 5 A Representative List of Excipients Currently Under Evaluation in New Vaccine Formulations

Excipient type	Name	Clinical status	References
Polymer	PLG	Phase I	21
Polymer	Chitosan	Phase I	22
Adjuvant	CpG ODN	Phase II	23
Adjuvant	MPL adjuvant	Phase III	24

Abbreviations: MPL, monophosphoryl lipid A; PLG, poly(lactide-co-glycolide); CpG ODN, CpG oligodeoxynucleotide.

sodium dodecyl sulfate polyacrylamide gel electrophoresis, spectrophotometric analysis, and colorimetric analysis.

SELECTION OF EXCIPIENTS FOR NEXT GENERATION VACCINES

Next generation vaccine formulations will comprise several antigens that will include glycoconjugates, recombinant proteins, plasmids, oligonucleotides, peptides and additional adjuvant molecules for enhanced immunogenicity. These complex formulations will need a rational selection of stabilizers, preservatives, and buffers. Most paramount in this selection is that the stability of all components of the vaccine should be such that the potency of the final formulation is maintained. Enhanced shelf life is another parameter that would dictate formulation development in vaccines. New vaccine modalities such as DNA vaccines are currently being explored using charged PLG microparticles (21) as delivery systems. These and similar novel delivery technologies will be essential components of some of the next generation vaccines.

SUMMARY

In conclusion, excipients to be used in vaccines must be very carefully selected and justified to the regulatory authorities because these are to be used in millions of healthy subjects. Their safety and compatibility with other vaccine components are of prime importance. Because vaccines of the future will be more and more complex, the need for suitable excipients is also likely to grow. It must be ensured that the excipients do not compromise the immunogenicity of the vaccine and accord maximum stability upon long-term storage.

ACKNOWLEDGMENTS

We would like to thank all members of the formulation group in vaccines research for their suggestions and help.

REFERENCES

1. Vogel FR, Powell MF. A compendium of vaccine adjuvants and excipients. In: Powell MF, Newman MJ, eds. Vaccine Design: The Subunit and Adjuvant Approach. New York: Plenum Press, 1995:141–228.
2. Edelman R. Adjuvants for the future. In: Levine MM, Woodrow GC, Kaper JB, Cobon GS, eds. New Generation Vaccines. New York: Marcel Dekker, Inc., 1997:173–192.
3. Gupta RK. Aluminum compounds as vaccine adjuvants. Adv Drug Del Rev 1998; 32:155–172.
4. Ramon G. Sur la toxine et surranatoxine diphtheriques. Ann Inst Pasteur 1940; 38:1.
5. Jefferson T, Rudin M, Pietrantonj S. Adverse events after immunization with aluminum containing DTP vaccines: systemic review of the evidence. Lancet 2004; 4:84–90.
6. Ulrich JT, Myers KR. Monophosphoryl lipid A as an adjuvant: past experiences and new directions. Pharm Biotechnol 1995; 6:495–524.
7. Ulrich JT. MPLr immunostimulant: adjuvant formulations. In: O'Hagan DT, ed. Vaccine Adjuvants: Preparation Methods and Research Protocols. Totowa, NJ: Humana Press Inc., 2000:273–282.

8. Kensil CR. Saponins as vaccine adjuvants. Crit Rev Ther Drug Carrier Syst 1996; 13:1–55.
9. Kensil CR, Kammer R. QS-21: a water-soluble triterpene glycoside adjuvant. Exp Opin Invest Drugs 1998; 7:1475–1482.
10. Krieg AM, Yi AK, Matson S, et al. CpG motifs in bacterial DNA trigger direct B-cell activation. Nature 1995; 374:546–549.
11. Klinman DM, Barnhart KM, Conover J. CpG motifs as immune adjuvants. Vaccine 1999; 17:19–25.
12. Singh M, Ott G, Kazzaz J, et al. Cationic microparticles are an effective delivery system for immune stimulatory CpG DNA. Pharm Res 2001; 18:1476–1479.
13. Ott G, Barchfeld GL, Chernoff D, Radhakrishnan R, van Hoogevest P, Van Nest G. MF59: design and evaluation of a safe and potent adjuvant for human vaccines. In: Powell MF, Newman MJ, eds. Vaccine Design: The Subunit and Adjuvant Approach. New York: Plenum Press, 1995:277–296.
14. Cataldo DM, Van Nest G. The adjuvant MF59 increases the immunogenicity and protective efficacy of subunit influenza vaccine in mice. Vaccine 1997; 15:1710–1715.
15. Higgins DA, Carlson JR, Van Nest G. MF59 adjuvant enhances the immunogenicity of influenza vaccine in both young and old mice. Vaccine 1996; 14:478–484.
16. O'Hagan DT, Ott GS, Van Nest G. Recent advances in vaccine adjuvants: the development of MF59 emulsion and polymeric microparticles. Mol Med Today 1997; 3:69–75.
17. Gregoriadis G. Immunological adjuvants: a role for liposomes. Immunol Today 1990; 11:89–97.
18. Bungener DT, Huckriede A, Wilschut J. Virosomes as an antigen delivery system. J Liposome Res 2000; 10:329–338.
19. Cleland JL, Jones AJS. Excipient stabilization of polypeptides treated with organic solvents. U.S. Patent 1994, 5589167.
20. Gribbon EM, Sen S, Roser BJ, Kampinga J. Stabilization of vaccines using trehalose. In: Brown F, ed. New Approaches to Stabilization of Vaccines Potency. Basel Kargar; Dev Biol Stand 1996; 228:692–700.
21. Singh M, Briones M, Ott G, O'Hagan DT. Cationic microparticles: a potent adjuvant for DNA vaccines. Proc Natl Acad Sci 2000; 97(2):811–816.
22. McNeela EA, Connor D, Gill J, et al. A mucosal vaccine against diphtheria. Vaccine 2000; 19(9–10):1188–1198.
23. www.coleypharma.com (accessed on Feb 3, 2005).
24. www.corixa.com (accessed on Feb 3, 2005).

19
Polymeric Excipients for Controlled Release Applications

Mahesh V. Chaubal
Baxter Healthcare, Round Lake, Illinois, U.S.A.

INTRODUCTION

The emergence of controlled release (CR) technology as an effective way to enhance patient compliance and extend the life cycle of a drug has led to the need for novel ways of controlling the drug release profiles. Polymers present a logical and simple approach to control the release of drugs. The use of polymers in pharmaceutical preparation dates back to 3000 B.C.E, with references in ancient Indian medical text (1). The use of polymers for oral CR was reported in the modern era, in 1930s, with the use of shellac in aspirin tablets. However elevation of this technology to its current commercial status was catalyzed in the 1970s and 1980s, with a rising need for minimization of toxic side effects and for life cycle management of drugs.

As shown in Figure 1, polymers are capable of providing sustained release of an encapsulated drug, within its therapeutic window. This leads to reduced peaks and valleys typically associated with immediate release dosage forms. Typically natural polymers or their derivatives (such as cellulose and methyl cellulose) as well as synthetic nondegradable polymers [such as poly(vinyl pyrrolidone) and polymethacrylates] are used for oral CR applications. Due to this generic status of polymers, the emphasis in novel oral CR systems is geared more toward the mixing and matching of polymers and fabrication of the solid dosage form, rather than design of novel polymers. On the other hand, use of polymers for injectable and/or implantable CR systems is relatively new and more infrequent. In such situations, the polymer must not only permit CR of the drug, but also be biocompatible and nontoxic. Several drug delivery applications also require the polymer to be biodegradable—degrading into by-products that are safe and can be cleared from the body.

Thus polymers serve as key excipients in oral and parenteral CR formulations. Other excipients used in sustained release dosage forms have been covered in other chapters within this book. For example, parenteral CR dosage forms involving polymers would still have other excipients as discussed in the chapter on injectable excipients (Chapter 16). Similarly oral dosage forms will require consideration of other excipients depending on the nature of the drug, as discussed in Chapter 12. This chapter reviews some of the promising polymers used in this application.

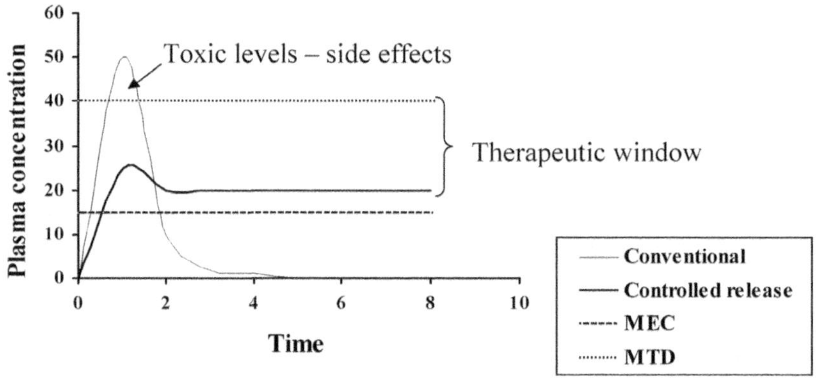

Figure 1 Comparison of typical pharmacokinetic profiles seen for conventional versus controlled release formulations. *Abbreviations*: MEC, minimum effective concentration; MTD, maximum tolerable dose.

ORAL DRUG DELIVERY

Typically, oral CR formulations utilize natural or synthetic polymers that provide CR of the therapeutic agent via diffusion and/or erosion mechanism. Erosion-controlled CR formulations involve the use of water-soluble polymers that encapsulate the active ingredient in specific patterns (layers, films, three-dimensional structures, etc.). The active ingredient is released over time as the polymer dissolves in the gastrointestinal (GI) tract. The rate of release is controlled by mixing and layering hydrophilic polymers with varying dissolution kinetics as well as by the use of innovative fabrication designs. Generally, the polymers used in oral CR systems are nonabsorbable biopolymers such as cellulose derivatives or hydrophilic gums. Alternatively, the polymer could be insoluble, and the drug release is governed by diffusion or osmosis. The diffusion could be Fickian or zero-order depending on the type of polymers and excipients used. In reality, except for osmotically driven systems, most oral CR formulations utilize a combination of dissolution and diffusion. Although the drug transport is via diffusion, the rate-controlling step for drug release is the dissolution of the polymer. An example of this concept is the TIMERx® (Penwest Pharmaceuticals) formulation used for CR of nifedipine. This formulation consists of a blend of locust bean gum and xanthan gum, which form aggregates via noncovalent interactions (2). The composition of these two polymers can be varied to obtain the desired drug release profile. The SODAS (Elan) formulation consists of the drug covered by layers of water-soluble and insoluble polymers in 1 to 2 mm beads. The drug is released via diffusion, wherein the rate of diffusion is controlled by the types of polymers and excipients used. Examples of commercial products utilizing this technology are Ritalin LA®, a sustained release formulation of methylphenidate hydrochloride, and Cardizem SR® that encapsulates diltiazem as the active ingredient.

Dissolution- and diffusion-based technologies can be utilized to obtain various novel profiles of drug release. For example, a repeat-action tablet can be created wherein the tablet provides two immediate release doses, with a time interval. This technology is used in Proventil Repetabs® to provide two doses of albuterol (2 mg) separated by six hours. Acacia and carnauba wax are the polymers used in this specific case to provide two immediate release doses over a defined period.

Polymers can also be utilized to obtain targeted delivery into certain regions of the GI tract. Table 1 provides typical transit time for food in various parts of the GI tract. Targeting to the GI tract can be achieved by a number of approaches. Enteric coating polymers are the simplest of examples, wherein the polymer dissolves in a specific portion of the GI tract (as a function of pH) resulting in targeted delivery. However, these formulations are typically pH dependant and thus depend on the fed or fasted state of the patient. More sophisticated pH-independent systems are also being developed. For example, drugs used for the treatment of ulcerative colitis, a form of inflammatory bowel disease, which affects the colon, are most effectively delivered directly to the colon. A novel formulation of prednisolone consisting of a combination of ethyl cellulose and a starch derivative to encapsulate the drug is being tested in the clinic (3). The polymer coating is susceptible to degradation specifically by enzymes present in the colon, thus leading to colon-targeted delivery. Azopolymers are examples of excipients, which are also useful for colon targeting. The drug is attached to these polymers via an azobond, which is degraded specifically by enzymes in the colon, thus leading to targeted delivery (4). Another approach to targeted delivery is delaying the release of the drug until it reaches the region of interest within the GI tract. This concept has been employed in a commercially available formulation of aminosalicylate for colon-targeted delivery. The most successful formulation utilizing the delayed release technology is AstraZeneca's Prilosec, which is a delayed-release formulation of omeprazole.

Delayed release can also be applied for chronotherapeutic dosing, whereby the drug is administered in sync with the body's internal circadian rhythm. Verelan PM is an example of chromotherapeutic formulation that utilizes the combination of water-soluble polymers (povidone and gelatin) in conjunction with insoluble polymers (shellac).

Besides the technologies mentioned above, several other strategies have been applied to obtain CR of orally active drugs. One of the examples is the ion exchange resins-based formulation. In this case, the drug is bound to ion exchange resin particles via electrostatic interactions. As the pH and ionic strength of the physiological fluids in the GI tract change (with time and location), the drug gets desorbed from the resin, providing a sustained release effect. Examples of marketed products utilizing this technology include duromine, which contains the basic drug phentermine,

Table 1 Description of the Gastrointestinal Tract and the Implication for Drug Delivery

Region	Length (m)	Transit time (hr)	pH	Approaches to drug delivery
Stomach	0.2	1–3	1.5–5	Enteric coating may be required for drugs that are sensitive to acidic conditions. If the stomach is the target region for absorption, superporous hydrogels may be used for gastroretention
Small intestine	7	3–5	5–8	Specific biopolymers such as tomato lectin may be used to exploit its selective binding characteristics to small intestine epithelium
Large intestine	1.5	4–16	5.5–7	Contains specific enzymes that may be exploited for drug release. Polymers that are specifically degraded by such enzymes include saccharide-containing polymers or azopolymers

complexed onto an anionic resin, and MS Contin suspension, which uses a negatively charged polystyrene sulphonate resin. Besides CR, ion exchange resins are also used for taste masking (because the drug will not display its taste characteristics in a bound state) as well as for enhanced dissolution and as a powder processing aid (5).

Another novel use of polymers in oral drug delivery is for gastric retention formulations, utilizing superporous hydrogels. Such hydrogels are prepared by the addition of sodium bicarbonate during polymerization reaction. This leads to the generation of carbon dioxide, which leads to bubbles that escape leaving a porous structure. These formulations swell rapidly (within 20 minutes) to a size that is bigger than the pylorus in the GI tract. This swelling prevents the oral solid dosage form from being removed from the GI tract during a turnover cycle. Poly(vinyl pyrrolidone) is one of the polymers being explored for this application (6). Other hydrogels possessing bioadhesive and mucoadhesive properties also find applications in oral drug delivery due to their ability to increase residence time of drugs in the GI tract (7). Table 2 lists some of the polymers commonly used as excipients in oral drug delivery formulations.

As seen in Table 2, majority of polymers used in oral drug delivery are derived from natural sources. While plant-derived materials may generally be acceptable based on their safety and toxicology profile, a significant concern exists for animal-derived excipients. Concerns regarding bovine spongiform encephalopathies and transmissible spongiform encephalopathies transmission risks have been discussed elsewhere in this book. Such concerns have led to research and development of synthetic replacements for animal-derived polymers. For example, a recombinant form of gelatin is being explored for drug delivery applications to replace natural bovine-derived gelatin (8). Similarly, a carrageenan and hydroxypropyl starch formulation has been developed as an alternative to gelatin-based soft elastic capsules.

Biodegradable Polymers for Oral Delivery

The use of biodegradable polymers, especially polylactic acid (PLA), in oral solid dosage forms has been reported in the literature. PLA has been used as a matrix for phenobarbital tablets (9). Similarly, the use of polylactide as a matrix for oral dosage form of naproxen has also been reported (10).

A novel use of biodegrabale polymers in oral drug delivery was demonstrated by Mathiowitz and coworkers, who showed that fumaric acid–based polyanhydrides had bioadhesive properties useful for increased gastroretention (11). Poly(fumaric anhydride) is a rapidly degrading polymer that degrades to a component of the Kreb's cycle, fumaric acid. The rapid surface-erosion type degradation leads to a high concentration of carboxyl groups on the surface of the dosage form. It is hypothesized that these highly acidic groups enhance bioadhesiveness of the matrix, thus increasing its residence time in the GI tract.

PARENTERAL DRUG DELIVERY

Early development of polymers in injectable drug delivery primarily involved PLA and poly(lactic-co-glycolic) acid (PLGA) due to the prior use of these polymers in biomedical applications as sutures. Besides the safe and biocompatible nature of these polymers, their ease of availability made them ideal first candidates for screening parenteral CR formulations. Some of the early biodegradable polymer-based products for injectable sustained release used these polymers. However because

Table 2 Examples of Polymer-Based Technologies Used for Oral Drug Delivery

Technology	Company	Polymers	Approach	Product example
Diffucaps	Eurand	Povidone, HPMC, ethyl cellulose	Combination of immediate and sustained release	Metadate CD
SODAS	Elan	Ammoniomethacrylate copolymers, povidone	Combination of immediate and sustained release	Avinza
Andrx proprietary technology	Andrx	Candellila wax, methacrylic acid copolymer, hypromellose	Extended release	Altoprev
Ion exchange	Celltech	Sulfonated styrene-divinylbenzene copolymer	Controlled Release	Delsym
Gastric retention	Depomed	Povidone, PEO	Extended release	Proquin XR
TIMERx®	Penwest Pharm-aceuticals	Xantham and locust gum	Extended release	Procardia XL
Microtrol	Shire Labs	Povidone, PEG, microcrystalline cellulose	Extended release/immediate release	Carbatrol
Contramid	Labopharm	High-amylose starch	Sustained release	Tramadol
Softgel	Cardinal	Gelatin	Microemulsion	Neoral

Abbreviations: HPMC, hydroxypropyl methyl cellulos; PEO, polyethylene oxide; PEG, polyethylene glycol.

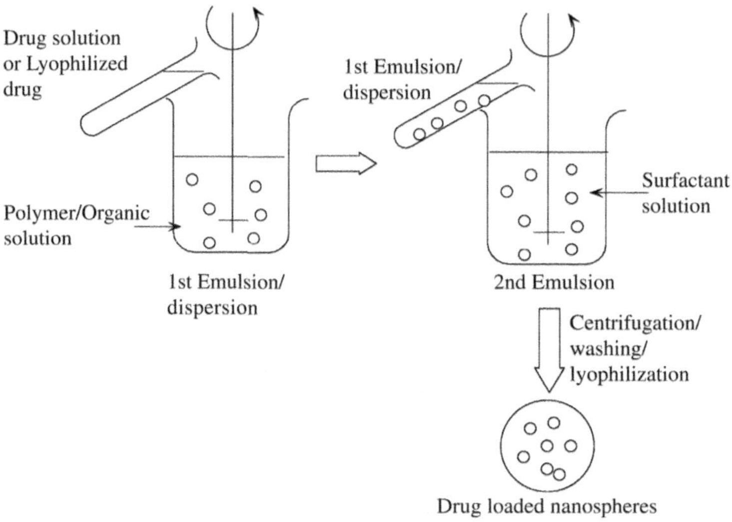

Figure 2 Schematic of the emulsion-based process typically used for preparation of polymeric nanoparticles.

the field has expanded to other areas such as medical devices, interest in novel polymers with unique functionalities has increased.

Polymers for parenteral drug administration can be used as matrices that are injected or implanted. Ease of administration requires the polymer to be in an injectable form, such as microspheres, nanoparticles, or reversible gels. Microspheres can be fabricated with the drug enclosed within, using emulsification-based techniques, as depicted in Figure 2 (12), or spray drying. The same techniques can be used for the preparation of nanoparticles. Other high-energy processes such as high-pressure homogenization have also been explored (13).

Polymer Gels

Certain types of polymers possess an ability to transform from flowable liquid form to viscous gels as a function of increasing temperature. Such polymers typically possess an alternating block copolymer structure, AB or ABA type, wherein A and B represent hydrophobic and hydrophilic segments within the polymer backbone. On increasing the temperature the polymers undergo micellization, leading to phase transition from a flowable liquid state to a gel state. For instance, copolymers of polyethylene glycol (PEG) and PLGA have been found to possess thermoreversible properties (14). The copolymer formulations are soluble in aqueous systems and form a flowable solution under cold conditions. As the temperature is raised to approximately 37°C, the polymer solution undergoes rapid gelation to form a semisolid depot.

Another approach to obtain injectable gels is using concepts of solvent-induced gelation (15). Common biodegradable polymers such as PLA and PLGA are soluble in a range of pharmaceutical grade organic solvents. When an organic solution of such polymers is injected, the organic solvent diffuses away, and the water-insoluble polymer precipitates to form a semisolid gel. This concept has been applied in a product Eligard®, used for sustained release of leuprolide acetate. The product utilizes PLGA polymer dissolved in *N*-methyl pyrrolidone as the matrix. By modifying the ratio of solvent:drug:polymer, a one-month, three-month, and four-month sustained release leuprolide acetate products have been developed.

NOVEL POLYMERS FOR DRUG DELIVERY

Polyesters, specifically polylactides and poly(lactide-co-glycolide)s have played a critical role in the development of polymer-based CR technologies. The biocompatibility and the well-established safety profiles of PLA and PLGA polymers have made them the polymer of choice for CR applications. However the off-patent status of these polymers makes them freely available for research in industry as well as academia. This has led to a vast number of patents covering various applications of these polymers within the drug delivery sector. Due to these issues, very limited scope remains to utilize these polymers to reformulate generic, off-patent drugs.

Another driver for novel polymer research was the increasing complexity of polymeric drug delivery systems. An ideal polymer for these applications should serve the following requirements:

1. It must be biocompatible and degradable (i.e., it should degrade in vivo into smaller fragments, which can then be excreted from the body).
2. The degradation products should be nontoxic and should not create an inflammatory response.
3. Degradation should occur within a reasonable period of time as required by the application (this may vary from days to months).
4. Based on the needs of certain application, the polymer should demonstrate versatile mechanical properties [e.g., stent coatings require polymers to be elastomeric and microsphere processing require them to have high glass transition temperature (T_g)].

No single polymer can match all of the above criteria. This has led companies to develop application-specific polymers and/or series of polymers that may have the structure–property variability to encompass all potential applications. As listed in Table 3, several properties of the polymer have a direct effect on its degradation kinetics and consequently on the drug release profile. Hence novel polymers that

Table 3 Lists of the Key Properties That Have a Direct Effect on the Polymer as a Release-Modifying Agent

Property	Effect on degradation kinetics
Chemical linkages	The type of hydrolytic linkage determines rate of degradation. For example, anhydride bonds are known to degrade faster than ester bonds
Molecular weight	Higher the molecular weight, slower is the degradation rate
Morphology	Porous forms (higher surface area) may be more susceptible to hydrolysis due to enhanced access for water penetration
Crystallinity	Higher crystallinity leads to slower degradation
Water uptake	Water uptake leads to faster degradation due to a better access for water to attack the polymer chains
Polymerization conditions	Use of catalysts, reaction temperature, etc. may affect the degradation properties of the polymer
Chain defects	Chain defects are often associated with faster degradation. Lesser the uniformity in structure, higher is the rate of hydrolysis

do not have prior history in biological applications undergo extensive regulatory scrutiny. As with any novel excipitent, significant physicochemical and biological testing is required to prove the safety of the polymer. Some of these challenges have been discussed in chapter of this book. This section describes some of the leading biodegradable polymers that are being developed toward commercialization.

Poly(α-hydroxy acid)s

The poly(α-hydroxy acid)s series of polymers includes PLA and PLGA as well as other polymers such as polycaprolactone and poly(butyric acid) (Fig. 3). Most of the drug delivery research has been carried out using PLA and PLGA polymers due to the fact that they can have degradation times in the less than one-year time frame. Because the lactide monomer possesses two chiral carbons, polylactide can be obtained using D-lactide (the D-,D-cyclic dimer), L-lactide (the L-,L-cyclic dimer), and mesolactide (the D-,L-cyclic dimer) or DL-lactide (a racemic mixture of D- and L-lactide). A typical reaction scheme has been depicted in Figure 4. The poly(L-lactide) (PLLA) is crystalline and possesses excellent mechanical properties. However it is very slow degrading, making it unsuitable for typical drug delivery applications. Most of the PLLA applications involve long-term therapies in tissue engineering. Poly (DL-lactide) on the other hand is an amorphous polymer that is brittle and leads to degradation time frames that are suitable for drug delivery applications. Adding increasing proportions of glycolide into PLA lowers T_g and generally increases polymer hydrophilicity. PLGA copolymers generally remain amorphous as long as the glycolide content remains within the range of about 0% to 70% (molar fraction). In contrast, poly(L-lactide-co-glycolide) is amorphous when the glycolide content is 25% to 70%. The most rapid degradation rate (i.e., two months) is observed in PLGA copolymers containing 50% glycolide.

The degradation products of PLGA polymers, lactic acid and glycolic acid, are components of the Kreb's cycle and hence are well tolerated by the body. Furthermore, the degradation rates of PLGA copolymer can be varied from two weeks to greater than a year, simply by adjusting the ratio of lactide to glycolide. Glycolide homopolymers are highly crystalline and poorly soluble in most organic solvents. On the other hand, lactide homopolymers could be crystalline in case of L-lactide or amorphous for DL-lactide. Copolymerization of DL-lactide with glycolide monomer leads to a polymer that is amorphous as well as possesses increased hydrophilicity as compared to the DL-lactide homopolymer. The PLGA copolymer also possesses more chain defects due to the randomness of copolymerization. Due to these aspects, PLGA 50:50 (50% DL-lactide and 50% glycolide) is the fastest degrading polymer in the series, with degradation time in the order of week. On the contrary, polymers made from L-lactide demonstrate degradation times that are in the order of years. This wide range of degradation kinetics facilitates the use of these polymers for a variety of

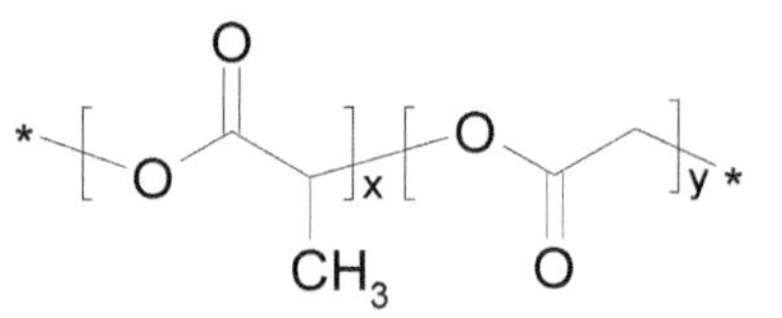

Figure 3 Chemical structure of poly(lactic-co-glycolic) acid polymer. The "x" component represents lactic acid and "y" component represents glycolic acid. For other poly(hydroxy acids), the side-chain methyl group is replaced by other alkyl groups.

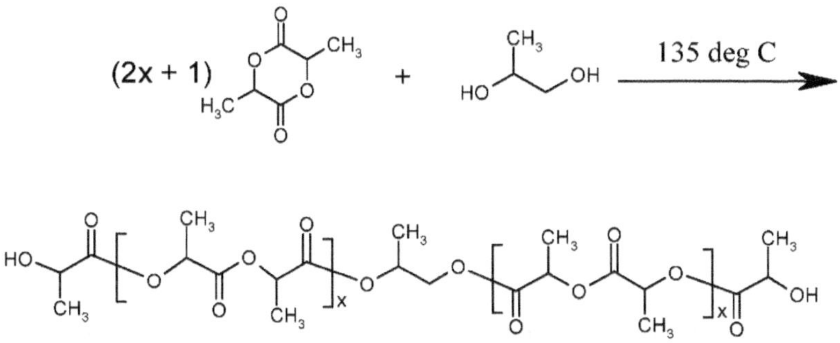

Figure 4 A typical reaction scheme for the synthesis of polylactides. By varying the ratio of the initiator (in this case propylene glycol) to lactide, the molecular weight of the final polymer could be controlled as desired.

drug delivery applications (16). Polyesters typically allow water penetration and hence undergo bulk erosion (degradation in the bulk of the matrix). This may be viewed as a limitation in certain applications, where maintenance of the matrix shape is desired, as in the case of medical device applications.

Besides lactic acid- and glycolic acid-based polyesters, polymers based on other ester-based cyclic monomers such as trimethylene carbonate and ε-caprolactone have also been tested for applications in drug delivery (Table 4). ε-Caprolactone-based polymers demonstrate enhanced flexibility with glass transition temperatures lower than room temperature. Copolymers of ε-caprolactone and lactide/glycolide demonstrate a wide range of properties in terms of degradation rates, glass transition temperatures, and elasticity. Due to the elastomeric properties, these copolymers have been studied for applications as coating materials for medical device applications.

Polyanhydrides

Polyanhydrides were first developed by Carothers and coworkers in the early 20th century for applications in the textile industry. The interest in these polymers waned soon thereafter because of their instability. However it was the poor hydrolytic stability that made these polymers attractive candidates for drug delivery applications (17).

Table 4 A Partial List of Marketed Drug Delivery Products Utilizing Polylactic Acid or Poly(lactic-co-glycolic) Acid Polymers

Product trade name	Polymer	Active ingredient	Indication
Nutropin Depot	PLGA	Human growth hormone	Growth deficiencies
Lupron Depot	PLA	Leuprolide acetate	Prostate cancer, endometreosis
Zoladex	PLA	Goserelin acetate	Prostate cancer, endometreosis
Trelstar Depot	PLGA	Triptorelin pamoate	Prostate cancer
Sandostatin LAR	PLGA-glucose	Octreotide	Acromegaly
Atridox	PLA	Doxyclycline hyclate	Periodontal disease

Abbreviations: PLA, polylactic acid; PLGA, poly(lactic-co-glycolic) acid.
Source: Chaubal M. Polyactides/glycolides—excipients for injectable drug delivery and beyond, drug Deliv Technol 2002; 2:34–36.

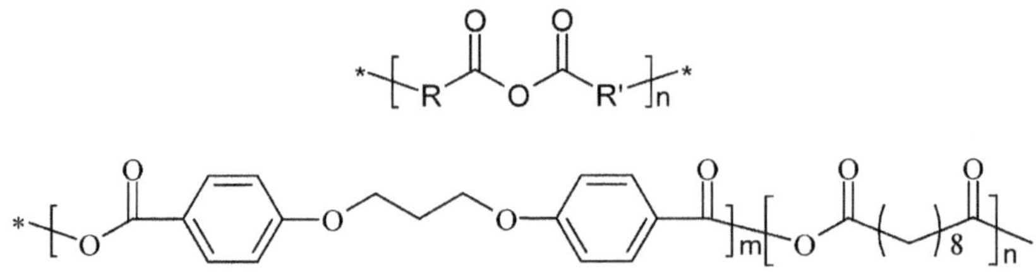

Figure 5 General structure of polyanhydrides. R and R′ can be varied to modify degradation kinetics and profile. The lower frame shows the structure of P(CPP-SA), a polyanhydride copolymer, used in the Gliadel product. *Abbreviations*: SA, sebacic acid; CPP, carboxyphenoxypropane.

Presently two series of polyanhydrides have undergone active clinical development. The first series involve sebacic acid (SA, hydrophilic component) and carboxyphenoxypropane (CPP, hydrophobic component). By varying the ratio of SA to CPP, the degradation rates of the copolymer could be significantly varied (Fig. 5). The commercially developed polymer in this series is P(CPP:SA 20:80), developed by Guilford Pharmaceuticals as a matrix for Gliadel, a commercial product for the sustained release of carmustine, for the treatment of brain tumors (18). A second family of polyanhydrides that has also been developed for applications in drug delivery is the copolymer of fatty acid dimer and SA. In this case, the fatty acid dimer acts as the more hydrophobic, slower degrading moiety. A poly(erucic acid dimer:SA 50:50) copolymer has been used for the sustained release of gentamicin, an antibiotic used for the treatment of osteomyelitis (19).

The property that makes polyanhydrides unique is their surface hydrophobicity. Due to this high hydrophobicity, polyanhydride matrices do not facilitate water absorption. Consequently, hydrolytic degradation is restricted to the surface—a property that is termed as surface erosion. This type of degradation allows for zero-order release of drugs, i.e., the drug release profile is independent of the residual concentration of the drug in the matrix.

Polyphosphoesters

Polyphosphoesters are a novel class of polymers being studied for applications in drug delivery and tissue engineering (Fig. 6). The individual components can be

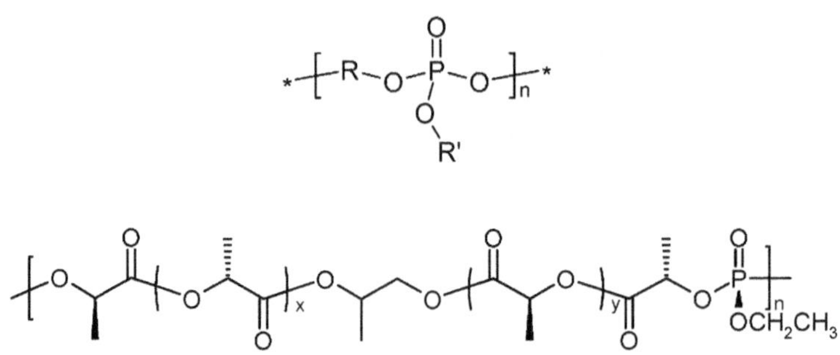

Figure 6 (*Top*) General structure of Polyphosphoesters. R and R′ can be varied to obtain polymers with varied physicochemical properties ranging from gels, to elastomers, to amorphous polymer particles. (*Bottom*) Example structure of a polylactide-co-ethylphosphate copolymer.

modified to obtain polymers with a wide range of properties and useful morphologies including injectable gels, elastomeric films, and amorphous solids (20).

The phosphate groups impart several unique characteristics to this polymer series. It makes the polymer more soluble in common organic solvents. It also acts as an internal plasticizer, making the polymer more flexible. Finally the phosphate groups impart hydrophilicity to the polymer, thus giving the surface lower fouling characteristics via reduced protein adsorption.

Polyphosphazenes

As seen in Figure 7, polyphosphazenes contain alternating phosphorus–nitrogen double and single bonds and side-chain functionalities that can be varied to obtain various series of polymers (21). Polyphosphazenes are synthesized by the reaction of poly(dichloro phosphazene) with organic nucleophiles such as alkoxides, aryloxides, or amines. The side-chain functionalities can be modified to obtain a wide range of properties including water solubility and degradability. Water-soluble polyphosphazenes have attracted special attention due to the possibility of encapsulating biopharmaceutical drugs via a completely aqueous process (22,23). Water-soluble polyphosphazenes gel in calcium chloride solutions enabling microencapsulation via a completely aqueous process. Such aqueous processing of polyphosphazenes allow their utility for the encapsulation of sensitive drugs such as proteins and vaccines (24).

Poly(orthoester)s

Poly(orthoester)s (POEs) developed in the 1970s were the first series of polymers developed specifically for applications in CR technologies. The first hydrolytically labile polyorthoesters were synthesized by polycondensation of 1,6-hexanediol or 1,4-cyclohexane dimethanol with an orthoester, diethoxy tetrahydrofuran (Fig. 8). However, due to the high hydrophobicity of POEs, the early generations demonstrated very slow degradation rates. To overcome the slow degradation, encapsulation of acidic entities, which could catalyze the degradation process, was proposed. However, such acidic components would leach out, and hence this concept did not meet significant success. The POE-IV was created with latent acid groups (such as lactic or glycolic acid) present on the polymer backbone (25). As the lactic/glycolic acid segments degrade, more acidic functionalities are generated, which autocatalyze the overall degradation of the polymer. The degradation rates could be varied by changing the amount of the internal autocatalysis moiety on the polymer backbone.

POEs have also demonstrated a variety of morphological characteristics. For example POE with cyclohexyldimethanol as a monomer unit is present at room temperature as a viscous paste rather than an amorphous solid. Such an embodiment

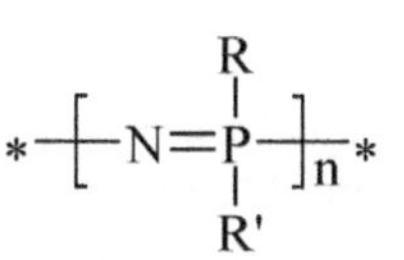

Figure 7 General structure of polyphosphazenes. R and R′ functionalities can be varied to create a library of polymers with differing structural properties.

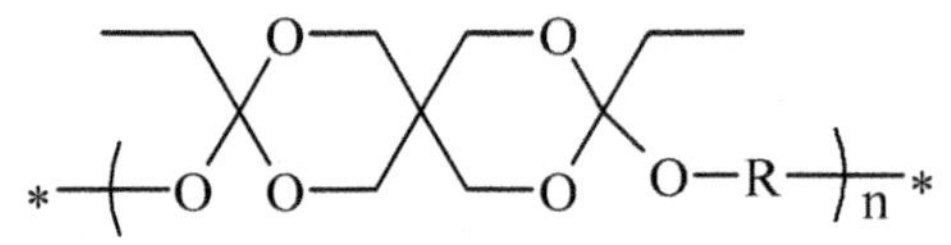

Figure 8 General structure of polyorthoesters. The functional group R can be varied to impart different properties to the polymer. Presence of lactic acid oligomer segments at R catalyzes degradation of the polymer.

may have unique applications in protein delivery wherein the protein could be incorporated by simple physical mixing or for applications in intraocular delivery.

Pseudopoly(amino acid)s

Pseudopoly(amino acid)s, developed by Kohn and Langer (26), have been investigated as implantable, degradable materials for medical applications. Inherent to all pseudopoly(amino acid)s are nonamide bonds incorporated into the backbone of amino acid–derived polymers. Contrary to conventional poly(amino acid)s, pseudopoly(amino acid)s are readily soluble in organic solvents and processable by conventional melt-fabrication techniques. The ease of fabricating pseudopoly(amino acid)s into fibers, films, rods, microspheres, porous foams, or other configurations appropriate for medical devices is one of their major advantages over conventional poly(amino acid)s. Thus, these materials combine the inherent nontoxicity and biocompatibility of amino acids with outstanding material properties usually only found in industrial engineering plastics.

Particularly noteworthy are the tyrosine-derived polycarbonates (27), a family of polymers based on alkyl esters of desaminotyrosyl-tyrosine. The lead polymer in this family is poly[desaminotyrosyl-tyrosine ethyl ester (DTE) carbonate], a polymer derived from desaminotyrosyl-tyrosine ethyl ester. Other polymers in this series of tyrosine-derived polycarbonates are poly[desaminotyrosyl-tyrosine butyl ester (DTB) carbonate], poly[desaminotyrosyl-tyrosine hexyl ester (DTH) carbonate], and poly [desaminotyrosyl-tyrosine octyl ester (DTO) carbonate], where the letters B, H, and O indicate the presence of butyl, hexyl, or octyl ester pendent chains, respectively.

Controlled intracranial release of dopamine, a drug for the treatment of Parkinson's disease, from a poly(DTH carbonate) matrix was one of the first medical applications investigated for tyrosine-derived polycarbonates. Poly(DTH carbonate) has a relatively low processing temperature and its backbone is structurally related to dopamine, which seemed to improve dopamine incorporation into the polymer. In experiments in vitro, dopamine was released from the matrices at a fairly consistent rate of 1 to 2 mg/day over a prolonged period of 180 days (Z. Dong, M.Sc. Thesis, Rutgers University, 1993). Although this release rate is within the therapeutically useful range, no subsequent studies of this release system in vivo were reported.

Tyrosine-derived polyarylates represent the first combinatorially designed library of biomaterials. Overall, 112 individual polymers were recently synthesized in such a way that the polymers exhibited continuous, incremental gradients of chemical, material, and biological properties (28). Tyrosine-derived polyarylates are strictly alternating copolymers of a diacid component and a diphenol component. Compared to tyrosine-derived polycarbonates, the polyarylates degrade faster, are more flexible, and encompass a wider range of physicomechanical and biological properties. The major advantage of their combinatorial design is that structure–property correlations established at the outset may be effectively used to help tailor material properties to specific applications.

Polymers for Implantable Drug Delivery

While nonsurgical injections are desirable from a patient compliance standpoint, implantable drug delivery systems can be utilized especially when surgery is inevitable. Nondegradable polymers are stable and hence do not lead to pH changes in the formulation microenvironment (as opposed to degradable polymers where polymer degradation leads to drop in pH due to formation of acidic fragments). Another advantage of nondegradable polymers is that formulations based on such systems typically have drug release exclusively by single mechanisms (diffusion and osmosis). Consequently drug release profiles can be accurately predicted and modeled. Furthermore, by fabricating the polymer implant to a specific geometry, the release profile can be tailored. For example, it was shown that hemispherical implants made out of nondegradable polymers provide a zero-order drug release (29). On the other hand, the nondegradability of polymers limits their use to applications involving very extended release or applications where the presence of an external entity is not considered a major issue. A commercial example of very long-term sustained release via nondegradable polymer implants is the Norplant implant, designed for sustained delivery of levonorgestrel for up to five years.

Nondegradable polymers are also useful as matrices for ocular implants. This application requires the polymer to be hydrophilic, to minimize local tissue irritation. Need for ocular implants stems from the challenges posed to conventional ocular medicines (i.e., eye drops) such as rapid dilution, tear washout, poor patient compliance, and limited bioavailability. Ocular implants from hydrophilic polymer matrices that provide localized sustained release may overcome the above limitations. The first polymeric sustained release product to reach the market was Ocusert®, a pilocarpin sustained release ocular implant developed by Alza. Ocusert® has the drug reservoir as a thin disc of pilocarpine–alginate complex sandwiched between two transparent discs of microporous membrane fabricated from ethylene–vinyl acetate copolymer. The microporous membranes permit the tear fluid to penetrate into the drug reservoir compartment to dissolve pilocarpine from the complex. Pilocarpine molecules are then released at a constant rate of 20 or 40 μg/hr for a four- to seven-day management of glaucoma.

Toxicology Assessment of New Polymeric Excipients

Successful development of new polymeric excipients depends on obtaining appropriate toxicological data on the safety and biocompatibility of such excipients. Implant applications have other relevant guidelines developed by United States Pharmacopoeia (USP) for testing of the polymer safety and tissue irritability. One such example of a test is USP Biological Reactivity Tests, in vivo <88>, which include the systemic injection test, the intracutaneous test, and the implantation test. Such guidelines may be of relevance when developing a polymer excipient for parenteral CR applications. Recent guidelines from the Food and Drug Administration describe the type of data package required in preclinical development of a new excipient. Shive and Anderson (30) have described the biodegradation, biocompatibility, and tissue/material interactions studies conducted on lactide-based copolymers. Duncan et al. (31) have described the preclinical data obtained for a polymer–drug conjugate developed for oncolytic applications. Kelly et al. (32) described the investigation on subchronic toxicity of poly(vinyl alcohol), for oral applications.

SUMMARY

Polymeric excipients are an essential part of formulation development for CR products. Polymer-based oral CR products are widely prevalent in the market. Such products utilize natural polymers (such as gelatin and starch) or nondegradable synthetic polymers (such as polystyrenes and polyacrylic acid). Degradable polymers are also being explored for oral drug delivery. While oral products use established polymers, novel polymers are being explored for emerging applications in parenteral delivery. It is anticipated that because new polymeric excipients are developed for medical device applications (such as drug-eluting stents), such polymers will find applications in pharmaceutical products as well.

REFERENCES

1. Filbey JA, Charles SA. Shellac to PEGylation: 2500 years of polymers in controlled release drug delivery. Drug Delivery Tech 2005; 5(6):64–69.
2. Baichwal A, Neville D. Advanced drug delivery technology infuses new life into product life-cycles. Business Brief Pharmatech 2002:1–5.
3. Hanauer SB, Sparrow M. COLAL-PRED Alizyme. Curr Opin Invest Drugs 2004; 5(11): 1192–1197.
4. Sakuma S, Lu ZR, Kopeckova P, Kopecek J. Biorecognizable HPMA copolymer-drug conjugates for colon-specific delivery of 9-aminocamptothecin. J Control Release 2001; 75(3):365–379.
5. Chaubal MV. Synthetic polymer-based ion exchange resins: excipients and actives. Drug Deliv Tech 2003; 3(5).
6. Omidian H, Rocca JG, Park K. Advances in superporous hydrogels. J Control Release 2005; 102(1):3–12.
7. Peppas NA, Sahlin JJ. Hydrogels as mucoadhesive and bioadhesive materials: a review. Biomaterials 1996; 17(16):1553–1561.
8. Olsen D, Yang C, Bodo M, et al. Recombinant collagen and gelatin for drug delivery. Adv Drug Deliv Rev 2003; 55(12):1547–1567.
9. Kader A, Jalil R. Formulation factors affecting drug release from poly(lactic acid)(PLA) microcapsule tablets. Drug Dev Ind Pharm 1999; 25(2):141–151.
10. Giunchedi P, Gavini E, Bonacucina G, Palmieri GF. Tabletted polylactide microspheres prepared by a w/o emulsion-spray drying method. J Microencapsul 2000; 17(6):711–720.
11. Thanos CG, Liu Z, Reineke J, Edwards E, Mathiowitz E. Improving relative bioavailability of dicumarol by reducing particle size and adding the adhesive poly(fumaric-co-sebacic) anhydride. Pharm Res 2003; 20(7):1093–1100.
12. Meng FT, Ma GH, Qiu W, Su ZG. W/O/W double emulsion technique using ethyl acetate as organic solvent: effects of its diffusion rate on the characteristics of microparticles. J Control Release 2003; 91(3):407–416.
13. Calvor A, Muller BW. Production of microparticles by high-pressure homogenization. Pharm Dev Technol 1998; 3(3):297–305.
14. Zentner GM, Rathi R, Shih C, et al. Biodegradable block copolymers for delivery of proteins and water-insoluble drugs. J Control Release 2001; 72(1–3):203–215.
15. Southard GL, Dunn RL, Garrett S. The drug delivery and biomaterial attributes of the ATRIGEL technology in the treatment of periodontal disease. Expert Opin Invest Drugs 1998; 7(9):1483–1491.

16. Vert M, Schwach G, Engel R, Coudane J. Something new in the field of PLA/GA bioresorbable polymers? J Control Release 1998; 53(1):85–92.
17. Shieh L, Tamada J, Chen I, Pang J, Domb A, Langer R. Erosion of a new family of biodegradable polyanhydrides. J Biomed Mater Res 1994; 28:1465–1475.
18. Gopferich A. Biodegradable polymers: polyanhydrides. In: Mathiowitz E, ed. The Encyclopedia of Controlled Release. New York: Wiley, 1999:60–71.
19. Li LC, Deng J, Stephens D. Polyanhydride implant for antibiotic delivery—from the bench to the clinic. Adv Drug Deliv Rev 2002(Oct 16); 54(7):963–986 (review).
20. Mao H-Q, Kadivala I, Leong KW, Zhao Z, Dang W. Biodegradable polymers: poly (phosphoester)s. In: Mathiowitz E, ed. The Encyclopedia of Controlled Release. New York: Wiley, 1999:45–60.
21. Allcock HR, Pucher SR, Scopelianos AG. Poly[(amino acid ester)phosphazenes] as substrates for the controlled release of small molecules. Biomaterials 1994; 15(8):563–569.
22. Payne LG, Andrianov AK. Protein release from polyphosphazene matrices. Adv Drug Deliv Rev 1998; 31(3):185–196.
23. Payne LG, Jenkins SA, Andrianov A, Roberts BE. Water-soluble phosphazene polymers for parenteral and mucosal vaccine delivery. Pharm Biotechnol 1995; 6:473–493.
24. Chaubal MV, Gupta AS, Lopina ST, Bruley DF. Polyphosphates and other phosphorus-containing polymers for drug delivery applications. Crit Rev Ther Drug Carrier Syst 2003; 20(4):295–315.
25. Heller J, Barr J, Ng SY, et al. Poly(orthoesters)—their development and some recent applications. Eur J Pharm Biopharm 2000; 50(1):121–128.
26. Kohn J , Langer R. Polymerization reactions involving the side chains of alpha-L-amino acids. J Am Chem Soc 1987; 109:817–820.
27. Ertel SI , Kohn J. Evaluation of a series of tyrosine-derived polycarbonates as degradable biomaterials. J Biomed Mater Res 1994; 28:919–930.
28. Brocchini S , James K, Tangpasuthadol V, Kohn J. Structure-property correlations in a combinatorial library of degradable biomaterials. J Biomed Mater Res 1998; 42:66–75.
29. Hopfenberg HB. Controlled release from erodible slabs, cylinders, and spheres. In: Paul DR, Harris FW, Gould RF, eds. Controlled Release Polymeric Formulations. A.C.S. Symposium Series, Chapter 3. Washington, DC: American Chemical Society, 1976.
30. Shive MS, Anderson JM. Biodegradation and biocompatibility of PLA and PLGA microspheres. Adv Drug Deliv Rev 1997; 28:5–24.
31. Duncan R, Coatsworth JK, Burtles S. Preclinical toxicology of a novel polymeric antitumour agent: HPMA copolymer-doxorubicin (PK1). Hum Exp Toxicol 1998; 17(2):93–104.
32. Kelly CM, DeMerlis CC, Schoneker DR, Borzelleca JF. Subchronic toxicity study in rats and genotoxicity tests with polyvinyl alcohol. Food Chem Toxicol 2003; 41(5):719–727.

20

Emerging Excipients in Parenteral Medications: The New Paradigm

Shireesh P. Apte
Chemologic LLC, Mansfield, Texas, U.S.A.

Sydney O. Ugwu
NeoPharm Inc., Waukegan, Illinois, U.S.A.

INTRODUCTION

Emerging excipients are increasingly becoming integral parts of sophisticated drug delivery systems rather than functioning as "stand-alone" chemical entities. Such systems require complex manufacturing methodologies and precise chemical/clinical engineering protocols to achieve the desired morphology of the drug–excipient recipe—mere mechanical incorporation of drug and excipient into a formulation rarely achieves the desired outcome.

The possibility of treating fatal and severely debilitating diseases such as cancer or Alzheimer's disease as chronic conditions requiring the lifelong administration of therapeutic agents necessitates that such agents be exquisitely specific in their pharmacological effects. Cytotoxic or neuroprotective agents must preferably be delivered to predetermined sites in the body. A need to deliver drugs (genes) to more specific (and increasingly inaccessible) cellular therapeutic targets may eventually necessitate such delivery systems.

1. Mimic natural cellular trafficking mechanisms.
2. Emulate the evolutionary egregious targeting mechanisms of certain pathogens; the so-called "killers as healers" model.
3. Evade degradation or misdirected delivery by the body's myriad built in defense mechanisms.

Intuitively; such delivery systems or excipients would best achieve this objective if they were to be composed of natural products or their modifications; an overwhelming presence of excipients under this category (discussed in Ref. 1) provides credence to this observation. The substitution of natural products comprising complex proteins, antibodies, chimera, or toxins in lieu of "stand-alone" simple inorganic molecules as excipients for parenteral drug delivery represents a paradigm shift in the introduction of emerging excipients in the therapeutic armamentarium.

The increasing investigative use and projected inclusion of a variety of natural products including naturally occurring polymers and their derivatives into drug delivery systems provides compelling evidence for such a paradigm shift. In addition, complex inextricable interactions between such "excipients" and drugs dramatically serve to diminish the molecular and functional boundaries between the two in emerging drug delivery systems.

The International Pharmaceutical Excipients Council defines excipients as "substances, other than the active drug substance of finished dosage form, which have been appropriately evaluated for safety and are included in a drug delivery system to either aid the processing of the drug delivery system during its manufacture, protect, support, enhance stability, bioavailability, or patient acceptability, assist in product identification, or enhance any other attributes of the overall safety and effectiveness of the drug delivery system during storage or use."

The definition above implies that excipients have a purpose in formulation. This contrasts with the old terminology of "inactive excipients," which hints at the property of inertness. From the literal interpretation of this definition, an excipient can include diverse molecules or moieties such as replication incompetent/selective viruses (adenoviral or retroviral vectors), bacterial protein components, antibodies, bacteriophages, fusion proteins, molecular chimera, etc. Furthermore, individual components of a drug delivery system frequently enhance the attributes of safety and efficacy of such a system while producing a pharmacological response only when acting in conjunction with other components (2–8). Frequently, the excipients used are themselves not entirely devoid of pharmacological activity. As parenteral drug delivery systems become complex and sophisticated, with diverse fields such as gene delivery, immunomodulation, sustained and targeted release, the differentiation between "excipient" and "active" is likely to become increasingly blurred.

What distinguishes a drug from an excipient? Is it the concentration at which it is administered or the indication for which it is administered? For an entity to be classified as a drug, is it necessary for it to *cause* a pharmacological response or is it merely sufficient to be able to *elicit* one? Furthermore, does the entity need to elicit a response from (i) another chemical entity, (ii) endogenous molecules/proteins, or (iii) an exogenously administered radiation [infrared (IR), ultrasound, magnetic field, etc.]? If the reason for administration of a drug delivery system is to enhance the effectiveness of a subsequently administered drug/delivery system, then the former could be literally interpreted as being an excipient. Therefore, the *context* in which the drug or delivery system is used should also be considered.

CONCENTRATION OF THE CHEMICAL ENTITY

Higher concentrations of a chemical entity may cause a pharmacological response whereas lower concentrations are incapable of doing so. For example, edetate calcium disodium (9) at a formulated concentration of 20% is indicated as a drug for the treatment of lead poisoning and lead encephalopathy while it can be found commonly present as an excipient in concentrations ranging from 0.01% to 0.1% in parenteral formulations to prevent oxidation of the active ingredient.

Oral administration of exogenous compounds can induce tissue-specific expression of previously transfected (and integrated) genes and transcription factors. Such compounds typically cause conformational changes in the transcription/activating factors that in turn cause their binding to the promotor linked to the transgene.

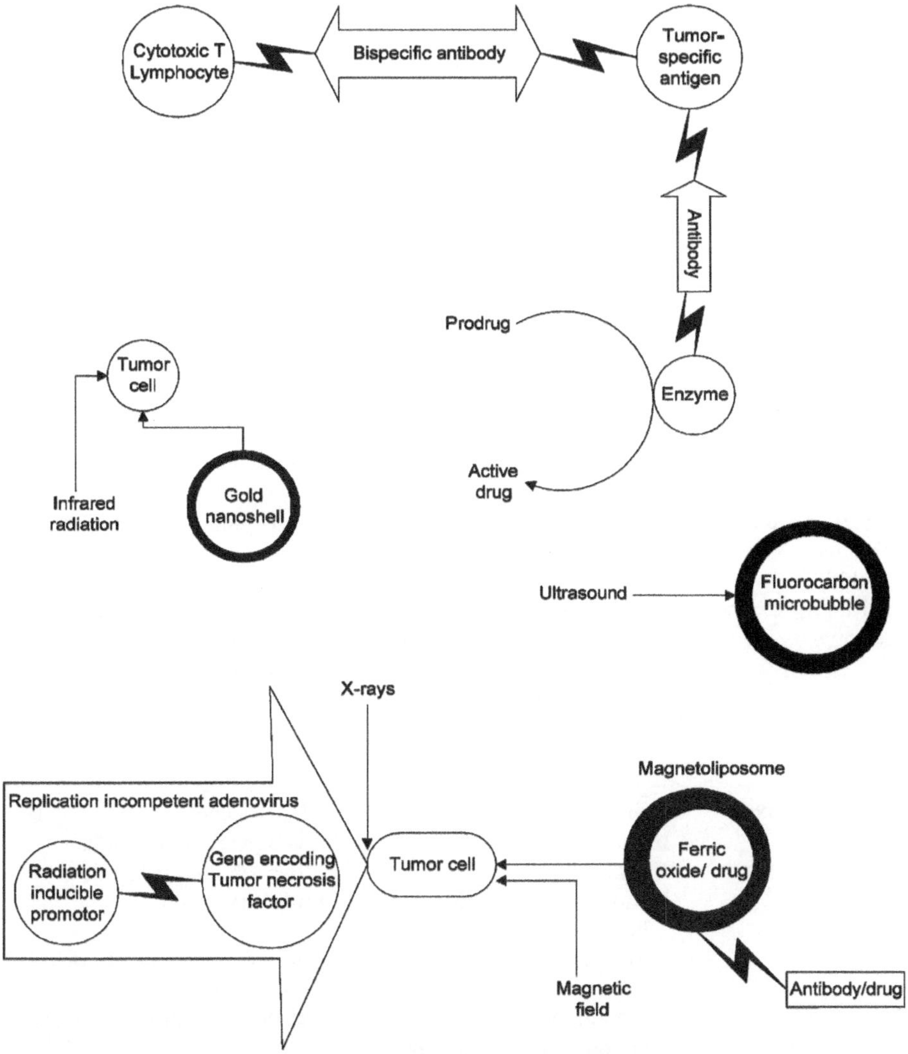

Figure 1 Increasingly complex drug-delivery systems erode definitive differentiation between "drug" and "excipient."

Such binding can cause either an expression or a repression of gene transcription. Mifepristone (RU486, GeneSwitch®) and muristerone A are examples of such exogenous compounds capable of inducing gene transcription typically at much lower doses than those that are responsible for their primary pharmacological response (Fig. 1) (Table 1).

INDICATION FOR WHICH THE CHEMICAL ENTITY IS ADMINISTERED

The indication for which a chemical entity is administered may also differentiate its designation as a drug or an excipient. For example, tyloxapol is an oligomer of octoxynol 9 (Triton X-100) that is used as an active agent in respiratory distress

Table 1 Summary of Emerging Excipients

Chemical entity	Excipient use	Therapeutic use
Siderophores	Antibacterial delivery	Antituberculosis, contrast agents for MRI
Keyhole limpet hemocyanin	Hapten carrier, vaccine adjuvant	Antiproliferative in several cancers
Tranexamic acid	Solubility enhancer	Antifibrinolytic agent, hemophilia
Fullerenes	Radioisotope delivery to tumor cells, MRI contrast agents	Antibacterial, neuroprotective, inhibition of HIV activity
Bactericidal permeability increasing protein	Increases antibiotic activity	Septic shock, antiangiogenic
Tocophersolan PEG succinate, PEG hydroxystearate	Antioxidant, solubility enhancer	Attenuation of multi drug resistance
Fluorocarbon-filled, phospholipid-coated microbubbles	Assessment of response to angiogenesis inhibitor therapy	Ultrasound contrast agent
Trans-splicing ribozymes or ribozymes	Expression of enzyme capable of converting subsequently administered prodrug to active form Suppression of multidrug resistance–associated proteins	Expression of cytotoxic or apoptosis inducing proteins in cancer cells
Mifepristone (RU486)	GeneSwitch®	Abortificient
Muristerone A	Inducible gene transcription	Hypoglycemic, anabolic, hypocholesterolaemic
Phospholipid	Passive drug targeting, liposomes	Lung surfactant
Tyloxapol	Surfactant	Mucolytic, respiratory distress syndrome
Melittin	DNA targeting, endosomolytic, nuclear localization	Cytotoxic, antiproliferative
Salmonella typhimurium	Expression of enzyme capable of converting subsequently administered prodrug to active form	Cytotoxic
Transposons	Nonviral gene vector	
Porphyrins	Tumor-specific targeting	Cytotoxic
Polymer platinate	Platinum tumor delivery	Cytotoxic
SiRNA		

Abbreviations: MRI, magnetic resonance imaging; PEG, polyethylene glycol; SiRNA.

syndrome (10) and as a mucolytic in cystic fibrosis (11). It is also a ubiquitous excipient found in parenteral disperse systems as a surfactant for increasing the kinetic stability of the dispersion and for enhancing the dendrimer-mediated transfection (12). Phospholipids are an integral part of liposomal drug delivery systems, acting as passive targeting agents. They are also added to surfactant-replenishing intratracheal suspensions to mimic the surface tension–lowering properties of a natural lung surfactant.

Hyaluronic acid is a drug that is injected into the knee to restore the viscosity of the synovial fluid in patients suffering from osteoarthritis (12). It is also used as a viscoelastic material in a viscosurgical device (13) and as an excipient in a dry powder nucleic acid composition comprising a cationic lipid–DNA complex for delivery to the lung (14). Tranexamic acid, an antifibrinolytic agent, is used as a drug in adjunctive therapy in haemophilia and some other bleeding disorders (15). It has also been used as an excipient in recombinant reteplase injection, presumably to increase the solubility of the non-glycosylated plasminogen activator (16).

Some excipients such as tocophersolan polyethylene glycol succinates (17) or polyethylene glycol hydroxystearates (18) that are used as antioxidants or to increase the solubility of hydrophobic drugs also possess the pharmacological property to attenuate the P-glycoprotein–mediated multidrug resistance (19,20).

Pure carbon spheres of C_{60}—the fullerenes—react avidly with free radicals with a higher antioxidant ability than the naturally occurring vitamin E (21). Endohedral fullerenes have been shown to be capable of encapsulating a variety of atoms such as radiotracers or noble gases, thereby making them efficient excipients in the delivery of radioisotopes to cancer cells or in magnetic resonance imaging (MRI) (22).

Fullerenes have also been shown to possess myriad biological effects such as inhibition of HIV, antibacterial activity, and neuroprotection (21). Therefore, whether or not fullerenes are classified as drugs or excipients depends upon the indication for which they are used.

Keyhole limpet hemocyanin (KLH) is an immune stimulant and a hapten carrier, derived from a circulating glycoprotein of the marine mollusk *Megathura crenulata*. KLH has significant antiproliferative effects in vitro against several types of cancers (23). It has also been conjugated to a variety of immunogens and used as a vaccine adjuvant (24).

Melittin is a cytotoxic, cationic membrane lytic component of bee sting venom. Melittin complexed with PEI-DNA substantially increased the levels of reporter gene expression even in slowly dividing or confluent cells. The complex accelerated the time of transgene expression, thereby supporting the possibility of venom mediating a dual effect, combining endosomolysis with functional nuclear localization (25). The incorporation of a matrix metalloproteinase 2 (MMP2) target sequence into a melittin–avidin complex showed strong cytotoxic activity against MMP2 positive cancer cells (26).

Microbes acquire iron by utilizing very specific, low molecular weight iron chelators called siderophores. The resistance of bacteria to previously effective antibiotics can be circumvented in part by covalent coupling of antibiotics to siderophores. Some synthetic siderophores have been found to possess significant antibacterial activity themselves. They have also been found to have considerable potential as nontoxic, organ selective MRI contrast agents (27).

A bactericidal permeability increasing protein found inside human neutrophils has been investigated as a potent endotoxin neutralizing agent in the treatment of septic shock. It has also been shown to enhance the activity of antibiotics, suggesting a potential use in treating antibiotic-resistant infections (28).

Historical reviews have revealed a number of clinical observations in which cancers were reported to regress in patients with bacterial infections. At least some of the anticancer effects of Streptococcal infections are mediated through stimulation of the host immune system. An attenuated strain of *Salmonella typhimurium*, not devoid of the ability to accumulate preferentially within the extracellular components of tumors, has been further modified by chromosomal insertion of an *Escherichia coli* cytosine deaminase gene which when expressed converted 5-fluorocytosine to 5-fluorouracil (29).

ELICITATION OF A PHARMACOLOGICAL RESPONSE

Antibody-directed enzyme prodrug therapy involves administration of a cell/tissue-specific antibody conjugated to an enzyme. This is followed by the administration of a non-cytotoxic substance (a prodrug) that is converted in vivo by the previously administered enzyme into its active form (2,3). In such an instance, the antibody–enzyme complex does not by itself cause a pharmacological response; rather, it *elicits* that response when another chemical entity (the prodrug) is administered. Therefore, the antibody–enzyme complex would qualify as an excipient. A similar argument can be made for gene-directed enzyme prodrug therapy, where a chimera containing a transcriptional regulatory DNA sequence capable of being selectively activated in mammalian cells is described (5). Such a sequence can be linked to another sequence that encodes an enzyme in the target cells. In this case, the chimera would qualify as an excipient. A variation of this approach is nucleic acid–triggered catalytic drug release, where the catalytic component linked to an oligonucleotide (ODN) that is complementary to the triggering ODN would be considered as the excipient (30).

Trans-splicing ribozymes have the potential to be used for the delivery of new gene activities in living cells, conditional upon the presence of a chosen mRNA species (31). Ribozymes could conceptually be constructed that are capable of splicing a mutant human enzyme coding sequence into tumor cell–specific mRNA. This would be followed by administration of a cytotoxic prodrug that is a substrate for this enzyme. Ribozymes could also cleave multidrug-associated proteins in another unrelated mechanism. In these instances, the trans-splicing ribozyme or ribozyme would function as an excipient. Alternatively, the ribozyme could splice an apoptosis-inducing or angiogenesis-inhibiting protein coding sequence into tumor cell–specific mRNA. In this instance, the ribozyme would function as a drug (Fig. 2).

Bispecific antibodies have been developed against malignant cells that recognize a tumor-specific antigen on such cells. Their other "arm" recognizes cytotoxic T-cells, one of the body's endogenous effector mechanisms against malignant cells (6). Such antibodies *recruit* immune cells for tumor cell elimination. In the absence of T-cells (such as would be expected to occur in immunocompromised subjects), such bispecific antibodies would be ineffective in causing a pharmacological response. Therefore, they would qualify as an excipient rather than as a drug (Fig. 3A).

Gold nanoshells are concentric, spherical nanoparticles consisting of a dielectric core and a gold shell. By varying the relative thickness of the core and shell layers, the plasmon-derived optical resonance of gold can be dramatically shifted from the visible region into the IR over a wavelength range that spans the region of highest physiological transmittivity (7). In a murine tumor model, the nanoshells *in conjunction* with exposure to near-IR (NIR) radiation induced cell death whereas exposure to either NIR light or nanoshells alone did not (8). The nanoshells would hence fall under the category of "excipient" as previously elucidated in this chapter (Fig. 3B).

Intravenously injected, fluorocarbon-filled, phospholipid-coated microbubbles by themselves have little or no therapeutic value unless they are made acoustically

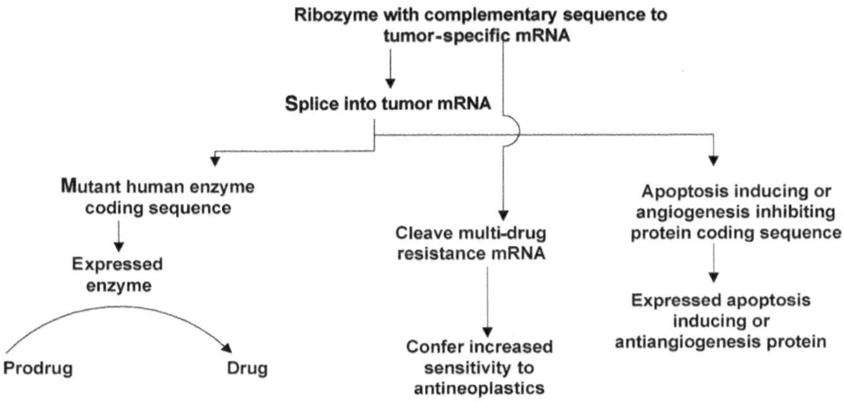

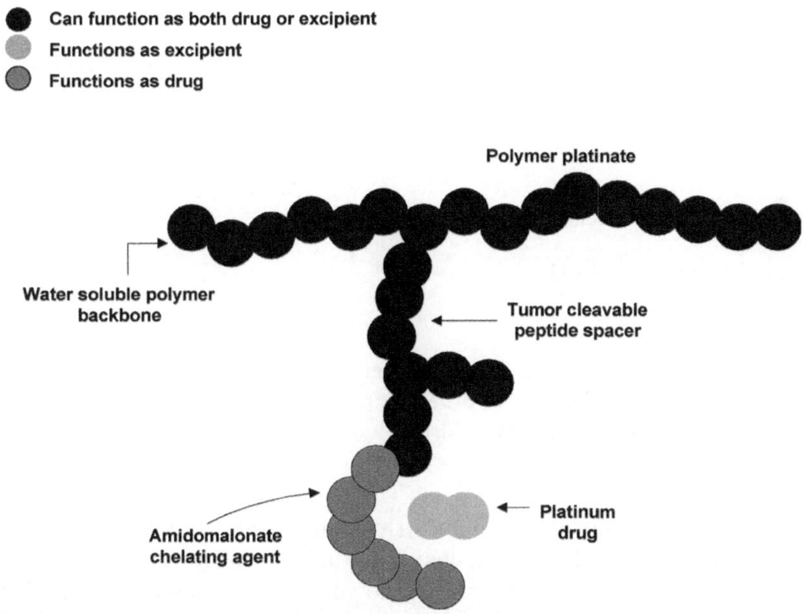

Figure 2 A chemical entity can function as either a "drug" or "excipient" depending upon its mechanism of action, requirement of other entit(ies) to achieve pharmacological response, or modulate the effectiveness of another "drug."

reflective, using exogenously applied ultrasound (32). The microbubbles and ultrasound acting in tandem cause a dramatic increase in contrast. Furthermore, the rate of contrast that can be measured with the above system appears to correspond directly with tissue perfusion and can therefore be used to enable a more rapid assessment of response to angiogenesis inhibitor therapy (Fig. 3C). In the above instance, the microbubble/ultrasound diagnostic system itself assists in the effectiveness of another (antiangiogenic) drug or delivery system, thereby designating it as an "excipient." Molecules that are administered as part of photodynamic therapy require the application of exogenously administered visible light for activation and pharmacological activity. Such molecules (Verteporfin, porfimer sodium) can therefore also be classified as "excipients" (33).

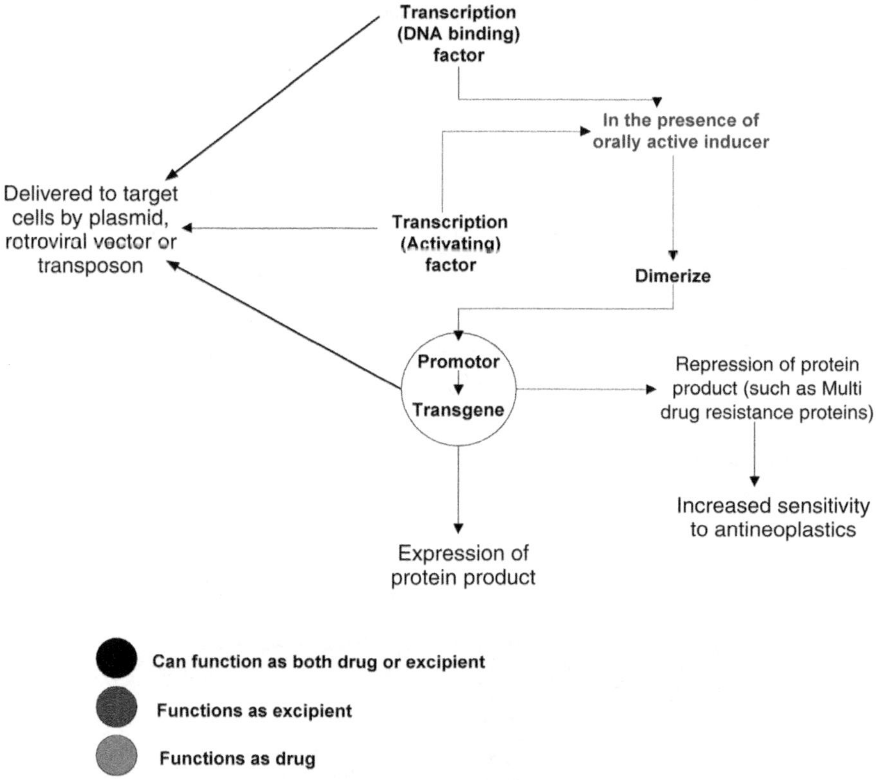

Figure 3 A chemical entity can function as either a "drug" or "excipient" depending upon its mechanism of action, requirement of other entit(ies) to achieve pharmacological response, or modulate the effectiveness of another "drug."

Fe_3O_4-containing magnetoliposomes (with or without antibody/drug conjugation/encapsulation) have been targeted to tumors using exogenously applied magnetic field. A high-frequency magnetic field can then be subsequently and specifically applied to the tumor site, thereby generating cellular hyperthermia (Fig. 3D) (34).

A replication-deficient adenovirus has been used to deliver a gene construct, which contains within its promoter region a radiation-responsive element. Upon irradiation with conventional doses of X rays, this construct initiates transcription of the gene coding for the toxic cytokine, tumor necrosis factor–α (35). The above is an exquisite example of how emerging delivery and gene therapies are fast blurring the distinction between an "active" and an "excipient" (Fig. 3D).

EXCIPIENT–DRUG CONJUGATES

Whether chelation (as opposed to covalent bonding) of an active moiety to an "excipient" qualifies the entire construct as a new chemical entity (NCE) is a moot point. A case in point is polymer platinate. This compound consists of a polymer backbone, hydroxypropylmethacrylamide, linked to a polypeptide spacer. The peptide is in turn linked with an aminomalonate chelating group, which chelates the platinum compound (Fig. 2).

Although it is generally accepted that covalent bonding or chelation of an "excipient" (eNCE) to a "drug" (dNCE) seems to qualify the new construct as a NCE, little or no part is played by the eNCE in either the causation or the elicitation of a pharmacological response in the vast majority of drug delivery systems. The authors have therefore classified the eNCE part of the construct as "excipient" in this discussion. Clinical experience and safety data from such NCEs may serve as supporting evidence to justify the usage of the eNCE as a "stand alone" excipient in another formulation or as an eNCE in another (eNCE + dNCE) drug delivery system. This may provide another avenue through which new excipients may be formally introduced into parenteral medications without explicit prior toxicological testing.

NATURAL PRODUCTS INCLUDING NATURALLY OCCURRING POLYMERS AND DERIVATIVES

These substances occur ubiquitously throughout the plant and animal kingdoms. Individual saponins derived from the South American tree *Quillaja saponaria* (36,37), KLH—a nonheme copper containing protein found in anthropods (38), MPL®, a monophosphoryl derivative of the lipid A molecule found in gram-negative bacteria, LeIF® (Leishmania elongation initiation factor), a protein produced by the parasite Leishmania (39), ricin, a potent immunotoxin obtained from the seeds of castorbean plants (40), and squalene, an isoprenoid found in large quantities in shark liver oil (41) have been used or investigated as vaccine adjuvants. Intranasal or ocular formulations of insulin containing a deacylated saponin derivative as a surfactant showed a dose-dependent hypoglycemic response in rats (42).

Albumin, gelatin (43), deoxycholic acid, sesame oil (44), and gangliosides (45,46) are substances that occur naturally in the body, and hence these would be exquisitely suited as excipients.

The coupling of naturally occurring polyamines such as spermine or its derivatives to bile acids allows the formation of facial amphiphiles, resulting in promising transfection agents (47,48). Spermidine has also been used to inhibit liposomal lipid peroxidation (49). Antifreeze glycoproteins, which are synthesized by fish living in polar regions (149), can be used to inhibit leakage from liposomes undergoing thermotropic phase transitions during lyophilization (12). Hyaluronic acid can be used as an excipient in a dry powder nucleic acid composition comprising a cationic lipid–DNA complex for delivery to the lung (148). It has also been injected into the knee to restore the elasticity and viscosity of the synovial fluid in patients suffering from osteoarthritis (50). Sphingomyelins occur in the myelin sheaths of nerves and can be used as components of liposomes (51). A glycoprotic biopolymer excreted by a new gram-negative species of bacteria, *Pseudoalteromonas antarctica* NF3, can be used to coat liposomes to protect the bilayers against the action of nonionic detergents that may be employed during manufacture (52).

Histones are highly basic, small, compact proteins, with a high affinity for DNA. They occur naturally, attached to the DNA of cell nuclei by ionic linkages. Their classification is based on the relative amounts of lysine and arginine. The galactosylated, lysine-rich histone H1 was found to be superior to the H2–H4 histones as a DNA carrier for liver gene delivery (53).

Heparin and heparin-like polyanions are known to have significant stabilizing effects on proteins. Heparin itself has been used as a constituent of the solution used to reconstitute lyophilized proteins (54) while enoxaparin is a low molecular weight fragment of heparin that was found effective in preventing heat-induced protein agglomeration (55). Phosvitin, a phosphoprotein isolated from egg yolk and phytic acid and a naturally occurring phosphorylated carbohydrate that has been shown to suppress the growth of epithelial cancers (56), was also effective in stabilizing protein (55). Polygalacturonic acid (57) derived from the hydrolysis of pectin has the potential to be used as a contrast medium in MRI. *N*-Acetyl [$Phe^8(CH_2–NH)$ Arg^9] bradykinin (Cereport®) (58) is a modified bradykinin that increases the permeability of the blood brain barrier to enable delivery of drugs to the brain (59,60).

Cyclodextrins (61) are naturally occurring clathrates obtained by the action of *Bacillus macerans* amylase on starch to form homogeneous D-glucopyranose–linked units. Their fluoro analogs (62) can be used to encapsulate extremely hydrophobic compounds. Phosvitin can be conjugated to galactomannan to yield an excellent emulsifying agent (63). Dipyridoxyl phosphate and its derivatives can be used as paramagnetic metal chelating agents in NMRI contrast agent compositions (64–66). The depolymerization kinetics of hyaluronan (HA) can be modified by direct coupling of β-cyclodextrin to HA carboxylic acid groups. The degree of substitution can be used to advantage to modify the release kinetics of entrapped drugs (67). Streptavidin or avidin genetically fused to an antibody specific for the transferrin receptor can transport biotinylated drugs across the blood brain barrier (68,69).

The expanding area of gene delivery is likely to witness the incorporation of many novel natural products as "excipients." An example is listeriolysin O, a thiol-activated protein. This is a bacterial component (70) of a targeted gene delivery system (71) that when combined with a polynucleotide and a binding agent can lyse the endosome of the targeted cell, thereby causing the internalized polynucleotide to be released into the cytoplasm. Linamarase is a plant enzyme that hydrolyzes the cyanogenic glucoside substrate, linamarin, into glucose, acetone, and cyanide. Retroviral vectors targeted to intracerebral gliomas that carry the linamarase gene cause a marked sensitization to the innocuous substrate, linamarin, followed by cell death (72). Ortho ester lipids contain a pH-sensitive ortho ester linkage that causes lipid headgroup cleavage on exposure to the mildly acidic environment of the endosome (73). As the ortho ester hydrolysis ensues, a lactonization process liberates a single-chain detergent–like lipid to destabilize the endosomal membrane.

The transferrins are a group of homologous, nonheme, iron-binding glycoproteins widely distributed in nature. The transferrin receptor is highly elevated on some tumors, including gliomas and haematopoietic tumors, and therefore suited to gene delivery (74). Gene transfer to K562 haematopoietic leukemic cells was achieved with a transferrin–polycation (poly-L-lysine or protamine) conjugate (75); melanoma cells were transfected with the gene for interleukin-2, resulting in a successful tumor vaccine (76).

Bacterial viruses capable of binding mammalian cells expressing the growth factor receptor ErbB2 and undergoing receptor-mediated endocytosis have been engineered to package the green fluorescent protein (GFP) reporter gene driven by the cytomegalovirus promoter. After application to cells, GFP expression occurred only in cells overexpressing ErbB2 (77).

Bacterial pore-forming proteins, such as α-hemolysin, secreted by *Staphylococcus aureus* can be modified so that pore formation is activated by chemical, biochemical, or physical triggers. Such hemolysins, when targeted to tumors, could increase tumor permeability, and hence susceptibility to various cytotoxic drugs (78).

Palmitic acid is conjugated to glucuronic acid to form a reticuloendothelial system-avoiding liposome delivery system (79). Phospholipids such as phosphatidyl choline or phosphatidyl ethanolamine are used as constituents of lipid complexes or liposomes (80). In Table 2, a partial list of the names and uses of excipients in this category is presented.

Porphyrins are chromophores, occurring in diverse molecules, such as hemoglobin and chlorophyll. They have a propensity to accumulate in tumors (81). Boronated porphyrins or their derivatives have great potential as boron carriers in boron neutron capture therapy (82).

Transposons are sequences of DNA that can move around to different positions within the genome of a single cell. They are also called "jumping genes" or "transposable genetic elements." Helper-independent transposon–transposase vectors have been constructed, which contain on single plasmids (i) a synthetic salmonid *sleeping beauty* transposon containing the transgene of interest and (ii) a *sleeping beauty* transposase expression cassette. These nonviral gene vectors have

Table 2 Natural Products Including Naturally Occurring Polymers and Derivatives

Name	Use
Squalene, squalane	Vaccine adjuvant
Phosvitin	Protein stabilizer
Phytic acid	Protein stabilizer
Phospholipids	Liposomes
Spermidine	Inhibitor of lipid peroxidation
Hyaluronic acid	Viscoelastic
Sphingomyelin	Component of liposomes
Biopolymer	Protective coating for liposomes
Fibronectin	Protein stabilizer
Spermine–bile acids	DNA transfection agent
Deacylated saponin	Surfactant in intranasal or ocular delivery
a-Hemolysin	Increasing susceptibility of tumor cells to cytotoxic drugs
Galactosylated histone H1	Liver gene delivery
Sialic acid	
Galacturonic acid or poly-galacturonic acid	Contrast medium in MRI
N-acetyl [$Phe^8(CH_2$—$NH)Arg^9$] bradykinin	Transport across the blood brain barrier
Enoxaparin, heparin	Protein stabilizer
Cyclodextrins	Solubilizer for hydrophobic molecules
Phosvitin–galactomannan conjugate	Emulsifier
Pyridoxal phosphate or derivatives	Paramagnetic metal chelating agent in NMRI
Recombinant fusion streptavidin-Mab protein	Transport across the blood brain barrier
Hyaluronic acid–cyclodextrin	Modification of depolymerization kinetics and release
Listeriolysin-O, ortho ester lipids	Endosomal escape
Transposons	Nonviral gene delivery
Porphyrin derivatives	Boron neutron capture therapy
Fusion-associated small transmembrane protein, TAT, VP22, Antp	Fusogenic liposomes, increase transfection efficiency
Palmityl-D-glucuronide	RES-avoiding liposomal drug delivery
Transferrin	Ligand in receptor-mediated gene delivery
Linamarase	ADEPT

Abbreviations: NMRI, nuclear magnetic resonance imaging; Antp, antennapedia; RES, reticuloendothelial system; ADEPT, antibody-directed enzyme prodrug therapy; TAT.

efficiently facilitated the long-term transgene expression in mouse liver and have partially restored blood-clotting capability in hemophilic mice (83). Hybrid vectors that combine the integration capability of mariner transposons with the transduction efficiency of recombinant adenovirus vectors have also been constructed and shown to be effective in introducing transgenes into the host chromosome (84).

Fusion-associated small transmembrane (FAST) proteins have been isolated from nonenveloped reoviruses that are capable of inducing fusion and promoting syncytium formation (85). Incorporation of such fusogenic FAST proteins into liposomes may promote liposome–cell fusion for the intracellular delivery of drugs, proteins, or genes. These fusogenic liposomes may increase transfection efficiencies of purely nonviral delivery systems. More fusogenic peptides such as HIV-1 TAT protein, HSV VP22, and the Drosophila Antennapedia homeotic transcription factor are known (86).

CONCLUSIONS

As the molecular mechanisms of disease become increasingly better understood, so does the realization that the possibility of relatively benign and therapeutically effective intervention requires ways to specifically target such uniquely identified mechanism(s) and/or pathway(s). Such targeting methods include—in large part—the recruitment of component(s) of the precise generic and specific targeting mechanisms employed by pathogenic organisms. Such components typically include proteins, enzymes, polymers, coding sequences, and in some cases the mutated organism(s) themselves. Properties such as evasion from host defense mechanisms, target detection, penetration, integration into the host genome, expression of the transduced genetic drug/enzyme product, and destruction of the targeted host cell(s) can be imparted at will into drug delivery systems by site/sequence-specific and/or morphology-specific incorporation of such "natural products."

Emerging excipients are more likely to be the products of gene coding sequences than inorganic molecules and be integral components of drug delivery systems than merely be mechanically incorporated into a formulation. They are more likely to possess pharmacological activity that is independent of their excipient functionality, thereby allowing their use as a "drug" in certain indications and as an "excipient" in others. These attributes constitute a new paradigm in the definition, function, and use of emerging excipients in parenteral medications (87).

REFERENCES

1. Lee J, Park YS, Kim YI, et al. Synthesis of a porphyrin having eight carborane cages. Bull Korean Chem Soc 1999; 20(11):1371–1372.
2. Smith GK, Banks S, Blumenkopf TA, et al. Toward antibody directed enzyme prodrug therapy with the T268G mutant of human carboxypeptidase A1 and novel in vivo stable prodrugs of methotrexate. J Biol Chem 1997; 272(25):15804–15816.
3. Vector consisting of a transcriptional regulatory DNA sequence linked to a DNA sequence encoding β-lactamase for enzyme prodrug therapy. Glaxo Wellcome, Inc., Research Triangle Park, NC. Patent No. WO9719180A2, issued 5–29–97.
4. Kukowska-Latallo JF, Chen C, Eichman J, et al. Enhancement of dendrimer-mediated transfection using synthetic lung surfactant exosurf neonatal in vitro. Biochem Biophys Res Commun 1999; 264(1):253–261.

5. Baeuerle PA, Kufer P, Lutterbuse R. Bispecific antibodies for polyclonal T-cell enlargement. Curr Op Mol Ther 2003; 5(4):413–419.
6. Oldenburg SJ, Jackson JB, Westcott JB, Halas NJ. Infrared extinction properties of gold nanoshells. Appl Phys Lett 1999; 75(19):2897–2899.
7. Hirsch LR, Sershen SR, Halas NJ, Stafford RJ, Hazle J, West JL. Nanoshell mediated near infrared photothermal tumor therapy. 2003 Summer Bioengineering Conference, June 25–29, Key Biscayne, FL.
8. Unger E, Matsunaga TO, Schumann PA, Zutshi R. Microbubbles in molecular imaging and therapy. Medicamundi 2003; 47(1):58–62.
9. Calcium Disodium Versenate. 3M Pharmaceuticals, Patient package insert.
10. Exosurf® Neonatal for intratracheal suspension. Glaxo Wellcome, Research Triangle Park, NC.
11. Supervent® Aerosol. Discovery Laboratories Inc., Doylestown, PA.
12. Orthovisc®. Anika Therapeutics, Woburn, MA.
13. Viscoat®, Provisc®. Alcon Inc., Fort Worth, TX.
14. Compositions and methods for nucleic acid delivery to the lung. Inhale Therapeutic Systems Inc., San Carlos, CA. U.S. Patent No. 5994314, issued 11–30–1999.
15. Cyclokapron®. Kabi Pharmacia, Patient package insert.
16. Pharmaceutical preparations containing plasminogen activators. Boehringer Mannheim Gmbh, Mannheim, Germany. U.S. Patent No. 5747030, issued 8–21–96.
17. Solutol® HS 15. BASF Aktiengesellschaft, 2002.
18. Dintaman JM, et al. Inhibition of P-glycoprotein by D-alpha-tocopheryl polyethylene glycol 1000 succinate (TPGS). Pharm Res 1999; 16(10):1550–1556.
19. Coon JS, et al. Cancer Res 1991; 51(3):897–902.
20. Method for antibody targeting of therapeutic agents. Immunomedics, Inc., Morris Plains, NJ. U.S. Patent No. 5851527, issued 12–22–98.
21. Hallahan DE, Mauceri HJ, Seung LP, et al. Spatial and temporal control of gene therapy using ionizing radiation. Nat Med 1995; 1(8):786–791.
22. Ros TD, Spalluto G, Prato M. Biological applications of fullerene derivatives: a brief overview. Croat Chem Acta 2001; 74(4):743–755.
23. Slovin SF, Ragupathi G, Adluri S, et al. Carbohydrate vaccines in cancer: immunogenicity of a fully synthetic globo H hexasaccharide conjugate in man. Proc Natl Acad Sci USA 1999; 96:5710–5715.
24. Ivics Z, Hackett PB, Plasterk RH, et al. Cell 1997; 91:501–510.
25. Holle L, Song W, et al. A matrix metalloproteinase 2 cleavable melittin/avidin conjugate specifically targets tumor cells in vitro and in vivo. Int J Oncol 2003; 22:93–98.
26. Cunningham C, Nemunaitis J. A phase I trial of genetically modified Salmonella typhimurium expressing cytosine deaminase (TAPET-CD, VNP20029) administered by intratumoral injection in combination with 5-fluorocytosine for patients with advanced or metastatic cancer. Human Gene Ther 2001; 12(12):1594–1596.
27. Levin M, Quint PA, Goldstein B, et al. Recombinant rBPI21 as adjunctive treatment for children with severe meningococcal sepsis: a randomized trial. RBPI Meningococcal sepsis study group. Lancet 2000; 356(9234):961–967.
28. Kohler U, Ayre BG, Goodman HM, et al. Trans-splicing ribozymes for targeted gene delivery. J Mol Biol 1999; 285:1935–1950.
29. Cortes ML, deFelipe P, et al. Successful use of a plant gene in the treatment of cancer in vivo. Gene Ther 1998; 5(11):1499–1507.
30. Shmulevitz M, Duncan R. A new class of fusion-associated small transmembrane (FAST) proteins encoded by the non-enveloped fusogenic reoviruses. EMBO J 2000; 19(5):902–912.
31. Ogris M, Carlisle RC, et al. Melittin enables efficient vesicular escape and enhanced nuclear access of nonviral gene delivery vectors. J Biol Chem 2001; 276(50):47550–47555.
32. Visudyne®. Novartis Ophthalmics Inc., Duluth, GA and Photofrin®. Axcan Scandipharm Inc., Birmingham, AL.

33. Cagle DW, Kennel SJ, Mirzadeh S, et al. In vivo studies of fullerene based materials using endohedral metallofullerene radiotracers. Proc Natl Acad Sci USA 1999; 96:5182–5188.
34. Apte SP, Ugwu SO. A review and classification of emerging excipients in parenteral medications. Pharm Technol 2003.
35. Shinkai M, Suzuki M, Iijima S, et al. Antibody-conjugated magnetoliposomes for targeting cancer cells and their application in hyperthermia. Biotechnol Appl Biochem 1994; 21:125–137.
36. So H-S, et al. Effect of a novel saponin adjuvant derived from Quillaja saponaria on the immune response to recombinant Hepatitis B surface antigen. Mol Cells 1996; 7(2):178–186.
37. A component of Theratope® vaccine in phase III clinical trials for use against metastatic breast cancer. Biomira Inc., Edmonton, Canada.
38. Corixa Corporation. Seattle, WA.
39. MDX-RA®, an immunotoxin conjugated to an antibody designed to prevent secondary cataracts, phase III clinical trials, enrollment stopped. Medarex, Annandale, NJ.
40. Allison AC. Squalene and squalane emulsions as adjuvants. Methods (Duluth) 1999; 19(1):87–93.
41. Pillion DJ, et al. Structure-function relationship among Quillaja Saponins serving as excipients for nasal and ocular delivery of insulin. J Pharm Sci 1996; 85(5):518–524.
42. Abbokinase® (Urokinase for injection). Abbott Laboratories, Abbott Park, IL. Acthar Gel® (Corticotropin injection). Rhone Poulanc Rorer, Collegeville, PA.
43. Solganol® (Aurothioglucose suspension). Schering Plough Corp., Madison, NJ.
44. Angelova MI, et al. Interactions of DNA with giant liposomes. Chem Phys Lipids 1999; 101(1):123–137.
45. Allen TM, et al. Biochim Biophys Acta 1990; 1029:91–97.
46. Walker S, et al. Cationic facial amphiphiles: a promising class of transfection agents. Proc Natl Acad Sci USA 1996; 93:1585–1590.
47. Patent # WO9529186. University of Princeton, 1996.
48. Awasthi S, et al. Inhibition of liposomal lipid peroxidation by spermidine. J Microencapsulation 1992; 9(2):237–242.
49. Hays LM, et al. Antifreeze glycoproteins inhibit leakage from liposomes during thermotropic phase transitions. Proc Natl Acad Sci USA 1996; 93:6835–6840.
50. Sphingosomes for enhanced drug delivery. Inex Pharmaceuticals Corp., Canada. U.S. Patent # 5741516, issued 4–21–1998.
51. De La Maza A, et al. Biopolymer excreted by Pseudoalteromonas antarctica NF3, as a coating and protective agent of liposomes against dodecyl maltoside. Biopolymers 1999; 50(6):5795–5788.
52. Chen J, et al. Galactosylated histone mediated gene transfer and expression. Human Gene Ther 1994; 5:429–435.
53. Zhang MZ, et al. A new strategy for enhancing the stability of lyophilized protein: the effect of the reconstitution medium on keratinocyte growth factor. Pharm Res 1995; 12(10):1447–1452.
54. Chong MW, et al. Stabilizers against heat-induced aggregation of RPR 114849, an acidic fibroblast growth factor (aFGF). Int J Pharmaceut 1998; 167:25–36.
55. Vucenik I, et al. Novel anticancer function of inositol hexaphosphate: inhibition of human rhabdomyosarcoma in vitro and in vivo. Anticancer Res 1998; 18(3A):1377–1384.
56. Natural polymers as contrast media for magnetic resonance imaging. ImaRx Pharmaceutical Corp., Tucson, AZ. U.S. Patent # 5368840, issued 11–29–94.
57. Method for increasing blood-brain barrier permeability by administering a bradykinin agonist of blood-brain barrier permeability. Alkermes, Inc., Cambridge, MA. U.S. Patent # 5112596, issued 5–12–92.
58. Emerich DF, et al. Enhanced delivery of carboplatin into brain tumors with intravenous Cereport® (RMP-7): dramatic differences and insight gained from dosing parameters. Br J Cancer 1999; 80(7):964–970.

59. Cloughesy TF, et al. Intra-arterail Cereport® (RMP-7) and Carboplatin: a dose escalation study for recurrent malignant gliomas. Neurosurgery 1999; 44(2):270–278.
60. Pharmaceutical formulation. Supergen Inc., San Ramon, CA. U.S. Patent # 5804568, issued 9–9-98. NDA filed for MitoExtra®, a proprietary reformulation of the anti-cancer drug, Mitomycin.
61. Fluorocyclosaccharide drug delivery systems. University of Alberta, Edmonton, CA. U.S. Patent # 5739121, issued 4–14–98.
62. Khan MAS, et al. Molecular mechanism of the excellent emulsifying properties of phosvitin-galactomannan conjugate. J Agric Food Chem 1999; 47(6):2262–2266.
63. Dipyridoxyl phosphate NMRI contrast agent compositions. Salutar, Inc., Sunnyvale, CA. U.S. Patent # 5091169, issued 2–25–92.
64. Marafie, et al. Complexes of vitamin B6 XX: equilibrium and mechanistic studies of the reaction of pyridoxal-5′-phosphate with pyridoxamine-5′-phosphate in the presence of copper(II). J Inorg Biochem 1989; 37:7–16.
65. Matsushima, et al. *N*-Pyridoxylidenehydrazine-*N*′,*N*′-diacetic acid. III. Formation of metal chelates in solution. Chem Pharm Bull 1989; 35(12):4695–4699.
66. Soltes L, et al. Cyclodextrin derivative of hyaluronan. Carbohydr Polym 1999; 39(1): 17–24.
67. Li JY, et al. Genetically engineered brain drug delivery vectors: cloning, expression and in vivo application of an anti-transferrin receptor single chain antibody-streptavidin fusion gene and protein. Protein Eng 1999; 12(9):787–796.
68. Penichet ML, et al. An antibody-avidin fusion protein specific for the transferrin receptor serves as a delivery vehicle for effective brain targeting: initial applications in anti-HIV antisense drug delivery to the brain. J Immunol 1999; 163(8):4421–4426.
69. Portnoy, et al. Capacity of listeriolysin O, streptolysin O and perfringolysin O to mediate growth of Bacillus subtilis within mammalian cells. Infect Immun 1992; 60:2710–2717.
70. Use of a bacterial component to enhance targeted delivery of polynucleotides to cells. University of Connecticut, Storrs, CT and Targetech, Inc. a wholly owned subsidiary of the Immune Response Corporation, Carlsbad, CA. U.S. Patent # 5728399, issued 3–17–88.
71. Wagner E, et al. DNA-binding transferrin conjugates as functional gene delivery agents. Bioconjug Chem 1991; 2:226–231.
72. Quilimmune-M®. Malaria vaccine adjuvant, phase I trials completed. Aquila Biopharmaceuticals Inc., Framingham, MA.
73. Schwarze SR, Hruska KA, Dowdy SF. Protein transduction: unrestricted delivery into all cells?. Trends Cell Biol 2000; 10:290–295.
74. Cotton M, et al. Transferrin-polycation mediated introduction of DNA into leukemic cells. Proc Natl Acad Sci USA 1990; 87:4033–4037.
75. Zatloukal K, et al. Somatic gene therapy for cancers: the utility of transfer infection in generating tumor vaccines. Gene 1993; 135:199–207.
76. Poul MA, et al. Targeted gene delivery to mammalian cells by filamentous bacteriophage. J Mol Biol 1999; 288(2):203–211.
77. Bayley H. Building doors into cells. Sci Am 1997:62–67.
78. Namba Y, et al. Liposomal modification with uronate, which endows liposomes with long circulation in vivo, reduces the uptake of liposomes by J774 cells in vitro. Life Sci 1992; 50(23):1773–1779.
79. Abelcet®, amphotericin lipid complex injection. Liposome Co., Princeton, NJ.
80. Ambisome®, Daunoxome®, liposomal amphotericin B and daunorubicin citrate injection. Liposomal lurtotecan in phase I clinical trials. Gilead Sciences Inc., Foster City, CA.
81. Barth RF. A critical assessment of boron neutron capture therapy: an overview. J Neuro Oncol 2003; 62:1–5.
82. McFadden DW, Riggs DR, Jackson BJ, et al. Keyhole limpet hemocyanin, a novel immune stimulant with promising anticancer activity in Barrett's esophageal adenocarcinoma. Am J Surg 2003; 186(5):552–555.

83. Zhang L, Sanker U, Lampe DJ, et al. The Himar1 mariner transposase cloned in a recombinant adenovirus vector is functional in mammalian cells. Nucleic Acids Res 1998; 26(16):3687–3693.
84. Ma Z, Taylor JS. Nucleic acid triggered catalytic drug release. PNAS 2000; 97(21): 11159–11163.
85. Nantz M, Zhu J. Ortho ester lipids. U.S. Patent No. 6200599, 2001.
86. Gomez JDF, Sansom MSP. Acquisition of siderophores in gram negative bacteria. Nature Rev Mol Cell Biol 2003; 4:105–116.

21
Excipient Manufacturing and Good Manufacturing Practices

Irwin Silverstein
IBS Consulting in Quality LLC, Piscataway, New Jersey, U.S.A.

INTRODUCTION

Pharmaceutical customers desire excipients that are produced with minimal variation in quality and performance characteristics from lot to lot as well as within the lot. In addition, pharmaceutical users do not want changes in excipient performance to cause them registration or regulatory issues. Registration issues can arise as a consequence of an adverse regulatory inspection of an excipient manufacturer. Regulatory issues for the drug product manufacturer can result from a significant change that changes the excipient impurity profile or excipient stability and which then alters the drug product impurity profile or stability profile. Thus the excipient customer wants a product that uniformly meets specification and performance criteria and is produced in adequate conformance to regulatory requirements.

This chapter will deal with the objective of manufacturing excipient ingredients to appropriate good manufacturing practices (GMP) requirements, as stipulated by the United States Pharmacopeia (USP) (1) and the International Pharmaceutical Excipients Council excipient GMP guide (2). It is beyond the scope to address the many quality techniques for minimizing variation in excipient quality. However this chapter will address the issues concerning assurance that all excipient material within each batch meets compendial or manufacturer's specification.

Quality Systems for Excipients

Manufacture of excipients in compliance with excipient GMP requirements is based upon an International Organization for Standardization (ISO) 9001 compliant quality system as the foundation to which additional GMP requirements are added. If the manufacturer of the excipient chemical has achieved certification to ISO 9001, then the specific GMP requirements can be added to the quality system already in place. Otherwise, all manufacturing-related ISO 9001 requirements should be implemented in an ISO compliant manner. Therefore such ISO requirements as a sound document control system as well as internal audit, calibration, and preventive maintenance and training programs should be in place. Also required under ISO and excipient GMP is adequate work instructions, release procedures, identification, traceability from raw materials to product and vice versa, and validation of the manufacturing process. With adequate compliance to these ISO 9001 requirements, the implementation of an excipient GMP compliant quality system is facilitated.

This chapter will focus on those additional GMP requirements for the manufacture of excipient ingredients under an ISO 9001 compliant quality system. While certification to ISO 9001 is unnecessary for compliance to excipient GMPs, many of the manufacturing-related requirements as noted above are essential for adequate GMP conformance. With the numerous ISO 9001 compliance resources available, this chapter will assume the reader has the knowledge to implement those relevant general ISO 9001 requirements.

Diversity in Production of Excipients

Excipients are produced by a diverse group of manufacturers who range from pharmaceutical companies to specialty chemical manufacturers, food producers, and even mineral quarries. These manufacturers are involved in a diverse range of chemical processing and usually produce excipients as a small component of their product mix.

Chemical companies, including some petrochemical producers, generally produce excipient grades of such classes of synthetic organic chemicals as solvents, polymers, alcohols, esters, surfactants, etc. These companies also usually produce these same chemicals in grades suitable for other markets such as those servicing personal care, foods, and industrial. It is quite common for these other applications to dwarf the excipient grade in terms of both volume and dollar value.

Food processors also supply excipients to the pharmaceutical market. Generally these excipients are either foods, such as sucrose, starch, oils, etc. or food-based derivatives such as various modified starches, cellulose derivatives, etc. These products originally found application as ingredients in processed food products.

Companies that manufacture food additives will often also produce an excipient grade. Many excipients first found application in processed foods where their demonstrated safety has made them attractive for use as pharmaceutical excipients. Examples of food additives also used in drug products include artificial sweeteners, antioxidants, and inorganic salts.

Mineral product companies also provide excipients to the industry. Such companies may produce these inorganic chemicals by quarrying rock and separating the desired chemical entity such as talc, sodium chloride, or sodium carbonate. Alternatively sometimes the inorganic excipient is produced synthetically. The market for excipient grade minerals is dwarfed by other applications for these products in the industrial market.

The importance in appreciating the diverse nature of excipient manufacturers and their processes is necessary for an understanding of the difference in excipient GMP requirements when compared to those for the drug product or active pharmaceutical ingredient. The manufacturing of drug products essentially involves the blending of ingredients and their formation into a dosage form. The production of active pharmaceutical ingredients typically involves the purification, derivatization, or synthesis of natural substances. However the manufacture of excipients may combine some or all of these processes ranging from isolation and purification to chemical synthesis and blending. The diversity of excipient manufacturers and their processing is the main element making excipient production unique from other aspects of the pharmaceutical industry.

As a consequence of the uniqueness of excipient manufacture as discussed above, the starting point for applying GMP principles, the manufacturing environment, manufacturing equipment, manufacturing processes, and other applications for nonpharmaceutical grades of the material all impact the application of GMP requirements.

Good Manufacturing Practices for Excipients

Implementation of excipient GMPs to the manufacturing process begins with a determination as to when full excipient GMP requirements must be applied. This decision begins with a review of the process flow diagram. Working back from the packaging of finished excipient, the manufacturer must determine at which step the final molecule is formed or where the final purification occurs. Generally whichever comes later in the process is the first point at which full GMP requirements must be applied.

Several examples will help to illustrate this point. For the manufacture of a polymeric excipient, full GMPs are applied at the polymerization step because there is usually no significant purification after the polymer is formed. For a small molecule excipient such as a solvent or alcohol, full GMP principles must be applied no later than the start of purification, which is usually a distillation step. Finally for a salt or inorganic powder, full GMPs are applied at the formation of the salt or its purification step such as crystallization.

It is important not to limit application of GMP principles solely from the point in the manufacturing process where FULL GMP requirements must be applied. Selective portions of excipient GMPs should be applied earlier in the manufacturing process as appropriate with the objective of producing an excipient with minimal variation in quality or performance characteristics. Some examples of earlier application of appropriate GMP requirements are presented below.

Processing

Excipient manufacture runs the full range of industrial chemical manufacturing processes. However, these production processes generally fall into one of two broad categories: batch or continuous processing. Often the manufacture of an excipient involves a combination of both batch steps and continuous steps. Whether wholly batch, continuous, or some combination of the two, there should be clear and complete manufacturing instructions available for the workers to follow. Also there should be sufficient records to show that the batch was produced according to those instructions. Finally, those records should be completed contemporaneously with completion of the processing step.

Excipient production facilities should be maintained in a good state of repair and with good housekeeping so as to present an acceptable appearance to regulatory inspectors. While much chemical processing occurs outdoors and is acceptable, wherever possible the facility should be free of peeling paint, loose insulation, rust, etc. Buildings should be watertight with screens on windows and doors to prevent entry by flying insects. Tools and portable equipment should be properly stored and the area kept free of trash.

Pharmaceutical excipients must be made using appropriate manufacturing instructions, equipment, raw materials, utilities, test procedures, and records. It is inappropriate to offer as excipient grade any material not produced in conformance to GMP requirements. Because it is not permissible to upgrade a lot from a lesser grade to an excipient grade, it is important to start all batches, which might be sold to the pharmaceutical industry using a GMP compliant quality system. Therefore the manufacturing instructions and production records should be labeled as excipient grade from the start of production. At any time during processing or after the lot is packaged, it is permissible to downgrade the lot to a lesser, nonpharmaceutical grade for whatever reason.

Production Records

All production records are controlled documents. Each blank record should have a unique identifier on each page; which is generally the product name or code. Each page should also have an identifier such as the issue date that enables verification that the production record is current. Also each page of the record should have its page number along with the total number of pages in the record. Blank production records should be issued to the production unit only as needed to control the number of copies, which will facilitate migration to a newer version when the production record is updated.

Production records must be completed in ink, preferably black or blue ink. Records should be completed in conjunction with the completion of each step. Records should never be completed before the task has been performed nor should tasks be completed with the entries made to the record at a later time. A space for an entry on a record should never be left blank. All notations expected on the production record must be either completed or the record should indicate why no entry was necessary. Completed production records should be reviewed by a supervisor to ensure that the record is complete with no blank spaces and that appropriate entries have been made. The supervisor acknowledges his/her review by signature or initials and date.

It is important to properly correct erroneous entries on GMP documents such as production records. Improper correction of errors can raise the concern of regulators that the change was not proper and will thus jeopardize the validity of the document. Any change to a document or record must be done using ink to draw a single line through the erroneous information and the individual making the correction must then add his/her initials and date. This provides traceability of the change back to the author. It is also beneficial for the individual to annotate the reason the correction was made. It is not acceptable to attempt to obliterate the erroneous information with ink or whiteout.

Training of Employees

Employees should be qualified for the jobs they perform. They should receive appropriate training that should be documented. Classroom training should be documented through retention of the sign in sheet as well as with an outline of the training, a copy of the presentation slides, or other suitable documentation. For on-the-job training, the trainer should complete a checklist confirming that all important tasks have been shown to the employee. There should also be a record showing the effectiveness of the training, which can be the completion of a written exam, performance of the task, or other confirmatory means.

Employees performing GMP-related tasks should be trained periodically in the principles of GMP that apply to their operation. At least biennially they should be reminded of the documentation practices noted above as well as their responsibility to follow procedures as prescribed. Employees should also be encouraged to inform their supervisor of any incorrectly performed operations. Finally they should be informed that it is necessary to report to their supervisor any illness, especially open lesions, that they have, which may contaminate the excipient during the performance of their responsibilities. Employee GMP training should also be documented.

Control of Raw Materials

Even though full GMP compliance may begin later than receipt of raw materials, it is important to treat raw materials in a GMP compliant manner. Raw materials should be purchased only from suppliers approved by the Quality Unit. For raw materials whose quality is important to conformance of the excipient to compendial or specification requirements, or to performance expectations, the supplier approval process should be a combination of site visit and an evaluation of raw material quality. For other raw materials, it would be adequate to confirm that the raw material supplier can meet the purchasing specification.

Where raw material quality is important to excipient quality or conformance, regardless of whether the raw material is provided in discrete packages, by bulk transport, or via pipeline, supplier approval should include an understanding of the quality system under which it was produced. A visit to the site will provide the excipient manufacturer with an understanding of the impurities and potential contaminants that may be present in the raw material. Site visits also give an appreciation of the possibility of mix-ups at the supplier, especially during packaging and labeling operations. With this knowledge, the excipient manufacturer can tailor their raw material approval-testing program to the potential risks to the quality of the raw material.

Raw materials must always be approved by the Quality Unit before use by production. Each lot of raw material should be sampled and the laboratory should perform at least an identity test in addition to verifying from the supplier Certificate of Analysis (COA) that the lot test results conform to the excipient manufacturer's specification. Upon approval, the status of the lot is changed from unapproved or quarantine to approved or available. The raw material lot status can be identified by use of approval labels on the container or pallet, movement of the raw material lot to the approved section of the warehouse, or by changing the lot status in a computerized inventory system.

Raw materials must be stored in a manner that protects their quality. While it is acceptable to store raw materials outdoors, there are risks that should be recognized. The most pronounced risks are that the container label will become unreadable or that water will enter the container. Because many raw material labels are printed on demand, their labels are often susceptible to degradation by sunlight and water. If the label becomes difficult or impossible to read, an error can be made such as using the wrong material, using an unapproved lot of the proper raw material, or recording an incorrect raw material lot number on the production record.

It is a recognized phenomenon that water can enter a drum, whether of plastic or steel construction, if the drum is stored upright unprotected from precipitation. Water collecting on the top of the drum can be sucked into the drum by the partial vacuum created as a result of temperature changes to the drum's contents. If the drum's contents are not tested again before use, wet raw material might be used in the process.

Oftentimes, bulk raw materials are stored in storage tanks. Usually these tanks have a vent so that the tank is allowed to "breathe." It is important to ensure that the vent opening is protected from precipitation and also to entry by birds or insects into the tank. A mesh screen on a gooseneck vent provides suitable protection. Sometimes the raw material is sensitive to environmental conditions and must be kept under a blanket of inert gas such as nitrogen or maintained at a fixed temperature. Wherever there is a requirement for an inert blanket or temperature control, there should be a device that allows for confirmation the storage tank is maintained under the specified environmental conditions. These conditions should be periodically recorded.

Excipient GMPs require the use of a suitable quality of water in the manufacture of excipient ingredients. Potable water is required wherever water comes into contact with the materials during processing especially the finished excipient molecule. It is preferable to receive potable water from a municipal authority, which would then have responsibility for demonstrating the water meets Environmental Protection Agency (EPA) requirements for potable drinking water. Otherwise the manufacturing site would have to self-certify conformance to the EPA standard by periodic testing of the water. Whether purchased from a municipality or self-certified, water should be periodically tested, at least annually, at the point(s) of use to demonstrate that it meets the microbiological requirements for potable water. Water that is used in the process for such operations as external heating or cooling need not be of potable quality.

Oftentimes water is further purified using deionization or reverse osmosis to meet the requirements for chemical processing. There must be specifications for the quality of this purified water and periodic testing to demonstrate conformance. The water fed to these operations must also be of potable quality unless the purified water is used for noncontact purposes as discussed above. If the purified water is stored prior to use, it must be stored so as to prevent the build up of microbial organisms. There are many techniques suitable to control microbes such as treatment with ultraviolet light or ozone and circulation. Whatever control method is used, it should be demonstrated that it effectively prevents microbial build-up.

Preventing Contamination

The manufacture of pharmaceutical ingredients must be protected from contamination beginning with the charging of raw materials through packaging the excipient into the market container. Generally contamination arises from two sources: exposure of the excipient to the environment during the production process and flaws in the equipment.

Generally during the synthesis of an excipient, the risk of environmental contamination only arises during the charging of raw materials, additives, or processing aids, or during the filling of the market container. While steps should be taken to prevent environmental contamination of the excipient at all stages of production, protection is particularly important after the final molecule has been formed or after the final purification, whichever step is later. At this point in the processing especially where final purification has already occurred, contaminants in the excipient will remain.

Prevention of environmental contamination of the excipient batch during the charging of materials can be done by keeping environmental contaminants out of the production vessel. Airborne contamination arises from dust such as from other chemical processing, dirt such as from nearby fields, insects, and fumes from processing. Flying insects can be controlled through the use of screens on doors and windows, insect lures, and traps. Keeping other airborne contaminants such as dust, dirt, and chemical fumes from blowing into the production vessel can be accomplished by enclosing the production equipment in a controlled environment room (discussed later in this chapter) or by developing an air curtain over the vessel opening. An air curtain can be constructed by installing a hood over the vessel where it is open to the air. Often it is satisfactory to use air, which has been filtered through an efficient air conditioner filter directed through the hood with sufficient velocity to keep dust from entering.

Efforts to prevent environmental contamination of the finished excipient must be more rigorous. Because this generally occurs in the packaging area, this operation should be conducted in an enclosed room. The packaging area should be constructed of washable walls, floor, and ceiling. It has been suggested that the room should be maintained under positive pressure using air passed through a 2 μm filter operating at 95% efficiency with about 20 air changes per hour. The packaging room should be cleaned with sufficient frequency to prevent cross contamination from materials present from prior packaging operations.

A less common risk of contamination arises from the raw material. Contaminants can enter the vessel during charging of raw materials from dirty packages. It is therefore important to wipe down the outside of drums if the raw material is to be poured into the vessel. An alternative technique for adding raw materials is to siphon or draw the raw material from the drum into the vessel, which poses reduced risk of contamination. Where the raw material is provided in bags or boxes, care must be exercised to prevent the paper, cardboard, or plastic liner from entering the vessel. Finally it must be recognized that sampling of raw materials for Quality Control (QC) testing can also result in contamination to the excipient through the raw material. Therefore sampling devices must be protected from contamination, even if the device is cleaned after each use. Samplers must be stored in a manner so that they do not become dirty prior to use.

The other significant risk of excipient contamination comes from the operating equipment. Excipients may become contaminated from equipment in two ways: cross contamination from other uses of the equipment and contamination from the equipment itself. Excipients produced in equipment dedicated to the production of the pharmaceutical ingredient and its other grades such as food, technical, cosmetic, etc. is inconsequential. This contamination is usually limited to incidental carryover from the batch of nonpharmaceutical grade to excipient grade and the chemical composition of each grade is often indistinguishable. In a similar way, the production equipment may be used to produce the excipient in multiple grades differentiated by such properties as particle size, molecular weight, or viscosity. Again incidental carryover from grade to grade does not present an unacceptable situation as long as it is kept to a minimum. However cross contamination between the excipient and other molecules of different composition is unacceptable. Therefore, whenever equipment is used to produce multiple chemical moieties, the manufacturer must have adequate procedures for cleaning between products. The cleaning of equipment is further discussed under cleaning validation.

The second way in which equipment can lead to excipient contamination is from equipment construction. Production equipment should not be reactive under the conditions used for producing the excipient. Also the materials used for the operation of the equipment such as lubricants, grease, coolants, etc. should not be allowed to contaminate the excipient. To prevent this, devices such as traps, seals, and preventive maintenance must be employed. However where the risk of such contamination to the excipient remains even with proper measures, the use of food grade materials should be carefully considered.

Qualification of Manufacturing Equipment

From the point in the process where full GMP compliance begins, the excipient should be produced in qualified equipment using a validated manufacturing process and testing should be done using validated methods. Where production equipment is not dedicated, validated cleaning methods are also needed. Full GMP compliance is required no later than the final excipient purification step or the manufacturing step where the excipient molecule has been synthesized, whichever occurs later in the process.

Qualification or validation begins with a protocol or plan that describes fully the approach, including the scope, description of equipment, utilities, test methods, acceptance criteria, etc. The approved protocol is then executed by performing the requisite number of replicates as specified in the protocol and gathering the indicated data. The final step is the preparation and approval of a report containing the findings of the activity against the acceptance criteria found in the protocol.

As summarized above, the procedure for qualifying equipment begins with a protocol that describes the qualification activity. Qualification of equipment begins with an installation qualification, which is followed by operation qualification and concludes with performance qualification.

Installation qualification is an exercise that shows the equipment has been installed properly, as specified either by the equipment manufacturer or by the purchaser. Operation qualification demonstrates that the equipment operates as intended. The operation of the equipment is compared to the equipment manufacturer's specification or the excipient manufacturer's design specification. Finally, the performance qualification shows that the equipment performs as intended. Where production equipment is involved, performance qualification usually involves running a trial substance such as water or a production batch.

As noted, excipient manufacture should take place using qualified equipment and a validated process. Generally excipient equipment has been in place for many years so that classical methods of qualification, which is done as new equipment is commissioned, are inapplicable. To retrospectively qualify the installation, operation, and performance of equipment, it is suggested to rely on historical records. For installation and operation qualification, a protocol is prepared that illustrates how maintenance and production records will be used to support the hypothesis that the equipment was installed properly and is operating as intended. Then the protocol is executed by reviewing the maintenance and production records for the supporting data. Finally a report is prepared that includes the data from the records, which support the conclusion that the installation and operation of the equipment conforms to protocol requirements. It is suggested that maintenance and production records for a minimum of one year but preferably five years be reviewed.

Performance qualification can be done on equipment even if it has been in use for some time. The performance qualification protocol should describe how the performance of the equipment is to be demonstrated. If this involves a trial batch, then the composition of the batch should be described. The protocol should also describe how the equipment would be used to process the trial material. Finally the protocol should delineate the expected output that will show the equipment performs as intended. The final elements of the performance qualification protocol are tests that show the equipment operates properly at the limits of its performance specification and that the equipment has the specified production capacity. Performance qualifications require a minimum of three consecutive batches.

Process Validation

With the completion of the installation qualification, operation qualification and performance qualification, the equipment is considered qualified and process validation can proceed. A master validation plan should be developed that

- Describes the validation technique; retrospective, concurrent, or prospective;
- Indicates the processing steps that require validation, and
- Establishes a schedule for completing each validation.

Usually the master validation plan covers the processing steps that impact excipient quality from the point in the process where full GMP compliance begins.

The most accepted validation method is prospective. This validation approach relies on completion of the validation before commercial production begins and requires the manufacture of at least three consecutive batches during protocol execution. The batches are evaluated for conformance to the protocol requirements; a report is prepared and approved. Then the lots are released for sale and production commences using the validated process. For excipient manufacture, where the material has been produced for quite some time, this approach is usually inappropriate.

Concurrent validation is another accepted approach to executing the protocol. In this method, again a minimum of three batches is produced consecutively. However, as each batch is produced, it is evaluated against the protocol requirements and if in conformance the batch is released. However the test requirements for concurrent validation are usually more stringent than for prospective validation so as to minimize the risk of releasing a batch whose quality or performance might be unacceptable. Therefore concurrent validation will normally require testing that goes beyond routine finished product analysis. This additional testing can include performance tests, measures of physical or chemical properties not normally evaluated, tests using methods with improved precision or lower detection limits, or development of the impurity profile. Once the designated number of batches has been produced consecutively, a report is prepared and approved and validation of the process step has been completed.

The validation approach excipient manufacturers prefer is retrospective because of the preponderance of production data they have available to support the hypothesis that the process operates reliably. In order for a process or process step to be considered for retrospective validation, it should have operated for at least one year with

1. No significant process change,
2. No significant equipment change, and
3. High process capability ($C_{pk} \geq 1$)

If these criteria are met, the retrospective validation protocol should require the evaluation of a minimum of 30 consecutive batches for conformance to production requirements and quality measures. Again the data is gathered and a report is written.

Once all of the processing steps that are in the Master Validation Plan have been validated, i.e., their validation reports have been approved, a summary report can be prepared, supporting the conclusion that the process is validated.

Cleaning Validation

The final type of validation applicable to manufacturing is cleaning validation. The purpose of cleaning validation is to confirm the hypothesis that the cleaning technique is effective in removing manufacturing residue from the designated equipment. Cleaning validation requires that workers clean the equipment in a repetitive manner and thus there must be detailed instructions as to how they should clean the equipment as well as records to confirm the instructions were followed.

Cleaning validation begins with a protocol that describes the

1. Equipment to be cleaned,
2. Cleaning procedure,
3. Cleaning materials,
4. Test method for evaluating the cleanliness of the equipment,
5. Cleanliness acceptance criteria,
6. Identification of the most difficult area of the equipment to clean,
7. Identification of the most difficult product to clean from the equipment (where the equipment is used for multiple products), and
8. Number of replicates to show the cleaning technique has been validated.

Usually the most difficult aspect of cleaning validation is in determining how to evaluate the efficacy of the cleaning method. Equipment should be sufficiently clean so that the incidental carryover in the first batch after cleaning presents an acceptable risk to excipient quality and performance. Once this determination has been made, it is possible to calculate the maximum amount of residue carried over into the excipient batch. Then a calculation can be made as to how much residue can be left on the equipment surface, assuming the residue is uniform throughout the equipment.

Evaluation of the cleanliness of the equipment involves measuring the residue left on a known area of the equipment. This is done by marking off a known area and then swabbing the area with a good solvent for the residue. Laboratory measurements quantify the amount of residue in the swabbed area after cleaning so that the quantity of residue left in the equipment can be estimated. If the estimate of the residue is below the maximum calculated from the risk analysis above, then the equipment has been adequately cleaned.

An alternative to determining the quantity of residue left in the equipment is to monitor the effluent of cleaning solution for the presence of residue. If it can be shown that the residue is readily soluble in the cleaning solution and the test method is sufficiently sensitive, the acceptance criteria for cleanliness might involve washing until the residue drops below the quantifiable limit of the test method or reaches an acceptably low steady state in the effluent.

Once the cleaning validation has been completed, equipment can be cleaned in a routine manner. This involves records to show that the cleaning followed the proscribed instructions. Also there should be an evaluation of the efficacy of cleaning which can involve documentation of a visual inspection.

Process Monitoring and Control

In-process monitoring of the production process is usually a combination of on-line measurements and sampling. Any instrument used to make a measurement whose result either indicates the quality of the output of the production step or is used to adjust operating parameters for the purpose of affecting product quality must be made using a calibrated instrument. The calibration program requires that each instrument

- Be uniquely identified,
- Have a schedule for calibration,
- Have a calibration procedure,
- Be calibrated against NIST traceable calibration standards,
- Be calibrated by qualified individuals, and
- Have a record of each calibration that includes, in addition to the information denoted above, the findings of the calibration.

It is also beneficial to tag the instrument so that it can be labeled with the date the next calibration is due. This allows the operator to verify the instrument is within its calibration interval.

Sampling and Testing

In-process sampling should be done in accordance with a written procedure. The Standard Operating Procedure (SOP) or production instruction should describe

- When in the processing, a sample is to be taken,
- Where from the equipment the sample is to be taken,
- How the sample is to be taken such as draining a fixed quantity from a sampling point or recirculation through a sample loop for a fixed interval before taking the sample,
- What sampling device is to be used for taking the sample and how the device is to be cleaned,
- The quantity of sample to be taken,
- A description of the container into which the sample is collected, and
- The content of the sample label.

The results of in-process sample testing are important to confirming the proper operation of the equipment. Therefore it is important to ensure the sampling device, wherever used, is clean. If cleaning after each use is not feasible or it is difficult to keep it clean until next used, then the sample device must be stored properly to protect it from environmental contamination.

In-process samples are usually tested at-line or in the Quality Control laboratory. If the testing is performed by production personnel, the following requirements should be met:

- There should be documented production personnel training in the method.
- Training of production staff should be done by qualified trainers.
- Instructions for testing should be readily available.
- Test records should be properly completed.
- There should be periodic evaluation of the quality of testing by the production personnel through techniques such as observation by qualified individuals of the production personnel performing the test or assessment through production personnel making the measurement on a reference sample.

If the sample is tested in a Quality Control laboratory, then the testing should be done to laboratory GMP requirements. If the QC test result is reported back to production, it should be done in writing. Usually this involves a remote printer operated from the laboratory but it can be accomplished through other means such as facsimile or e-mail.

There should be a sampling plan for finished excipient testing. The sampling plan should be in accordance with the in-process sampling requirements outlined above. It is important that if only one sample from the finished batch is taken, that it be representative of the excipient. This is usually not possible if the lot is not homogeneous. In that event, it is important to establish a sampling plan that assures sufficient samples are taken so that with all samples in conformance, the entire excipient batch is known to be conforming.

Deriving a sampling plan for nonhomogenous excipients begins by sampling every excipient package. The variation in the test results from each package when compared to the specification limits will guide the preparation of a sampling frequency adequate to assure that the excipient batch is conforming. The more centered within the specification range that the process operates, the less frequently the excipient packages in the batch should be sampled.

In the event that one or more samples from the finished excipient batch are found nonconforming, at least a portion of that batch has been shown not to meet excipient grade requirements. Downgrading the entire batch to a lesser grade, such as food, personal care, or industrial, if it meets those specifications, is the most conservative disposition for the batch. However it would be acceptable to cull out all material from the last package whose test result was conforming to the first subsequent package that provides once again a conforming test result. Under this scheme, only the packages from which nonconforming results were obtained are culled from the excipient lot. The consequence of this approach is excipient lots of varying quantity.

A suitable remedy to nonhomogeneous batches is blending to uniformity. Such blending operations must be conducted using detailed procedures with records to demonstrate the blending instructions were followed. Blending operations must be validated to confirm the hypothesis that when the instructions are followed, a homogeneous batch is the result.

Certificate of Analysis

Where one sample representative of the excipient batch is used for QC testing, the results of those tests are reported on the COA. Where multiple samples from the finished excipient batch are taken and individually tested, the matter of reporting results on the COA must be addressed. There are two approaches for dealing with multiple batch sampling. The first approach is to report test results from a specified sample on the COA. The procedure might therefore identify, for instance, that either the least conforming test result, indicative of the least conforming excipient sample, be reported on the COA or the average of the test results from the individual samples is reported.

The alternative approach is to prepare a composite sample from the individual batch samples and from that take a sample for finished excipient release testing. This approach requires combining equivalent quantities from the individual batch samples into a container. Once all of the material has been added from the individual samples, the composite sample must be blended to make it homogeneous. Once the composite has been adequately blended, a representative sample must be taken for finished excipient testing. These test results are then reported on the COA.

Packaging and Labeling

Packaging and labeling operations present a significant risk to the quality of the excipient or its regulatory compliance with GMP requirements. The risk to excipient quality results from excipient exposure during packaging operations to airborne contaminants as well as mold, yeast, and humidity. To protect the excipient, the filling of finished excipient containers should take place in an enclosed area, which is maintained under positive pressure using filtered air. This area should be kept clean to prevent the excipient from becoming contaminated. Therefore the walls, floor, and ceiling should be washable to allow frequent cleaning. Measures should be taken to ensure the package exterior is free of extraneous dust and dirt before being brought into the packaging room such as wiping down containers or removing shrink-wrap covering the packages. Contact packaging should not be opened until brought into the packaging room. If the contact packaging is a bag or if the excipient is packaged into a bag, which is then placed inside a drum, the bag should not be opened until it is inside the packaging room. Containers received closed should not be opened until they are in the controlled environment packaging room; otherwise they will be exposed to airborne contamination. Filled containers and bags must be closed or sealed while still in the controlled environment.

Employees involved in packaging operations should wear appropriate attire. Where the excipient is exposed during packaging and might become contaminated by the packaging operator, the operator should wear clean clothing or a disposable outer garment. The operator should also wear head covering such as a hard hat or hairnet and where necessary, a covering for facial hair. Where the operators' hands may come into contact with the excipient, the employee should wear clean gloves, preferably disposable gloves. Finally if there is a risk from the operator breathing on the excipient, the employee should wear a face mask.

Proper packaging and labeling procedures ensure regulatory compliance to GMP requirements. It is necessary to limit packaging operations in the same area to one at a time. The packaging room should be used to package a single batch of excipient so as to prevent mix-ups that might occur from labeling multiple batches. Prior to the start of packaging, the room should be inspected to ensure there is no packaging or labeling materials leftover from prior operations, especially printed labels with lot numbers or lot labeled excipient containers. Once the room has been inspected, the packaging and labeling materials for the upcoming operation can be brought into the area.

Excipient labels are typically either printed on demand or purchased preprinted. Where labels are purchased, the Quality Unit should approve their receipt by matching the incoming label content against an approved reference label. Where labels are printed on-demand, the Quality Unit, or their designate, should compare a printed label against the approved reference label to assure the printed content is correct. The approved reference label is one that has been carefully reviewed by the organization responsible for label content and design. The initials and date of the responsible individual affixed to the reference label provide confirmation that the label is an approved reference.

Label usage must be reconciled to assure there are no stray labels. Whether printed on-demand or preprinted, labels should be issued to the packaging operators with a label reconciliation form. The form will indicate the label by reference identification and if the labels were printed on-demand will indicate the printed labels have been reviewed and approved by the Quality Unit or their designate. The reconciliation form will also indicate the total number of labels issued. As the labels are used, they will be accounted for on the form. Thus there should be an entry for the number of labels affixed to excipient packages, the number of labels damaged during application and thus discarded, the number of labels left unused and thus destroyed, and finally one label, which is attached to the packaging or batch record for reference. The sum of these should equal the total number of labels issued.

Preferably labels are applied either right before or after the container has been filled. One consideration is the ability of the label to adhere to the filled package, which may now have excipient on the exterior. However if containers are labeled after filling, procedures should ensure there is never more than one container of packaged excipient lacking a label at any time so as to prevent mix-ups. Alternatively, at the start of the operation, labels can be affixed to the minimum number of containers that are expected to be packaged so as not to have any unused lot labeled excipient packages after the entire batch has been packaged. In the event that there are unfilled labeled packages, they will have to be discarded.

The containers are filled in the packaging room and then sealed or at least closed in the room. If the weight of the container must be adjusted to meet the net weight tolerance, it must be done in a similarly controlled environment. The excipient used to adjust the container net weight must be from the same lot. Once the containers are at least closed if not sealed, the excipient package can be moved outside the controlled environment packaging room for final sealing. Once the package is permanently closed, a tamper evident device should be affixed. This is especially necessary when the outer package is a box or drum. For a box, a security tape is usually sufficient as long as the tape is personalized and not generic. Where the container is a drum, security tapes or seals on the clamping ring are needed. With the ever-present risk of product tampering, the use of tamper evident seals should be evaluated so as to enable recognition by the recipient that the container was opened.

Packaging of an excipient batch usually results in some excess quantity of material insufficient to make a full excipient package. There are a number of options concerning what to do with this excess excipient. It is acceptable to

1. Reprocess the material into a subsequent batch of excipient. The material should be reprocessed into the batch prior to the purification step. The quantity and batch identity of the reprocessed material should be noted on the batch record into which the excess material was added.
2. Combine the residual excipient from multiple batches into a new batch. This can be done either by combining the residual excipient and reprocessing it through the purification step or by putting the residual excipient in a blender to make a new batch that is homogeneous. The blending requires the use of a validated blending operation.
3. The residual quantities of excipient can be downgraded to a lesser grade and reprocessed or sold as nonexcipient grade.

Quality Release

The final step in producing an excipient batch is Quality Release, which involves the verification that all necessary records are properly completed for the batch in question. Upon completion of this task and with approval from the Quality Unit, the lot of excipient is suitable for sale.

The Quality Unit or their designate should review the records of the following operations:

1. Production batch record or logbooks
2. Packaging record
3. Label reconciliation record
4. QC test record

The objective is to verify that each record has been properly completed with signatures or initials and time where stipulated and that it has been reviewed by supervision as required. The Quality Unit can perform the review by having the records assembled for their review or they can perform the review of the records at the various departments wherein the records are maintained. Upon completion of the review and with acceptable findings, the Quality Unit then changes the lot status, or directs the lot status be changed to indicate the lot is available for sale.

Storage

The last step at the manufacturing facility is storage of the finished excipient. The warehouse should be properly maintained so that the packaged excipient is not exposed to water or sunlight, which might cause deterioration of the package or its label. The warehouse must be well lighted and organized to facilitate finding the lot that is specified on the shipping paperwork. Where computer quarantine of unreleased excipient is used, it is especially important that the location of excipient lots in the warehouse be accurate. Also the warehouse must be kept clean which requires that containers be kept at a sufficient distance from adjacent containers and perimeter walls for effective cleaning around stacked containers.

The excipient must be stored following the manufacturers recommended storage conditions. Where temperature or humidity is specified, it is necessary to provide appropriate controls in the warehouse along with records to show compliance. If normal warehousing conditions are appropriate, the warehouse can be kept under ambient conditions and no monitoring of temperature or humidity is required.

It is also important that the area where finished excipient is stored be under an insect and rodent control program. Typically, the site contracts with an exterminator to perform monthly inspections for evidence of insect or rodent activity. The exterminator should only use Food Drug and Administration approved materials in controlling these pests. The exterminator should tour the site following a proscribed path usually defined on a site map. The exterminator should provide a report of his/her findings to the site. This report can be in the form of a notation on the paperwork left with the facility host by the exterminator as a record of their visit.

A final note on excipient manufacturing is needed. If an excipient package is to be sampled for any reason and then the excipient package is to be resealed and the material sold as excipient grade, the sampling must take place under the same environmental conditions as the original packaging operation. It is usually inappropriate to open the excipient package and sample the contents in the warehouse.

SUMMARY

This chapter is meant as a survey of the manufacturing issues in the production of pharmaceutical excipients. As such many important decision points have been discussed. It is a basic tenet of GMP compliance and a principle of auditing that if it is not written, it did not happen, which is a basic assumption used by pharmaceutical regulatory auditors. It is therefore essential that decisions reached in relation to GMP compliance be documented. While proper documentation presenting the logic behind a compliance decision will not ensure that regulatory authorities will agree with the decision reached, it will demonstrate to the regulators that the issue was given consideration. As such, the consequences of an adverse finding by the regulator will usually be mitigated and the regulatory action may be more limited.

REFERENCES

1. USP < 1078 > Good Manufacturing Practices for Bulk Pharmaceutical Excipients.
2. The IPEC Good Manufacturing Practices Guide for Bulk Pharmaceutical Excipients, 2001.

22

Excipient Quality Assurance: Handling, Sampling, and Regulatory Issues

Raafat Fahmy
Office of New Animal Drug Evaluation, Center for Veterinary Medicine, Food and Drug Administration, Rockville, Maryland, U.S.A.

Stephen W. Hoag
School of Pharmacy, University of Maryland, Baltimore, Maryland, U.S.A.

INTRODUCTION

One of the Food and Drug Administration's (FDA) key mandates is to ensure that drug products are safe and effective, and for excipients, the pharmacologically inert ingredients that constitute most of a drug product, a key safety issue is purity. Unfortunately, there are many well-known examples of patient harm that resulted from excipient impurities. For example, in 1996, the CDC reported that 86 children in Haiti died when given a pediatric acetaminophen syrup containing glycerin contaminated with diethylene glycol (1). This glycerin was apparently manufactured in China (China would not let investigators verify this) and shipped through Europe to Haiti, using a supply chain with several participants. The product was shipped with no traceability and with insufficient controls and documentation needed to avoid or even reduce manipulation and alteration of the material. This tragic example illustrates what can happen when the components of a drug product are improperly handled. The fact that this incident happened in Haiti may lead some manufacturers to say it could not happen here in the United States, but as the excipient business becomes more global and more of a commodity business, increased care will be required to ensure that similar incidents do not happen.

Section 501 [21 U.S.C. 351] of the Federal Food Drug and Cosmetic Act (FFDCA) act describes the conditions under which a drug or device may be deemed to be adulterated. One of the requirements of the act is that the drug be manufactured under current Good Manufacturing Practices (cGMP). Pursuant to this provision of the federal FFDCA act, the regulations that follow describe the intent of this provision, which is to ensure the finished drug product meets the expected safety, identity, strength, purity, and quality characteristics that it purports or is represented to possess.

For finished drug products, cGMP under 21 CFR, Parts 210 and 211, refers to the finished drug products. Section 501(a)(2)(B) of the FFDCA requires that all drugs be manufactured, processed, packed, and held in accordance with cGMP. Any pharmaceutical dosage form must be manufactured according to the current GMP regulations. The pharmaceutical manufacturer is responsible for producing a safe and effective product that includes the quality of components: the active ingredient(s), the excipients, and the packaging materials.

In addition to regulatory issues, a manufacturer will also have compendial and noncompendial excipient requirements that are of critical importance to the manufacturability of a particular drug product. For example, a particular particle size distribution, viscosity grade, or hydrate may be needed for the successful and reproducible manufacturing of a safe and effective product. This chapter focuses on the issues needed to assure that the excipients used in drug product manufacturing are really what they are intended to be, i.e., that they meet manufacturing specifications and regulatory requirements. This chapter is divided into three sections: (i) the regulation of excipients, (ii) sampling for testing, and (iii) the use of near infrared (NIR) and chemometrics for excipient identification (ID) testing. To fully cover all aspects of excipient acceptance is beyond the scope of this chapter, and this presentation should not be thought of as a comprehensive treatise on the subject; however, many of the key issues are discussed below. In addition, it should also be noted that the authors have directly quoted federal regulations and the United States Pharmacopeia (USP), where applicable.

REGULATORY ASPECTS OF EXCIPIENT QUALITY ASSURANCE

As a part of the pharmaceutical manufacturers' quality systems, all the manufacturers of the finished dosage forms should qualify their raw materials suppliers; the manufacturers can use several avenues such as auditing, testing, certificate of analysis (COA), questionnaires, or third-party audits to qualify their suppliers and assure the quality of their excipients. Recently, the International Committee on Harmonization (ICH) has produced the Q7A guideline describing the GMP standards for active pharmaceutical ingredients (APIs). In the European Union, GMP guidance may also be made applicable to certain excipients, where the excipient plays a critical part in assuring the safety and effectiveness. The excipient manufacturers can use the specific guidance, issued by International Pharmaceutical Excipients Council of Americas (IPEC-Americas) for defining the GMP standards required in their manufacturer. However, GMPs used by excipient manufacturers should be compatible with the quality systems and GMPs used by pharmaceutical manufacturers.

CERTIFICATE OF ANALYSIS

The goal of the COA is to assure that the materials meet the expected quality criteria as per the Code of Federal Regulation (CFR) 211.84(d)(2). "Each component shall be tested for conformity with all appropriate written specifications for purity, strength, and quality. In lieu of such testing by the manufacturer, a report of analysis may be accepted from the supplier of a component." The COA is also discussed in the USP, see General Chapter <1078>. The COA should be attached to any excipient shipment, and is always generated by the original manufacturer of the excipient and must include the following information:

- Supplier's name and address
- Name of product-U.S.P/National Formulary (NF) designation

- Lot number
- Production date
- Specification and acceptance criteria of the product
- Test method used and reference to the analytical procedures
- Actual analytical results
- Supplier's signature and date

All the acceptance criteria and test results are best expressed numerically or qualitatively (e.g., clear, colorless solution), as appropriate. The use of terms such as "conforms" or "meets specification" is discouraged.

When the specification for a raw material excipient is compendial and conforms to the monograph standard, a citation to the appropriate official compendium needs to be provided. The excipient specification is expected to be identical to the compendial monograph and full monograph testing will be performed on each batch of excipient by the excipient's manufacturer. At a minimum, the drug product manufacturer must perform an appropriate ID test [21 CFR 211.84(d)(1)], and for materials held in inventory, full monograph testing is expected once a year. However, when there are specific safety concerns relating to an excipient, testing in addition to an identity test would be warranted.

When the specification for a compendial excipient differs from the compendial monograph (e.g., additional tests, different analytical methods, or different acceptance criteria) the test results will be accepted from the excipient manufacturer's COA. However, the excipient should still conform to the monograph in an official compendium if there is such a monograph; otherwise, justifications must be provided, and labeling needs to be changed to state plainly that the article does not meet the compendial requirement.

The drug product manufacturer must establish the reliability of the supplier's analyses through appropriate validation of the supplier's test results at appropriate intervals per [21 CFR Part 211.84(d)(2)]. The reliability of the analyses need not be established at the time the application is submitted. However, the specification should indicate that the tests that will be performed once the reliability of the supplier's results has been established in accordance with cGMPs, prior to marketing the drug product.

A compendial excipient should comply with the current revision of the official compendium cited. Therefore, when citing an official compendium, the version of the compendium should not be included in the citation. For example, the NF should be cited rather than NF 20. Certain General Chapters in the USP contain a statement that the text of the USP is harmonized with the corresponding texts of the European Pharmacopoeia and the Japanese Pharmacopoeia. However, where a difference appears, or in the event of a dispute, the result obtained from the USP procedure is conclusive.

Additional guidance is available in:

ICH: *Q6A Specifications: Test Procedures and Acceptance Criteria for New Chemical Substances and New Drug Products.*

Functional and Noncompendial Testing

It is important to assure consistency from lot-to-lot for the desired functionality. While the compendial monograph focuses on chemical characteristics and demonstrates the safety of the raw material, some excipients require functionality tests to assess their performance in the finished drug product. These tests reflect the physical characteristics of the excipients such as particle size or density per 21 CFR Part 211.84(d)(4). These tests may not be included in the COA unless the drug product manufacturer specifically requests them from the supplier.

ADDITIONAL GUIDELINES

USP <1078> provides the following guidelines for the excipient manufacturer and the purchaser to use in establishing standards for excipient materials provided.

Contract Review

The manufacturer and user should mutually agree upon the excipient specifications. The manufacturer must have the facility and process capability to consistently meet the mutually agreed upon specifications of the excipient(s). Subcontracting or significant changes to a supplier's audited process that could affect the physical, chemical, or functionality of the excipient in a final dosage form should be immediately communicated or pre-approved as mutually agreed upon between customer and supplier.

Purchasing

The purchaser should verify that the supplier of raw materials, components, and services for the manufacture of excipients has the capability to consistently meet the agreed-upon requirements. This may include periodic audits of the vendor's plant, if deemed necessary. Purchasing agreements should contain data clearly describing the product ordered, including where applicable, the following:

- The name, type, class, style, grade, item code number, or other precise identification traceable to the raw material specification.
- Drawings, process requirements, inspection instructions, and other relevant technical data, including requirements for approval or qualification of product, procedures, process equipment, and personnel. These requirements also apply to selection and control of subcontractors. Subcontractors include toll manufacturers and contract laboratories.

Control of Customer Supplied Products

The manufacturer should establish and maintain procedures for verification, storage, and maintenance of customer supplied products intended for incorporation into the customer's excipients.

RECEIPT, SAMPLING, TESTING, AND APPROVAL OF RAW MATERIALS

While the COA is the excipient manufacturer's responsibility, once the material is received, it is the drug product manufacturers' responsibility to verify the product and ensure that it is properly tested, handled, and stored. Upon receipt of a shipment, each lot of excipient will be withheld from use until the lot is sampled, tested, or examined according to the written procedures. The quality control (QC) personnel will examine each container for (i) manufacturer's name, (ii) manufacturer's lot number, (iii) leaks or spills, (iv) contamination, (v) breached containers, (vi) proper labeling, and (vii) material safety data sheet and determined material hazards.

Representative samples of each shipment must be collected for testing as required by § 211.84. (b). The number of containers to be sampled depends on the (i) component variability, (ii) confidence level, (iii) degree of precision desired, (iv) past quality history of the supplier, and (v) quantity needed for analysis and reserve samples. For hazardous or highly toxic raw materials, where on-site testing may be impractical, suppliers' COA should be obtained, showing that the raw materials conform to specifications. In addition, the identity of these raw materials should be confirmed by examination of containers and labels. The lack of on-site testing for hazardous raw materials should be documented.

Each bag or container of raw materials should be identified with a unique code, lot, or receipt number. This code should be used in recording the disposition of each lot. Raw materials will be held under quarantine until they are sampled, tested, and released. Raw materials should be carefully handled and stored to avoid any contamination or cross contamination. When bagged and boxed raw materials need to be stored, it must be done so in adequately cleaned buildings that are free of infestation by rodents, birds, insects, and other vermin, and the building should be maintained. A controlled environment may be necessary to avoid microbial contamination or degradation caused due to exposure to heat, air, or light. When the raw materials are stored outdoors, the containers should be adequate for the outdoor storage.

Representative samples of each lot will be collected for testing in accordance with the established procedure. The number of containers to sample and the sample size should be based upon appropriate criteria as required by § 211.84 (3) (e.g., raw material variability, confidence levels, degree of precision desired, past quality history of the supplier, and the quantity needed for analysis). Sample containers should be properly labeled. The label should include, at minimum, information about the sample name, manufacturer name, sample size, and date and hour of sampling. Raw material containers selected for sampling should be opened, sampled, and resealed in a manner that prevents contamination of their contents and of other raw materials.

Materials are held in quarantine until the QC approves the product. Once the product is approved, the QC inspector will stamp the material as released and the product is transferred from quarantine to the available inventory area for use. If the QC lab rejects the raw material, it will be properly identified and moved into a separate indentified area. The system should be designed to avoid using the rejected materials in the operation. All the acceptable raw materials should be stored under appropriate conditions; the drums will be rotated so that the oldest stock is used first. Raw materials should be reevaluated, as necessary, to determine their suitability for use (e.g., after prolonged storage or after exposure to heat or high humidity).

Additional guidance is available in the following:

21 CFR 211.113(b): Control of microbiological contamination
21 CFR 211.122: Materials examination and usage criteria

EXPIRATION OR RETEST DATING

To ensure safe and effective finished drug products, the excipients must be stable. Some excipients may be stable and may not require extensive testing, while others may be less stable and require more scrutiny. A retesting or expiration date should be identified on the container label and the COA of the raw material at the time of use. Expiration or retest dates should relate to any storage conditions stated on the label and should be supported by appropriate stability studies.

Packaging and Labeling Control

Storage conditions should be specified on the label of raw material containers. Detailed written procedures describing the receiving, handling, identifying, storing, and testing of the raw materials should be in place. Raw material container labels should include material name, supplier's name, lot number, storage conditions, retesting date, and any other cautions or hazards. Labeled storage conditions should comply with standard definitions for "Controlled Room Temperature," "Cold," or "Freezer," as defined in the USP or guidelines of the ICH. Statements of specific storage conditions should be used instead of more general terms such as "room temperature," when it is critical for maintaining the quality of the raw material. The label on each container should also include any warnings to protect the contents from excessive heat, freezing, light, or moisture. Any labeling or packaging materials that do not meet the specifications should be rejected.

RECORDS

Records for excipients should include the following [information as indicated in 211.184]: (i) The identity and quantity of each shipment of each lot of components, drug product containers, closures, and labeling; the name of the supplier; the supplier's lot number(s) if known; the receiving code as specified in § 211.80; and the date of receipt. The name and location of the original manufacturer, if different from the supplier. (ii) The results of any test or examination performed including those performed as required by [21 CFR Part 211.82(a) and Part 211.84(d)] and the conclusions derived. (iii) An individual inventory record of each component, drug product container, and closure, and, for each component, a reconciliation of the use of each lot of such component. The inventory record should contain sufficient information to allow determination of any batch or lot of drug product associated with the use of each component, drug product container, and closure. (iv) Documentation is required of the examination and review of labels and labeling, for conformity with established specifications in accordance with the [21 CFR Part 211.122(c) and 211.130(c)]. (v) The system should be able to document the disposition of rejected components, drug product containers, closure, and labeling.

Each batch of excipients is identified by the receiving code bar as specified in 21 CFR Part 211.80. The code bar will be used in recording the disposition of each lot. Each lot is indentified as to its status (i.e., quarantined, approved or rejected) and the date of receipt. As stated in CFR211.180, all components, production, control and distribution records that are associated with any drug product will be retained for at least one year after the expiration date of the drug product and should be available for any authorized inspection.

TRACEABILITY

With the globalization of the supply system and more open markets, traceability is required to enable everyone in the drug manufacture's supply chain to control and investigate any problems that may arise in one of their products, which may be caused by or linked to any excipient. A traceability system is required by cGMP. Cell excipients should be traceable and the drug product manufacturer may need to trace the origin of the excipient to assess whether it is appropriate for the intended use or not.

ANALYTICAL PROCEDURES

Analytical procedures and references should be available if the analytical procedure used is in the current revision of an official compendium or another FDA-recognized standard reference (e.g., AOAC International Book of Methods), and the referenced analytical procedure is not modified. However, the validated analytical procedures for novel excipients should be provided.

Additional guidance is available in the following:

ICH: *Q2A Text on Validation of Analytical Procedures*
ICH: *Q6B Specifications: Test Procedures and Acceptance Criteria for Biotechnological/Biological Products.*

LABORATORY CONTROLS

Laboratory controls should include the establishment of scientifically sound and appropriate specifications, standards, sampling plans, and test procedures to ensure that raw materials and containers conform to established standards of quality and purity. Specifications, standards, sampling plans, test procedures, or other laboratory control mechanisms, including any minor changes, should be updated by the appropriate organizational unit and reviewed and approved by the QC unit. Laboratory controls should be followed and documented at the time of performance. Deviations from written specifications, standards, sampling plans, test procedures, or other laboratory control mechanisms should be documented and justified.

Procedures should be established to determine conformance to appropriate written specifications for the acceptance of each lot of raw material, containers, intermediates, and APIs. Such procedures should also cover appropriate sampling and retesting of any materials used in the manufacturing or holding of an intermediate or API that are subject to deterioration or degradation. Laboratory test samples should be representative, properly handled, and adequately identified.

A program should be in place for calibration and qualification of each instrument, apparatus, gauge, and recording device at suitable intervals. The program should contain specific directions, maintenance schedules, limits for accuracy and precision, and provisions for corrective action in the event that accuracy and/or precision limits are not met. Instruments, apparatus, gauges, and recording devices not meeting established specifications should not be used. Secondary laboratory reference standards (e.g., production lots that are further purified and qualified in the laboratory) should be appropriately prepared, identified, stored, and tested, as necessary, to ensure their suitability for use. The suitability of each lot of secondary reference standard should be determined prior to use by comparing against a primary reference standard.

Analytical reagents used in testing the excipients should be prepared and labeled following established procedures. Retest or expiration dates should be used, as appropriate, for analytical reagents, or standard solutions. Analytical methods should be validated unless the method employed is set forth in the current revision of *the United States Pharmacopeia/National Formulary, Association of Official Analytical Chemists (AOAC), Book of Methods*, or other recognized standard references, or detailed in the Drug Master File or approved New Drug Application and are used unmodified.

Laboratory Records

As per 21 CFR Part 211.194, laboratory records should include the following information. All data should be included in laboratory records. The sample should contain the quantity, lot number or a distinctive code, the date the sample was taken, and the date the sample was received for testing. All methods used to test this sample should be recorded, and any other additional statements should be recorded. A statement of the weight of the sample should also be included. All the specific components should be indicated through any type of preferred data analysis, in addition to the calculations which are to be submitted. The date and a signature of the person who completed the test should also be recorded and submitted.

All records are to be kept. If modified, a reason and a summarized statement should be given as to why the modification is taking place to verify that the current modifications support the validity of the testing. Keep complete records of labortory standards. In addition, complete records should be kept of periodic calibration of laboratory instruments, apparatus, recording devices, and gauges.

EXCIPIENTS OF HUMAN OR ANIMAL ORIGIN

Excipients of human or animal origin should be identified. The genus, species, country of origin, source (e.g., pancreas), and manufacturer or supplier should be clearly indicated. Furthermore, for excipients derived from ruminant materials, the application should state whether the materials are from bovine spongiform encephalopathy countries as defined by the U.S. Department of Agriculture (9 CFR 94.11).

Guidance is available from the FDA on *The Sourcing and Processing of Gelatin to Reduce the Potential Risk Posed by Bovine Spongiform Encephalopathy (BSE) in FDA-Regulated Products for Human Use*. The potential adventitious agents should be identified, and general information regarding control of these adventitious agents (e.g., specifications, description of the testing performed, and viral safety data) should be provided in this section. Details of the control strategy and the rationale for the controls should be provided.

Additional guidance is available in the following:

ICH: Q5A Viral Safety Evaluation of Biotechnology Products Derived From Cell Lines of Human or Animal Origin

ICH: Q6B Specifications: Test Procedures and Acceptance Criteria for Biotechnological/Biological Products

SAMPLE COLLECTION

Acquiring a representative sample from a bulk powder lot is a difficult procedure that requires special consideration; the basic procedures for acquiring a representative sample are discussed below. It should be noted that every situation requires techniques that are appropriate for the given population to be sampled. The methods presented here are applicable to the sampling of static powders stored in midsize bulk containers such as 1 ton "super sacks," 50 kg drums, or 50 pound bags. These methods are not necessarily applicable for the sampling of liquids, large storage containers such as train cars or silos, and in-process systems such as a blender or a moving conveyer belt. In addition, the procedures described here are for particles in the approximately 1 to 1000 μm size range; significantly smaller or larger particles require special procedures not covered below.

Before sampling any excipient there must be a documented sampling plan in place. The purpose of a sampling plan is to obtain a representative sample of a population from which reliable inferences about the population sampled can be drawn to a certain level or degree of confidence. Acquiring a representative sample from a lot is critical to all future analyses; without a representative sample all further analyses and data interpretations about this lot are in doubt. A perfect sample[a], as defined by statisticians, is a sample in which every member of the sample population has an equal probability of being in the sample. In addition, the sampling procedure needs to be reproducible, i.e., if the sampling protocol was repeated there should be a high probability of obtaining similar results. At first glance, this seems like a simple requirement, much like flipping a coin or drawing colored marbles from an urn. However, sampling real-world powder systems that do not contain well-defined discrete units and have a propensity to segregate can make obtaining a perfect sample very difficult. Thus, to obtain representative samples, one must carefully develop a sampling plan that accounts for and mitigates the segregation tendencies of a particular powder system. Thus, the goal of this section is to outline the steps that are needed to develop a sampling scheme or plan for a particular system, which is consistent with good sampling practices.

The primary difficulty with acquiring a representative sample is that the measurement sample, typically a few milligrams to grams, must be withdrawn from a large population on the order of hundreds to thousands of kilograms; i.e., the few milligrams analyzed in a laboratory must be taken from a large population of particles in a warehouse in such a manner that the measurement sample is representative of all the particles in the lot. Any bias or error in the sampling process will cause all future inference to be in error. Over the years, methods have been developed and refined to ensure that the measurement sample is representative of the whole population (2,3). A typical strategy is shown in Figure 1: the strategy is to sample in stages, starting with the initial gross or primary sample, which is withdrawn directly from the received containers. In the laboratory, the gross sample must be reduced in size until it is the appropriate size for measurement. This should be done in a manner that minimizes the introduction of sampling errors. Randomness is the key to reducing the sampling error because it assures that every element of the sample has an equal probability of being included in the sample; however, due to the segregating or nonrandom nature of powders, there are many pitfalls that can cause bias and contribute to the sampling error. Following the flow chart shown in Figure 1 and the subsequent discussion will help to minimize sampling errors.

DEVELOPING A SAMPLING PLAN

Step 1: Sample Size Selection

Sample size selection can either be statistically based or arbitrarily chosen. Arbitrary sampling plans are plans in which the number of samples is arbitrarily set at some

[a] This definition neglects systems where different members of a population have different probabilities of occurrence.

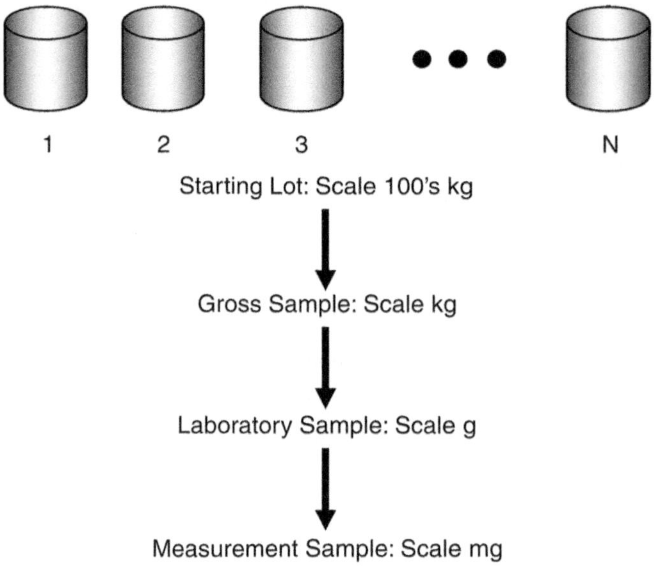

Figure 1 Overall sampling strategy for reducing 100 kg to measurement sample of mg.

number between 1 and N. Statistically based sampling plans are based upon statistical principles and depend upon the population's heterogeneity and intrinsic variability. Statistically based plans offer many advantages over arbitrary sampling plans and are generally preferred. For example, statistically based plans are more efficient, allowing the collection of a sufficient number of samples to obtain the desired degree of certainty without collecting too many or too few samples. For example, the $\sqrt{N} + 1$ sampling plan often collects too few samples for small populations and too many samples for large populations. However, the primary advantage of arbitrary sampling is that sample size can be arbitrarily chosen with little forethought or a priori knowledge of the population to be sampled.

Statistical plans are preferred but not always necessary, for example, when determining identity, a nonstatistical plan will probably be sufficient; however, when determining a variable attribute such as particle size, it is better to use the statistically based sampling plans. In addition, if the population to be sampled is homogenous, using statistically based sampling methods can actually reduce the number of samples needed, which saves time and money. Statistically based samples are extensively used in other industries and are the subject of many standards set by AOAC, American National Standards Institute (ANSI) , American Society for Testing and Materials (ASTM), ICH, ISO 9000, and International Union of Pure and Applied Chemistry, to name a few. Thus, if you are going to use a statistically based method, please refer to the appropriate standard reference, such as ANSI/ASQ Z1.9–2003 for bulk materials or ANSI/ASQ Z1.4–2003 for multiunit or discreet populations; these standards are put out by the American Society for Quality (4).

Determining the sample size is often a difficult decision, and sometimes has to be made without sufficient information. If there is uncertainty, sometimes a quick small informal test of the system may help decide what is the best way to proceed.

Step 2: Sample Collection

Primary Sample

To collect a representative primary or gross sample, see Figure 1. First, the appropriate container or containers must be selected from the population of M containers. Second, a representative sample must be withdrawn from each of the selected containers.

Step 3: Container Selection

To avoid bias and other sampling errors, the containers to be sampled must be randomly selected. To make a random selection, first number all containers in the lot, and then use a random number table (or computer generated set of random numbers) to choose from which container or containers to withdraw the samples. See Appendix I for an example on the use of random number tables.

For systems with a large number of small units, it may not be practical to assign each unit a number, for example, withdrawing a sample from a thousand units[b] (e.g., tablets or capsules, or small unit packages). In these situations, practical compromises must be made; however, some effort must be made to insure randomness. For example, the top layer of tablets could be different from the tablets beneath. Ideally, one should carefully mix the tablets such that they are not damaged or spread them out, and then have another operator withdraw the sample in a manner analogous to closing their eyes and withdrawing colored marbles from a black bag or urn.

Step 4: Withdrawing Sample from a Container

There are many methods available for obtaining a sample from a powder system (5). Unfortunately, most of these methods involve either setting the powder bed in motion or in-process sampling. Due to concerns about cross contamination and containment of potentially toxic materials, most of these methods are impractical for the bulk sampling that is required for the cGMP and FDA regulations described above. Hence, most of the sampling done in the pharmaceutical industry is static sampling done via either (i) scoop or grab sampling (ii) or stratified sampling typically done using a sampling thief. The choice of either of these methods is dictated by the distribution of the attribute being sampled in the container; see below for further discussion.

Container Types

The three most popular container types are the bag, drum, and super sack. Generally bags are closed and not resealable; thus, special sampling thieves, sometimes called bag triers, have been designed to puncture the bag (Fig. 2). If the system to be sampled is heterogeneous, the samples should be obtained from the bottom, center, and top of the bag, and depending on how the bags are stacked on the pallet, they should also be sampled from the front and the back. When sampling from bags, particular attention should be given to the corners, because they can disproportionally trap certain types of particles such as fine particles. If no bag thief is available, then a knife can be used to cut open the bag for sampling. When sampling from a bag, the

[b] It would be far better to take this sample in-process to ensure randomness, but in-process samples are beyond the scope of this section and may not always be possible.

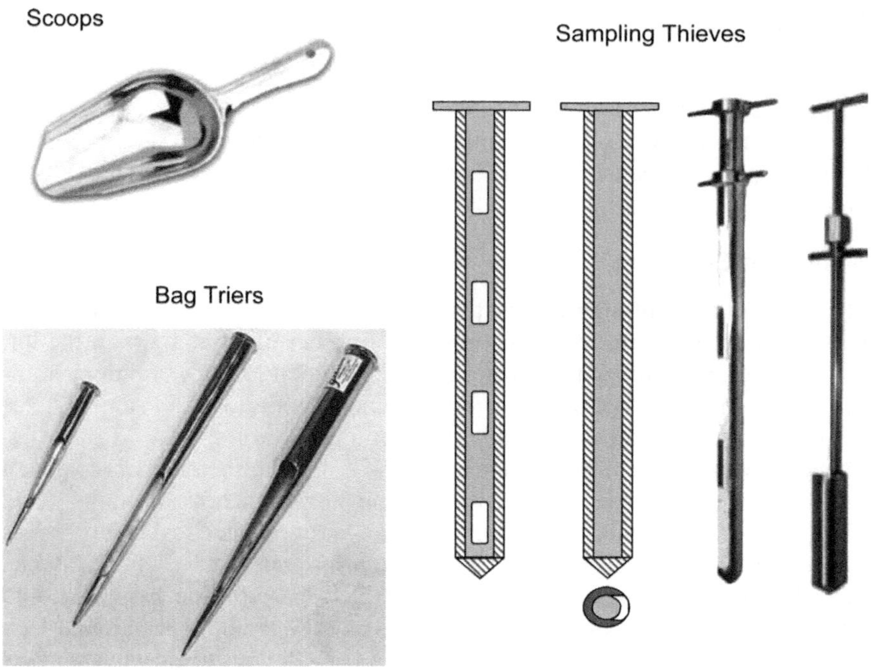

Figure 2 Sampling devices commonly used in the pharmaceutical industry.

external surface should be sufficiently cleaned so that the sample is not contaminated and foreign material is not introduced into the bulk material. Once the sample has been taken, a compatible material should be placed over the hole in the bag and then this patch should be fixed with an appropriate adhesive tape. When sampling from drums, either a scoop or sampling thief may be used, depending upon the heterogeneity of the system. The term "super sacks" is a trade name for large sack containers that can hold from hundreds to thousands of pounds of material, and they are usually placed on a pallet so as to be lifted by a fork truck. These sacks usually have a fill spout on the top and a discharge spout at the bottom. As mentioned before, for homogeneous systems, scoop sampling is appropriate, but given the large size of these super sacks, if there is any concern about the heterogeneity of the material, then a thief should be used, i.e., the large size of these supper sacks makes the use of the thief more important for representative sampling than in the case of a drum or bag.

Step 5: Sample Handling

The samples collected can either be assayed individually or combined, and then a subset of the gross sample is assayed as depicted in Figure 1 and described below. Sample increments should be combined on a clean, dry surface or in a suitable container or bag. All containers that the sample comes into contact with should be inert and not chemically or physically react with the sample. In addition, accurate sample labeling and records must be kept. If the sample is divided, then the retained portion should be kept for possible future analysis and labeled with at least the date, gross field sample number, lot number, and/or reference number from bill of lading. The material to be tested should be labeled with at least the date, gross field sample number, lot number and/or reference number from bill of lading, a unique test sample number, sample weight, and number of reduction iterations (if applicable).

SAMPLING EQUIPMENT AND INFRASTRUCTURE

Sampling Equipment and Facilities

All equipment used for sampling should be made of inert materials, kept scrupulously clean and dry, and stored under clean conditions. Adequate washing facilities should be provided to insure cleanliness.

The facilities should be sufficient to prevent open containers from becoming contaminated by other materials and the operator.

Also, the facilities and sampling equipment should be sufficient to prevent cross contamination by other materials, products, and general environmental contaminates. Where possible, the sampling should be done in a quarantined area separate from other materials, especially, those that have already passed inspection. For rejected materials, procedures and facilities, i.e., storage space, to handle these materials should be in place.

Sampling of materials that require dedicated environments such as sterile components, etc. must be done in a manner so that these environments are not compromised. The equipment used for sampling should not compromise these special environments (e.g., sterility) and should be maintained in a manner that does not compromise these special environmental conditions.

Sampling Health and Safety

It is the responsibility of the operator to read and understand relevant health and safety information (e.g., material safety data sheets) before sampling. The operator must wear appropriate protective clothing for the materials to be sampled; the operator should be protected from the material and the material should be protected from the operator. If specific safety precautions and equipment (e.g., respirator) are required, then the operator must be properly trained in the use of the specified equipment and procedures.

The facilities where the samples are to be taken should be safe with proper access, ventilation, light, etc. Also, fine powders can explode and precautions must be taken to prevent this. All safety concerns should be explicitly mentioned in the sampling plans.

Primary Sampling Process

There should be written standard operating procedures (SOP) in place before sampling. Before sampling, the operator should have all the equipment, containers, and labels needed for sampling; this information should be included in the SOP. All labels should be applied at the time of sampling.

Before removing the sample, the operator should examine the population to be sampled and look for visual signs of heterogeneity and/or overt damage of the material, for example, different appearance, colors, contaminates, obvious particle size differences, prior damage due to water, different lot numbers on drums, etc. The sampling plan should have contingency plans for any materials with gross visual differences. Any sampling plan should take into account prior history with a material and its supplier. For example, if you have received 100 lots that met your specifications, then the sampling plan can be relaxed; however, if the last lot failed or you are dealing with a new supplier, more scrutiny may be needed.

After the sample is collected, the sample should be stored in sealed containers that do not interact with the sample, and should be kept from being contaminated and from contaminating other materials. Also, the sample containers should protect and be protected from things that degrade the sample, such as air, moisture, light, heat, etc. Sample containers that separate, such as jars with lids, should be labeled on all parts to avoid mix-ups. The samples should be stored in secure facilities that have environmental conditions that do not promote excessive sample degradation before analysis. Samples should be properly disposed of after analysis and not be returned to their bulk material.

Counterfeiting

Unfortunately, counterfeiting and alterations of raw materials and finished product can occur; operators should be aware of this possibility and be on their guard for products that are likely to be counterfeited. This is an area that falls outside of the normal sampling plans, because there could be the intent to deceive the purchaser by hiding counterfeit materials within the lot, e.g., putting counterfeit drug products in the bottom of a drum and covering this material with the genuine product. Materials purchased on the secondary market and not directly from the manufacturer have historically been more likely to be counterfeited, and one should be on guard for this possibility—especially for high value products that are more likely to be counterfeited.

SPECTROSCOPIC TECHNIQUES FOR SAMPLE QUALIFICATION

Product identity testing is one of the most important tests to be done on received materials, and in recent years, advances in NIR spectroscopy and better computer software have led to the situation in which excipients can be identified in real time. Traditional ID tests are done in an analytical laboratory; the process of collecting a sample, sending the sample to the laboratory, waiting for the laboratory to perform the test, and waiting for the results to be sent back to the quarantine area for a decision to be made can take days. Therefore, real-time ID testing can save a significant amount of time. In addition, this rapid, low-cost sample analysis will allow a greater number of samples to be measured at the same cost, thus increasing the test reliability and accuracy. This is a significant advantage over traditional laboratory ID techniques such as those described in the USP monographs; however, the many advantages of NIR should not be thought of as a replacement for compendial tests. In addition, it should be noted that, even with NIR methodology, the fundamental underlying principles of good sampling practices still apply and must be followed to obtain a representative sample. This section discusses this technology and the practical aspects behind product ID technology using NIR. The main focus will be on data analysis; however, it should be noted that NIR can do much more than identify products (6).

Near Infrared Spectroscopy

The electromagnetic spectrum can be divided into distinct ranges from low-energy long-wave radiation to high-energy short-wave X rays and beyond (Fig. 3). From a spectroscopic point of view, the basis of these different ranges is the nature or type

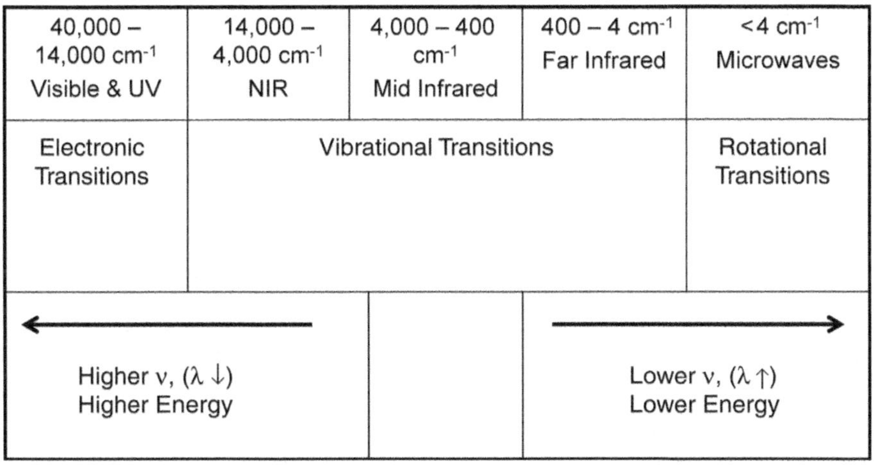

Figure 3 Electromagnetic spectrum.

of molecular interactions between the light of a given wavelength and matter. For example, UV radiation typically interacts with the electronic structure of a molecule. The infrared region (IR) region of the electromagnetic spectrum is next to the visible region, and IR radiation primarily interacts with the molecular vibrations of a molecule. This region of the electromagnetic spectrum goes from approximately 800 to 300 μm, and can be divided into three regions: near, middle, and far IR.

Traditional laboratory chemical organic chemistry identity studies are done using mid-IR/-FTIR, which is in the range of 2.5 to 50 μm (2.5–15 μm are most often used). This region can very accurately ID materials in the laboratory, but the difficulty with this region of the electromagnetic spectrum is that the mid-IR absorbance is very strong and sample preparation must be done so as not to saturate the instrument. However, in the NIR region, the absorption is much weaker and thus sample preparation is not as critical, because there is less of a concern about instrument saturation. With NIR, measurements can be made by sticking a probe in the sample, and for some types of packaging, the analysis can even be done through the packaging. In addition, the signal can be transmitted up to 10 m via a fiber optic cable. All of these factors make NIR ideal for a manufacturing facility. Thus, when working in the analytical laboratory, mid-IR is a popular method, and in the manufacturing facility, NIR is often the instrument of choice.

Typically, a NIR spectrometer measures the amount of light absorbed at each wavelength. An idealized instrument is shown in Figure 4; here a light source illuminates the sample, but before the light reaches the sample it passes through a wavelength selector, which removes all wavelengths except the wavelength of interest. Then a photo-detector measures the amount of light reflected back or transmitted through the sample at each wavelength. The typical result of this measurement is shown in Figure 5. A review of the different types of instruments can be found in the following texts by Burns (7) and Skoog (8). As seen in Figure 5 the spectrum is quite complex, which until recently had been a barrier to the use of NIR. However, with the advent of chemometric-pattern ID methods such as hierarchical cluster analysis, K-nearest neighbor, and Soft Independent Modeling of Class Analogies (SIMCA), NIR has become much more commonly used (9). A detailed discussion of the many patterns of ID methods is beyond the scope of

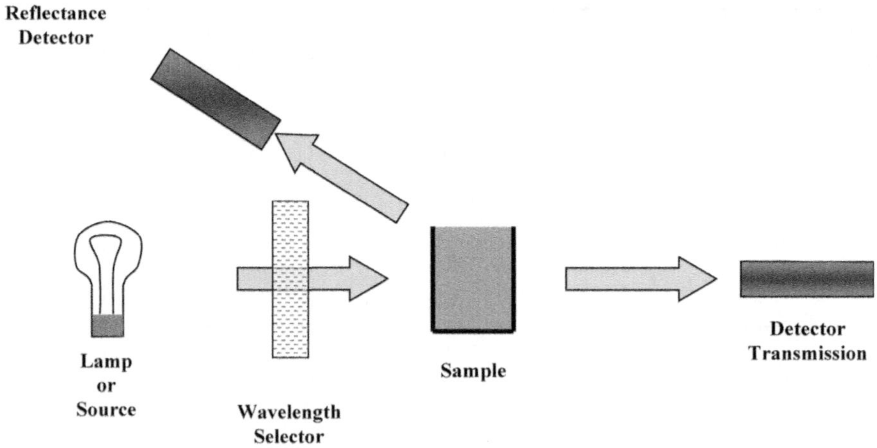

Figure 4 Configuration of a near infrared spectrometer.

the present chapter; however, there are many good books on chemometrics that can describe the methods available and their appropriate use (9–11).

DATA ANALYSIS

The basic steps involved in identifying a sample using NIR will be illustrated using the SIMCA method, which is one of the popular methods for product ID. The

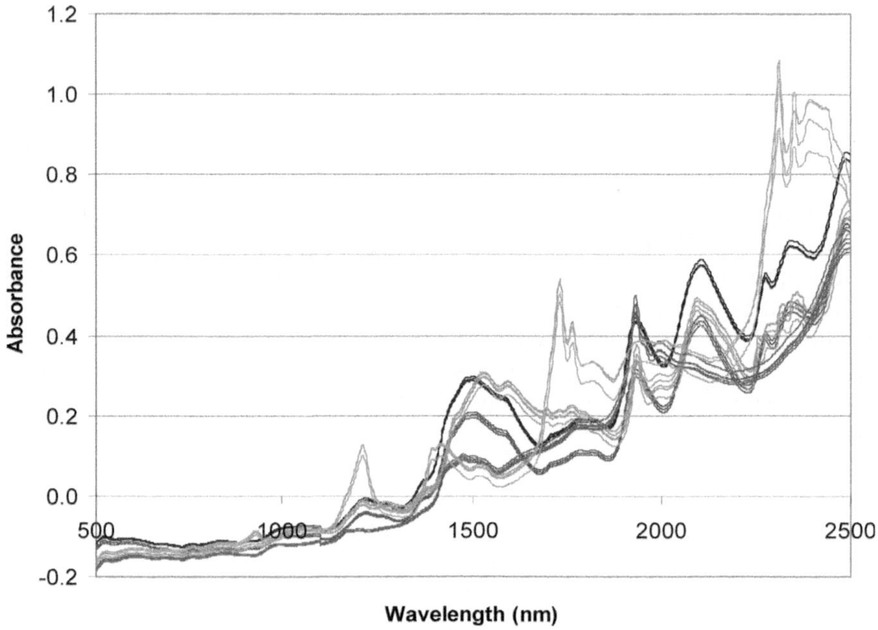

Figure 5 Scans of commonly used excipients for tablet manufacture. Key–MCC (Avicel PH101),—MCC (Avicel Ph200), Lactose (Foremost 325),—dibasic calcium phosphate (di-tab) and—Mg Stearate.

SIMCA method relies on a pattern-recognition technique called principal component analysis (PCA).

To begin with, it is worth talking about typical data structures used for chemometric analysis of NIR spectra. Typically, the unprocessed spectral data are put in a design matrix where each column contains the absorbance at a given wavelength (the number of columns equals the number of wavelengths measured) and each row contains the spectrum of an individual sample (the number of rows equals the number of samples).

$$X_{ij} = \begin{bmatrix} x_{11} & x_{12} & \cdots & x_{1j} & \cdots & x_{1m} \\ x_{21} & x_{22} & \cdots & x_{2j} & \cdots & x_{2m} \\ \vdots & \vdots & & \vdots & & \vdots \\ x_{n1} & x_{n2} & \cdots & x_{n3} & \cdots & x_{nm} \end{bmatrix} \quad (1)$$

This choice of columns and rows is completely arbitrary and switching the columns for the rows will yield equivalent results, but for the sake of consistency, this analysis will use the row column convention stated above. This data structure has been illustrated diagrammatically in Figure 6. In Figure 6, there are two hypothetical excipients: excipient A has an absorption peak at 2100 nm and excipient B has an absorption peak at 1400 nm. In this hypothetical example, absorbance measurements were taken at 100 nm intervals from 1000 to 2500 nm; thus, there are 16 variables corresponding to the absorption data at the 16 wavelengths and there are two samples one for each excipient.

The basic problem is taking the spectra shown in Figures 5 or 6, which the human eye can readily tell apart, and convert these patterns into a form that the computer can readily analyze. The human eye can readily see peak and shape differences, but when there are hundreds of materials that must be identified, it can become a very complex task to differentiate all the different spectrum. In addition, the amount of variability must be accounted for, because you would not want to reject a lot which is in fact a good lot, but is slightly different from the previous lots. Conversely you would not accept a material that is close to the material you want but is out of specification. PCA and SIMCA will help you judge and quantify these differences.

The basic idea is to take the spectral data shown in Figures 5 or 6 and replot them using the variables, i.e., columns. Thus, going back to our hypothetical

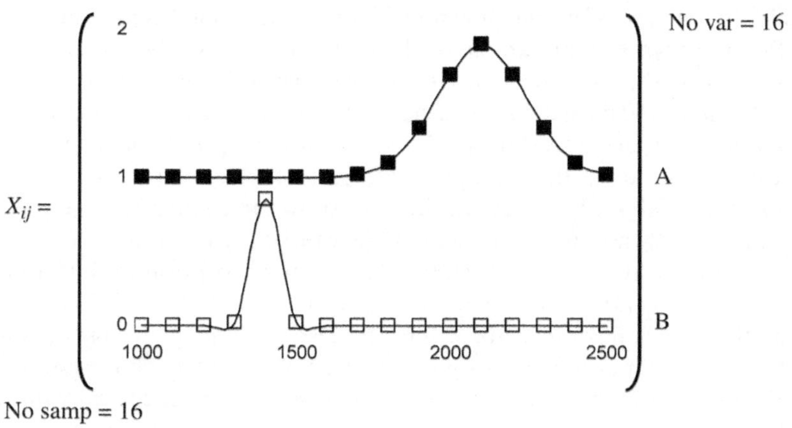

Figure 6 Representation near infrared spectra in matrix form.

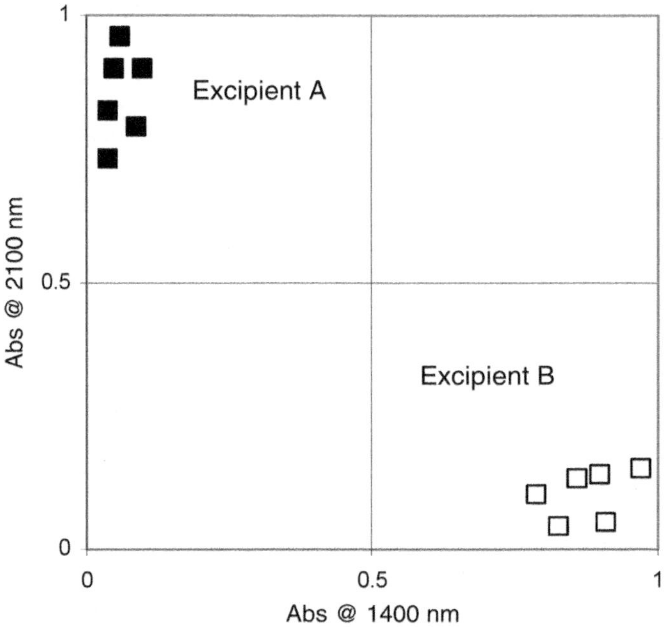

Figure 7 Replotting spectra in Figure 6 using the absorbance at 1400 and 2100 nm variables. Solid symbols (■) excipient A sample, open symbols (□) excipient B samples.

example of excipients A and B shown in Figure 5, the data is replotted as shown in Figure 7. In this illustration (16 dimensional graphs with one dimension for each variable are hard to draw in excel), the spectra shown in Figure 5 are reduced to two wavelengths or variables, absorption at 2100 and absorption at 1400 nm, which are the absorption peaks for the compounds A and B, respectively. In other words, in this example, each spectrum (row of the design matrix) is reduced to two numbers, which are plotted in Figure 7 on the coordinate system variable 1 (abs @ 2100 nm) as the y-axis and variable 2 (abs @ 1400 nm) as the x-axis.

The type of graph represented in Figure 7 is fundamentally different from the typical spectral graph shown in Figure 6. The traditional graph plots the absorption intensity against the measurement variable/wavelength; this set of points is used to represent a sample. In the type of graph shown in Figure 7, each point represents one sample or row. Because each row or sample of the design matrix is represented by a point, this type of plot is said to represent the row space of the design matrix. In Figure 5, there are 25 spectra and 1050 measured wavelengths for each spectrum. The representation of these data in row space would have 25 points in a 1050 dimensional space.

One consequence of this is that samples with similar characteristics tend to group together when plotted in row space. In the data shown in Figure 7, there were 12 samples, 6 of excipient A and 6 of excipient B. As can be seen in Figure 7, excipient A samples have high absorption at 2100 nm and low absorption at 1400 nm; thus, excipient A samples tend to cluster at the upper left hand side of Figure 7, whereas excipient B samples have low absorption at 2100 nm and high absorption at 1400 nm and tend to cluster at the lower right hand side of Figure 7. Thus, samples that are clustered close to each other in the row space have similar numerical values in the design matrix. This fact allows numerical similarity in the design matrix to be determined by proximity or distance in the row space. In other words, numerical distances in the row space can be used as a basis for assessing chemical similarity, and in our example excipient ID.

It should be noted that these distances in the row space do not necessarily have physical meaning, and care must be taken that the measurements taken really relate to the properties one is trying to distinguish. Fortunately, for excipient ID, NIR measurements are very telling of chemical identity. However, factors such as particle size or viscosity grade may be more difficult to distinguish via NIR; in addition, the ID of contaminants such as in the glycerin example given at the start of this chapter could be more difficult to identify using NIR.

PRINCIPAL COMPONENTS ANALYSIS

As shown in Figure 5, there are 1050 absorbance measurements at 1050 wavelengths, and dealing with this number of variables in the row space is practically difficult and cumbersome. PCA analysis is one of the factor analysis methods (12) that uses mathematical manipulations to simplify the dataset. Given the typical spectral range of a typical vis-NIR spectrophotometer from 400 to 2500 nm, it is obvious that each excipient will have certain wavelengths that are very telling of the excipient and other wavelengths that do not yield much information. In other words, certain wavelengths will contain significant variation[c] while other wavelengths contain very little variability and, consequently, very little information.

PCA finds a smaller number of factors that describe the majority of the variability or spread in the dataset. Using these factors often called principal components, the row space is transformed or mapped into a new coordinate system in which the principal components, which can combine the information from several variables, redefine the axes based upon the factors, and these new axes describe the degree of variation or spread in the dataset.

To illustrate these points, we can examine the greatly simplified diagram shown in Figure 8A. In this figure, the hypothetical data is taken from the lower right hand side of Figure 7. The 24 samples are plotted using their two variables, i.e., each sample is represented by a point in the plane defined by these two variables. On this graph, the dataset can be described by the original variables or the coordinates of two new axes, PC1 and PC2; notice that the PC1 axis aligns itself with the greatest spread or variability in the dataset. The first principal component (PC) describes the maximum amount of variability possible. The second PC describes the second most variability in the dataset and is perpendicular to PC1. Figure 8B illustrates some of the common terms used to describe PCA analysis. The new axes called principal components can be used to describe a point's location in the new coordinate system and these coordinates are called scores (see grayed line in Fig. 8B). A plot of score values on the PC axes is often called a score plot, and score plots give you information on the relationship between samples.

In addition, the PC that can be used to describe the variability in the dataset is made up of combinations of the original variables and it would be useful to know which variables have the most influence on the PC and which is typically related to how important a variable is. As with the PC, the variables that describe the most

[c] As used in this context, variation refers to an excipient's attributes that influence the spectra and not just random noise.

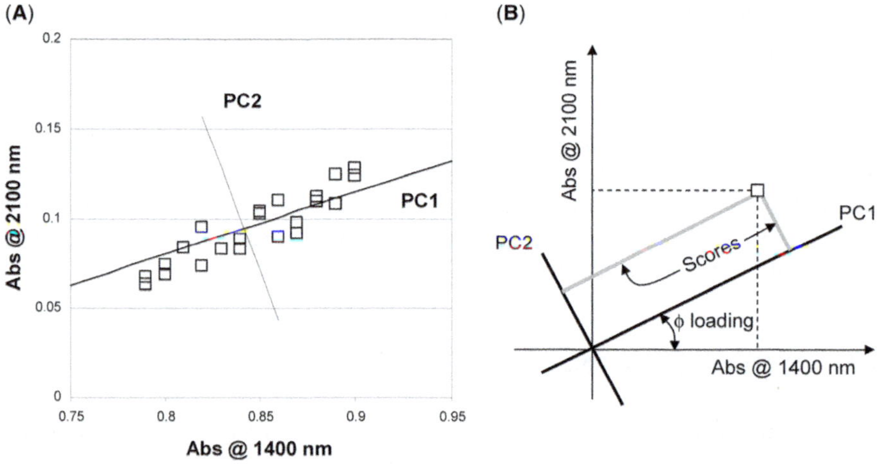

Figure 8 (**A**) With the addition of a few more data points, this figure shows an enlarged view of lower right hand side of Figure 7. (**B**) Showing the new in principal component axes PC1 and PC2 and the old original variable 1 (axes Abs @ 1400 Principle nm) and variable 2 (Abs @ 2100 principle nm).

variability in the dataset contain the most information. In our example shown in Figure 8A, the original variable axis most parallel to PC1 contains the most information. For the example data shown in Figure 8A, variable 1 Abs @ 1400 nm contains the most information about excipient B. The influence a variable has on a PC can be assessed by the cosine of the angle between the original variable and the PC. Cosine ranges between –1 and 1. The cosine of 0° and 180° are 1 and –1, respectively, and these values would indicate that the PC is parallel to the original variable, whereas a value of 0 would indicate that the variable is perpendicular to the PC and there would be no relationship between this variable and the PC. The cosine values are called the loadings, and are used to examine the variables.

Figure 9 shows the PCA of the data presented in Figure 5. This score plot gives a great deal of valuable information that can be used to identify an excipient. The graph shown in Figure 9 plots the scores for PC1 on the *x*-axis and the scores for PC2 on the *y*-axis. The score plot takes the 26,250 data points (25 samples × 1050 absorbance measurements) shown in Figure 5 and reduces them down to 25 points. The score plot shows that the samples have five distinct groupings that correspond to each excipient. It is the analysis of these distances between the clustering of samples in the PC space, which can be used to identify the different excipients.

It should be noted that on these graphs the percentage of the total variability accounted for by a particular PC is given. It is important to keep track of this information because it gives insight into how many factors are needed to adequately model the dataset. For example, if the intrinsic variability of the measurements is 2%, and on the graph the total variability described by all the PCs used to make the model totals to 85%, then there probably are not enough PCs to accurately model this dataset, and there is a lot of variability in the dataset that must be accounted for. However, if the intrinsic variability is 15%, and your PCs total to 99%, then you are probably modeling noise in the data. When your model includes random noise, the PCA model will not be robust because random noise is not reproducible; this situation is called overfitting the data. The mathematical details of the PCA calculations are left to the interested reader (9–13).

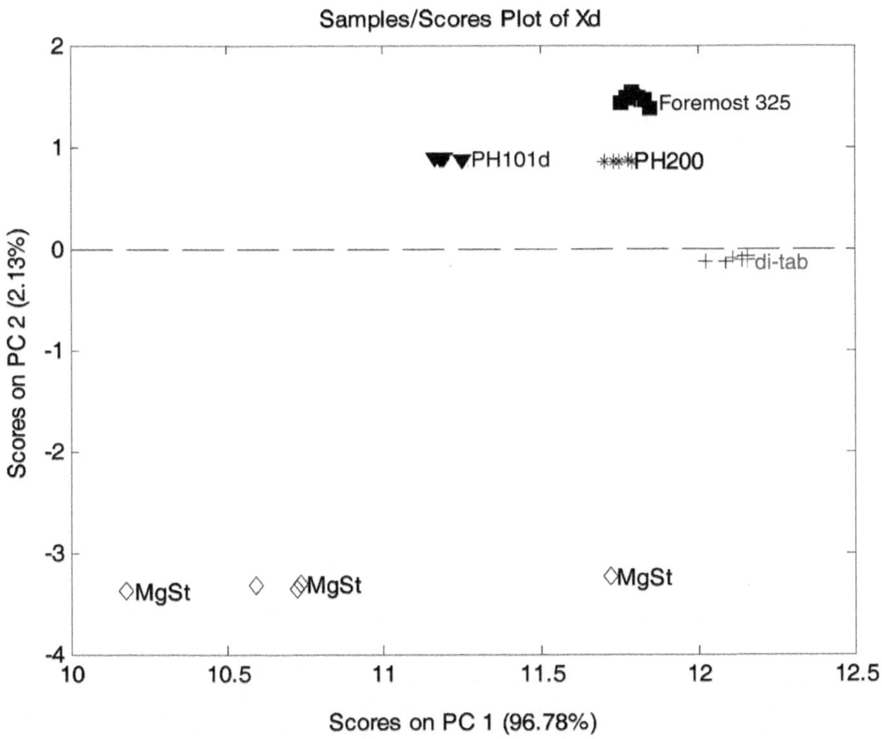

Figure 9 PCA analysis of the scans shown in Figure 5. PCA model calculated using Preprocessing: MSC (mean) Normalize with two PC.

As discussed above, PCA models attempt to describe the overall variation in a dataset with as few factors as possible, and determine what variables are most influential. The application of this method gives an analysis of the type shown in Figure 9. The analysis shown in Figure 9 was done without any knowledge of what samples belonged to what class, and there is no guarantee that a single model will be able to differentiate between all the different classes. This is especially true for materials that are chemically similar, but have different physical grades such as particle size or viscosity. In this context, the term "class" means a collection of samples that are defined as being similar (9). Thus, to make the PCA useful for ID, class information must be incorporated into the model, and a wider range of models must be used.

The SIMCA method has been developed to overcome some of these limitations. The SIMCA model consists of a collection of PCA models with one for each class in the dataset. This is shown graphically in Figure 10. The four graphs show one model for each excipient. Note that these score plots have their origin at the center of the dataset, and the blue dashed line marks the 95% confidence limit calculated based upon the variability of the data. To use the SIMCA method, a PCA model is built for each class. These class models are built to optimize the description of a particular excipient. Thus, each model contains all the usual parts of a PCA model: mean vector, scaling information, data preprocessing, etc., and they can have a different number of PCs, i.e., the number of PCs should be appropriate for the class dataset. In other words, each model is a fully independent PCA model.

Once a PCA model has been built for each material in the library, then the spectrum from an unknown sample can be run through the PCA model and its

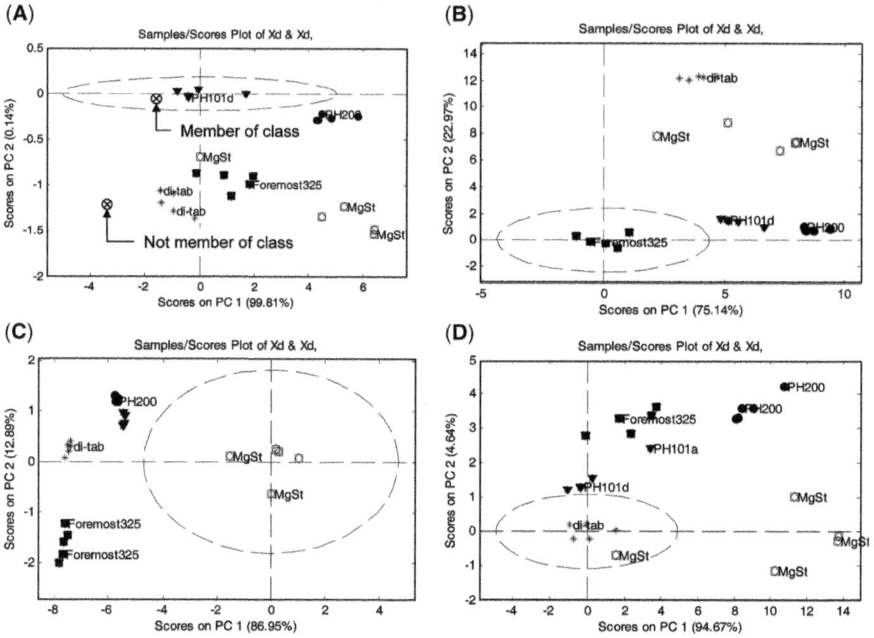

Figure 10 SIMCA analysis of the scans shown in Figure 5. (**A**) SICMA model for Avicel PH 101, (**B**) SIMCA model for Lactose, (**C**) SIMCA model for Mg Stearate and (**D**) SIMCA model for di-tab. *Abbreviation*: SIMCA, Soft Independent Modelling of Class Analogies.

position on the score plot can be calculated. If the unknown sample has a position that is close to the average of the other members of the class, then one can be relatively assured that it belongs to that class. However, if the sample is far from the average value, then one can be relatively assured that it does not belong to that class. For example, in Figure 10A, if an unknown spectrum plots at a point inside the 95% confident limit, then the SIMCA model would assign that spectrum to the Avicel PH101 class, and if the unknown spectrum plots at a point outside the 95% confidence limit, then the model would not assign it to Avicel PH101. By determining the distance an unknown spectrum is from the center of each class model allows one to compare the unknown to all the excipients in the library. This way the identity of an unknown sample can be determined.

ACKNOWLEDGMENTS AND NOTICES

The authors acknowledge the useful comments given by Albinus (Abi) M. D'Sa, Ph.D.

This information reflects the personal opinions of the authors and does not reflect official policy of the FDA or USP.

REFERENCES

1. CDC. Fatalities Associated with Ingestion of Diethylene Glycol-Contaminated Glycerin Used to Manufacture Acetaminophen Syrup—Haiti, November 1995–June 1996. MMWR 1996; 45(30):649–650.
2. Petersen L, Dahl CK, Esbensen KH. Representative mass reduction in sampling—a critical survey of techniques and hardware. Chemom Intell Lab Syst 2004; 74:95–114.

3. Petersen L, Minkkinen P, Esbensen KH. Representative sampling for reliable data analysis: theory of sampling. Chemom Intell Lab Syst 2005; 77:261–277.
4. http://www.asq.org/.
5. Sastry KVS et al. Solid-solid operation and equipment. In: Perry RH, Green DW, Maloney JO, eds. Perry's Chemical Engineeer's Handbook. New York: McGraw-Hill, 1997:19–1–19–65.
6. Tatavarti AS et al. Assessment of NIR spectroscopy for nondestructive analysis of physical and chemical attributes of sulfamethazine bolus dosage forms. Pharm Sci Tech 2005.
7. Burns DA, Ciurczak EW. Handbook of Near-Infrared Analysis: Practical Spectroscopy. 2nd ed. New York: Marcel Dekker, 1992.
8. Skoog DA, Leary JJ. Principles of Instrumental Analysis. 4th ed. Fort Worth: Saunders College Publishing, 1992:700.
9. Beebe KR, Pell RJ, Seasholtz MB. Chemometrics A Practical Guide. New York: John Wiley & Sons, Inc., 1998:348.
10. Kramer R. Chemometric Techniques for Quantitative Analysis. New York: Marcel Dekker Inc., 1998:203.
11. Brereton RG. Chemometrics Data Analysis for the Laboratory and Chemical Plant. Chichester, England: John Wiley & Sons Ltd., 2003:489.
12. Malinowski ER. Factor Analysis in Chemistry. 3rd ed. New York: John Wiley and Sons Inc., 2002:414.
13. Martens H, Naes T. Multivariate Calibration. Chichester: John Wiley & Sons, 1989:419.Chowhan Z. Segregation of particulate solids, Part 1. Pharm Technol 1995; 19:56, 58, 60, 62, 64, 66, 68, 70.Chowhan Z. Segregation of Particulate Solids, Part II. Pharm Technol 1995; 19:80, 82, 84, 86, 88, 90, 92, 94Allen T. Particle size measurement. In: Scarlett B, Jimbo G, eds. Powder Technology Series. Vol. 1. 5th ed. London: Chapman & Hall, 1997:525Smith PL. A primer for sampling solids, liquids, and gases: based on the seven sampling errors of Pierre Gy. In: Gunst RF, ed. ASA-SIAM Series on Statistics and Applied Probability Society for Industrial & Applied Mathematics. Philadelphia: 2001:96Muzzio F et al. Sampling and characterization of pharmaceutical powders and granular blends. Int J Pharm 2003; 250(1):51–64Muzzio FJ et al. Sampling practices in powder blending. Int J Pharm 1997; 155(2):153–178Venables H, Wells J. Powder sampling Drug Dev Ind Pharm 28 2 2002 107–117Jillavenkatesa A, Dapkunas SJ, Lum L-SH. Particle Size Characterization Special Publication 960–1. National Institute for Science and Technology, ed., 2001Hewa Saranadasa. The square root of N plus one sampling rule. Pharm Tech 2003; 27:50–62

APPENDIX I: RANDOM NUMBER TABLES

1. Number all containers in the lot to be sampled.
2. Pick the random number table page or generate the random numbers.
3. Choose a direction to read, i.e., up–down, right, or left.
4. To get a starting point, close your eyes and pick a point on the table.
5. Read numbers off table in that predetermined direction.
   If the digits that you selected are between 01 and *N*, this is the first sample element.
      If not, keep going in the preset direction until you find a suitable number.
6. Continuing to move in the preset direction until you find the next number between 01 and *N*. That is your second element. Continue in this manner until you have enough samples, see Table A-1 for an example.

Table A-1 Random Number Table

$N = 500$, $n = 50$	$N = 100$, $n = 10$
6977	69**77**
8**377**	8377 ⇐ Sampling w/o replacement
3**034**	30**34**
9903	99**03**
6955	69**55**
5**483**	54**83**
5733	57**33**
0**126**	01**26**
4**329**	43**29**
3776	37**76**
etc.	etc.
Containers to be sampled: 377, 34, 483, 126, 329	Containers to be sampled: 77, 34, 3, 55, 83, 33, 26, 29, 76

Note: Population size = N; Sample size = n.
Source: From Refs. 14–22.

APPENDIX II: SAMPLING AND SEGREGATION PRINCIPLES

To acquire a representative sample, one must develop and implement a suitable sampling plan. A good sampling plan includes (i) population determination and sample size selection and (ii) sample collection procedure and sample size reduction method. In addition, one needs an infrastructure to maintain the integrity of the samples and sampled materials. To begin with, a brief introduction to the sampling theory and terminology is in order.

The process of selecting n objects for a lot of size N is called sampling. The n objects are the sample and the total lot size N is the population from which the sample is taken. This lot may be in one mass or distributed among M separate containers. Randomness is critical to sampling because a random sample has minimum errors; ideally all the error should be due to the intrinsic variability of the material and the variability due to sampling should be minimized. An attribute is the property of interest that is going to be measured, be it particle size, composition, etc.

Population Heterogeneity

The key to getting a good sample is understanding and accounting for the degree of heterogeneity in the system that is to be sampled. This heterogeneity of a particle system arises from two sources: the intrinsic, constitutive, or compositional heterogeneity and the spatial distribution heterogeneity. The intrinsic heterogeneity of the powder system reflects the fundamental differences in the particles, for example, particle size, but it should be noted that this concept applies to any attribute that is measured. In other words, this heterogeneity is associated with the fundamental variation of the powder system. On the other hand, distribution heterogeneity is the distribution of these particles in space (or time in the case of a process). For example, small particles are preferentially in the lower portion of the powder bed. This type of situation can arise due to powder bed segregation and is common in some particle systems with a broad particle size distribution. In other words, the particles may not be uniformly distributed throughout the lot. This spatial heterogeneity introduces variation in the sample and is a source of variation that contributes to the total variation. Together these sources of heterogeneity give rise to the sampling error, which dictates how variable the samples will be, how large the sample size should be, and how hard it will be to obtain a representative sample.

It should be noted that the intrinsic or compositional heterogeneity is a function of the powder system and is a fundamental unalterable characteristic of the material. Thus, the intrinsic heterogeneity is the minimal variance a system can have. The difference between the true state of the system and what is actually measured is called the "fundamental error." When all the other sources of error are added to the intrinsic heterogeneity, this gives us the fundamental error and it is our goal to minimize these other sources of error. Thus, knowing where the error comes from can help to minimize these errors.

To successfully withdraw a sample from a bulk container that is representative of the population, one needs to have an idea of the population's homogeneity, i.e., how segregated the system is. Knowing what factors can accentuate segregation and the patterns of segregation that are likely will help one account for segregation in a powder bed and take better samples. There are many factors that can affect the degree of powder bed segregation. For segregation to occur, sufficient energy needs to be put into the powder bed to induce motion between particles. When the amount of supplied energy is sufficient, segregation can occur via three modes—in the powder bed, on the free surfaces of a powder bed, and when the powder bed is fluidized. These modes are illustrated in Figure A-1.

Within the powder bed, percolation, also called sifting segregation, and the rising of course particles via vibration can occur. With sifting segregation, the smaller particles acting under the influence of gravity can more easily move down into the void spaces between the larger particles when the particle bed is perturbed. The net effect of these movements is that the smaller particles percolate down into the powder bed; i.e., the top of the powder bed will have a higher proportion of larger particles. A common example of sifting segregation is the unpopped popcorn kernels that are found at the bottom of a bag of popcorn. In addition, when larger particles are mixed with smaller particles and the bed is vibrated, larger particles can, in effect, float to the top of the powder bed.

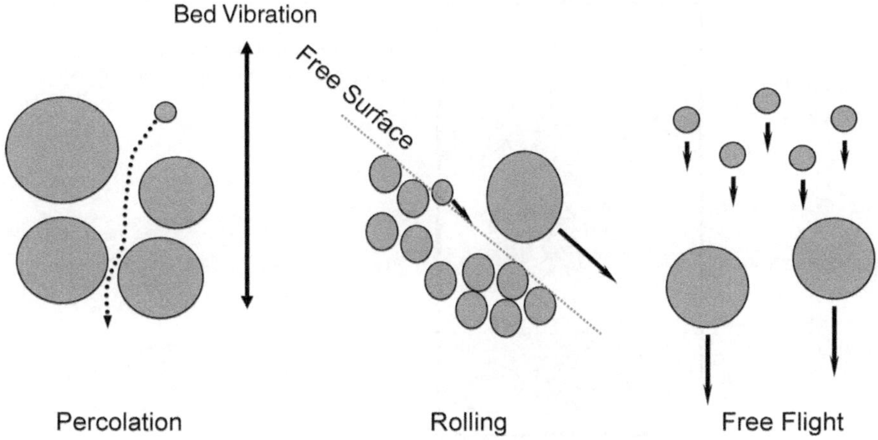

Figure A-1 Illustration of three modes of particle segregation.

For free surfaces, segregation can occur anytime a free surface that particles can roll down is created, i.e., segregation can occur on any nonlevel surfaces where there is the relative movement of particles. When particles roll down these free surfaces, larger particles tend to tumble farther down the surface than the smaller particles (Fig. A-1). For example, if a conical heap or pile is formed in the middle of a hopper during loading, the larger particles are more likely to roll farther down the heap toward the outer edge of the hopper. This creates a situation in which the smaller particles tend to be in the center of the hopper and the larger particles toward the outer wall of the hopper. The formation of these free surfaces can be a major factor in segregation.

When the powder beds are fluidized, a large amount of air is incorporated into the powder bed, and when this air is moving, the air velocity may exceed the terminal velocity of the smaller particles. When this happens, the fines are suspended in the airstream while the coarse particles settle out. The fines eventually settle on top of the powder bed, forming a top layer that has a higher concentration of fine particles. This type of segregation is sometimes called "elutriation segregation." This type of segregation can occur when a powder is discharged from a hopper, or is poured into the top of a hopper and a large volume of air is displaced.

In summary, for a highly segregating system, the powder bed could have an idealized spatial distribution of particles similar to that shown in Figure A-2, where, due to elutriation segregation, there could be a layer of fine particles on the top followed by a layer of larger particles caused by percolation segregation, and the radial distribution with larger particles toward the outer wall is due to rolling segregation.

In general, the primary factors affecting segregation are particle size and size distribution, density, and shape and shape distribution; of secondary importance are surface roughness, surface coefficient of friction, moisture content, and container shape and design. Experience has shown that of all these factors, particle size is the most important, and subtle differences in particles can be sufficient to cause a measurable segregation. It should be noted that if the attribute of interest is associated with particle size, then this attribute will segregate along with the different particle sizes. For example, if a granulation is made in which the larger particles contain more drug

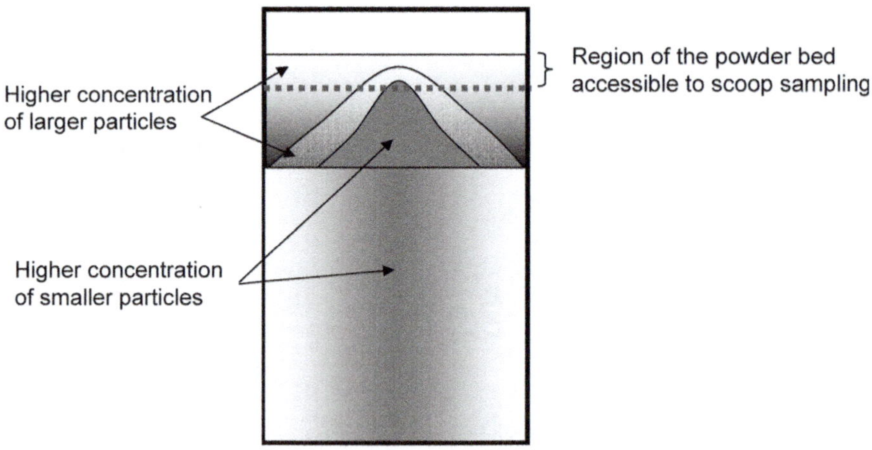

Figure A-2 An example of a powder drum that has undergone significant segregation.

than the smaller particles, then drug content could be very prone to segregation, i.e., drug content will show segregation patterns similar to that for particle size.

Segregation can drastically increase the sampling error because it decreases the probability of certain particle types from being in the sample. In addition, it should be noted that the powder bed can already be segregated upon receipt of material, and that poor sample handling can also cause segregation. To avoid further segregation during sample handling, the operator should avoid situations that promote segregation such as pouring where the powder forms a sloping surface, pouring into the core of a hopper, vibrations, shaking, stirring (unless done to promote mixing); in addition, the use of mass flow hoppers is preferred.

For segregating systems to ensure every element of the population has an equal probability of being in the sample, there are two basic strategies: (i) a sampling thief and (ii) sampling from a moving powder stream. A sampling thief is a long spear-like probe that can be inserted into the powder bed, and once inserted it can collect powder samples from points adjacent to the spear (Fig. 2). By using a sampling thief, particles from almost any point in the powder bed can be included in the sample. The second method relies upon the golden rules of sampling, which state that (i) a powder should always be sampled when in motion and (ii) the whole stream of powder should be taken for many short increments of time in preference to a part of the stream being taken for a longer time. For example, if the container to be sampled were emptied onto a conveyer belt, all the material would pass by a single point, which could then be sampled. Thus, no matter how segregated the system is, the collection of the powder at random time points ensures that every particle has an equal probability of being included in the sample. The second condition of the golden rule accounts for material segregation on the conveyer belt: by collecting the entire stream, one gets a cross section of all the particles, no matter how much segregation occurs on the conveyer belt.

APPENDIX III: SAMPLING CONSIDERATIONS AND TOOLS

Homogenous Systems

For powder systems where the attribute of interest is randomly distributed throughout the container, scoop sampling is adequate. Typical scoops are shown in Figure 2; scoop sampling is a straightforward procedure where the operator, after randomly selecting the containers to be sampled as described above, opens a container and then scoops out a sufficient amount of material from the top of the powder bed and then seals the container. If a thin layer of material on top of the powder bed is suspected of being different from the bulk, then samples should be taken from a point below this top layer. For example, with elutriation segregation, a thin layer of fine parties may lie on top of the powder bed; in this case the operator should dig down into the powder bed to avoid sampling from this layer. The scoop should be large enough so that no material is lost during handling; lost material may result in sample bias. In other words, the use of a heaping scoop with material rolling off the scoop should be avoided. The advantages of scoop sampling are convenience and cost, and with highly potent materials, low-cost disposable scoops can be used to minimize cross contamination.

Heterogeneous Systems

If there is a spatial distribution of the attribute of interest, this method is prone to error and potentially significant errors. Scoop sampling is a nonprobabilistic method because only the most accessible fraction of the container is sampled. This is illustrated in Figure 2, where only the material in the top layer can be reached with a scoop. For example, a sample from the top outer edge of the drum shown in Figure A-2 could be biased because, in this example, the larger-sized particles are preferentially toward the top and outer edges of the drum. Hence, the smaller particles will not have an equal probability of being in the sample. As a result, the smaller particles will be underrepresented in the sample, and any analysis of particle size will not reflect the true particle size of the original population.

For heterogeneous systems, the initial primary sample is the most difficult to obtain; the use of a sampling thief, sometimes called a grain probe or sampling spear, is needed. The advantage of a sampling thief is that much more of the powder bed is accessible because the sampling thief can sample from different points in the powder bed, which helps to reduce sampling bias. There are many different types of sampling thieves available; the most commonly used sampling thieves in the pharmaceutical industry are shown in Figure 2. They are the (i) concentric sleeve with slotted chambers, (ii) concentric sleeve with a groove, sometimes called open-handled, (iii) end sampler, and (iv) core sampler. Each type has it own unique operating procedures, which are described below.

a. The concentric sleeve with slotted compartments is probably the most popular type of sampling thief used in the pharmaceutical industry. This type of sampling thief consists of two concentric tubes or cylinders where the inner tube is divided into compartments (Fig. 2). This design enables the detection of differences in the attribute of interests across the depth of the container. To collect a sample the compartments are closed and the sampling thief is inserted into the powder bed with the collection zone openings facing upward. The handle is turned to open the sample zones, and the handle is moved up and down with two quick short strokes to help fill the compartments. The sampling thief is then closed and removed from the powder bed. Visual inspection of the powder bed through its depth should be done before emptying the sampling thief. The powder from the individual compartments can be combined on a clean surface or in a collection container. In certain situations the material from each compartment may be analyzed separately, i.e., without mixing.
b. The concentric sleeve with groove is an open-handled probe where the inner tube is not divided into compartments. The probe is inserted into the powder bed with the grove open and then closed by rotating the outer sleeve before withdrawing the sampling thief from the powder bed. The probe's contents are emptied from the handle end by holding the probe upright and letting the sample slide out from the handle, which is more convenient than the slotted thief. However, this makes visual inspection for material inconsistencies by depth more difficult.
c. The end sampler is a type of probe that has a single entry zone at the bottom of the sampling thief. These types or probes are often used to sample slurries. Often the end-sampling zone is larger then the rest of the sampling thief, which is a disadvantage, because the larger the probe is the more it perturbs the powder bed, which can introduce sampling bias.
d. The core sampler has a hollow outer cylinder that has a tapered outer wall on the open end. This probe is inserted into the powder bed, and the intrinsic cohesion of the particles keeps them from flowing out when the probe is withdrawn. The contents of the cylinder are then emptied into a clear container.

General Considerations

To obtain reliable and reproducible results the particle size of the powder should be larger than the 2 to 10 μm range or else the powder will be too cohesive and not flow properly into the sampling thief. In addition, particles larger than about one-third the width of the slot will also give poor results. Samples should be taken from several sites throughout the container. The probe should be long enough to penetrate at least three-fourth of the depth of the powder bed; this insures that material from all depths appear in the sample. The choice of sites should be dictated by an understanding (often subjective) of the degree of heterogeneity in the powder bed, which may have been caused by handling or movement during transport. Sampling plans can either call for the insertion of the probe at random locations and random angles or at predetermined locations and angles. For example, the plan may call for the probe to be inserted at the center and two locations near the edges. Also, many operators recommend that the probe always be inserted at a 10° angle from vertical, which increases the range of locations sampled.

Some of the disadvantages of sampling thieves include the facts that the procedure is more labor intensive because the probe has to be physically inserted into the powder bed, often multiple times, and then the contents of the probe must be emptied, and then the probe must be thoroughly cleaned, and for settled powder beds, the sampling probe can be difficult to insert. In addition, the sampling probe can introduce the following errors: fine particles can lodge in between the inner and outer tubes, particles can fracture, fines can compact and not flow well into the sampling chambers, segregation can occur during flow into the sampling zone, and the act of inserting the probe can disrupt the powder bed by dragging powder from the top layers of the bed down through the bed.

APPENDIX IV: PRIMARY SAMPLE–SIZE REDUCTION

As mentioned before, typically the primary sample consists of multiple samples taken from containers and mixed together. To obtain an analysis sample (Fig. 1), the primary sample has to be reduced to a size appropriate to the analysis method. Primary sample–size reduction is an often-overlooked aspect of a sampling plan, but just as important because the same factors that cause segregation in a container will cause segregation in the primary sample, and any bias in the primary sample–size reduction method will lead to erroneous results. The advantage of the secondary sample is that the mass has been reduced to the point where it is much easer to obtain a representative unbiased sample, because every element in the powder bed is readily accessible, which makes it is easy to adhere to the golden rules of sampling. Generally speaking, the measurement sample is either wet or dry: the choice is dictated by the analysis method requirements. For example, the Coulter Counter needs the samples to be uniformly suspended in an electrolyte, whereas other methods such as sieving are typically done with dry powders.

Dry Analysis Methods

There are many laboratory devices available for the reduction of the primary sample to an analysis sample. The three most important methods used in the pharmaceutical industry are (i) scoop sampling, (ii) cone and quartering, and (iii) the spinning riffler or rotary sample divider (Fig. A-3).

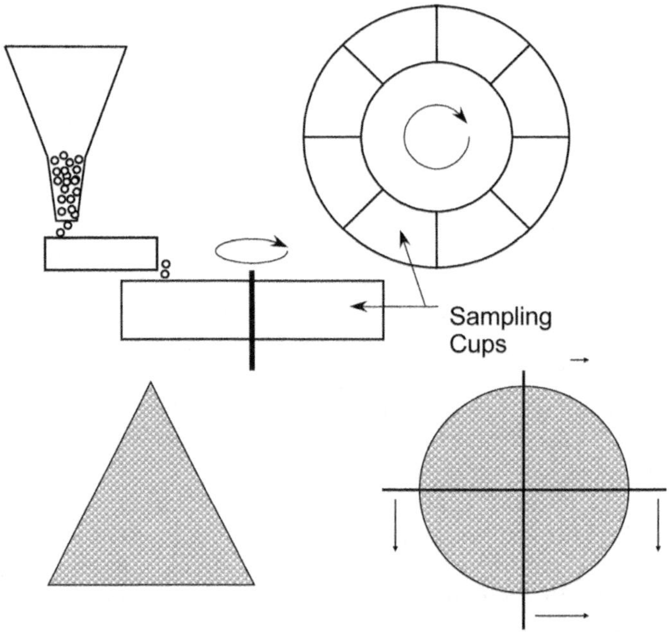

Figure A-3 Sample dividing.

a. Scoop sampling is done as described previously but generally with a smaller scoop or spatula. Great care must be taken when removing material from the primary sample because this material could be highly segregated due to handling. Scoop sampling is appropriate for homogenous or cohesive powders, but if the powder is prone to segregation, scoop sampling can introduce significant errors. There are several serious disadvantages of scoop sampling. The first is its dependency on the operator to decide where to scoop the material and what quantity of the sample to extract, which can introduce operator bias. Second, with scoop sampling, there is a natural tendency for operators to withdraw the sample from the free surface, which is highly prone to segregation and not representative of the bulk. Third, one should avoid creating a heap where rolling segregation can occur, and when sampling, make sure no material falls off the edges of the spatula or scoop because this could bias the sample. Ideally the operator should make some attempt to mix the primary sample before using the scoop, but this can also exacerbate segregation problems and should be done with great caution.
b. Cone and quartering is done by pouring the primary sample into a symmetric cone on a flat surface. The cone is then flattened by a flat surface such as a spatula and divided into four identical quarters (Fig. A-3), thus reducing the amount of material by one quarter. This procedure can be repeated until the desired sample size is obtained. The theory of this method is that by creating a symmetric cone all the segregation processes also occur symmetrically around the cone; hence the symmetry is used to mitigate the effect of segregation. In practice, it is very difficult to actually make a symmetric powder cone and the method becomes very operator dependent and often unreliable. Operator differences in how the heap is

formed and subdivided can lead to poor precision and significant errors. In addition, if the method is done more than once, errors can propagate each time the cone and quartering is done. Some experts in the field do not recommend this method.

c. A spinning riffler (Fig. A-3) has a series of containers mounted on a circular holder. The circular holder rotates at a constant speed and the sample is loaded at a constant rate into the containers via a vibratory chute, which is feed by a mass flow hopper. Once the material has been divided among the different holders, an individual holder can be removed for testing or for further sample division. The angular velocity of the circular holders and the amplitude of the vibratory feeder can be controlled to accommodate powders with different flow properties. The holder velocity and feed rate should be adjusted so that the containers fill uniformly and a heap does not form on the vibratory feeder. Spinning rifflers are available in different sizes, enabling subdivision of powders from a few milligrams to hundreds of grams. The only drawback of the spinning riffler are the time it takes to process the sample and the capital expense. Despite these minor disadvantages, the spinning riffler is by far the best method for subdivision of free-flowing powders.
d. It should be noted that when dividing a sample, if the sample has agglomerated, the agglomerates should be broken apart by a suitable technique such as sieving before dividing the sample.

Wet Analysis Methods

Wet analysis methods require dispersing the sample in a liquid suitable for analysis and then withdrawing an aliquot using a syringe or pipette. Effective secondary sampling requires making a stable[d1] homogenous suspension; factors to be considered include sample solubility in the dispersion vehicle, aggregation of the sample, the use of suspending agents, deaggregation of primary particles in the dispersion vehicle, etc. Even though a uniform suspension is created, the sample should be homogenized, typically by shaking, immediately before withdrawing the sample with a syringe or pipette. The diameter of the syringe or pipette should be sufficiently large so that it does not exclude particles, nor should it be prone to clogging. It is recommended that the largest particle diameters do not exceed 40% of the syringe or pipette tip diameter. If for practical reasons, the amount of material from the primary sample is too large, the sample size must be reduced before a suspension is made; to do this, use the methods described in the Dry Analysis Methods section.

As a precaution, enough sample should be retained to repeat all tests a minimum of five times.

23
Excipient Distribution

Victoria M. Shaheen
Mutchler Pharmaceutical Ingredients, Inc., Harrington Park, New Jersey, U.S.A.

INTRODUCTION

"Oh, you're just a distributor." If I had a nickel for every time I heard that when a prospective customer approached our booth at a trade show, let us just say I wouldn't have to be working. Many assume that the role of distributor can be relegated to that of middleman, reseller, or intermediary: one who lacks the expertise of those who actually manufacture products—"Oh! So you just buy products and resell them to someone else." These gross oversimplifications do little to explain the scale, complexity, and critical role that a distributor plays in the movement of goods and services from manufacturer to end user (the supply chain), particularly in the pharmaceutical industry.

Distribution of goods was once considered a low-level operation in the supply chain, almost a nuisance, ranked with warehousing and freight transport in its tedium, and was relegated to a secondary or tertiary function after production or operations. It was not focused on as an important part of business strategy, and was often appointed less able staff than other functions such as manufacturing or operations (1). In the last century, however, it has been discovered that distribution is a critical determinant of the success of a business and must be paired equally with manufacturing when allocating focus and resources in order for a company to achieve optimal profitability. In the physical transfer of goods from producer to consumer, the distributor plays a number of important roles—from reducing marketing costs of the producer to providing a specialized high level of customer service to the consumer or user. The distributor may also handle a series of related functions such as transport, handling, storage of inventory, and order processing.

In the pharmaceutical industry, distributors fall under intense scrutiny, as critical players in the supply chain. From the early development of many drug products to production and commercialization, distributors are integral parts of the entire process— supplying equipment, active pharmaceutical ingredients (APIs), and excipients—the focus of this chapter.

Traceability of excipients is a major focus of concern during the development and production of drug products. Knowing with certainty exactly where the excipients were manufactured and how many times they change hands and are repacked, handled, etc. is critical information that relates directly to the safety of the finished product. It is also a key element that is scrutinized by the Food and Drug Administration (FDA) in its broad regulation of the pharmaceutical industry.

Unfortunately, there have been tragic events in the past, which have brought the need to increase controls on the excipient supply chain. In 1996 in Haiti, for example, over 80 children died after the administration of a cough syrup in which one of the key ingredients (glycerin) was impure and contaminated with levels of diethylene glycol high enough to be lethal. A full investigation was launched by regulatory authorities and it was found that the glycerin had changed hands so many times that the original source was never determined (2). This tragedy and another like it in India (3) have really brought to the forefront of pharmaceutical manufacturing the importance of traceability of excipients as well as the critical role of vendor qualification when choosing ingredients for products destined for human consumption, and continue to do so. They are also reminders of how necessary it is that these efforts to increase supply chain integrity include intensifying the focus on those who distribute goods and services into the pharmaceutical industry.

THE EVOLUTION AND SPECIALIZED ROLE OF THE PHARMACEUTICAL DISTRIBUTOR IN THE PHARMACEUTICAL INDUSTRY

There are so many different types of distribution and variations amongst the different types in the chemical and pharmaceutical industries that it would be impossible to outline all of them in a short chapter. The focus of this chapter however is the "pharmaceutical excipient distributor." A specialized distributor, the excipient distributor can supply just about every ingredient in a drug delivery system other than the API. Their business is conducted either through contractual relationships with excipient makers, or they may also buy and sell excipients on an ad hoc basis, based on excipient user needs. On a very basic level, there are primarily two types of pharmaceutical excipient distributors: those engaging strictly in warehousing and distribution, and those who further process the material by repacking, sampling, micronizing, or performing analytical testing or any other activities involving exposed material.

To understand the role of such distributors in the pharmaceutical industry, it is important to have a cursory knowledge of how such a type of distribution came about. It is logical to assume that excipient distribution was born of chemical distribution, because many excipients are general chemicals that are manufactured under a special set of circumstances, or within the recommendations of an established quality system such as those put forth in the International Pharmaceutical Excipient Council's (IPEC) Good Manufacturing Practices (GMPs) for Bulk Pharmaceutical Excipients guidelines. For the purpose of this text, three types of distributors are discussed: chemical distributors, chemical distributors offering excipients, and the highly specialized pharmaceutical excipient distributor.

Chemical Distributors

Chemical distribution has had a long presence throughout the economic development of the United States, the origins of which developed from the goal of increasing the efficiency by which a manufacturer marketed their goods and services (4). From the standpoint of chemical manufacturers, it is more efficient to have a lot of firms selling their products. Many general chemicals are commodities, and as a result carry a fairly low margin. The more points of distribution there are for these products, the lower the fixed sales cost to the manufacturer (sales management and staff, travel, etc.). In some cases, it is more beneficial for a manufacturer to pay a distributor a commission, thereby all but eliminating those costs. When business declines, the distributor bears the overhead burden of costs accrued in warehousing, administration, and sales. Those costs become variable to the manufacturer because they are absorbed by the distributor.

Another factor to be considered in the distribution of general chemicals is that many chemical distributors do not necessarily have to have a high level of expertise regarding the application of the products in consumables, nor do they need to understand with any great depth of complexity the characterization of the chemicals that they market. This is partially due to the fact that they have such a broad range of distribution into so many applications that it is difficult to develop expertise in all areas. Additionally, the return on such an investment in training may be minimal, because the industrial applications of many of these chemicals do not necessarily warrant such expertise on the part of the distributor. It is often difficult to quantify the benefits of a highly trained and knowledgeable sales force in the chemical distribution industry as it relates to the generation of revenue.

Although the business practices and general quality and safety of the chemical distribution industry are monitored, it is on a different scale than that of the pharmaceutical industry. Overall, there is no regulatory agency that applies to all chemicals equally, because it would be impossible to provide standardized industry regulation (e.g., water is a chemical, but is much more loosely regulated than, say, toluene or hydrochloric acid because of the dramatically different risk levels or relative toxicities presented by each chemical). The entire industry is of course regulated by the Occupational Safety and Health Administration (OSHA), a part of the U.S. Department of Labor (5). OSHA's mission is to continually ensure the safety of the U.S. work force by setting the standards by which safe and healthy workplaces can be achieved. Beyond the broad scope of OSHA and more specifically relevant to the area of chemical distribution, however, is the National Association of Chemical Distributors (NACD). A condition of membership in this organization is that all member companies commit to adhere to the guidelines put forth in the Responsible Distribution Practices Guideline, which encompass all phases of the distributive process. NACD provides instruction on how to achieve responsible distribution, as well as a forum for communication within the industry. NACD is also a resource for consumers and chemical manufacturers, because all member companies are listed. A growing trend in the industry is to only conduct business with companies listed in the NACD membership log (6).

Chemical Distributors Offering Excipients

Many large chemical distributors service a range of industries. They may market their products for use in cosmetics, food, or other industrial applications. Because many excipients are multifunctional, it is not unreasonable to assume that the same ingredient, albeit a different grade, marketed to the food or cosmetic industry may also be marketed to a pharmaceutical drug maker. At this point, it is important to consider the level of service that will be offered by such a distributor, as well as the integrity of the grades offered. In other words, if a large distributor is offering many grades of the same ingredient [Food Chemical Codex (FCC), or food-grade; United States Pharmacopoeia/National Formulary (USP/NF) or pharmaceutical grade, etc.] and they are engaging in any further processing of the excipient such as analytical testing or repacking, it is important to determine whether the original excipient is manufactured according to IPEC GMP guidelines, and not just tested to meet the pharmacopeial standards and labeled as such. According to IPEC, testing excipients that are not manufactured in accordance with a GMP quality system to show that they meet the specifications put forth in the USP/NF and labeling them "USP" or "NF" grade is not an acceptable practice. Ingredients entering a pharmaceutical drug product are subjected to a higher level of scrutiny than those entering other industries, both from a regulatory and a technical standpoint. In qualifying a large chemical distributor who also happens to sell excipients, the levels of service and attention that will be delivered are not the only considerations. It is also important for a user to understand the extent to which such a distributor understands the regulations a user must comply with when choosing their excipients. It is simply not acceptable for a large distributor with analytical testing capabilities to test excipients for compliance with the NF monograph and then label them as such, although this is sometimes done. These considerations are especially important when the annual volumes of a required excipient for a drug product are low. Most excipients, by their nature, are not very costly relative to the cost of the end product. They are certainly not as profitable as APIs, and thus many times, small volume, high maintenance excipient sales can be cumbersome to a large distributor who may be selling truckload quantities of the same ingredient into the food industry with significantly *less* of a regulatory burden.

This is not meant to imply that excipient users should not consider sourcing and purchasing products from large chemical distributors. In fact, some of these distributors are configured in such a way that they have special business units designed *specifically* to service the pharmaceutical market. These business units are managed separately from the other groups (such as food, cosmetic, automotive, etc.) and they typically try to provide the high level of service required by the pharmaceutical industry. They are generally staffed by individuals having at least a cursory knowledge of the applications of their marketed excipients in pharmaceutical formulation development, and a basic understanding of chemical and physical properties such as particle size distribution, bulk density, moisture content, etc. It is also important that the pharmaceutical business unit of a chemical distributor can understand and effectively communicate the different grades of excipients that may be available, excipient incompatibilities, and other important information regarding the use of the excipients. Additionally, by virtue of their size and reach in the industry, these types of distributors may have valuable contractual relationships with excipient manufacturers, whereby a user is actually required to purchase the material from such a distributor, and cannot buy directly from the excipient maker.

A further advantage to a large chemical distributor offering excipients may be a broad range of distribution points. A large firm is much more likely to have warehousing across the country/globe, allowing them to service a large geographical territory without the burden of unmanageable freight and transportation costs.

Pharmaceutical Excipient Distributors

Truly a "niche" distributor, the pharmaceutical excipient distributor places intense focus on a number of considerations in servicing the pharmaceutical industry. It is a very tightly regulated industry and one that requires extremely high levels of service—from technical, regulatory, and logistical standpoints.

This type of distribution can again be broken down to either the strictly warehousing and distribution outfits or those who engage in further processing of the material. It should also be noted that many excipient manufacturers also distribute excipients, but these types of organizations fall a bit outside the scope of this chapter. For the purpose of this section, Mutchler, Inc. pharmaceutical ingredients can be used as a model for focused pharmaceutical excipient distribution. Mutchler, Inc. distributes only factory-sealed excipients almost exclusively to the pharmaceutical industry, with great emphasis on technical proficiency and compliance with industry standards. Mutchler does not repack, sample, test, or handle exposed material. Nor are they brokers, which is to say that they take title of the material and inspect the goods prior to shipment to the customer, which includes an inspection of the integrity of the sealed packaging and confirmation that the tamper-evident seals are intact, and that the Certificates of Analyses (C of A) examined reflect the lot numbers received in a particular shipment.

Although it is the responsibility of an excipient user to qualify their excipient suppliers, that burden is shared with the distributor as well. A pharmaceutical distributor must provide the user with the assurance that the excipients supplied are unadulterated, have not been tampered with, and are free from contamination. To ensure this, a distributor must have warehouse and internal controls in place, comply with current regulatory standards, and should be an active NACD member. Traceability of ingredients is another critical need that must be met by pharmaceutical distributors. This can be achieved by providing the manufacturer's original packaging, documentation, and manufacturing site address, confirmed by the manufacturer.

Distributors are often only as good as the compliance of their suppliers, and in order to meet the aforementioned considerations, they must exercise diligence when evaluating and pursuing supplier representation. The pharmaceutical excipient distributor must be responsible for obtaining the most current documentation available on the products from the suppliers, and should seek out suppliers that are forthcoming about the origin of materials and the quality systems to which they adhere. Additionally, the suppliers chosen and represented by the distributor should provide the distributor with the assurance that they have internal controls in place to ensure the integrity of their materials. Periodic audits/visits by the distributor may be conducted at the suppliers' manufacturing facility to assure compliance with industry standards, but because part of the function of an excipient distributor may be to arrange an audit on behalf of a user and accompany them, it is often unnecessary for the distributor to audit their suppliers alone.

SPECIALIZED MODEL OF THE PHARMACEUTICAL DISTRIBUTOR

The pharmaceutical excipient distributor, in order to service demands of the pharmaceutical industry, generally performs many more functions than just the movement of excipients along the supply chain. A number of these functions and their benefits to excipient users are outlined below.

Supplier Relationships and Liaison Service

In an industry where quality, service, and rapid problem resolution dominate, the relationships that an excipient distributor fosters with their suppliers and customers are critical. The nature of distribution dictates that a particular distributor may purchase excipients from the same supplier to market to several users. This places the distributor in a good position to cultivate this relationship with the excipient maker on behalf of each user, because the total volume purchased by the distributor will generally be greater than that purchased by any one user. This is one of the many ways that an excipient distributor is in an excellent position to advocate the interests of each individual user.

Essentially, a user should be able to contact their excipient distributor and request a particular excipient, and the distributor should perform all of the sourcing activities, such as elucidating the exact grade of material required, if not already specified; provide documentation generated by the manufacturer [product specifications or data sheets, typical C of A, material safety data sheets (MSDS)]; and obtain samples of representative lots of the excipient free of charge for the user.

The manufacturer liaison function of the excipient distributor also relieves the user of many troubleshooting activities and quality investigations that may have to be taken up with the excipient makers. For example, if the analytical testing of an excipient, which a user performs, yields results indicating that the excipient is not meeting the specifications put forth in the appropriate compendium or on the C of A, a skilled and experienced pharmaceutical excipient distributor knows which avenues to pursue with the excipient maker to investigate the out-of-specification result and perhaps obtain a written explanation from the maker and/or replacement material. Further, a more serious quality issue such as the discovery of foreign matter (FM) in a factory-sealed package of excipient can also be handled by the distributor, who should have an established protocol for initiating investigations of their suppliers, and generating the appropriate corrective action documentation that provides the user with the information that the introduction of the FM has been identified, and controls have been added to the manufacturing/packaging process to ensure that such a deviation in quality will not occur again.

Document Control

Drug makers (excipient users) generally require much more information on the excipients they purchase than what is presented on an MSDS, C of A, or a product specification sheet. The regulatory environment dictates that as much information as possible be gathered about an excipient, at the time of sourcing and beyond. Such documents may include letters outlining the origin of materials (whether the excipient is derived from animal, vegetable, or mineral sources), shelf life, and recommended retest/reevaluation procedures, quality system statements (is the manufacturer operating according to the recommendations put forth in the IPEC GMP guidelines? Are they adhering to another quality system?), analytical test methods used to evaluate the excipients, and allergen/hypersensitivity statements. The list is by no means comprehensive, but an excipient distributor will generally have most of these form letters on file for each specific excipient or group of excipients to easily provide to purchasing customers. In the case of special or customized requests for documentation specific to a particular drug maker, the distributor can generally use their aforementioned relationship with excipient makers to obtain such documentation quickly and pass it on to the user.

Technical and Regulatory Services

It was previously noted that some excipient distributors offer additional services to users, which involve further processing of the excipients. These services can consist of analytical testing, repackaging to smaller or larger container sizes, or manipulation of the particle size of the material, such as milling or micronizing. These types of activities are more typical of larger distributors and especially of excipient manufacturers who also distribute excipients. In these cases, it is imperative that the facilities where these types of functions are performed are operating under the appropriate quality systems. Repackaging should be performed under GMP conditions, laboratory testing should be executed according to the recommendations put forth in Good Laboratory Practices guidelines, and so forth.

Excipient sourcing and vendor qualification generally requires that the user perform an audit of the excipient maker. These audits may range from four hours to two days, and generally include a thorough review of the excipient maker's facility, quality systems, record keeping detailing the adherence to such quality systems, and security. A pharmaceutical excipient distributor generally understands the need for and utility of these audits and can many times offer assistance in scheduling these and many times may accompany the user on the audit. This serves the dual purpose of providing a high level of customer service and can be an educational experience for the distributor because they will gain a better understanding of how the excipients are manufactured, as well as what users are looking for regarding safety and quality of the excipients.

PHARMACEUTICAL-ORIENTED CUSTOMER SERVICE

Whether onsite or centralized, the customer service department of an excipient distributor is one of the most critical interfaces between the distributor and their excipient user customers. In the pharmaceutical industry, time and accuracy of order processing are often critical components of a transaction, and time, in particular, always seems to be in short supply! Servicing the pharmaceutical industry requires great attention to detail, in the early stages of excipient sourcing and procurement, all the way through order processing, transport, and delivery. The degree to which a customer service department achieves this attention to detail can certainly be a key determinant in the success of the business. A distributor marketing exclusively to the pharmaceutical industry gives customer service personnel the opportunity to sharpen their ability to meet specialized high maintenance requests for service: e.g., hard-to-find excipients, special high functionality excipients, specific lot numbers required for certain orders, and short lead times.

TECHNICALLY TRAINED/PROFICIENT SALES STAFF

The sale of excipients is generally a bit more complicated than presenting a product to a user, obtaining a purchase order, and shipping the material. Although it is not necessarily a requirement that the staff understand formulation development, product applications, and chemical/physical properties, it certainly yields a significant advantage to those who do possess such understanding. A technical staff automatically elevates the level and depth of service that can be provided to the user and certainly aids in the aforementioned evolution from chemical distributor to that of the specialized pharmaceutical excipient distributor.

In the pharmaceutical industry, the level of user need varies tremendously. Some large global pharmaceutical manufacturers have very efficient research and development (R&D), sourcing, and purchasing units, but this is by no means representative of the industry as a whole. The level of expertise and experience of customers calling in for excipients varies greatly. It is at this point where a distributor can often find themselves in the position of educating the user about the excipients, from either a technical or regulatory standpoint.

The technical department of an excipient distributor should be staffed with individuals with at least a basic science background. This allows them to more fully comprehend the chemical and physical attributes of the excipients they are marketing, the implications of analytical test results, the ability to spot "out of spec" results on a C of A, as well as the applications of the excipients in various dosage forms or delivery systems. Further, they may be able to make educated recommendations about which particular excipients or grades may solve unique formulation problems. Understanding how particle size can affect blend uniformity, how moisture can interact with an API, or how excess mixing with too much magnesium stearate can be the focus of formulation failure can be valuable information for a distributor to have in order to effectively market their excipients. Very specialized excipient distributors may even go to the lengths of having their technical/regulatory customer service and/or sales personnel trained in tableting. There are a few three- or four-day courses available through various organizations or companies that offer training on blend technology, excipient characterization, tableting, coating, etc.

Of course, technical departments of excipient distributors offering additional services involving further processing or handling of exposed material must be staffed with technicians such as analytical chemists, microbiologists, etc., unless they are outsourcing these functions to contract laboratories.

WAREHOUSING, LOGISTICS, AND MATERIALS MANAGEMENT

It is understood that as a supplier of excipients into the pharmaceutical industry, a distributor is responsible for warehousing the material in a facility where it can easily be delivered and will not be at risk of contamination or adulteration of any kind. The practices of warehousing, logistics, and materials management depend on the type of distribution that is being carried out (warehousing and distribution vs. further processing of the excipient, such as repacking, analytical testing, or sampling services). An excipient distribution warehouse is not just a place to retain inventory, but must be managed and secured to ensure the quality and integrity of the materials being stored. Adequate pest control, air handling, temperature conditions, security, etc. must all be in place and closely monitored. Although most excipients are quite stable, some require special storage and handling conditions. The warehouse used by excipient distributors should be able to accommodate these requirements and be able to offer protection from the elements.

It is common practice in the United States for distributors to outsource warehousing and transport functions. Although this is a totally acceptable practice, it is important that the excipient distributor perform periodic inspections of the warehouse facility and be in continual communication with warehouse staff. Warehouses storing excipients destined for pharmaceutical products must be kept clean, dry, orderly, pest free, and secure.

PHARMACEUTICAL EXCIPIENT DISTRIBUTOR STRENGTHS

The existence of pharmaceutical-focused distributors with a "value-added" service philosophy can provide tremendous advantages to the user, especially in the United States, which is said to bear the burden of R&D by a large margin over other areas of the world. The following sections outline several advantages that can be gained by purchasing excipients from a specialized pharmaceutical excipient distributor.

High Service Levels (Logistics, Technical, and Regulatory)

First, a pharmaceutical excipient distributor should be committed to providing a very high level of service to their pharmaceutical customers. The pharmaceutical industry is highly detail focused, and the requirements of service can often surpass what a large-volume manufacturer or less specialized chemical distributor is willing or able to provide.

The distribution of excipients is not known for being easy or a "quick sell." The sales process can be long and laborious, with payoffs coming *years* after the initiation of sourcing projects, if at all. That is why it is important for a specialized excipient distributor to maintain enough existing business to sustain itself in order to offset the arduous pursuit of new business. The cost of prescription and over-the-counter drugs is prominent in the social and political agendas in the United States, and drug makers are under pressure to reduce costs while still producing quality and safe drug products. Additional costs in the supply chain are intensely scrutinized by drug makers, and it is here that it is crucial for distributors to leverage their costs by providing extremely high levels of service. This pressure to provide excellent service to foster customer loyalty and retain business really works in favor of the excipient users.

Broad Technical and Regulatory Knowledge of Excipient Industry and Pharmaceutical Company Requirements

As previously discussed, a highly specialized pharmaceutical excipient distributor who only supplies the pharmaceutical industry will be much better prepared to accurately serve that particular industry.

Due to the chemically and physiologically inert nature, narrow toxicity range, and general safety of most excipients, the regulations governing them are a bit more relaxed than API regulations, and as a result are often less clear. However, it is extremely important that drug makers follow the guidelines put forth by the FDA to the best of their ability when sourcing, purchasing, and using excipients. Interpretations of these guidelines vary almost from user to user, making it difficult for some distributors to satisfy all of the individual regulatory requirements of the excipient users. In many cases, where excipient regulations are unclear, a user will default to API regulations as a guideline. It is not unusual for distributors to be asked for "debarment statements" or access to "Drug Master Files," both of which are pieces of information related specifically to APIs. It is here where an excipient distributor would benefit greatly from developing proficiency in interpreting the regulatory guidelines that govern excipients. An excellent step in achieving this would be for a distributor to join the IPEC (7), a global federation of three independent regional industry associations headquartered in the United States (IPEC-Americas), Europe, and Japan. Each association's mission is to focus its attention on the applicable law, regulations, science, and business practices in order to work together on excipient safety and public health issues as they relate to excipients.

Can Supply Excipient Volumes Be Too Small to Be Leveraged with a Manufacturer?

In many cases, it is absolutely critical that if a distributor is committed to servicing the pharmaceutical industry they are able to supply small quantities for product development. This includes free samples, single unit orders (one bag, drum, carton, etc.) to provide partnership with the drug maker from delivery system development through formulation, pilot batches, and scale-up to commercial production.

Many large excipient manufacturers have volume or invoice minimums that must be met to purchase material. Small drug makers or development companies may not have the budgets, warehouse space, or logistical resources to purchase the required minimum quantities, nor can they manage the higher maintenance small volume purchases. Although it carries a certain degree of risk, an excipient distributor is much better suited to purchasing the minimum quantities of an excipient such as a pallet or truckload, and selling off the smaller volumes to individual accounts. Additionally, some distributors have long-standing business relationships that span decades, with excipient manufacturers. In these cases, an excipient maker may be willing to overlook volume or invoice minimums for a distributor in order to get their materials into R&D projects without executing the logistically complicated single unit orders.

Ability to Consolidate Diverse Materials and Offer Value-Added Services

An observable advantage of purchasing excipients from a pharmaceutical excipient distributor is the ability of that distributor to consolidate the purchase of many excipients on behalf of the user. By this rationale, the distributor becomes an extension of the excipient user's sourcing and procurement departments. This presents obvious time-, resource-, and cost-saving opportunities to the user, because they can allocate their resources to other functions, while the distributor assembles all necessary documentation and executes all orders with the various excipient makers.

As the costs of managing excipient inventory are becoming more burdensome to drug makers, some excipient distributors are implementing vendor-managed inventory (VMI) programs, based on models used by excipient makers to manage key customers inventory as a value-added service. There are many variations on how VMI programs can be configured, but a basic model involves a distributor holding inventory that is designated and reserved for specific customers. Material can be sampled and tested by the user and held in the distributor's warehouse facility. Then, smaller and more frequent releases can be shipped to the user and transferred directly to production. These types of programs, of course, carry a certain amount of liability and require that the distributor have specially allocated customer service and logistical and warehouse personnel. Naturally, this type of VMI, where a distributor is holding tested, quality assurance (QA)-released material in their warehouse, may not be appropriate when a distributor is using public warehousing. That is not to say that some form of VMI cannot be implemented by distributors using public warehousing. The objectives of VMI are to reduce material testing on the part of the user, reduce inventory and carrying costs, and to increase manufacturing space for the excipient user. The distributor in this sense becomes an extension of the user's QA and storage operations and thus offsets some of the costs of pharmaceutical production while at the same time helping to increase efficiency.

If Purchasing and Logistics Resources Are Tight, a Distributor Can Fill in the Deficits

As the last section implies, a pharmaceutical excipient distributor can greatly complement the procurement and logistical functions of an excipient user by taking on as much of those responsibilities as the user is willing and able to relinquish. An excipient distributor would typically be proficient in order processing, can arrange for transportation, and would be responsible for the storage of materials.

Distributors Can Be Helpful in Development Projects When Multiple Excipients Are Being Tried, and Formulation Challenges Exist

An experienced pharmaceutical excipient distributor understands the basics of excipient sourcing patterns during formulation development, pilot batch manufacturing, clinical trial supply manufacturing, and scale-up processes to commercial production. Such a distributor knows that samples and single unit orders (one bag, drum, or carton) are critical elements of these processes but are by no means guarantees for future orders or increased volumes. In these cases, a distributor will provide multiple lot number samples of ingredients for the purposes of excipient characterization and vendor qualification and may also provide several different grades or brands.

A highly specialized excipient distributor with a technically trained staff can actually provide some degree of partnership to product development groups as they are developing their formulations and making R&D or pilot batches. Often, if a distributor has exclusive representation of a particular excipient or group of excipients, the makers will provide technical trainings during which the technical sales personnel of the distributor learn the application potentials of the excipients in great detail and the physical and chemical characterization of the excipients and will become educated enough about how the excipients should be used to provide instruction and guidance to the user's formulation scientists.

Broad Market Views of and Experience in Both the Excipient Industry and the Finished-Dose Pharmaceutical Industry Are Useful Tools to Share with Users

An excipient distributor's history and experience servicing the pharmaceutical industry can add significant value to the end user beyond the aforementioned high levels of service. High quality and consistently available excipients are the most highly sought after in the industry. An experienced and historically established excipient distributor can provide users with the assurance that the continuity of supplied material will not be interrupted, and that manufacturers that the distributor chooses to represent are committed to supplying the pharmaceutical industry with high quality, safe excipients.

PHARMACEUTICAL EXCIPIENT DISTRIBUTOR CHALLENGES

As this chapter has hopefully illustrated, distribution is not at all a secondary or tertiary function, especially when it comes to servicing the pharmaceutical industry. Distributing excipients to drug makers presents some unique challenges, and sometimes a distributor is at a disadvantage compared to excipient makers in meeting these challenges. When choosing a distributor, it is important to evaluate whether or not they are configured in such a way that these disadvantages or challenges are minimized.

General Chemical Distributors May Lack Technical and Regulatory Knowledge and Resources to Deal with Technical and Regulatory Detail-Intensive Pharmaceutical Companies

Not all distributors are committed to maintaining a highly technically trained and regulatory-trained staff, nor is it necessary that they do, depending on the scope of industries that they service. The less a distributor is focused on the pharmaceutical industry, the less likely they are to be able to meet the high demands of excipient users. Depending on a distributor's business model (committed only to the pharmaceutical industry, or supplying greater volumes to other industries with less investment of resources?) an excipient user may encounter customer service issues if their business is not the focal point of the distributor's goals. Further, distributors not committed to the pharmaceutical industry may not have a comprehensive understanding of all of the documentation required by drug makers, and may not have the turn-around time required by the pharmaceutical industry in fulfilling requests for information.

Change Control Notification

Any changes, major or minor, in the process, raw materials, or site involved in the manufacture of an excipient can have an impact on a pharmaceutical formulation in its early stages of development, or during the commercial production of a drug product. For this reason, many drug makers have strict standard operating procedures that they adhere to in order to better control the effects that these changes may have on the safety, efficacy, or manufacturing process of their drugs. Under the umbrella of "change control," generally excipient makers supplying the pharmaceutical industry commit to providing users adequate notification of such changes in product, process, or site and support users with samples, data indicating functional equivalency between the excipient pre- and postchange, or any other information required by the user to document and adopt the change in their internal systems. The nature of excipients is that they are multifunctional and may be used in many other industries. Unfortunately, not every excipient maker is committed to servicing only the pharmaceutical industry and may not fully understand the implications of such changes. For example, many users will have to gather as much prechange material as possible and may have lengthy testing procedures to qualify postchange material depending on the nature of the change. This issue becomes especially challenging for distributors supplying excipients from makers that they may not have contractual relationships with. In some cases, the excipient maker may not provide a distributor with the appropriate notification of an impending change, in turn limiting the notice that the distributor can provide to users. When the change is announced (or in the worst case, *discovered*), a conscientious pharmaceutical excipient distributor will scramble to gather any prechange material to satisfy immediate user needs while trying to give users as much time as possible to qualify postchange material. This presents obvious challenges, and could be avoided through better communication between excipient makers and distributors.

Distributors May Be at a Disadvantage Regarding Request for Pricing/Quotations

As a cost-savings initiative and/or method of reducing the number of suppliers a company uses, many large drug makers have implemented electronic "request for pricing or request for quote" (RFP/Q) programs. Although there are many variations, generally a large list of excipients used by the drug maker is sent to all potential suppliers for their review and opportunity to quote on the business. The process is fairly new, but already excipient distributors are at a distinct disadvantage regarding these types of programs. Distributors have the most difficult job providing price quotations, because they generally have so many items to quote on. Often, for the sake of consistency and ease of interpretation, the rules of the RFP/Q dictate that it may count against them if a participant in the bid does not submit a quotation for every item. In these cases, exclusive distributorships or contracts are not considered in the configuration of these programs, and as a result distributors may be penalized by not quoting on every item, even if contractual relationships do not allow them to quote on or supply these particular items. There are actually some RFP/Qs that have the contingency that if a firm does not quote on every item, they will be dismissed from a preferred vendor list. This puts distributors in a very difficult position, because contractual relationships prohibit them from quoting on or supplying certain brands of excipients if they exclusively represent one manufacturer. In these cases, an honest distributor has no control of how other firms will choose to complete the RFP/Q to be considered as a supplier. The ramifications of this are obvious. If RFP/Qs are going to be used as a tool in the pharmaceutical industry and are going to be issued to distributors in the interest of having them bid on certain pieces of business, then the nature of distributorship (wide range of products, exclusive representations, etc.) must be considered in order to receive the most comprehensive and accurate data in return.

Small Volume Orders: A Necessary Evil

The issue of small excipient volumes is raised again as a challenge that excipient distributors face. This is really a dual issue, because a specialized excipient distributor will be committed to servicing product development groups with the small volumes they need to not only qualify and characterize excipients, but also to produce their R&D or abbreviated new drug application (ANDA) submission batches. It is an obvious advantage to begin supplying a user at this stage in hopes that some of these projects make it to commercial production. However, the cost and logistical resources that these types of transactions require can be enormous—from shipping and handling, to having to purchase more material than what is needed to satisfy a volume minimum a manufacturer might have, to trying to bundle shipments of many small units if a user is taking advantage of a distributor's function as a consolidator of miscellaneous items. Many times, the price a distributor must charge for these small shipments is disproportionate to larger volume pricing, and may give the mistaken impression that it is more expensive to work with a distributor.

In the race to be the first to market a drug product, or in the midst of quality or supply issues, it is not uncommon for a distributor to receive phone calls on a Friday afternoon asking for a Monday delivery of excipients, because many excipient users do not necessarily follow a Monday through Friday, 9:00 A.M. to 5:00 P.M. schedule. This is especially true of product development groups in small, privately owned companies. It is not uncommon for formulation development to continue into the weekend. It has already been noted that small volume shipments are more difficult than large volume ones, and, further, arranging for these materials to be available on short notice can certainly compound the challenge.

Some Distributors Configured for Efficiency and Standardization, Not the "Special Cases," Which Are the Norm in Pharmaceutical Industry

Because excipient regulations are not as well defined as they are for APIs, many drug makers utilize their own quality and regulatory departments to attempt to develop policies to which they can adhere to ensure compliance with FDA standards. As a result, there are many interpretations of exactly how to qualify and procure excipients. This lack of standardization leads to confusion and adds time and energy into the excipient sourcing and procurement process. Some drug makers wish to qualify three different lot number samples of an excipient and require one of those same lot numbers to produce their ANDA submission batches, others require 10 different lot number samples. When the time comes to begin purchasing the excipient, some actually will ask the distributor to perform lot selection of the manufacturer's available stock. Others still may only take material from one lot at a time, which can be a challenge depending on the excipient maker's standard batch size and frequency of production.

All of these scenarios are typical of servicing the pharmaceutical industry. Drug makers work hard to ensure that their systems are optimized to produce high quality, safe, and effective medicines. They have determined the best way to achieve this, and without standardization, a distributor may get a number of unique or special requests for documentation, lot selection, packaging, etc., depending on the customer. It becomes a challenge to a distributor to maintain the high level of service, and speed of order processing when having to use customer service and logistical resources to meet these special needs.

Supply Chain Integrity Compromised by Some Distributors (Labeling, Manufacturer Site Disclosure, etc.)

As tragic events have illustrated in the past (2,3), not all distributors conduct business as carefully and as transparently as they should when marketing products into the pharmaceutical industry. A few bad examples have cast a shadow on a very necessary and hardworking member of the excipient supply chain. Again, the multifunctional nature of excipients and the fact that so many are used in other industries with more relaxed regulations and requirements may tempt a distributor with analytical testing capability to test material manufactured to yield a grade for a different application than pharmaceuticals [Cosmetic, Toiletry, and Fragrance Act or FCC] to meet the USP or NF and sell the material at a much higher price to the pharmaceutical industry. Industry trade associations, whose primary function is to interpret and help users adhere to guidelines the FDA recommends when using excipients, have come right out and called this practice unacceptable.

Custom-Designed or Custom-Manufactured Excipients Are Needed

There are some customer requirements for ingredients other than APIs, which present great challenge to distributors and are sometimes better handled directly by the manufacturer. These are cases where a custom-designed or -manufactured ingredient is required. For example, makers of specialized coatings or coating systems for tablets, particles, etc. generally conduct business directly with a user, because such a high level of expertise is required to instruct the user on the properties and application of such coating systems. Further, many flavors and colors that are needed for certain formulations, while not considered APIs, are generally customized to meet a very tight specification developed by the user, especially in the cases of generic manufacturers attempting to formulate the generic version of a branded pharmaceutical product. In these cases, it makes little sense for a distributor to be involved, because the services they offer may not offset the extra time it may take for the distributor to interpret the customer specification and communicate it to the manufacturer.

Holding Inventory to Satisfy Pharmaceutical User Need and Minimizing Risk

In nearly every industry, inventory management presents a challenge. Holding inventory carries a certain amount of risk and ties up capital. Ensuring excipients are available at the right time and in the right quantities to satisfy user demand can be especially challenging for a distributor, because it is impossible for an excipient distributor to have a working knowledge of the production planning of all their suppliers.

Once a distributor purchases an excipient from a maker, that excipient only gains sales value and the potential to recover the cost of purchasing and marketing it when a user actually places an order for it. Bringing in large amounts of inventory without standing or blanket orders from users can place strain on the cash flow of a distributor.

Maintaining too much excipient stock versus not having enough to fill orders in a timely manner is a delicate balance that excipient distributors must maintain. Knowing how much material to keep inventoried is especially challenging in the pharmaceutical industry without standing or blanket orders. In the pharmaceutical industry, forecasts for material can change with a day's notice, especially in the cases of drug makers utilizing centralized purchasing or purchasing departments that are not physically located at the manufacturing site. Depending on the level and speed of the internal communication of the drug maker, enormous requirements for excipients may appear overnight. Conversely, even when a user has a rigid forecasting system and is able to place standing or blanket orders with their distributor, unforeseen circumstances may cause the need for a material to diminish greatly and unexpectedly. These circumstances include the unanticipated poor sales performance of the finished drug product, production problems, or quality issues.

ANTICIPATING, ESTIMATING, AND COMMUNICATING ACCURATE LEAD TIMES

When an excipient distributor is sourcing a material for a user for the first time or even for subsequent orders on a nonstock item, a key element of information that must be communicated to the user is the availability of the excipient and the time it will take between when an order is placed and when the excipient is actually delivered to the user site. The definition of availability can be variable, especially if the distributor happens to represent international excipient makers, who may have the material in stock overseas, requiring additional lead times of anywhere from two to eight weeks, depending on the service of the supplier and customs, and, now as a result of the Bioterrorism Act, FDA intervention with entry into the United States.

Lead time is also affected by order cycle time, which includes the time it takes to transmit the order to the supplier, process the order, assemble the material, and transport them to the customer (1). Time between order and delivery may be affected by a number of factors.

This is all assuming that the materials arrive at their final destination fully intact. In the event of breakage, damage, contamination, etc., the distributor must have protocols in place to initiate corrective action investigations as well as a well-defined policy on the return of damaged shipments.

REGULATING PHARMACEUTICAL DISTRIBUTORS

Currently, the regulatory environment in the pharmaceutical industry is such that pharmaceutical distributors, in addition to excipient users and makers, are falling under increased scrutiny. There are a few organizations attempting to address this issue specifically regarding excipient distribution, by developing and publishing guidelines with the goal of helping distributors achieve compliance with current industry best practices and generally engage in good business practices that ensure that excipients delivered into the pharmaceutical market are of the highest quality and, of course, safe. IPEC is at the forefront of these efforts.

When attempting to develop guidelines to aid a distributor, it is important to consider exactly the type of distribution that is being engaged in, because there are different levels of risk to each type and therefore differing levels of compliance. A distributor that engages in strict distribution and warehousing will be subject to different levels of controls than a distributor who engages in further processing of the material, such as repacking, sampling, etc.

In the past, the role of distributor has been taken for granted, garnering public attention only when there was an interruption of the supply of goods or an adverse event involving distributed material. However, with today's rapidly expanding pharmaceutical market, the growth of the fast paced and high demand generic industry, and the trend towards outsourcing supply and services, one can only conclude that the distributor will continue its emergence as a critical link in the pharmaceutical supply chain.

ACKNOWLEDGMENTS

I would like to offer my sincere thanks to Glenn, Laurie, and Dwight Mutchler for providing me as well as the pharmaceutical industry as a whole with an outstanding model of how excipient distribution should be executed. Their philosophies of customer service and their method of pharmaceutical excipient distribution, conducted with a high degree of integrity and transparency, influenced much of the content of this chapter. I would also like to thank Linda Herzog for her contribution in configuring the outline used as a framework for this chapter.

REFERENCES

1. Brewer SM, Button KJ, Hensher DA, eds. Handbook of Logistics and Supply Chain Management. New York: Pergamon, 2001.
2. O'Brien KL, et al. Epidemic of pediatric deaths from acute renal failure caused by diethylene glycol poisoning. J Am Med Assoc 1998; 279(15):1175–1178.
3. Singh J, et al. Diethylene glycol poisoning in Gurgaon, India, 1998. Bull World Health Organ 2001; 79(2):88–95.
4. McKinnon, Alan C. Physical Distribution Systems. New York: Routledge, 1989.
5. www.osha.gov.
6. www.nacd.com.
7. www.ipec.org.

Index

For Product Safety Concerns and Information please contact our EU representative GPSR@taylorandfrancis.com Taylor & Francis Verlag GmbH, Kaufingerstraße 24, 80331 München, Germany

Batch number: 10731069

Printed by Printforce, the Netherlands